Studies in Modern History

General Editor: J.C.D. Clark, Joyce and Elizabeth Hall Distinguished
Professor of British History, University of Kansas

Titles include:

James B. Bell
THE IMPERIAL ORIGINS OF THE KING'S CHURCH IN EARLY AMERICA
1607–1783

Jonathan Clark and Howard Erskine-Hill (*editors*)
SAMUEL JOHNSON IN HISTORICAL CONTEXT

Bernard Cottret (*editor*)
BOLINGBROKE'S POLITICAL WRITINGS
The Conservative Enlightenment

Richard R. Follett
EVANGELICALISM, PENAL THEORY AND THE POLITICS OF CRIMINAL
LAW REFORM IN ENGLAND, 1808–30

Andrew Godley
JEWISH IMMIGRANT ENTREPRENEURSHIP IN NEW YORK AND
LONDON 1880–1914

Philip Hicks
NEOCLASSICAL HISTORY AND ENGLISH CULTURE
From Clarendon to Hume

Mark Keay
WILLIAM WORDSWORTH'S GOLDEN AGE THEORIES
DURING THE INDUSTRIAL REVOLUTION IN ENGLAND, 1750–1850

William M. Kuhn
DEMOCRATIC ROYALISM
The Transformation of the British Monarchy, 1861–1914

Kim Lawes
PATERNALISM AND POLITICS
The Revival of Paternalism in Early Nineteenth-Century Britain

Marisa Linton
THE POLITICS OF VIRTUE IN ENGLIGHTENMENT FRANCE

Nancy D. LoPatin
POLITICAL UNIONS, POPULAR POLITICS AND THE GREAT REFORM
ACT OF 1832

Karin J. Mac Hardy
WAR, RELIGION AND COURT PATRONAGE IN HABSBURG AUSTRIA
The Social and Cultural Dimensions of Political Interaction, 1521–1622

Robert J. Mayhew
LANDSCAPE, LITERATURE AND ENGLISH RELIGIOUS CULTURE,
1660–1800
Samuel Johnson and Languages of Natural Description

Marjorie Morgan
NATIONAL IDENTITIES AND TRAVEL IN VICTORIAN BRITAIN

James Muldoon
EMPIRE AND ORDER
The Concept of Empire, 800–1800

W.D. Rubinstein and Hilary Rubinstein
PHILOSEMITISM
Admiration and Support for Jews in the English-Speaking World,
1840–1939

Julia Rudolph
WHIG POLITICAL THOUGHT AND THE GLORIOUS REVOLUTION
James Tyrrell and the Theory of Resistance

Lisa Steffen
TREASON AND NATIONAL IDENTITY
Defining a British State, 1608–1820

Lynne Taylor
BETWEEN RESISTANCE AND COLLABORATION
Popular Protest in Northern France, 1940–45

Doron Zimmerman
THE JACOBITE MOVEMENT IN SCOTLAND AND IN EXILE, 1746–1759

Studies in Modern History
Series Standing Order ISBN 0–333–79328–5
(*outside North America only*)

You can receive future titles in this series as they are published by placing a standing order. Please contact your bookseller or, in case of difficulty, write to us at the address below with your name and address, the title of the series and the ISBN quoted above.

Customer Services Department, Macmillan Distribution Ltd, Houndmills, Basingstoke, Hampshire RG21 6XS, England

Landscape, Literature and English Religious Culture, 1660–1800

Samuel Johnson and Languages of Natural Description

Robert J. Mayhew
University of Wales
Aberystwyth, UK

palgrave
macmillan

First published 2004 by
PALGRAVE MACMILLAN
Houndmills, Basingstoke, Hampshire RG21 6XS and
175 Fifth Avenue, New York, N.Y. 10010
Companies and representatives throughout the world

PALGRAVE MACMILLAN is the global academic imprint of the Palgrave
Macmillan division of St. Martin's Press, LLC and of Palgrave Macmillan Ltd.
Macmillan® is a registered trademark in the United States, United Kingdom
and other countries. Palgrave is a registered trademark in the European
Union and other countries.

ISBN 0–333–99308–X hardback

This book is printed on paper suitable for recycling and made from fully
managed and sustained forest sources.

A catalogue record for this book is available from the British Library.

Library of Congress Cataloging-in-Publication Data

Mayhew, Robert J. (Robert John), 1971–
    Landscape, literature and English religious culture, 1660–1800 : Samuel Johnson
    and languages of natural description / Robert J. Mayhew.
        p. cm. — (Studies in modern history)
    Includes bibliographical references and indexes.
    ISBN 0–333–99308–X
    1. Johnson, Samuel, 1709–1784—Knowledge—Natural history.
    2. Landscape in literature.   3. English literature—Early modern,
    1500–1700—History and criticism.   4. English literature—18th century—
    History and criticism.   5. English language—Early modern, 1500–1700—
    Rhetoric.   6. English language—18th century—Rhetoric.   7. Nature—
    Religious aspects—Christianity.   8. Christianity and literature—
    England.   9. Natural history—England—History.   10. Description
    (Rhetoric).   11. Nature in literature.   I. Title.   II. Studies in modern history
    (Palgrave (Firm))

    PR3537.L3M28 2004
    828'.609—dc22   2003066389

10   9   8   7   6   5   4   3   2
13 12 11 10 09 08 07 06 05
Transferred to digital printing 2005

Contents

Acknowledgements

This book is the product of over ten years of studying the nexus of landscape, literature and religion, in the course of which I have incurred many debts to institutions and individuals. First, I would like to thank the late J.D. Fleeman for encouraging a geographer to think he could write a meaningful contribution concerning Samuel Johnson. His rigour and tolerance are an example to all scholars. I also acknowledge the support and guidance of Jack Langton who aided me at all stages in the production of this book. Jonathan Clark and David Livingstone examined the doctoral thesis from which this book springs, and I am grateful to both for their consistent support and encouragement over many years. Paul Langford made helpful comments on Chapter 8 which forced me to alter and improve it and to develop a new conclusion. Roey Sweet made useful suggestions for Chapter 3, which have improved it considerably.

The Economic and Social Research Council supported me for the first three years of my work on this book. St John's College, Oxford then generously supported my work by the award of a North Senior scholarship. The staff of the Upper Reading Room of the Bodleian kindly endured years shuttling books to me whilst both the Cambridge University Library and the National Library of Wales have provided the resources with which this book has reached fruition.

Finally, I must thank my partner, Yvonne, and my son, Samuel. This work began long before I knew either of them, but it would have no meaning if they were not with me now: *coelum, non animum mutant*.

Part I

Historiography and Landscape Studies

The opening chapters of this work examine the question of how to study landscape depictions from the past. Chapter 1 offers a critical review of landscape studies over the past thirty years, focused in the main on the methodological and historiographical frameworks which commentators from a range of disciplines – notably history, art history, geography and criticism – have deployed in their attempts to come to a deeper understanding of the meaning of landscape depictions as a 'window' on the societies that produced them. Looking in particular at works studying English landscape depictions in the 'long' eighteenth century, it is suggested that there has been a critical consensus in such studies that landscape must be understood in the light of a broader historical context. Yet Chapter 1 shows that the contexts which have been deployed are partial and oversimplified, in good part due to an impoverished approach to historical method. This suggests two needs: first, to engage with recent revisionist scholarship about the eighteenth century in understanding what landscape 'meant'; and secondly, to develop a more rigorous and philosophically informed approach to historical method. This book as a whole responds to the first of these needs, and it does so in the light of an understanding of historical contextualism developed in Chapter 2. Chapter 2 draws on the philosophical writings of Michael Oakeshott, together with work by Bradley and Wittgenstein, to develop a more theorized understanding of what it means to put a landscape depiction – written or graphic – into historical context. The framework developed using these writers is given operational meaning through the work of Quentin Skinner. Put together, Part I, then, offers a critique of recent approaches to landscape studies as failing to live up to their historical rhetoric and then provides a framework through which the analysis of landscape can be genuinely historical.

1
Contextualizing Landscape History: Mainly with Respect to Eighteenth-Century England

[T]he history of the idea of landscape has to be traced in the works of poets and artists, for it is only in the present century that there has been any technical or academic discussion of the meaning of landscape as a concept.[1]

This begs the question of how to understand the historical meanings of works of art as they pertain to landscape. The quest for historical understanding has led to attempts to *contextualize* expressions relating to landscape. Contextualism will be taken in this chapter as the attempt to explain past statements, actions and events in terms of the social and intellectual categories which could have been invoked to explain them at the time, rather than in terms of subsequently created explanatory systems, a definition whose substance will be elaborated on in Chapter 2. The claim to be doing contextual research has powerful rhetorical appeal because it aims to tie an interpretation down to a clear body of historical data which is open to scrutiny in a way that criticism is not.

One of the features of landscape studies across a range of disciplines over the past fifteen years has been the convergence on claims to contextual sensitivity. I wish to assess these claims in the light of the definition of contextualism given above by dividing recent studies of landscape into two groups.[2] First, and in response to traditional humanistic work, there has been the joining of landscape studies to a broader (and largely Marxist) attempt to contextualize in terms of socio-economic history. Second, and more recently, there has been a more diffuse contextualization of the landscape as a text to be read or as a symbol.

The socio-economic contextualization of landscape studies

The socio-economic contextualization of the history of landscape ideas is based upon the belief that only in this way can we understand that history:

> it is possible and useful to trace the internal histories of landscape painting, landscape writing, landscape gardening and landscape architecture, but in any final analysis we must relate these histories to the common history of a land and society.[3]

This formulation was followed closely by Cosgrove, for whom 'landscape is a way of seeing that has its own history, but a history that can be understood only as part of a wider history of economy and society'.[4]

Implicit in both statements is the aspiration to tie together two narratives, the result being an historically grounded understanding of landscape ideas. Accepting that there is some need to connect the two narratives together, what is the nature of that linkage? Proponents of this form of contextualization move between the temptation to suggest a causal link and the tendency to speak of the two as simply being compatible. Cosgrove spoke of culture and landscape having to be 'homologous' with socio-economic conditions,[5] and argued that 'during that period [the Renaissance] many Europeans came to see nature in novel ways, ways that *corresponded* to new approaches to production on the land', speaking also of 'important historical *parallels*'.[6] It is unclear from the outset whether this linkage is a methodological demand or an empirical and historical hypothesis.[7]

If it is not simply to be assumed that the landscape narrative must be tied to a socio-economic one, two alternatives have been canvassed: either the chronology of the development of landscape representations has been tied to the development of the capitalist mode of production (an historical hypothesis); or the techniques of landscape representation have been taken to mean that it is necessarily implicated in the transition from feudalism to capitalism (an argument about the essence of landscape representation).

The empirical/historical hypothesis

The argument from chronological suggestiveness

The suggestion of some chronological correlation between the history of socio-economic change in Europe and the history of landscape representations was made most starkly by Cosgrove. His treatment is understandably sketchy given a scope of half a millennium of European history, but his basic argument is that the period of the capitalist transition in Europe is precisely one in which the status of land is uncertain. Its redefinition, from use value to exchange value, was a long and hard fought process ... For a long period land was *the* arena for social struggle.[8]

It was in this period that landscape representation emerged, its birthplaces being in Northern Italy and Flanders, these two areas also being the first to experience the transition to capitalism. Cosgrove summarizes that

in this dual significance of land during the struggle to redefine it in terms of capitalist relations is the key to the modern landscape idea and its development.[9]

By 1900, the transition from use value to exchange value being complete, the tension between the conceptions of land was diffused, the same period seeing the 'atrophy' of landscape.[10]

What has been developed is an argument from chronological suggestiveness; given that the narratives of the two histories are so alike in their commencement, sites of origin and temporal span, is it not likely that the connection is less than accidental? This argument has been deployed in a number of ways. Fitter links landscape sensibilities to commercialization rather than capitalism *sensu stricto*, which allows him to explain the pictorial naturalism of the Greeks and to extend the time frame of the connection of economy and landscape to two millennia.[11] On a more restricted timescale and with a more specific linkage, Bermingham also employs such an argument: 'the emergence of rustic landscape painting as a major genre in England at the end of the eighteenth century coincided with the accelerated enclosure of the English countryside'. She then begins 'with the assumption that the parallelism of these events is not an accident but rather a manifestation of profound social change...'.[12] A third form of the argument links the narrative of landscape representations to a less rigidly economic context, as in Mitchell's attempt to point to a correlation between imperialism and landscape.[13] Yet the historical approach needs to specify the linkage between the two chronologies, rather than assuming it. Chronological similitude, regardless of time span and the narratives juxtaposed, can be no more than suggestive.[14]

Even accepting some chronological correspondence, what is the relationship between landscape discourse and socio-economic change? Cosgrove's argument tends to suggest that either landscape representations passively reflected the battle over the status of land or that by retaining the concept of use value during the transition to exchange value, landscape representations were active in the transition, obscuring the social realities of changing conceptions of land in the interest of the owning classes. The uncertainty is heightened, for the reader at least, by the use of the language of intention (which strongly suggests landscape representations were active in the transition) at the same time as it is denied that the language of intention should be taken as such.[15]

A more detailed link is needed if the historical argument for the socio-economic contextualization of landscape studies is to be accepted.[16] There have been two ways in which the link has been further specified: by showing how landscape representations functioned in relation to socio-economic realities; and by taking a more limited period and showing that the putative connection still holds when examined in greater depth. I will look at the deployment of these arguments with particular reference to eighteenth-century England, which is seen as a crucial moment and case study in the transition to capitalism.[17]

The function of landscape representations

A complex of ideas has been suggested whereby landscape representations 'cover up' socio-economic realities, yet at the same time admit these realities in the form of characteristic absences in modes of representation. The starting point for such an analysis is given by Barrell, who sees his work as

> an attempt to study the image of rural life in the painting of the period 1730–1840, not exactly in the light of this new historiography [of E.P. Thompson *et al.*], . . . but taking advantage of the new freedom that Thompson's works have given us to compare ideology in the eighteenth century, as it finds expression in the arts of the period, with what we may now suspect to have been the actuality of eighteenth-century life.[18]

It is the gap between reality and representation which demonstrates the complicity of landscape representation in socio-economic change. Landscape was an idea under tension due to this gap: it was a 'realistic' portrayal of landscape, yet was so far from reality as to beggar belief. This tension provides a dynamic for the stylistic development of landscape painting: Barrell sees a whole sequence of forms of representation of the rural poor, each replacing the last 'when that image would serve no longer', due to its unbelievable representational conventions. Landscape painting was forced to shift to a discernibly English (as opposed to Arcadian) representation, and this

> committed the poets and painters to a continual struggle, at once to reveal more and more of the actuality of the life of the poor, and to find more effective ways of concealing that actuality.[19]

Given this function in a social formation, 'landscape is an ideology, a sophisticated "visual ideology" which obscures not only the forces and relations of production but also more plebeian, less pictorial experiences of nature'.[20]

Closely related is the idea that the very absence of socio-economic realities from landscape representations testifies to the interdependence of landscape representation and socio-economic transition:

> it is not often intended or explicit meanings that I shall be pointing to...but meanings that emerge as we study what can *not* be represented in the landscape art of the period.[21]

Solkin made a similar point in reference to Richard Wilson's work:

> we can only sense the discontent of the poor in the crevices of elite culture...Any serious attempt to comprehend Wilson's happy landscapes must take into account not only what they show but also what they leave out...[22]

The class-specific nature of landscape representation is thus reinforced by absences, and this establishes the linkage with the transition from feudalism to capitalism.

As a complex of arguments, this group fails to tie the two chronologies together. It is assumed that making *realistic* landscape representations demanded that they reflect or represent the *actual*. Whilst this could have been the case, it does not appear to have been so: all landscape painting and poetry were clearly recognised to belong to certain genres with their own conventions.[23] As a consequence, there need not be a 'tension' created by a gap between image and reality 'forcing' a sequence of representational changes. The gap is a broken rule, as representations were not primarily compared with social actualities. To picture a dramatic tension in landscape portrayal is to ignore the function of art in eighteenth-century England. The period saw only the beginnings of an appreciation of the possibility of using images as historical evidence for social conditions,[24] which strongly suggests that the tension found in landscape imagery is the product of an approach to history and art history not clearly articulated in the eighteenth century. Also 'many English buyers of landscapes tended to value them primarily as decorative objects, and only secondarily, if indeed at all, for their subject matter'.[25] Given this, it appears unlikely that purchasers of landscape art would demand that it represent social actualities. Or, if purchasers did desire a realistic picture, this was a demand for something which looked plausibly like the English countryside, rather than something representing social conditions in the actual countryside. Just as the purchasers' demands were vague, so any tension was unlikely to be pressing.[26]

Barrell himself in earlier work recognized the functional demands and nature of eighteenth-century landscape representations: the eighteenth-century eye looked 'over' not 'at' the landscape, and the phrase indicates how little...the eye could be engaged by its object. It indicates how much the impression made upon the eye was a general one.

This in itself would suggest that the actual conditions of the countryside would have created little tension for those buying landscape representations or looking at the actual countryside around them, a suggestion strengthened by Barrell's comment that

> they [the landowning classes] gave little evidence of caring that the topography of a landscape was a representation of the needs of the people who had created it.[27]

However reprehensible this may now be found, it invalidates as contextual arguments the views about the social functions of landscape imagery canvassed so far. The self-understandings of eighteenth-century elites, unlike those of twentieth-century academics, did not necessarily generate the tension alleged to drive the changing representations of the landscape.

A similar argument applies to 'absences' in landscape representations: they are in the main testimony to the fact that landscape representation was not intending to represent social 'realities'. Of course, it can still be said that such realities are absent, but from a contextual perspective the crucial yardstick is what the author, work and genre could be expected to represent, given the state of the landscape discourse. In any case, the existence of rural poverty was not simply masked but discussed by eighteenth-century writers on the landscape.[28]

The eighteenth-century context of landscape studies

For the plausibility of the account given of a gap between reality and representation, socio-economic contextualizations of landscape history all rely upon a limited range of authors, the result being a coherent view of the realities of eighteenth-century English society.[29]

There are three main elements to this view. First, the existence of something akin to Marxist classes is assumed.

> [A]n acquaintance with eighteenth-century writing, whether with the imaginative literature or with the literature concerned more directly with the discussion of social problems, will reveal that the 'poor' were indeed coming to be thought of as a class.[30]

Related to this, and secondly, is the notion of the poor or proletariat as a threat to the elite classes such that their discontent forced itself upon upper-class consciousness and culture. It is only for this reason that the tension Barrell speaks of as driving change in the depiction of the rural poor makes sense. He attacks the nostalgic view of the eighteenth century as an age of stability, drawing on the social history of E.P. Thompson *et al.*, focusing on riots and criminal law, and sees one of his aims as to 'look beneath the surface of the painting, and to discover there evidence of the very conflict it seems to deny'.[31] Thirdly, the eighteenth century is seen as a period of rapid commercialization and the development of 'capitalist' property rights. Thus,

> the Palladian country house and its enclosed parkland... represent the victory of a new concept of landownership, best identified by that favourite eighteenth-century word *property*.[32]

In sum, then, landscape studies of this variety have found a highly agreeable context within which to place themselves. Eighteenth-century English society is portrayed as class based and class conscious, with the seething discontent of the lower classes being either obscured or suppressed by draconian property laws eroding a moral economy. The English aristocracy is portrayed as simultaneously confident and fearful of a 'proletariat' which posed a structural threat to it. Within such a context the ideological function of landscape representations makes considerable sense: landscape representations betray the concerns and projects of elite groups, and can thus be expected to serve the needs of those groups in relation to class, suppression of conflict and the promotion of private property.

Yet this portrayal of eighteenth-century society is itself contentious and one aim of revisionist contextual history has been to undermine it. This is of considerable importance to socio-economic contextualizations of landscape history, yet it seems to have gone all but unnoticed. If it can be shown that the prevailing image of eighteenth-century society is distorted, then the foundation upon which previous landscape contextualizations have been built will appear less than stable, suggesting the validity of a project to reformulate the aims of contextual landscape history to focus on the concepts contemporaries could have held, rather than the social conditions under which they held them.

First, with respect to the transition from feudalism to capitalism, this is of course a highly controversial concept. Whilst this should not prevent its being used as the broad context for shifting attitudes towards the landscape, it does call for caution. Above all, the language of intention should not be invoked in relation to the transition as it has been by several authors writing on landscape:

> the success of the 'glorious revolution' provided the political conditions under which landowning and mercantile groups could, through the control of the Lords and Commons, jointly direct the English (soon British) social formation towards full market capitalism.[33]

The problem with such characterizations is that work on economic discourse in seventeenth- and eighteenth-century England[34] has suggested that there was no clear understanding of the economic system as separate from moral and political concerns until at least the time of Ricardo.[35] This work suggests eighteenth-century social theorists still grappling with manifest changes in the economy in the moral language of 'the passions and the interests'. As such the language of intention is inappropriate from a contextual perspective: how could people or classes consort to bring about that which could have no meaning in their self-understandings? Economics was part of a different division of knowledge in the period: debates over its legitimation were closely tied to denominational politics,[36] its language was one derived in good part from classical debates,[37] and the Tory view of the period was strongly opposed to the human calculus political economy was said to involve thanks to its Christian paternalism.[38]

Class should be a more concrete historical concept around which contextualization of landscape studies can occur, yet in fact, it turns out to be an equally contentious issue for eighteenth-century English history. When landscape historians refer to 'class', they allude generally to the Marxist sense of the term, which can be taken minimally to mean a group defined by a similar position in relation to the means of production and conscious of that position.[39] A tripartite division in the language of orders only emerged as a concept in the 1750s and 1760s, and the language of ranks and orders coexisted with that of class for a considerable time.[40] If the language of class does have meaning, it is only late in the century, and its continuing fluidity strongly suggests it is not an adequate organizing concept for our understanding of the function of landscape representations, at least as cognitized by the actors of the period. Therefore the notion of the suppression of the proletariat is not an adequate explanation for the development of landscape representations in the eighteenth century, at least as this development could have been understood at the time.

With respect to the law, this has been seen as giving powerful support to the view of an eighteenth-century England where the ruling classes were engaged in a vicious suppression of workers: 'the law was one of their [the ruling classes'] chief ideological instruments'.[41] Such a view was important to the socio-economic contextualization of landscape studies, supporting the general view of a ruling class project to suppress and sublimate threats to their supremacy. Just as the law was one gauge of this, so the unreality of landscape representations was another. Yet this interpretation of the role of eighteenth-century law has come under increasing pressure in the light of regional studies.[42] These have shown that all groups had recourse to the law to settle grievances, and that the previous emphasis upon a limited range of criminal law had led to a misleading picture of eighteenth-century legal practice as it affected people's lives *in toto*.

New work suggests an aristocratic and gentry attitude towards the poor far different from that of fearful suppression. As Hirschman said of Burke, his 'primary emotion toward the "lower orders" was not so much class antagonism and fear of revolt as utter contempt and feelings of total separateness'.[43] Barrell admits as much when he says: 'it seems in fact that the polite classes of the eighteenth century had no fear of such [egalitarian] notions making headway among the poor until the 1790s'.[44] This was a period when many of the most widely accepted conceptions of social organization were based on hierarchy.[45] As such, status differentials and occasional unrest did not demand the sort of sublimation so important to socio-economic contextualizations of landscape. The postulated class 'realities' behind the history of landscape representations betray more about twentieth-century assumptions than those of the eighteenth-century elite who bought the representations, the painters, poets and writers who created them, or indeed, of the poor who were or were not represented.

Finally, with respect to property in eighteenth-century England, Cosgrove says

> we know from the writers of social history how fierce were the battles to establish the notion of untramelled personal property in land over the still-powerful conception of common ownership and access to it, for example in England in the eighteenth century.[46]

Obviously, the notion of a transition from use value to exchange value is important to this particular approach's notion of the function of landscape ideas. Yet the chronology given is problematic in several senses. In the legal realm, the right to alienate property freely as an individual was established by the fourteenth century; as such, the use value of land was not of paramount concern to English land law from its earliest development.[47] Equally, notions of common rights and common land remained far more vibrant in the eighteenth century than has traditionally been thought, and, as Neeson shows, were defended by sections of the ruling orders until the 1790s.[48] In the realm of ideas of property, Cosgrove's reference to the 'notion' of untrammelled personal property opens up the issue of the Macpherson thesis, which supported the socio-economic contextualization of landscape by suggesting that the bourgeois revolution had been backed up by characteristically capitalist theorizations of land as property. Yet the Macpherson argument has been questioned by detailed work which suggests that in Locke's theory 'private and common ownership are not mutually exclusive but mutually related'.[49] Moving into the eighteenth century, further research suggests that no one defended a notion of untrammelled private property until the last decades of the century, largely because their thoughts, like Locke's, derived from a natural rights discourse.[50] All this means that it is not clear that the struggle to establish private property rights in land is a helpful context within which to locate the discourse of landscape representation in the eighteenth century. Until the end of the period, a coherent strand in debates about landscape imagery continued to defend a paternalistic ideal of the landscape which was built on customary rights and the notions of self-sufficiency which drove the defence of common land.[51] It is clear that just as the interactions between notions of common land and private property were complicated in the legal and the conceptual realm, and as the positions adopted about these matters cannot be mapped in any simple way onto the socio-economic situations of those engaged in debating these issues, so the connections between these debates and landscape imagery were complex, landscape by no means only having the potential to legitimate one of the two poles simply because it was primarily a discourse of interest to the wealthy. The function of landscape representations *vis-à-vis* notions of property was not monolithic, and this is not surprising, given the manifest complexity of the arguments about the status of property and land in eighteenth-century England.

The conclusion must be that whatever socio-economic context students of landscape history attempt to take as foundational proves unsettlingly mobile and contentious. In the light of these difficulties in connecting the history of the landscape idea to a socio-economic history, it is worth now investigating the attempts to show the two to be linked by their very nature.

The essential/necessary linkage hypothesis

There have been two related arguments put forward to suggest that landscape representations, by the very nature of their construction, are consonant with capitalist society. The first centres upon the perspectival techniques central to seeing the land as an ordered assemblage or landscape. The second focuses upon the existential categories of 'insider' and 'outsider', suggesting that landscape's attachment to the latter makes it an alienated and alienating vision, this being the product of capitalism.

Perspective, partiality and tendentiousness

In the first argument, as well as being a visual term, 'landscape was, over much of its history, closely bound up with the practical appropriation of space'.[52] Realist representation by perspective 'gives the eye absolute mastery over space... Visually space is rendered the property of the individual detached observer'.[53] Perspective itself helps to reinforce capitalist notions of private (individual) property, which are also naturalized by realist art. Moreover, the link between the appropriation of space visually and physically is more than metaphorical, the same perspectival techniques being used in the physical control and delimitation of territory, notably in the elevated prospect of the battlefield from which many of the techniques of landscape representation derive.[54]

It is suggested that the claims to 'realism' made by landscape representations produced according to the rules of linear perspective are in fact ideological for two reasons. First, linear perspective can only display one moment in time, and can only suggest the passage of time by certain conventional subject matters. Secondly, perspective is directed towards a single spectator:

> the claim of realism is in fact ideological. It offers a view of the world directed at the experience of one individual at a given moment in time ... it then represents this view as universally valid by claiming for it the status of reality.[55]

This is bolstered by a form of the argument from chronological suggestiveness:

> it is significant that the landscape idea and the techniques of linear perspective emerge in a particular historical period as conventions that reinforce ideas of individualism, subjective control of an objective environment and the separation of personal experience from the flux of collective historical experience.[56]

If we accept this argument, painted landscape representations become part of an individualist, bourgeois and capitalist way of seeing, such that there is no need to link the chronology of representations to that of a transition from feudalism to capitalism, or to the history of a specific social formation. The structure of the picture space ties it to capitalism. This argument can be extended to verbal representations of the landscape, since the entire idea of a project, controlling and organizing objects in the landscape, creates an 'idea' of landscape built upon the same perspectival assumptions.[57]

The argument given above should not, however, be accepted at face value, for it relies upon a number of inferences and analogies. First, whilst it is quite accurate to say that perspective was important to the appropriation and control of physical space, it is less clear why such control should be so closely connected with capitalist notions of space, for the attempt to accurately delimit space and property does not begin with the advent of capitalism, but is a far older demand.[58] Moreover, there is a *non-sequitur* being employed: even if perspective were crucial to the appropriation of space, and this was to be deemed peculiarly capitalist, it does not follow that any employment of perspectival techniques would be implicated in capitalism (to argue thus is to deploy the genetic fallacy).[59] Even the genetic fallacy does not work satisfactorily, because the origins of perspective are in Greek mathematics, which reinforces the point that perspective in and of itself is not inextricably intertwined with capitalism.[60]

Secondly, it is true that perspective focuses itself upon the individual who can then appropriate the scene. Space can thus be said to be the 'property' of the individual, but this form of property and appropriation of the land is categorically distinct from the physical appropriation of the land. Perspective was regarded as having its origin in the eye itself 'thus confirming its sovereignty at the centre of the visual world'.[61] As such, appropriation by one person would not prevent appropriation by another such that the property enjoyed by the sovereign eye would be quite different from that enjoyed over physical space. There is an analogy between perspectival and physical control of space, but it is only an analogy.[62]

Thirdly, the claims of perspective to be 'realistic' must be examined: perspective does not address all modes of experiencing the land, and is as such 'partial'. Yet the resultant portrayals are partial in the sense of being 'less than the whole', rather than 'tendentious'. Moreover, this does not amount to an ideology for the eighteenth century in the sense of a false consciousness, as the partiality of perspective was widely understood: a number of aestheticians pointed to the fact that a painting could only capture one moment in time, and the multiple perspectives of individuals were recognized.[63]

The claim that perspective renders representation fused with a capitalist way of seeing appears to be questionable, as it relies upon equivocation over the meaning of key terms in its argument. Once these issues are clarified, the resultant argument appears to be one of analogy, not synonymy. In this case, the argument is forced back to an attempt to render this link to capitalism more concrete by an appeal to history, an appeal we have already found to be unsatisfactory.

Outsiders, alienation and individualism

While socio-economic contextualizations have attempted to overcome an earlier humanistic view of landscape, existential notions of insiders and outsiders in the land have been retained. Landscape is seen as an ideology not simply because it claims the status of reality, but also because

> the experience of the insider, the landscape as subject, and the collective life within it are all implicitly denied. Subjectivity is rendered the property of the artist and the viewer – those who control the landscape – not those who belong to it.[64]

The aesthetic and perspectival cognition of landscape is seen by its very nature as the view of the outsider[65] because 'linear perspective directs the external world towards the individual located outside that space'.[66] Such a view is not open to the man who works on and in the land. The most extended treatment built upon these notions comes from Barrell's study of Clare's 'sense of place'. Barrell contrasts this form of knowledge which is only valid within a certain place with the bulk of eighteenth-century topographical poetry which sought to control and command the land and manipulate it into a landscape.

The argument, then, takes a similar form to that dealing with perspective: inasmuch as landscape must be a detached and organized view of a scene, it must be the view of an outsider, the alienated individual of a capitalist society, regardless of its content. It is by its structure part of the worldview, or way of seeing the land, of capitalism.

There are, however, difficulties with this argument. Any representation of the land will have to be detached and organized in some fashion. The attempt to capture the view of the insider will always be riddled with contradictions, precisely because that view is an unarticulated one.[67] As such, while the landscape way of seeing may be that of an outsider, it is not clear precisely what form of expression it is being contrasted with. And if any form of expression is that of an outsider, it is hard to see why landscape representation is peculiarly linked to capitalism.

Another assumption is that landscape by being an exclusively visual way of organizing and understanding the landscape, denies non-visual, less pictorial experiences of nature.[68] Yet this is to ignore many instances of non-visual landscapes. In John Clark, who wrote the first Board of Agriculture report for Herefordshire, 'the idea of [agricultural] richness is rather prevalent, and apt to overawe the mind by that self-sufficiency...what Clark finds oppressive is what he apprehends by taste and smell'.[69] Handel's soundscapes set 'him high among those artists of all time who have made Nature an important part of their subject matter'.[70] These examples suggest that landscape was not an exclusively visual concept, even in its periods of most rigid formulation, and that the argument based upon landscape's suppression of the non-visual is at best partial.[71] This is not surprising, given the generic traditions of landscape description derived from antiquity: the charms of landscape in standard exercises were 'distributed first among the five senses and then among the four elements'.[72]

If it is claimed that landscape as a way of seeing has tended to denigrate other understandings of the land, this evaluative hierarchy has been reversed in most discussions of insiders and outsiders. It is often intimated that the workers' view of the land is more 'real' than the distancing view of the aesthetic. Yet such a suggestion rests upon moral and ideological assumptions which are far from universally agreed upon.[73] This second attempt to show landscape representations to be intrinsically capitalist is also unconvincing.

Some historiographical issues relating to the socio-economic contextualization of landscape studies

The aim of a socio-economic contextualization of landscape is to ground representation in another chronology. Thus we move beyond enumerating the twists and turns of the landscape discourse to understand it as implicated in, and explained by reference to, something far broader. Yet it should be apparent that for this to hold, the two chronologies must have some reasonable degree of independence in their initial construction. If the chronological similarities are due to the application of some overarching theory of history, they will be a product of the historical method through which they have been organized, not of actual correspondences between the two histories being linked.[74] This problem arises when conjoining the history of landscape ideas with the history of socio-economic change. The socio-economic contextualization of landscape studies has been primarily a Marxist-inspired project.[75] As shown above, the context on which these Marxist readings of landscape history have drawn has been that established by Marxist historiography, creating the danger of historiographical self-confirmation replacing the empirical connection of landscape and society. This should not lead to the abandonment of contextual work in landscape studies, but suggests that instead of subordinating the history of landscape ideas to another history, landscape might be viewed initially as a relatively autonomous discourse, influenced by many others and yet forming a coherent object of study, in order to discover which discourses were connected with it. If the history of the landscape idea is recognized to be underdetermined by socio-economic context or indeed by any other context, then a space has been cleared for its study *sui generis.* Cosgrove argued that 'closing cultural history within the boundaries of its own discourse simply mystifies it'.[76] It would appear that its connection to another discourse can have a similar effect unless the worth of the connection is empirically demonstrated.

Another element of historiographical self-confirmation is to be found in the manner in which the history of the landscape idea has been constructed. It would appear that a basic chronology of the development of landscape ideas is accepted as an assumption, and then items which do not fit into this scheme are either ignored or reduced to regressive elements or anomalies. Thus Rosenthal makes the claim that 'British landscape painting is a product of the Restoration. It did, however have medieval origins'.[77] For some reason, these origins are removed to the status of precursors to a predeveloped notion of the correct chronology.[78] Anomalies in the tradition receive a similar treatment: Solkin speaks of six 'exceptional' works by Richard Wilson, because they are outside the tradition in which he wishes to categorize Wilson, while Rosenthal speaks of 'anticipations' and 'prefigurations' in the narrative of landscape he constructs, simply because pictures come at chronologically inconvenient times.[79] The clearest theorization is Hemmingway's: justifying his focus on certain landscape images he says

> underlying this...is a concept of value which appraises art objects in terms of their cognitive effects. Value is measured in terms both of the acuity and depth with which objects engage with the historical development of

the forms of representation involved, and with contemporary beliefs and social phenomena.[80]

This clearly suggests images being selected by virtue of the ease with which they can be connected with social issues: it can hardly be surprising if the theory of a parallelism of art and society is then confirmed!

Also common is the claim only to be studying certain historical 'moments' in a broader tradition. This is most clearly stated by Barrell:

> although I shall suggest that these painters I discuss may be seen in terms of a tradition, I have not tried to study that tradition as a whole, and have been content to discuss what I shall argue are its most important moments.[81]

Yet the highlighting of moments can go with the ignoring of the spans of time in between, such that simple linear histories are drawn up, these serving to lend justification to a distorted narrative.

Labelling work 'contextual' is desirable because of the rhetorical force thus acquired by alignment with the practice of history. Yet this appeal carries with it a commitment not to conflate moral or interpretative statements with statements about the past.[82] Partly because landscape studies relies so heavily upon images and literary representations, the traditional fare of criticism, it has been tempted to conflate the two.

One form of conflation is that of the moral with the historical, a good example being given by Cosgrove's description of Blenheim:

> entering it even today one is overwhelmed by the arrogant assertion of total control in the vulgar classicism of the house and the subjection of the valley floor to a lord's parkland. There is a military feel to this scale and ordering of nature.[83]

Whether we agree with such a statement is strictly irrelevant; what matters is that it is a different category of statement from an historical one. The other prevalent conflation of moral with historical statements is class based and has already been discussed. It is the idea of an 'authentic' working lifestyle which is opposed to the 'cultural mediocrity' of the eighteenth-century English polite classes.[84] Many contextual landscape historians deride traditional art history for its moral assertiveness,[85] and yet they practice the same sort of criticism.

More prevalent is the confusion of critical interpretation of a landscape representation with a statement of historical fact. Solkin's reading of the enclosure scene, *Moor Park, Hertfordshire*, is a good example: he says that

> the picture also transmutes the building of a fence, together with all its potent implications, into an act of nature. Instead of imposing itself

upon the scene, the fence seems almost to have sprung out of the landscape itself, confirming a territorial division already inherent in the disposition of water and foliage.[86]

Again, the value of such statements as criticism is not the issue; the problem is the juxtaposition of contextual work with interpretation, giving a rhetorical power to the latter by the overtly historical nature of the surrounding text.

The outcome of criticism being confused with contextual history is a certain arbitrariness of interpretation. To give an example, Barrell suggests that the pastoral imagery of the English landscape had to be increasingly 'inoculated' with the georgic imagery of hard work in the face of rural realities and rising tension in the mid-1760s.[87] Yet in the same decade Rosenthal suggests that a secure English ruling class was confident enough to take up a concern with the landscape 'as such', with georgic conventions petering out.[88] At the same time, Solkin has the sensual replacing the intellectual in landscape art, a response in part to the rise of individualism and the power and self-consciousness of the middle class, but also to the volatility of the first decade of George III's reign.[89] Of course, none of these claims are directly incompatible with each other (they are probably too vague for that to be the case), yet they do suggest considerable disagreements about the interpretation of what was occurring in the history of landscape representations and what to relate this to. Their only point of agreement seems to be the attempt to map this history straight back onto socio-economic change. The landscape discourse itself becomes secondary, a metatheory about the nature and causes of cultural production driving any interpretation and historical reconstruction of specific instances of cultural production. This appears to be an exercise in what Ricoeur has termed the 'hermeneutics of suspicion', 'an obsessive hunt for the "power" and "oppression" which lie concealed in traditional discourse'.[90] While this approach has its own rationale, its aims and methods are distinct from those of contextual historical research into the mentalities of past actors.

Conclusion

The notion of linking the narrative of landscape to a narrative of socio-economic transition is seductive. It seems to lend to landscape studies an aura of respectability by tying it in to a broader theory about the nature of socio-economic change and of cultural production. 'We can offer structure and coherence to historical understanding and place our detailed knowledge within a wider perspective'.[91] Yet the attempt to specify an empirical linkage between the two narratives has been largely unsuccessful. The failure of the project derives from the basic assertion made by advocates of this form of contextualization. To revert to the beginning of this discussion, Williams argued: 'in any final analysis we must relate these histories [of landscape] to the common history of a land and society'. That a final analysis would have to relate back to socio-economic history is an assumption for which no justification has been found in its use as a working hypothesis.

The symbolic contextualization of landscape studies

In the last decade a new approach to contextualization has emerged in landscape studies. There is no rigid distinction between socio-economic and symbolic approaches to contextualizing landscape studies, the difference being a shift of emphasis. A concern for the symbolic element was always present in the project of socio-economic contextualization but has now come to predominate.[92] The shift towards viewing landscape as a symbol has also been carried out largely by those students of landscape history who had previously engaged in socio-economic contextualization. In the attempt to analyze some of the elements of this more recent work on land-scape, I shall try to highlight the ways in which it is different from and similar to the previous socio-economic contextualizations from which it has grown.

The duplicity of landscape

Perhaps the most obvious way in which more recent work on the history of landscape representations sets itself apart from the writings discussed under the heading of socio-economic contextualization is by its greater willingness to recognize that the debates about landscape have at least a relative autonomy from socio-economic history. The physical nature of a landscape is now rec-ognized to influence ideological strategies of representation in a reciprocal or 'duplicitous' interaction. Daniels chronicles this change, and argues

> it is both possible and desirable to conserve both an ideological and an ontological interpretation [of landscape]...and to bring each critically to bear upon the other.[93]

Daniels gives a clear statement of the shift in emphasis:

> the project of combining the aesthetic with the social has often amounted to fixing images to literal conditions, translating them into concepts, reducing them to 'signifiers' of social forces and relations... I have attended to the social history of landscape images to unfold their range and subtlety, to amplify their eloquence. It is not so much a procedure of unmasking images, to disclose their real identity, as one of revisioning images, of showing their many faces, from many, shifting, perspectives.[94]

While social history is still attended to, its connection to landscape repre-sentations is a far less mechanical process. Socio-economic material may amplify our understanding of images, but it will not be invoked so casually (and causally) as explaining them.

Cosgrove has also moved away from the simple connection of the socio-economic to landscape representations in his recent work, his stress on the duplicity of landscape focusing more upon human ideals and imagination. This is not a new theme for him,[95] yet it has become more pronounced in both his theorization and practice. Recent work 'has changed the questions asked of the evidence, redirecting them towards symbolic rather than purely instrumental interpretation'.[96] Cosgrove always argued that in studying the landscape he was investigating the history of an idea. The shift has been from studying and usually explaining that idea in socio-economic terms, to relating landscape ideas to other ideas:

> here the geographer enters fields of study traditionally tilled by the humanities, because it is in philosophy, religious belief and practice, literature and the arts that cultures most directly express ideas and values about nature, the world, human life and how it is to be lived.[97]

The duplicity of landscape has led back, then, to a recognition of the specificity of landscape, that it forms its own discourse.

The contextualization of the landscape discourse

Given that landscape is granted a higher degree of specificity and autonomy, it follows that the process of its contextualization will be far more arduous than it was previously. Indeed, in socio-economic contextualizations, the question of what context to place landscape representations in could not arise. This assumption not holding for those who have accepted the duplicity of landscape, contextualization becomes a matter related to the specific image or representation under discussion, and the number of possible contexts for any given work multiplies.

There have been two main responses in the light of the far wider linkages between landscape representations and other discourses. First, Barrell has moved towards linkage to a broad discourse of eighteenth-century intellectual life, linking the theory of aesthetics to what Pocock has termed the discourse of civic humanism, a set of framing assumptions and terminology for eighteenth-century English discussion.[98] He has shown how in a number of specific cases the discourse of civic humanism and its interaction with an emergent language of commerce was relevant to understanding the pictorial conventions adopted by eighteenth-century artists. This approach, built upon a growing awareness of the autonomy of the history of landscape ideas, tends, then, to subsume them once more, this time under a broader intellectual structure. This is perhaps a more satisfactory approach than the cruder forms of socio-economic contextualization, acknowledging as it does the degree to which representations of landscape were ideas to be struggled over and fitted into an individual's intellectual world. And yet, the discourse of civic humanism underdetermines the discourse of landscape. There is a danger of returning to a process of linking landscape representations to another factor, that factor simply changing to civic humanism.[99]

The second option is that taken by Daniels, who summarizes his approach as the belief that 'running through many of the images I discuss are a variety of discourses and practices'.[100] Where Barrell links landscape representations to one broad discourse, Daniels links them to a variety of more specific fields of knowledge and beliefs.[101] As such, Daniels's procedure is perhaps better able to respect the specificity of the various moments in the history of landscape it chooses to focus on. His studies of Wright of Derby and Loutherbourg are good examples of this method, linking both to the consumer culture of the eighteenth century, to the scientific developments of the period and to the more mystical elements of the Enlightenment.[102] This does not mean that all individuals producing landscape images in this period have to be contextualized in the same manner: depending upon their range of intellectual interests, the elements relevant to placing a work may be totally different.

This approach, however, does run the risk of being drawn into overinterpretation. Eco argues that overinterpretation occurs where a suggestion transgresses the lexical-historical repertoire an individual could have drawn on.[103] Overinterpretation is a possibility, given the sheer range of discursive practices existent at any one time and the implausibility of the creator of a landscape representation attending to more than a limited number of these practices. Thus overinterpretation would take the form of an arcane science of cultural 'echoes' to numerous contemporary practices for which there was no evidence that the creator was aware. It is unclear, for example, whether the image of *Rain, Steam, and Speed – the Great Western Railway* offers 'a commentary on the ambition, financial as well as technological, it [Maidenhead bridge] represented', because no evidence has been brought to show Turner's concern for the issues he is supposed to be commenting on.[104] Daniels's approach is at its most effective as historical contextualism where his subject is shown to have been concerned for the subject he is said to be alluding to.

One other problem for landscape studies arises from Daniels's approach to contextualization. While his aim may be 'to show how landscape intersects with other forms of representation, verbal as well as visual, and other subject matter',[105] the danger is of following this process to the extent that the specificity of landscape is diffused in the welter of other discourses to which it is connected. Whilst there is no doubt that landscape does relate to numerous other issues, it is itself a point of concentration for these issues and recognizably its own coherent object of inquiry.

The claims of symbolic contextualization

Barrell has attempted to sketch the aims and methods being adopted by 'a new kind of approach...to the history of art'. Characterizing a collection of essays, he says they 'do not seem to me to belong within any established discipline' and that

> I would describe this kind of work as 'cultural criticism', except that as that term is more and more exclusively applied to the analysis of the modern and post-modern, it seems to leave out of account the concern with history exemplified in these essays.[106]

This summary suggests a certain ambiguity of aims which Barrell addresses directly in a collection of his own essays, where he says

> they are preoccupied with questions of cultural history, but they are not attempts to write a history of ideas, still less a history of real events, but rather of discursive representations. To say that is to say that they are necessarily as concerned with questions of meaning as of history...I try therefore to be a historian among literary critics, and a literary critic among historians.[107]

This ambiguity as to the nature of the project stems from its attempt to yoke together symbolic and contextual reasoning. The aim of calling something a symbol is to say that it stands for or represents a larger entity. Thus Daniels sees landscapes as symbols for broader myths of national identity such that 'they picture the nation'.[108] Yet the aim of recent contextualism has been almost the reverse: to build up a body of information about the intellectual and discursive milieu into which a specific text can be placed. The aim is to move away from having classic texts stand for an entire period and to understand them in the light of a more continuously evolving discursive formation. As such, there will always be conflicting pressures when symbolism and contextualism are put together. Cosgrove says:

> in seeking to describe and understand the cultural transformation of a part of the Venetian land empire in the middle years of the sixteenth century, I have found it helpful to use the undoubted genius of Palladio as an entry into the various discourses through which the transformation was effected and represented in landscape...I shall keep the architect firmly in context, using his work as a *leitmotif* for the cultural world in which he operated and which his designs so brilliantly articulated.[109]

But to the extent that Palladio is used as a *leitmotif*, he will become a bearer of attitudes and issues he was unconcerned with. To the extent that his genius is studied in its context it will be unable to bear the historical load that being a symbol would demand.

Related to this conflation of interpretation and contextualization is an uncertainty over whether to describe symbolic contextualism as distinctively historical. This expresses itself in simultaneous declarations that work stems from present-day concerns and that it takes into account historical discourses and practices. Both Daniels and Barrell stress that their work emerges from 'that very coherent decade' the 1980s.[110] The suggestion is clearly that the present is implicated in our study of the past. Yet at the same time, both assert the distinctively historical character of their work. Thus Barrell writes of the complexity of eighteenth-century discourses

> that mobility...is not at all the same thing as historical indeterminacy; each change of allegiance or identification is an anticipation of, or a response to, another, and takes its course according to a recoverable trajectory and logic.[111]

I have spoken in relation to socio-economic contextualizations of the problems resulting from the conflation of intellectual categories, but this approach appears to revel in this confusion (and in this sense does link with postmodernism). To the extent that interpretative work has different standards of practice from historical work, this confusion is problematic. These different standards are not merely limiting factors to be transcended, but they are characteristic 'forms of attention' within which structured argument and explanation can occur.[112]

Historical and historiographical reflections on symbolic contextualization

Implicit in newer writings on landscape history has been acceptance of the need for a broader approach to the use of historical sources. There has been some widening of the canon of writers and painters addressed. This is a move away from men standing as symbols for their age and of representations as 'anticipations' and 'exceptions' to predetermined trends which was so common. An example of widening the canon comes in Daniels's article on Loutherbourg. As he says,

> when *Coalbrookdale by night* does appear in texts in English art history, it is usually as a freak. In its style as well as its subject-matter, the painting does disrupt the conventionally rustic genealogies of English landscape art.[113]

By 'revisioning' the image in relation to a variety of eighteenth-century discourses, Daniels is able to rescue the picture from being an exception and contextualize it with respect to aspects of the eighteenth century outside the scope of the social history appealed to by socio-economic contextualization.

The broadening of the canon, coupled with the diversification of contexts appealed to mark a move to dissolve stereotypes about a period and can be seen as part of a dissatisfaction with taking 'culture' as an entity capable of characterization (and action).

Symbolic contextualization has exhibited a concern for the instability of interpretation of landscape representations over time. This history has emphasized 'the diversity, the incoherence, the loose ends, the unstable excess in the images it examines'.[114] Whilst the concern for the continuities and changes in perception and expression in landscape representations is not new,[115] the deconstructive tone is. In practice, this approach (as exemplified by Daniels's work on St Paul's and John Constable) has reinforced the idea that landscape images are not tied down to a specific political or socio-economic stance. It has also emphasized the contextual recoverability of the meanings ascribed to an image at any one time.

It would appear, then, that symbolic contextualization has moved a considerable distance towards a more nuanced view of history and a less whiggish historiographical model. Yet there are still valid reservations about certain commonly held historiographical assumptions which have been carried over from socio-economic contextualization. There has been a continuation of the belief in certain 'moments' standing as symbols for broader issues. Daniels characterises this as

> realising the historical momentum of images . . . specifying those episodes when pictures, texts or designs condense a range of social forces and relations, when images assume a high specific gravity.[116]

While such moments may indeed exist, to focus exclusively upon them will tend to give a somewhat distorted view of the degree to which landscape imagery is charged with social significance, and thus underplay the fact that landscape imagery also forms its own discourse with meanings beyond those of social condensation.

Due to their ambiguous fusion of history and meaning, the practitioners of symbolic contextualization still shift between moral and historical modes of argument. This attitude is demonstrated most clearly by Barrell who criticizes the discourse of civic humanism as

> a discourse which defined "man" – not man in general, as it is sometimes pretended, but man as opposed to women and even most men – as a "political animal".[117]

Whilst a twentieth-century perspective may agree with such statements, the transhistorical language of class and sexual politics is not useful: that men were trying to justify their actions coherently is the mainspring of Pocock's work on the discourse of civic humanism on which Barrell draws. This is part of an inability to countenance the ability of past discourses to accept inequality and related structures. Howkins is unable to treat the idea of paternalism as anything more than oppression which is contestable, and also believes that people could not honestly have believed such notions in the past (which seems ahistorical).[118] This failure means that a certain whiggishness remains, for all the historiographical improvements in contextual landscape studies:

much greater ingenuity and a higher imaginative endeavour have been brought into play upon the whigs, progressives and even revolutionaries of the past, than have been exercised upon the elucidation of tories, conservatives and reactionaries. The whig historian withdraws the effort in the case of the men who are most in need of it.[119]

2
Landscape History: An Essay in Historiographical Method

Emergent from the previous discussion of socio-economic and symbolic contextualization is the suggestion that there exists an epistemological space wherein to accept the specificity of landscape studies as its own discourse,[1] and yet at the same time not to abandon the openness to context which historical inquiry fosters. The aim, then, is a rigorous intellectual history: intellectual in its recognition of the vitality of the landscape idea, that it is not a mere cipher for something else; and historical in its adoption of a specific contextual method.

Elements of such an approach were implicit in the earlier criticisms of previous contextual landscape studies. These criticisms were related to a specific Oakeshottian view about the nature of historical inquiry, and to a group of ideas about how to convert this view into a practical method for intellectual history. It is to this view of the nature of history and to some resultant methodological injunctions that I now turn. The hope is thereby to provide a rationale for and pointers towards a more contextual approach to the history of landscape ideas, which will then be deployed in Parts II and III of the book.

The historical mode of inquiry

The distinctiveness of history

Across a broad range of disciplines, recent years have seen attempts to reaffirm that some form of inquiry exists which is distinguished by its sensitivity to historical context. Methodologically, this has been carried furthest by Skinner.[2] He highlights the way in which people have sought in great thinkers of the past doctrines on subjects on which they could not possibly have meant to contribute. If such a doctrine could not be found, this was frequently a cause for complaint. As he says, this strategy appears as a means to fix one's own prejudices on to the most charismatic names, under the guise of innocuous historical speculation. History then indeed becomes a pack of tricks we play on the dead.[3]

This fixing of prejudices has considerable rhetorical power and may therefore be of great use to present-centred discussion. Skinner's point is that to displace opinions onto an historical figure does not *per se* make a piece of writing historical: it results in 'exegetically plausible but historically incredible interpretations'.[4] This distinction between fixing a doctrine to a past figure and actually finding them to have held it has been taken up by many across the humanities and social sciences.[5]

The desire to avoid foisting our ideas onto past figures is pertinent to landscape studies in two ways. First, it is pointless to seek modern ways of seeing landscape in previous thinkers. To argue that Breughel's view of the landscape is that of an 'insider' appears largely meaningless in historical terms, as this is not a doctrine he could possibly have held.[6] Secondly, landscape studies have all too often claimed that individuals were 'responding' to events they could not possibly have cognitized, let alone formed a response to.[7] Thus to see the rise of landscape as part of an individualist, bourgeois way of seeing related to the triumph of capitalism is to impose categories on those who actually represented the landscapes in question they could not in principle have recognized. This adds to the smoothness of the narrative of landscape history, but it does so at the expense of playing tricks on the dead.

In the hope of avoiding this, there has been an attempt to make a division between the historical *meaning* of an action, idea or event and its subsequent *significance*, which was not in principle knowable to those enacting it.[8] Skinner called the conflation of these two forms of thinking about an event in the past, the 'mythology of prolepsis'. To follow his example, Rousseau is often seen as 'responsible' for the emergence of totalitarianism: this may indeed be the significance of his words to subsequent generations, but could not in principle be an account of his aim at the time of writing.[9] It is also on this basis that discussion of 'influences', 'anticipations' and 'prefigurements' has come to be recognized as inadequate in intellectual history.[10] An influence must be shown to be direct, a general similarity meaning little; a writer could not have been influenced by someone he had never heard of and the most general similarities do not amount to influences. Discussion of 'anticipations' is a conflation of meaning and significance, as a writer could not have meant to anticipate in his writings the ideas of a future writer. Recognition of this categorical division between meaning and significance has come in a variety of binary divisions of approaches to historical works.[11]

One element of a more fully contextualized history of landscape ideas must be the recognition of this division between meaning and significance.

Cosgrove may show the significance of the landscape idea in terms of its connection to the transition from feudalism to capitalism,[12] and Barrell may link representation to sexual politics and the discourse of the division of labour,[13] but this still leaves the meaning of landscape representations to those engaged in their production untouched.[14] It is by collapsing meaning into what is in fact significance that Cosgrove *et al.* have been able to claim to be doing contextual work. This is not to deny the interest of the work done on the significance of the landscape idea, but rather to show more clearly what that writing is or can be about, and what remains to be studied by a contextual method.

For Skinner, it is a basic methodological tenet for an approach to meaning as opposed to significance that

> no agent can eventually be said to have meant or done something which he could never be brought to accept as a correct description of what he had meant or done.[15]

Skinner has subsequently weakened this position somewhat, arguing that on rare occasions an agent may not have to fulfil these conditions to have meant something,[16] yet as a methodological injunction this is still vital to those who are attempting to understand an actor historically. Skinner argues that what count as sufficient reasons for holding an idea vary historically and culturally such that present-day cognitive discomfort with an idea is no gauge to the degree of sincerity with which an historical figure could have held that belief.[17]

In landscape studies, the focus being upon the significance of landscape representations, there has been a tendency to ignore or at least underplay what the actors themselves thought they were doing. The best example of this comes from Barrell who recognizes Gainsborough's belief that figures in the landscape are mere objects of colour, but overrides this statement to link his representations into the tension during which the pastoral was inoculated by the georgic.[18] Whilst this is indeed one possible significance, a more context-ual approach is duty-bound to consider whether the figures mean anything like what they are supposed to signify. As well as a willingness to accept the statements of landscape representers (unless they can be shown to be insin-cere), contextualization of their representations must be prepared to accept the *prima facie* evidence of people holding ideas different from our own. This is why the lapses in contextual sensitivity identified in both socio-economic and symbolic contextualizations are so important: they bespeak a failure in the historical imagination which must be taken as the mainspring of a con-textual inquiry motivated by the desire to avoid playing tricks on the dead.

Oakeshott and the historical mode of experience

It is in this context that Oakeshott's writings on what distinguishes historical inquiry are so important, for, straddling philosophy and methodology, he makes a clear case for the intellectual separateness and rationale of the historical mode of experience – the quest for meaning – as distinct from the practical mode which addresses significance.

For Oakeshott, the historical mode of experience is distinguished by its concern for 'the attempt to explain the historical past by means of the historical past and for the sake of the historical past'.[19] The fundamental confusion is the belief that any statement utilizing information about the past is, *ipso facto*, an historical statement.[20] The historical mode of experience is categorically distinct from the scientific and the practical modes of experience (in 1962 he added the aesthetic mode of experience[21]) though all make statements utilizing 'historical survivals' in various ways. Historical survivals are pieces of evidence about the past which have survived into the present; as such, even the historical mode is present-centred, but its interest is the present construction of an understanding of the past, the motivation for which is not linked to any present goal. By contrast, the scientific mode uses these survivals to achieve universal generalizations, yet this is rarely mistaken for history, few taking the claims of scientific history seriously.[22] Reasoning in the practical mode, however, which is distinguished by its concern for past facts for the sake of the present, has frequently been regarded as 'history'. Oakeshott is emphatic as to the categorial distinction of the two:

> Wherever the past is merely that which preceded the present, that from which the present has grown, wherever the significance of the past lies in the fact that it has been influential in deciding the present and future fortunes of man, wherever the present is sought in the past, and wherever the past is regarded merely as a refuge from the present – the past involved is a practical, and not an historical past.[23]

These 'modes' can be seen simply as ways of describing the world akin to Wittgenstein's language games which are forged by humans by usage, rather than as the ontological entities Oakeshott's idealism tended to envisage.[24]

The conflation of these categories Oakeshott called *ignoratio elenchi*: this is irrelevance by which 'a hybrid and nonsensical world of ideas is produced'.[25] It is on the basis of *ignoratio elenchi* that my previous criticisms of both socio-economic and symbolic contextualizations of landscape studies were based. Both laid some claim to belonging to the historical mode of experience by their rhetoric of contextualism. Socio-economic contextualizations, however, conflated historical and practical modes of experience by intermixing historical and moral considerations, and conflated historical and scientific modes by the hope of achieving general causal explanation in history. In symbolic contextualism, the conflation was of interpretation (the poetic mode) with the historical mode, that confusion being both more overt and celebrated. This admixing of categories does not lead to work recognizing no disciplinary boundaries, but to an elegant and erudite sterility which cannot be contested or consented to simply because it has categorially different aims and approaches within it. If the proponents of socio-economic and/or symbolic contextualism claim to avoid this problem of *ignoratio elenchi* by arguing that their claim to contextualize was not an attempt to adhere to the historical mode of experience, this is acceptable, but it leaves a space in landscape studies for such an historical understanding to be attempted.

Modes cannot simply be mixed at will, because they do not apportion sectors of experience between them; each provides a consistent way of seeing the whole.

> [A] mode for him [i.e. Oakeshott] qualifies the world "adverbially": it modifies an ongoing activity, enabling us to experience the world historically, practically, scientifically, poetically.[26]

It is for this reason that history is a totality outside of which nothing can stand; this unity of history is not a finding of history but a presupposition of the engagement of thinking historically. This has important consequences for the characterization of history and for the notion of explanation in history. First, as history is an understanding of the totality, it is nonsense to speak of forces acting on history. Here Oakeshott argued that the search for underlying structures is not bad history, it simply is not history at all. This approach is prevalent in Marxist history, and was transmitted to the socio-economic contextualization of landscape studies. The whole notion of linking the two narratives, as cause and effect, was shown previously to be inadequate empirically. More fundamentally, this project was doomed from the outset as an example of *ignoratio elenchi*, conflating the historical and the scientific.

> What, I take it, is fundamental to this conception is that we should be able to separate the cause and its effect, and endow each with a certain degree of individuality; but it is just this which is impossible while we retain the postulates of historical experience.[27]

Second, in the totality that is the historical mode, no specific can be privileged: 'nothing in the world of history is negative or non-contributory'.[28] As such, no event can be 'decisive', or a 'turning point'. It is in the light of this that we can be sceptical of the historical nature of the claims of symbolic contextualism to be studying moments of 'high specific gravity' in the landscape idea.[29]

A certain view of historical explanation follows from this: it is by full description that we explain things in history, for this is the only way to recognize that in history no survival has a different status. History is thus characterized by Oakeshott as a drystone wall: its total character is given by its parts which take on their character by their relation to other stones (events), the relation being one of contiguity, not a togetherness created by mortar (i.e., historical laws or underlying factors).[30] There is no archimedean point from which to explain history: understanding will come only by looking at the stones of history. It is in the light of this that the previously identified ambiguity in the symbolic contextualization of landscape between description and explanation is helpful: unlike other conflations, it is not *ignoratio elenchi*, but the reverse – an acceptance of the requirements of historical discussion.

Oakeshott's most extended treatments of the nature of history start with historical evidence and its limits. 'I take it, first, that history is concerned only with that which appears in or is constructed from record of some kind.'[31] This leaves Oakeshott with a pragmatic approach to what history is and the certitude of its findings:

> 'What really happened' (a fixed and finished course of events, immune from change) as the end in history must, if history is to be rescued from nonentity, be replaced by 'what the evidence obliges us to believe'. All that history has is 'the evidence'; outside this lies nothing at all.[32]

In the light of this, the evidential impatience of previous contextualizations of landscape representations appears misguided, and the search for absences which would not have been recognized as such by contemporaries, given the state of discourse at the time, appears positively unhistorical. History is a coherent way of seeing things, but it is not a revelation of the whole. If the 'reality' of social conditions eludes the evidence of landscape representations, an historical inquiry into these representations will be forced to ask other questions. If the meaning of landscape representations is partly internal to discourses of landscape, or to realities in intellectual rather than social history, this must be respected by historical inquiry. Whilst we may agree with Collingwood on the need to approach evidence with a question and torture it,[33] on some matters the evidence cannot speak and hence only screams the 'truths' the torturers already knew beforehand. We need to approach the evidence with 'that mixture of activity and submission we call curiosity'.[34]

For Oakeshott the past is different from the present, and thus demands attention in all its specificity and otherness if an approach is to be distinguished by its concern for the past for the sake of the past; yet this difference does not amount to an ontological otherness, history being a 'passage of differences', such that the past is not an entity to be opposed to the present.[35] As such, history is both approachable and other: the study of history is not vitiated by the ontological otherness of some monolith called 'the past'.[36]

Contextualism: a methodological approach to the historical mode of experience

Skinner and linguistic contextualism

'[F]or most historians, the contextual imperative has become ... the fundamental distinction between historical and non-historical studies.'[37] It can be seen as the methodological attempt to pursue the historiographical implications of an Oakeshottian view of history as a mode of experience.[38] Skinner characterizes the starting point for a contextual study:

> I am only pleading for the historical task to be conceived as that of trying so far as possible to think as they thought and to see things in their way. What this requires is that we should recover the concepts they possessed, the distinctions they drew and the chains of reasoning they followed in their attempts to make sense of their world.[39]

Skinner, Pocock and Dunn have all stressed the linguistic context as a manageable arena within which to pursue this inquiry. This context is specific to a particular discourse (they all focus upon the history of political thought) and within that discourse a variety of genres and idioms will emerge.

> Each of these languages, however it originated, will exert the kind of force that has been called paradigmatic ... That is to say, each will present information selectively as relevant to the conduct and character of politics, and it will encourage the definition of political problems and values in certain ways and not in others.[40]

Furthermore, if an individual wants at any given time (and place) to be understood as contributing to a debate, he will have to adopt a great deal of this language and its resultant problems and values. The author having learned to understand a concept within a certain discursive formation is bound to investigate with respect to that discourse. Language 'supplies the categories, grammar, and mentality through which experience has to be recognised and articulated'.[41] The question or insight an author generates must always be located in relation to something, and it is by looking at this discursive situation that the rationale of a specific 'performance' can be grasped.[42]

Specifying this context is the real methodological problem, and only the most general prescriptions will apply generally to such a linguistic contextualization.

> The historian pursues his first goal by reading extensively in the literature of the time and by sensitizing himself to the presence of diverse idioms.

To some extent, therefore, his learning process is one of familiarization, but he cannot remain merely passive.[43]

The activity required is the attempt to discover 'moves' and 'countermoves' whereby conventional assumptions were challenged by terminological innovations, by the deliberate use of familiar terms in unfamiliar ways, or by the use of a genre to subvert its own conventional message.[44] All this reinforces the point Skinner was motivated by in his earliest methodological statement, namely that neither text nor context by itself will yield historically satisfactory discussion. For similar reasons, Skinner's contextualism treads a middle path between the voluntarism which stresses authorial creativity and the determinism which speaks of the death of the author.[45]

By focusing on linguistic context, Skinner puts to one side the issue of truth in a foundational sense. The concern is not for the correspondence between reality and an historical statement, but for an identity between what we claim an author said and what he did in fact say:

this does not mean that such an explanation [i.e. an historical one] cannot include an account of why X thought it to be true ... nor even an account of why X thought it to be true though many with the same values as X ... would have been able to show conclusively why it was false. What explanation cannot give in purely historical terms is an account of why it *is* true or false.[46]

Linguistic contextualism is concerned with evidence which exists, and cannot make up 'moves' in a discourse for which it has no evidence:

it is a cardinal rule of the historiography which defines itself as the recovery of languages that we must reconstitute the languages we find and follow the implications of their discourse wherever these may lead.[47]

As such, it is concerned with what the evidence obliges us to believe, rather than with what actually happened.

If, as my previous argument has suggested, we can see landscape as an independent subject of inquiry, then the approach of linguistic contextualism would seem to have some claims as a methodology by which to study it within the historical mode of experience. Given that eighteenth-century England saw an ongoing discussion about landscape in various media, we can look to the moves and countermoves by which understandings about the nature of landscape and evaluations of various landscapes were reinforced, challenged and changed. Within such a project, it is important to isolate various genres of discourse which established the discursive frame within which authors operated and which they altered. It is necessary, for example, to distinguish between descriptions of paradisiacal landscapes and those of actual landscapes, between remote landscapes and frequently described ones. Furthermore, there is another set of considerations related to genre: the specific genre – poetry,[48] travel,[49] painting[50] – is important, as authors and painters worked within a given set of conventions which are as important to the representation used as the landscape being represented. None of these points is claimed to be particularly novel, but landscape has not previously been seen as its own discourse: internal histories of landscape were not discursive, and histories of the landscape discourse were always contextualized exclusively to something beyond landscape.

Landscape is a way of seeing which has to be acquired culturally. It is because this is the case that linguistic contextualism would appear such a potent approach: individuals have to learn the set of conventions related to landscape evaluation in order to communicate ideas about it, and can only then go about adjusting or altering prevalent perceptions of landscape. This changes our approach to landscape history, dispensing with the foundational truth of a representation as an issue: a linguistic contextual approach can only show the range of understandings an author could have achieved of landscapes within various genres, and how they relate to that range. The landscape discourse talks largely about itself; its response to social realities, or indeed ideological demands, is discursive, based upon its existent modes of expression. As such, there are no 'absences' in the landscape discourse, only the loss of certain pieces of evidence to further our picture of that discourse, whether due to actual loss or the condescension of history. An absence would imply something about which the landscape discourse should speak, yet this is to put mortar between the evidential stones: the landscape discourse in its various guises is evidence of itself and should be treated as such.

Caveats to linguistic contextualism

Bevir has suggested that linguistic contextualism should be seen as only one line of useful evidence for recovering the meaning of a text. Agreeing that writers must use language in ways which others of their time and place can understand, he argues that the shared conventions for both understanding and expression are far less determinate than Pocock suggested. Within a broad set of linguistic conventions, historians or contemporaries can approach a given text with a faulty 'prior theory', yet develop an adequate 'passing theory' by which understanding of the intention of the text is achieved. As such, a knowledge of the language of discourse being adopted is not important except in the most sketchy way.

On this basis Bevir, whilst accepting the importance of linguistic contextual work in elucidating meaning, argues:

> linguistic contexts have no greater claim on the historian than do other possible sources of evidence...we cannot specify in advance what evidence

either historians in general or any particular historian will have to consider in order to come to understand a text correctly.[51]

While linguistic contextualism has dammed

the tides of psychological and sociological reductionism...a new form of reduction has become evident, the reduction of experience to the meanings that shape it...a new form of intellectual hubris has emerged, the hubris of wordmakers who claim to be makers of reality.[52]

In fact, this is a re-emphasis of points the linguistic contextualists have made, not a rejection of their method: Skinner himself admitted 'we must be careful not to assume too readily that the business of interpretation need always be entirely a reading process'.[53] Confusion has ensued by trying to make Skinner's method more systematic than it is or could be: the resources and contexts drawn upon must be determined in each individual case, by the dictates of the discourse in which a specific performance is being contextualized, and by the exigencies of evidence.[54]

In a study of the landscape discourse, it is important to bear in mind that linguistic contextualism is not simply a focus upon the terminological. For to focus exclusively upon the linguistic would be to collapse that duplicity of landscape which first allowed recognition of the relative independence of the landscape discourse:[55] it would be to ignore the sensory pull of landscape, simply replacing the ideological with the linguistic. The extra-linguistic seems particularly relevant to landscape studies precisely because of its duplicitous nature. While landscape may be seen through the eyes of linguistic and intellectual conventions, there is also the possibility of the 'shock of the new', and the attempt to describe this: Fabricant discussing Swift and Barrell discussing Clare show how both subverted the conventional language of aesthetic appreciation.[56] While both were clearly making countermoves in a discourse, with Clare in particular struggling to break out of the demands eighteenth-century topographical poetry made upon him, both were making these moves in response to specific landscapes. To argue that both made moves within a discourse would be as incomplete as it was true, for it overlooks a descriptive element vital to giving the actor a reason for making his move.

The contextualization of landscape studies, then, must look beyond language to visual, experiential, social, biographical and intellectual contexts in order to describe and thus contextualize more adequately, the specificity of a discussion of landscape. The landscape discourse can also be illuminated by reference to closely related discourses, perhaps most obviously that discussing the natural world in general, from which it was so poorly distinguished throughout the eighteenth century.[57] An initial concern to distinguish the ways in which an author is distinctive within the landscape discourse is then more fully investigated by moving beyond the linguistic. This is not so much explanation as the more complete understanding of an observed position within a debate. This would appear the most satisfactory way to recognize both the specificity of landscape, and that its autonomy is not tantamount to isolation from other discourses.

Criticism and the claims of linguistic contextualism

To introduce the notion of putting a work into context, immediately leaves one open to the charge of making historical work an impossible dream:

> the relevant context may have to be defined so that it includes all 'the theoretical, literary, and religious traditions and other cultural resources that historians know to have been accessible to most well-informed members of a given society at a given historical moment.' In other words, the relevant community of discourse may have to include all of Western civilisation. And more....[58]

Contextualism is thus seen as the historical mutation of Laplace's dictum, the empty formalism that if we understood everything bar the text, we could sensibly locate the text in its environment.

This does not make contextual inquiry impossible in principle, but only reminds us of our inability to reach a final interpretation. It does not deny that the more contextual work that is done by generations of historians, the more easily we can locate the problem situation to which an author was responding through his work. Moreover, the criticism seems to ignore the rather modest claims Skinner *et al.* have made for their approach. Their aim is not certitude or the 'whole truth':[59]

> we must certainly be careful to avoid the vulgarity...of supposing that we can ever hope to arrive at *'the* correct reading' of a text, such that we may speak of having finally determined its meaning and thereby ruled out any alternative interpretations.[60]

The aim of linguistic contextualism has always been to delimit the possible range of historical meanings an utterance could have, and only then to assign relative plausibility. Context only shows what might have been communicated in a given situation, it is not a determinant.

It is clear that a linguistic contextual approach has to select the evidence it addresses as its relevant community of discourse, then has to select the relevant contextual evidence by which to deal with a specific performance within that discourse, and then develop a narrative description of its findings. Modern critics suggest that this means history cannot be objective,[61] and yet in this sense contextualism has never claimed to be objective.

[I]t is impossible to exclude criticism from history, and where there is criticism there is judgement...An event independent of experience, 'objective' in the sense of being untouched by thought or judgement, would be an unknowable...[62]

But this does not allow us simply to dispose of the notion of truth as what the evidence obliges us to believe.

The other main complaints relate to the impossibility or irrelevance of history as it is practised by the contextualists. Thus Femia, in his critique of Skinner's method argues 'all history is "contemporary history", dictated by the interests of the historian; study of the past is valuable only in so far as it throws light on present problems or needs'.[63] It is not clear on what grounds this claim of the impossibility of history as a mode concerned for the past for its own sake is made. The interest of an historian may be directed towards a certain topic for present-day reasons, and the historian also cannot but approach the evidence with presuppositions, but neither of these truths suggests that history as a concern for the past is impossible. The reason for studying a given topic need not determine the conclusions arrived at, just as it does not follow from approaching the archive with presuppositions that they will be confirmed.[64]

A second claim is that contextualism, even if it is practicable, is merely dusty antiquarianism which would 'render intellectual history gratuitously barren'.[65] Why it is irrelevant to show what great thinkers in history saw themselves as doing is not clear, especially when this forces us to see the contingency of ideas and beliefs we take for granted. But further, the relevance of the complaint of irrelevance must be considered; on any view of history as being distinguished by its concern for the past for its own sake, the complaint is itself irrelevant, *ignoratio elenchi*. The historical approach is a mode of experience distinguished precisely by its concern for the past: the claim of irrelevance amounts to practitioners of the practical mode of experience complaining that others are not utilizing the same mode. As such, it is not a criticism of history at all, but of the existence of ways of thinking other than that which generated the criticism. It makes no more sense than complaining that a polemicist has distorted the historical facts. Contextualism itself has been guilty of such a confusion when it has complained at interpretations which have been unhistorical but never claimed to be otherwise.[66]

The claim that contextualism is irrelevant also ignores that the aim is not to rule out other approaches to historical works;

I see no impropriety in speaking of a work's having a meaning for me which the writer could not have intended...I have been concerned only with the converse point that whatever the writer is *doing in* writing what he writes must be relevant to the interpretation, and thus with the claim that *amongst* the interpreter's tasks must be the recovery of the writer's intentions.[67]

Contextualism is the attempt to clear up what the writer could have been trying to say at the time he said it. As such some interpretations will be historically grotesque, but this is evaluative only of their status as history; it does not impinge upon their interest.

Part II

Landscape and Religion, 1660–1800: Preliminary Contexts

Parts II and III of the book will put into operation the contextual approach to landscape depictions we have established in Part I. Part II will establish both the heterogeneity and the coherence of discussions of landscape in the 'long' eighteenth century. Chapter 3 is a brief prospectus, whose aim is to give the reader some sense of the sheer diversity of meanings which landscape could have in eighteenth-century prose and poetry. It is suggested that any attempt to theorize landscape as a 'window' on intellectual life must recognize and deal with this diversity. The second part of the chapter suggests that it is the religiosity of eighteenth-century English society which inspired the diverse landscape discourses we have mapped out. In short, landscape discourses were entwined with religion, something which work on the meaning of landscape has at best occluded and at worst simply denied. Chapter 4 moves the argument forward, suggesting not simply that religion grounded landscape depictions, but further that different theological positions led to different ways of describing the landscape. This contention is made good by a study of the dominant form of Anglicanism in the 'long' eighteenth century, namely Latitudinarianism. Chapter 4 shows that Latitudinarian theology was at the root of many of the most celebrated landscape writings of the period. It is against this benchmark that the very different style of Samuel Johnson's writings will be counterpointed in Part III.

3
Diversity and Coherence in the Discourse of Landscape in the 'Long' Eighteenth Century: A Preliminary Survey

Having suggested that landscape must, as a working hypothesis, be taken *sui generis* and argued that this is best achieved by a focus on the historical languages of debate, in this chapter I seek, in consequence, to map out some of the intellectual issues which were connected to printed discourses about landscape in the eighteenth century (my concern is not with practical matters of landscape design). Having shown the diversity of themes treated in writings about landscape, I will suggest that the structure which drew these themes together derived from an intellectual and social foundation very different from that invoked by the recent work in landscape studies discussed in Chapter 1, namely theological discourse. I also briefly locate Johnson within this structure of debate as a prelude to the third part of the book. The whole argument is framed by a brief discussion of the chronology of the rise and fall of the set of discursive connections the language of landscape was connected with in the 'long' eighteenth century.

The religious origins of eighteenth-century discussions of landscape and nature

The seventeenth century saw the decline of one complex of ideas pertaining to the natural world.[1] How, then, was the natural world incorporated into the intellectual life of the 'long' eighteenth century? Hill has provided one answer, arguing that the close relationship between the Bible and the appreciation of landscape and the natural world, at both a conceptual and linguistic level, was severed by the Restoration nexus of Latitudinarianism and the Royal Society.[2] In fact, however, that nexus simply forged a new connection of landscape and religion, part of a broader strategy which dignified discussions of the natural world by connecting them to debates seen to be of greater significance.

The combined importance of Newtonian physics and rationalistic theology to interest in nature in the later seventeenth century has often been pointed out. In the English enlightenment,

> science formed part of the established Church's own armoury – in large part thanks to the type of education offered to intending clergymen at Cambridge and, to a lesser extent, at Oxford.[3]

As we will see in greater detail in Chapter 4, it is mistaken to see Latitudinarianism solely through the attacks it was subjected to by High Churchmen and the Evangelical revival as a tepid faith.[4] Its aim was to render religion respectable by giving it a rational basis. As Simon Patrick had put it 'nor will it be possible otherwise to free religion from scorn and contempt, if her priests be not as well skilled in nature as the people'.[5] The book of nature was an increasingly attractive defence of religion. From the outset, then, the purpose of looking at landscape and nature was a physico-theological one, the connection being codified by the institution of the Boyle lectures.

The 'grander aspects' of nature, known more commonly as the sublime, also emerged as landscapes worthy of interest for religious reasons. Newton's vast sensorium had to reflect God's immensity, and analogically this was best achieved by the vast in landscape and nature. Burke's *Philosophical Enquiry* has often been understood through the 'conservatism' of his *Reflections*, but Harris has shown it is better to see both as efforts to work through concepts of the divine origin of inequality.[6]

The interest in gardening, so important to the development of the discourse of landscape in eighteenth-century England, also had religious origins. A number of emblematic gardens reinforced the message of human transience, perhaps most notably The Denbies, which had a clock to chime every minute, as well as coffins, skulls, and contrasting pictures of a sinner and a man of virtue on their deathbeds.[7] Also significant were Pope's garden at Twickenham, with its stone of the five wounds of Christ at the grotto, and its crown of thorns at the river, and Young's inscription in his garden at Welwyn: *Ambulantes in horto audiebant vocem Dei.*[8] The importance of Eden as a justification of gardening should also be borne in mind: 'for God doubtless would never have placed him in a Paradise, had not a Garden of Pleasure been consistent with Innocence'.[9] What is more, the resultant design of English gardens connected back to the divine view of nature expressed in Newtonianism:

> The emphasis upon an "artful wildness" among British theorists of the 'natural' garden may be better understood once we sense the necessity, after Burnet and Newton, for scientists and divines alike to adopt a less simplistic view of the order of things, a view that would make Nature's

rudeness and variety and limitless extent fundamental elements in a more intricate, though by no means less real, harmony of creation.[10]

Thomas gives a wealth of evidence suggesting the religious origins of the close observation of, and concern for, nature. What he says of the sensibilities behind kindness to animals applies more broadly:

the debate on animals thus furnishes yet another illustration of that shift to more secular modes of thinking which was characteristic of so much thought in the early modern period. Yet the initial impulse had been strongly religious.[11]

Finally, the travel account gained an initial *raison d'être* from Christianity.

Christian missions had made contact with ancient cultures and religions in India, China, Japan and seventy-odd volumes of Jesuit records bore witness to them. We in England have domesticated two such volumes, *Robinson Crusoe* and *Gulliver's Travels*, which were written by Christians for Christians.[12]

What is more, religion played a role in the significance of journeys to the Holy Lands and of sacred geography.[13] In the domestic context, as well, regional studies were 'stimulated by the search for documentary support in the political–theological strife associated with the rise of Protestantism'.[14]

Rationalistic religion, by drawing the face of the land into debates about the proof of God, gave landscape writings an unprecedented role and importance in English intellectual life. As well as rationalizing religion by naturalizing it, this elevated the status of nature and landscape by intellectualizing them. We will see how this 'mainstream' of landscape appreciation was manifested in a literary context in Chapter 4.

'Interwoven with the vegetable world'[15]: the diversity of the landscape discourse

It would be erroneous to argue that because the accepted genres of writing and ways of seeing landscape in eighteenth-century England had their origins in religious discourse, therefore landscape was a religious phenomenon in all its subsequent manifestations. The continued importance of religion to the apprehension of landscape in the eighteenth century is not in doubt, but independent intellectual debates also developed. Once landscape and nature had been intellectualized, it is not surprising that they became sites for a far broader range of intellectual debates.

Gascoigne traces the gradual dissolution of the 'Holy alliance' of Newtonian science and religion in Cambridge, concluding the confidence that all fields of learning could be integrated by reference to common religious or ethical ends was undermined by the ever-growing barriers of expertise and formal qualifications which separated one discipline from another [in late eighteenth-century England].[16]

Similarly, the discourse *on* the sublime was overtaken by a discourse *of* the sublime, taking on a life less connected to its religious origins.[17] Gardening manuals and travel journals also built up modes of representation and intertextuality far removed from their starting points. Thus across the range of genres,[18] landscape came to be a vehicle for discussion of a large number of themes, moving out from religion to morality and the connected debates on polity and patriotism, these last in turn being related to an antiquarian and a utilitarian approach to the landscape. Equally, some sought to develop an approach distinguished by the concern for accurate description divorced from other discourses.

'Nature is Christian'[19]

Throughout the eighteenth century, landscape was commonly connected to various aspects of a Christian worldview, most notably to the Mosaic chronology of the Earth based on a concept of Nature

> as an event unfolding in time and involving human agents, it was properly seen as a play or poem with its own beginning, middle, and end – a drama whose every episode, from the fall of a sparrow to the bursting of a star, takes place within the Providential scheme of history leading from Genesis to Apocalypse.[20]

It was only late in the eighteenth century that Hutton began to develop a consistent alternative chronology for the earth,[21] so it is unsurprising that discussions of landscape and nature betray a continued belief in, and reference to, the Creation, Eden and the Fall of Man, the Deluge and Apocalypse.

The design argument that the world gives evidence of a Divine artificer was very common. Whilst Hume's *Dialogues Concerning Natural Religion* (1779) may be seen as destroying the conventional argument from design, their significance should not be antedated, for in the eighteenth century the design argument's strength was overwhelming. The argument covered all levels of nature appreciation. On the smaller scale, Shakespeare's words on sermons in stones were oft quoted, and Bowden could complain:

> Yet Man, unthinking Man! regardless sees
> Himself more dull, Herbs, Animals and Trees;
> Sees all the fair Creation round him rise,
> Sees Ants, and Flow'rs with unsurprised Eyes.[22]

More famously, Cowper wrote:

> — Not a flow'r
> But shows some touch, in freckle, streak, or stain,
> Of his unrivall'd pencil —[23]

Man had divine lessons to learn from nature's minutiae in the eighteenth century as Ruskin would in the nineteenth. The design argument was also used in relation to the large scale. The fullest expression of this was Blackmore's *The Creation*, of which Book I 'contains the proof of a Deity, from the instances of design and choice, which occur in the structure and qualities of the earth and sea', whilst Book II deals with the same theme with respect to the solar system.[24]

References to Eden were, unsurprisingly, most common in gardening treatises. This image was invoked in the anonymous *The Rise and Progress of the Present Taste in Planting Parks* where 'at his [Capability Brown's] command a new Creation blooms'.[25] It could also be reversed, the orderly beauty of the garden becoming a design argument by analogy for the existence of a divine creator:

> a Man that should meet with a Palace, beset with pleasant Gardens, adorned with stately Avenues, furnished with well-contrived Aqueducts, Cascades and all other Appendages conducing to Convenience or Pleasure...we should conclude the Man was out of his Wits, that should assert and plead that all was the Work of Chance, or other than of some wise and skilful Hand. And so when we survey the bare Out-works of this our Globe.[26]

References to the Deluge and its effects on landscape were also frequent. Travellers to the Middle East and Africa made reference to the possible effects of the Flood,[27] but even in Britain, Hanway could invoke this as an explanation of Stonehenge:

> if we contemplate them on a supposition of their having been once emboweled in the earth, just where they stood, the soil washed down from them by the deluge, it fills the soul with religious fear.[28]

The Apocalypse was also discussed: the earth may be beautiful, but it is finite, and hence the pleasures of this world, of nature and landscape, must be understood as minor and best dignified as evidence of a more important truth:

> — Man, 'tis true,
> Smit with the beauty of so fair a scene,

> Might well suppose th'artificer divine
> * Meant it eternal, had he not himself
> Pronounc'd it transient, glorious as it is,
> And, still designing a more glorious far,
> Doom'd it as insufficient for his praise.
> These, therefore, are occasional, and pass;
> Form'd for the confutation of the fool,
> Whose lyeing heart disputes against a God;
> That office serv'd, they must be swept away.[29]

The Apocalypse could also draw the prospect into denominational debate, as in Burnet's *Sacred Theory*. In book III he sought to suggest where the purifying fire would begin, arguing:

> if we can find some Part of the Earth, or of the Christian World, that hath more of these natural Dispositions to Inflammation than the rest; and is also represented by Scripture as a more peculiar Object of God's Judgments at the coming of our Saviour, we may justly pitch upon that Part of the World, as first to be destroyed.

He found such a combination in 'the Roman Territory, or the Country of Italy' which was well known to natural philosophers as 'a Store-house of Fire', and, as obviously to Burnet, 'seeing "Mystical Babylon," the seat of Antichrist, is the same Rome and its Territory' his conclusion was that 'there is a Fairness, on both hands [the natural and the providential] to conclude, that, at the glorious Appearance of our Saviour, the Conflagration will begin at the City of Rome'. Burnet's theory was not taken seriously in the scientific community for long, but his vision remained powerful.[30]

In addition to these fixed points of biblical chronology influencing the understanding of landscape, there was also the notion of God's continued presence as expressed through exceptional events such as earthquakes. Gibbons was uncompromising in his view of the London earthquake:

> Hints of Wrath th'Omnipotence behind
> — the God
> Who struck the angry blow was unador'd,
> As nature's hidden elemental War.

The only answer was 'And turn with speedy Penitence and Pray'r, / And Faith, on JESUS' bleeding Merits fix'd.'[31] This perception extended across the century, from Story in Jamaica in 1692 to Cowper of the Sicilian earthquake (1783).[32]

Two religiously connected metaphors of landscape and nature should be mentioned, given the frequency of their recurrence. First are the references throughout the century to the 'book of nature' as instructing us to look to something higher. Milton had said:

> To ask or search I blame thee not, for Heav'n
> Is as the Book of God before thee set,
> Wherein to read his wondrous Works, and learn
> His Seasons, Hours, or Dayes, or Months, or Yeares[33]

Given Milton's influence on nature poetry and perception in eighteenth-century England,[34] the continuity of this theme is unsurprising. Second was the conflation of prospects with heavenly prospects. The heavenly prospect was always said to be far wider than man's. But as we learn from the Book of Nature:

> — larger Prospects of the beauteous Whole
> Would gradual open on our opening Minds;
> And each diffusive Harmony unite,
> In full Perfection, to th'astonish'd Eye.
> Then would we try to scan the *moral World*.[35]

By the same token, to be in heaven is

> To view more glorious sights in realms of light,
> Than this dim world to mortal eyes can yield.[36]

for as perishable man's faculties cannot grasp, so the perishable earth cannot offer, the whole prospect.[37]

Yet the use of religion as a way of understanding nature and landscape (and *vice versa*) was not uncontroversial. Attacks came from High Churchmen, who stressed revelation as the route to an understanding of God.[38] The Evangelicals and Methodists also tended to a more sceptical position about the book of nature, all three drawing to some degree on Hutchinson's anti-Newtonian natural philosophy.[39] This is not to suggest that any of these positions ruled out the evidence of nature, but that nature was only a poor supplement to revelation, which led to a discernible downplaying of the connection of religion, nature and the earthly prospect.

Whilst many have sought the decline of religion in the arts of eighteenth-century England,[40] it would seem that in landscape discourse at least, there was no marked decline in discussions of the themes outlined above or debate as to the propriety of that mode of argument as the century wore on.

The moral landscape

Intimately interconnected with Christianity was the notion of moral lessons in the landscape. 'No age has been blamed more for its innumerable allegories than has eighteenth-century England',[41] and the landscape proved a pliant vehicle for such allegorization.

One of the most common themes was transience. Ruins were particularly conducive to this theme, redirecting the focus from the sensual landscape to lessons of the mind:

> Fall'n, fall'n, a silent heap; her heroes all
> Sunk in their urns; behold the pride of pomp,
> The throne of nations fall'n; obscur'd in dust;
> Ev'n yet majestical: the solemn scene
> Elates the soul —[42]

Warton's *Enthusiast* caught 'the moral strains from nature's muse', and Dodsley said that while 'we rave delighted' about nature we catch the sight of

> The early mower, bending o'er his scythe,
> Lays low the slender grass; emblem of Man,
> Falling beneath the ruthless hand of Time.[43]

Modelled on the Horatian *beatus ille* (happy man) theme and seen through the lens of Christianity, retirement was another important moral theme interlinked with landscape.[44] The ideal had broken down by the 1760s, such that a person had to be sociable to attain wisdom.[45] Yet the theme of retirement was not abandoned, but transformed: that the city was corrupt remained a commonplace, and the superiority of retirement to the man of taste, virtue and industry was rarely questioned. The critique of retirement was not of its moral aims, but its efficacy, suggesting that the country was most useful as a temporary retreat, before rejoining active life. Satire of retirement pointed out not the bankruptcy of a moral ideal, but of a method of obtaining it which had been swamped by fashion.

Another linkage between landscape and morality was seeing landscape as a visualization of the moral health of its inhabitants. This can be seen in travel accounts, where the state of the polity is reflected in the look of the land, and also in tours where morality, utility and beauty were interlinked,[46] but it is most apparent in country-house poetry where eulogy of owner, and house, garden and landscape were interchangeable, and the landscape frequently acted as praise of its owner. Thus Bolingbroke's Dawley Farm:

> See! emblem of himself, his Villa stand!
> Politely finish'd, regularly grand![47]

But a reply printed in the following month's *Gentleman's Magazine* shows how the trope could be reversed. If, as the admirer of Bolingbroke had said, Dawley Farm was really a villa not a farm:

> Let Dawley triumph in the builders art,
> And stand the emblem of the owner's heart;
> If the low title wrong the pompous frame,
> Observe the man! his fraud is just the same.[48]

If the connection of landscape to virtue was controversial, a weaker argument of particular importance in the defence of gardening took the same linkage in a negative form: even if gardening was not positively conducive to virtue, it was at least not encouraging vice. Lawrence defended his encouragement of gardening to the clergy as being designed 'not to make them envy'd by Magnificence, but to make them happy, by loving an innocent Diversion'.[49] Price justified the picturesque similarly: 'I have been desirous of opening new sources of innocent, and easily attained pleasures.'[50] Yet not even this could be accepted by all. In some instances it was believed that gardens and the landscape could actually set an inappropriate moral example. Thus Wesley's praise of the gardens of Stourhead was tempered by his disgust for the 'images of devils' in the temples where he defied 'all mankind to reconcile statues with nudities to common-sense or common decency'.[51] Samuel Horsley, the High-Church bishop, similarly opposed clerical farming by which 'the whole dignity and sanctity of his character will be obliterated'.[52]

There was, then, a debate over the relation of morals to the look of the land. On a theoretical level, this was reflected in debates about the independence or otherwise of aesthetics, those influenced by Platonism favouring a linkage of morals and aesthetics.[53] Yet there were equally powerful opponents of this linkage. Certainly, by the 1780s when Percival wrote an essay *On the advantages of a taste for the general beauties of nature* which argued that 'physical and moral beauty bear so intimate a relation to each other, that they may be considered as different gradations in the scale of excellence', the stiff response of *An attempt to shew, that a taste for the beauties of nature and the fine arts has no influence favourable to morals* was unsurprising.[54]

Polity, economy and rank in landscape discussion

For those who accepted the importance of tying morality and aesthetics together, landscape became a vehicle for ideals and arguments related to polity, economy and society.

In topographical poetry, from *Cooper's Hill* (1642) until the decline of the genre, the linkage between prospects and politics was commonplace. Viewing Windsor led Denham to a long passage in praise of the English monarchy. Denham also established a concern for reflections on trade with his musings on the Thames:

> Rome only conquer'd halfe the world, but trade
> One commonwealth of that and her hath made.[55]

However, the exact nature of the linkage was less easily determined: landscape could be deceptive, the attractive prospect tempting the viewer to ignore the infringement of liberty, or it could be reflective of the health of the governance of a realm. The theme of landscape as political deception was used in relation to Italy:

> But small the bliss that sense alone bestows,
> And sensual bliss is all the nation knows.
> In florid beauty groves and fields appear,
> Man seems the only growth that dwindles here.[56]

The reverse also held: the forbidding landscape would be attractive if it had liberty, a point exemplified by Switzerland.[57] Just as the landscape of Italy was deceptive, so the ruins in Rome were instructive:

> O Britons, O my countrymen, beware;
> Gird, gird your hearts, the Romans were once free.[58]

Landscape as a reflection of polity was one oft-invoked justification of travelling:

> 'I know no better way of estimating the strength than by examining the face of the country' such that 'when I perceive such signs of poverty, misery and dirt among the commonality of France, their unfenced fields dug up in despair; without the intervention of meadow or fallow ground... I cannot help thinking they groan under oppression.'[59]

Denham's association of trade and topography continued to be important. Defoe viewed the land through the filter of trade, a concern summarized in *Caledonia*:

> Fitted for Commerce, and cut out for Trade;
> The Seas the Land, the Land the Seas invade.

and:

> Nature, that well foreknows a Nation's Fate
> Thus fitted Caledonia to be great.[60]

Arguments of profit and trade also informed the literature on gardening. That the garden should be both profitable and pleasurable remained a common claim, manifesting itself later in the century in the *ferme ornée* and in the writings of improvers. Even Gilbert White described the quality of trees by the prices paid for them.[61]

The nature of these reflections in landscape discussions must be borne in mind. Landscape became a vehicle for generalized reflections on the importance of liberty and commerce, and the disastrous effects of tyranny. The nature of good governance, and how and whether it could be 'seen' in the land was at issue.[62] More exact links to high politics were far less frequent. That there were instances of this is clear: Knight's exchange with Repton after the French Revolution, and Mason's *Heroic Epistle to Sir William Chambers* are good examples.[63] The convenience with which these examples fit into the socio-economic model of contextualization should not, however, mislead as to their inability to represent the patterns of intention the landscape discourse as a whole displays.

Similarly with trade the concerns betrayed in agricultural treatises of the eighteenth century were not those encompassed today by the term profit.[64] Moreover, the discussion was informed by the generic precedents available from the agricultural manuals of Cato, Varro and Columella, the eighteenth century inherited a series of standard ways of approaching the subject of 'profit and pleasure'.[65]

'Class' issues connected to landscape discussion are also to be found. Praise of the aristocracy through praise of their houses and gardens was common, this being linked to a moral discourse about the individual owner and his family. It is hard to discover the consciousness of a rise of the middle classes in eighteenth-century England: what is evident is a concern for parvenues, which is important to debates about retirement and gardening.

> The wealthy Cit, grown old in trade,
> Now wishes for the rural shade[66]

A series of poems by Graves satirized these cits: in 'The Cascade', for example, Curio has an artificial cascade:

> Regardless of domestic matters,
> Curio plays on; the torrent patters

but his servant reminds him of his commercial origins and the absurdity of his aspirations to taste:

> 'Hold! hold!' cries Doll, with unfeign'd sorrow:
> Why Sir! – we are to brew tomorrow.[67]

Likewise in gardening, Chambers spoke in reference to Capability Brown of 'peasants [who] emerge from the melon grounds to commence professors'.[68] As with the middle classes, discussions of a lower class are notable by their absence. Nourse spoke of the 'common people': 'such Men are to be look'd upon as trashy Weeds or Nettles growing usually upon dunghills, which if touched gently will sting, but being squeez'd hard will never hurt us', but such an outburst is memorable for its singularity.[69] Crawford found Richmond Hill ruined by 'vile Plebeans, turbulent and loud',[70] but again this was unusual, the theme of rural retirement *from* such tumults being more common than the idea of tumult *in* the countryside. Issues of rank and subordination rather than class tension seem to be vital, being linked to overtly moral discourses about virtue, the virtues of ownership, and about fashion and taste.

The debate over patriotism and landscape

Closely related to the interleaving of political reflections in landscape writing was the theme of patriotism, the look of the land being woven into the character of a people and their country.[71]

The four nations forming Britain proved fertile ground for the linkage between landscapes and nationhood and for its contestation. One link was the eulogy of a land, as in *Devonia, an itinerary; written at the beginning of the present war*, where at the sight of Plymouth, the 'patriotic muse' exclaimed:

> From Instow northward, to the Dartmouth cliffs
> The sound is heard; forth issuing dauntless come
> Neptune's vice-gerunts, Mars' intrepid sons,
> Go on – ye sons of valour, and of fame; –
> Go on – our annals let your prowess grace,
> Rescue Britannia, and her wrongs revenge![72]

There was also the complaint that the British landscape had been ignored:

> Still shall we read of fam'd Versailles Bow'rs,
> Eternal Grotts, Cascades, and gilded Tow'rs;
> While unobserv'd our native stones remain,
> And Caverns yawn, and Mountains rise in vain?[73]

Also common were comparisons consistently to the advantage of one country. Smollett suggested that

> the Circus Maximus, by far the largest in Rome was not so long as the Mall; and I will venture to affirm, that St James's Park would make a much more ample and convenient scene for those diversions.[74]

Finally, there was the simplicity of Cowper's landscape patriotism:

> England, with all thy faults, I love thee still...
> I would not yet exchange thy sullen skies,
> And fields without a flow'r, for warmer France
> With all her vines...[75]

Yet the theme of landscape visualizing patriotism was also subject to subversion to create political attacks on the neglect or malevolence of other countries, a tactic used by Swift in his *Verses occasioned by the sudden drying up of St Patrick's Well* (1728). The patriotic landscape could visualize a xenophobic attack upon a nation. Thus where for Defoe Scotland was 'cut out for Trade', the same geography could be explained in less complimentary terms, as giving Scotland its other name:

Louseland; because its Promontories, Craggedness and Buttings-out into the Sea, with a Multiplicity of Nooks and Angles, have a kindred Appearance to the Legs and engrailed Edges of a Louse...Nor does the Comparison end here, for as that Vermin preys upon its Preserver and Fosterer, so does Scotland upon England.[76]

Neither eulogy nor denigration were uncontroversial tactics; on the contrary, the whole linkage between landscape and patriotism was questioned throughout the period. Baretti lambasted travel writers who fill

pages and pages with scurrilous narratives of pretended absurdities... very gravely insisting that these crimes and absurdities were not single actions of this and that individual, but general pictures of nature in the countries through which he has travelled.[77]

Patriotic paeans were also criticized:

Though patriots flatter, still shall wisdom find
An equal portion dealt to all mankind,
As different good, by Art and Nature given,
To different nations makes their blessings even.[78]

With Shandean associationism, Sterne satirized Smollett's mode of patriotic comparison:

The learned SMELFUNGUS travelled...but he set out with the spleen and jaundice, and every object he pass'd by was discoloured and distorted – He wrote an account of them, but 'twas nothing but the account of his miserable feelings.[79]

We can therefore see a set of discursive practices built around the connection between landscape and patriotism, whereby that connection was used, reversed and questioned. This is one exemplification of the fact that links between ideas of landscape and nature, and the society in which they existed were complex, and that debates in the former realm could generate their own momentum.[80]

Antiquarianism, historicism and landscape

Though often a prelude to political or religious reflections, the antiquarian and historical way of seeing landscape had its own coherence. This was shown clearly by Addison, who summarized the manner in which such an approach to landscape operated:

> I took care to refresh my Memory among classic Authors...I must confess it was not one of the least Entertainments that I met with in Travelling, to examine these several Descriptions, as it were upon the Spot, and to compare the Natural Face of the Country with the Landskips that the Poets have given us of it.[81]

This method of comparison was of lasting importance in Italy ('on classick ground', as Addison himself had put it).[82]

As Addison had Horace to guide him from Rome to Naples and made the return journey with Virgil,[83] so travellers to the Middle East frequently travelled with the ancients and the Bible. In seeking the site of Solomon's house and gardens Pococke rejected the 'village of Solomon' because 'it is a very bad situation, and there is no prospect from it', preferring a summit near the Greek convent of St George, reasoning that

> the summit of it commands a very fine view of the pools, Bethlehem, and all the country round; and this seems to be a situation for a house of pleasure worthy of the taste of Solomon; and it is probable that there were hanging gardens on the side of the hill; so as to answer this description exceedingly well.[84]

This is a good example of the antiquarian way of seeing: topography, reasoning and (in this case scriptural) reading interacting to create a consistent approach to landscape.

The same way of seeing was applied to Great Britain. Pococke made detailed notes on antiquities whilst touring England and Wales in the 1750s, as did Pennant in his *Tour of Wales* (1778). Such an approach was by no means confined to the genre of travel accounts. In topographical poetry, reflections on historical events occurring at a site were a commonplace, the best (and longest) example being *Edge Hill*, wherein over 150 lines were allotted to the Civil War fought there.[85] The historical approach also influenced gardening: important treatises on gardening attempted to reconstruct the gardens of the ancients,[86] and gardens themselves clearly appealed to certain classical and biblical texts.[87]

The antiquarian way of seeing landscape was not without its critics. Campbell attacked antiquarians so intoxicated with the love of the Ancients, that to support a Passage in Herodotus, to justify a fact related by Diodorus, or to make good somewhat in Pliny's Natural History, [they will] take a great deal of unnecessary Pains, and not only give a wrong Biass [*sic*] to their Thoughts, but which is still more extraordinary, to their Eyesight, so strongly are they possessed with a Desire of beholding Things not as they are, but as they were in the Times of those ancient Writers.[88]

Sterne irreverently summed up the complaint against those such as 'the great Addison who did it [travel] with his satchel of school-books hanging at his a—'.[89]

Agriculture, utility, beauty and landscape

The linkage between agriculture, fertility and utility to beauty was an important one. This is best known in Young and Marshall's tours,[90] but was part of a broader debate about the relation of utility and beauty in the apprehension of the land which continued throughout the century across a range of genres.[91]

In terms of landscape appreciation, this meant that the features highlighted and approved were different from those of ways of seeing discussed previously. There can be no doubt that Young, for example, was making a statement about beauty as well as utility; namely, that they are fused. As such, it is misleading to see Young simply as an apologist for 'capitalist farming', his remarks being a continuation of the ancient connection between beauty and use.[92] Moreover, improvement was viewed in a Christian context: Worthington quoted Ezekiel in which 'this land that was desolate is become like the garden of Eden', adding

> in this view, what think ye of the great improvements in husbandry of various kinds...And particularly, of that spirit of planting, and raising all manner of trees and shrubs; as well for use, as for ornament...?[93]

This connection of beauty to use was criticized in aesthetic theory

> however we may pursue beautiful Objects from Self-Love, with a view to obtain the Pleasures of Beauty, as in Architecture, Gardening and many other Affairs; yet there must be a Sense of Beauty, antecedent to Prospects even of this Advantage.[94]

The majority of aesthetic theorists over the century agreed in this particular, for in the theory of association, utility could be seen as beautiful only insofar as it was associated with pleasing ideas.[95]

The same division was in the ascendant in the appreciation of landscape. The connection between beauty and use was less common than the deployment of criteria of beauty derived from the sister arts,[96] most notably painting and poetry. The importance of poetry to landscape appreciation has already been mentioned in relation to antiquarianism, but it was of more general importance: Defoe cited *Cooper's Hill*, Baretti Dante, and Forster Shakespeare; despite differences in the discourses to which these writers habitually linked landscape, all could express themselves through the use of apt quotation.[97] The appreciation of landscape painting and the growing art of connoisseurship provided a fruitful parallel for the appreciation of landscapes,[98] and not only after the rise of the picturesque movement. In 1738 Herring was put 'much in mind of Poussin's drawings' by the Welsh mountains;[99] and it was only as the parallel became more elaborate after Gilpin that the analogy rather than synonymy of painting and landscape had to be reasserted.[100]

'Pure views' of landscape

In eighteenth-century discussion one possibility was to focus on landscape in a fairly 'unwavering' fashion. In other words, it was possible to write of these issues without connecting them to the other discourses outlined above. This is not to say that a 'naive' view, uninformed by prior knowledge, was possible, but that a relatively straightforward descriptive approach to landscape was one option.

Travellers frequently attempted faithful descriptions. One example is Wright's view of Vesuvius:

> We turn'd back to take a Survey of the Way we had come; and as we look'd upon the rough Currents we had pass'd along, their Surfaces, which seem'd so very irregular, when we were upon them, and like rude Heaps furl'd together at random, at that distance appear'd plainly to have form'd themselves into a perfect natural wavy Surface; which could only shew itself at such a distance as took off those Asperities, which distracted the Eye, and obstructed its appearing so at a nearer View, where the Eye cou'd not take in all together.[101]

Wright clearly recognized the difference between description and simile in this passage, was not attempting to render the scene in a painterly way, and was scrupulous in attempting to convey the appearance of the landscape from his survey point and how distance had affected that appearance. This 'will to verisimilitude' of travel accounts can be seen in a domestic context in Brown's description of Keswick.[102]

There was also an abiding concern for the minutiae of nature. Whilst the eighteenth century has perhaps become best known for a concern with general prospects, this was counterpointed throughout the century:

it was in the early modern period that the taste for small-scale flower-gardening gradually established itself as one of the most characteristic attributes of English life.[103]

In aesthetic theory, the priority to be given to the visual was a subject for debate,[104] and in discussions of landscape, whilst the visual may have been the most frequently invoked sense, other senses were also treated. Champion and Woodhouse both spoke in landscape poetry of the 'ravish'd ear', and once again Wright's travel account provides a good example, drawn from the Mediterranean near Genoa:

which came rolling to the Shore with such a Force, that the Sound it made resembled Thunder; the vast Waves with a grumbling at first, forcing Shoals of Pebbles along with them, which ended with a Rattling like that of a Thunder clap.[105]

Poets also showed that 'Nor rural sights alone, but rural sounds, / Exhilarate the spirit—':[106]

> Sweet was the sound when oft at evening's close,
> Up yonder hill the village murmur rose;
> There as I past with careless steps and slow,
> The mingling notes came softened from below;
> The swain responsive as the milk-maid sung,
> The sober herd that lowed to meet their young;
> The noisy geese that gabbed o'er the pool,
> The playful children just let loose from school;
> The watch-dog's voice that bayed the whispering wind,
> And the loud laugh that spoke the vacant mind,
> These all in sweet confusion sought the shade,
> And filled each pause the nightingale had made.[107]

The value of pure description was debated in poetry and travel writing. In both genres, opinion was divided between those who stressed the value of description as truth and those who saw mere description as unedifying. Hurn asked

> Shall distant towns that rural fame ne'er know,
> Because thy waters near no palaces flow?
> Or must thy banks less beautiful appear,
> Because no scepter'd hero wanders there?
> No —

He later went on:

> For nought so lowly, as esteem'd so small,
> But still Description lives, the same in all.[108]

The same belief in the value of description can be seen in the prefaces to several travel journals. Forster signalled his intention to describe Cook's voyage 'without the assistance of fictions' going on more aggressively to assert the value of unencumbered descriptions: 'the learned, at last grown tired of being deceived by the powers of rhetoric, and by sophisticated arguments, raised a general cry after a simple collection of facts'.[109]

The defence of description had affinities with the argument for the independence of aesthetics from moral discourse. Once that argument was accepted, the linkage between natural description to other discourses came to appear as a juxtaposition rather than a discussion of fused issues, and the call for unencumbered description then made sense. But just as that separation was the subject of controversy with respect to utility and morality, so was it in this case. The topographical poet Maude justified his 'ingrafting upon the natural stock of rural description' because

> objects of inanimated nature ... are found in part common to all countries; and few have features so peculiarly striking and dissimilar, as to mark them out for any great length of description ...

A purely descriptive poem from this perspective was 'on grovelling themes'.[110] In travel writing the same points were made by Fielding:

> if the customs and manners of men were every where the same, there would be no office so dull as that of a traveller: for the differences of hills, valleys, rivers; in short, the various views in which we may see the face of the earth, would scarce afford him a pleasure worthy of his labour.[111]

With more humour but the same intent, Swift mocked those who are 'big with description' and Sterne wrote:

> "Now before I quit Calais," a travel writer would say, "it would not be amiss to give some account of it." – Now I think it very much amiss – that a man cannot go quietly through a town, and let it alone, when it does not meddle with him, but that he must be turning about and drawing his pen at every kennel he crosses over, merely o' my conscience, for the sake of drawing it...[112]

Conclusion: diversity and debate, continuity and changeability in the discourse of landscape

Writers in the eighteenth century could frequently shift their positions in relation to debates about the use of landscape according to the convenience with which a 'move' fitted into their argumentation. For this reason the same authors can be used to exemplify many of the discursive connections suggested above. Individuals made these connections with facility and no sense of contradiction:

> I took the coach-road, over Bagshot Heath, and that great forest, as 'tis called, of Windsor. Those that despise Scotland, and the north part of England, for being full of waste and barren land, may take a view of this part of Surrey, and look upon it as a foil to the beauty of the rest of England; or a mark of the just resentment showed by Heaven upon the Englishman's pride . . .[113]

Defoe's first sentence is basically descriptive, but the next sentence has elements of the debate over patriotism in Britain, of Defoe's Whiggish narrative wherein the health of the land is indicative of that of the polity after 1688, of the utilitarian equation of waste with ugliness, and of the notion of the land reflecting divine judgements.[114] This sense of the interplay of many discourses through the landscape is by no means confined to travel accounts. The descriptive poem has been described as

> a sort of genre-of-all-trades; it may embrace topographical, pastoral, didactic, narrative, political, and practically every other sort of stock poetic interest.[115]

As such, landscape discussion reflects a point increasingly recognized as the periodization of Augustan neo-classicism being overtaken by romanticism is deconstructed: that

> wherever we pause, we are bewildered by the diversity that surrounds us: not alone in the conflict of opinion but shot through the very texture of every considerable author's or artist's work.[116]

Landscape, then, was a vehicle for the expression of opinion about a wide variety of subjects. Its status as a vehicle was continually being asserted, questioned and extended in a series of debates which reveal the landscape discourse to have had a dynamic of its own. Recent scholarship has tended to exaggerate the prevalence of political references. The highlighting of one discursive context coupled with the studied neglect of others has resulted in a distorted picture of the scope and complexity of the intellectual conversation about landscape in eighteenth-century England.

Confusion from order sprung: coherence and structure in the landscape discourse

Having shown the diverse uses of landscape in eighteenth-century England, I now wish to suggest that this apparent chaos presupposed an ordered approach. Debates in the landscape discourse were structured by a hierarchy of values which had originated in the religious basis of the period's interest in the natural world. It is at the level of the presuppositions behind the arguments and connections made that 'the religious dimension in which all moved' is to be found,[117] these presuppositions being outlined by the authors themselves. Whilst the specificity of the landscape discourse devolved into contests, and political and personal debates, the 'rules of engagement' were agreed upon.

'Things material, moral ties and revelations laws': the *ancien régime mentalité* and landscape

The eighteenth century's ruling orders held a hierarchy of value in which debate was framed and which was determinate of the status of landscape and nature appreciation throughout the period. This can be called an *ancien régime mentalité* because its structuring and operation fit closely with the mentality of the eighteenth-century 'confessional state' in England outlined by Clark.

> It involved stress on the correspondence between hierarchy in the divine sphere and the hierarchy of creation, including human goverment, and in the other direction an anthropomorphic analogy between a man's natural body and the body politic . . . The argument from correspondence thus showed the naturalness of unified authority at the same time as it justified it morally. Order within creation was the result of a just harmony of ranks and a due arrangement of them.[118]

The correspondence of natural, human and divine hierarchies allowed not only the naturalization of the human and the divine, but, moving in the opposite direction, the infusion of societal and religious debates into discussions of nature and the appreciation of landscape. The hierarchy, then, was interpenetrative by analogy: 'it was assumed that God, expressing Himself in all creation, made the physical, moral and spiritual levels analogous to each other and to himself'.[119] This intellectual hierarchy of value, when turned to the natural world, placed the analogical knowledge it yielded of the divine at the highest level, followed by insights for man as a moral actor, and finally the concern for nature, landscape, animals and vegetables in their own right as aesthetic or scientific objects of interest.[120]

The clearest expression of the three levels of the hierarchy and their connection is found in the *Ode to the Genius of the Lakes in the North of England*:

> Objects three the duly wise
> With raptur'd eye explore,
> Things material, Moral ties,
> And Revelation's lore.
> As each unfolds a world complete,
> Where all that's fit and comely meet,
> So each, through binded laws
> And kindred charms, new light bestows...[121]

Many other writers of the period discuss parts of the hierarchy. One frequently expressed point was the superiority of morality to all that beauty can achieve: after discussing the beauty of nature, landscape, and the animal and vegetable creation, Spence continued, 'and yet all the Profusion of Beauty I have been speaking of, and even that of the whole Universe taken together, is but of a weak Nature in comparison to the Beauty of Virtue'. Similarly, Goldsmith's conclusion was that it is:

> Vain, very vain, my weary search to find
> That bliss which only centres in the mind.[122]

To look for happiness via place or travel is a chimera, the answer resting in the mind and virtue. As nature was inferior to morality and virtue, so it was also interleaved with moral lessons. Thomson sought to

> – meditate the Book
> Of Nature, ever open, aiming thence,
> Warm from the Heart, to learn the moral Song.[123]

That such an aim was seen as dignifying landscape discussions can be gleaned from two sources. First, prefaces justified the 'ingrafting' of moral themes in this manner. Jago endeavoured to make his topographical poem 'as extensively interesting as he could, by the frequent Introduction of general Sentiments and moral Reflections'.[124] Secondly, the praise given to topographical poets was based upon a positive evaluation of the introduction of more elevated themes:

> It is one of the greatest and most pleasing arts of descriptive poetry, to introduce moral sentences and instructions in an oblique and indirect manner, in places where one naturally expects only painting and amusement... it is this particular art that is the very distinguishing excellence of Cooper's Hill; throughout which, the descriptions of places and

images raised by the poet, are still tending to some hint or leading into some reflection, upon moral life, or political instruction; much in the same manner as the real sight of such scenes or prospects is apt to give the mind a composed turn.[125]

The last part of Warton's comments is important, suggesting that in the interpenetrative hierarchy of values, the linkage of reflections to natural scenes was seen not as a juxtaposition, but as a natural conjunction.[126]

A similar range of arguments applied to the connection between religion and nature. God is far above the sight of man in nature, being at the apex while landscape and nature are at the base. Yet the analogy built into the hierarchy also worked between the extremes, such that:

> – from a heav'n taught sense
> Of weakness, to embrace high truths reveal'd.
> Nature's fair volume, character of God
> Expressive, answ'ring imag'ry divine,
> Harmonious movements, when inchantment flies,
> Soft trilling to the concord of the heart,
> Still wou'd I read – [127]

This expressive character both raised the importance of landscape and nature to the highest level, and was evidence of the benevolence of God, who had given man this route to understanding the reality of his presence.

Positionality in the *ancien régime* discourse of landscape

Once this hierarchy is recognized, the structural order behind the debates enumerated in the previous section becomes clearer. The positions which this hierarchy established for the discussion of nature were: a position for those who discussed nature within this structure; the possibility of divorcing landscape and nature from religion and morality whilst maintaining the evaluative judgement of their lowly importance; and finally, the questioning of the entire structure.

Debates within the ancien régime *hierarchy (including nature)*

Given the hierarchy of intellectual importance within which the landscape discourse operated, the most common move was to 'dignify' landscape by its connection to social/moral or religious issues. As we shall see in the next chapter, writers influenced by Latitudinarian theology did this repeatedly. Thus landscape was made instructive rather than merely sensual.

The direct discussion of virtue, morality and religion via landscape is fairly simple to understand, given a hierarchy which valued moral and religious above natural discourse, whilst also accepting that nature revealed those two higher levels. This was not simply a question of adding dignity to their writings, but also based upon the genuine belief that such instruction was made visible in the landscape. This connection was so common because its directness made for ease of composition for the writer and of comprehension for the reader sharing the same presuppositions. Modern incomprehension of the consistent personification of landscape elements (to raise them to the human/moral sphere) and discomfort with the continued injection of the deity into landscape discussions cannot be taken as a gauge of the sincerity with which these statements could be made or the enthusiasm with which they could be read in eighteenth-century England. The presuppositions which guided political and intellectual debate at this time made moves such as personification, allegory and linkage to other discourses entirely acceptable and, indeed, to be expected if landscape was to focus as a centre of intellectual concern.[128]

It is in this context that the 'politicization' of landscape in eighteenth-century England is best understood. Reflections upon polity and the nature of good governance served to connect the landscape discourse to higher concerns with the well-being of man. This was not a crude ideological use of landscape, for this presupposes that nature and landscape could not analogically demonstrate social truths, which was not the opinion of eighteenth-century England. Moreover, the nature of the political language to which landscape was attached must be borne in mind: political discourse occupied a different position in the field of knowledge from that it holds today. It was understood as being 'for the honour or interest of the nation', not as partisan or factitious, and was thus a moral discourse.[129] Politics was also incorporated into the religious understanding of the English nexus of church and state: 'the great discovery which we constantly make and remake as historians is that English political debate is recurrently subordinate to English political theology'.[130] It is in this context, too, that patriotism in English landscape discourse can be understood: the Glorious Revolution was justified in political rhetoric as having established a perfect earthly polity. The discussion of landscape gained in worth to the extent that it could demonstrate such truths from the moral sphere. The discursive connections which have been uncovered between politics and landscape are, then, a subset of a broader strategy of elevating landscape.

The connection of landscape to issues of agriculture, utility and commerce can also be seen as an attempt to dignify discussions of nature: appreciation of landscape could only be beneficial or at least innocent, when connected to a concern for the well-being of those in the landscape. This view was by no means confined to the advocates of improvement.[131] It was because of the close link between the health of the polity, the well-being of the common people and the appearance of the land, that arguments about agriculture and improvement in the landscape were so often linked with political arguments. The discursive similarity is that in both cases the landscape was valourized if it could be rendered instructive in moral and social terms. The insistent concern for the health of the people in the landscape which this view imposed belies the notion of landscape debates as exclusively the concern of the outsider for the outsider's view: interest in the poor on the land could be insistent because paternalism was a dominant structure of feeling in England.[132]

Both the antiquarian and the picturesque ways of seeing related to this hierarchy by connecting the landscape to the actions of men, both historical and artistic (as we shall see for William Gilpin in Chapter 4). Landscape could be instructive by revealing sites of historical interest or simply spark by association reflections on historical events which were believed in eighteenth-century England to contain moral lessons. To view landscapes through great poets and painters was also to dignify them by association with moral and creative legislators, which lifted landscapes into the moral sphere from the sphere of nature. Of course, to view landscape through paintings and poetry was to impose hierarchical criteria in another sense as well, because the arts developed their critical analyses in response to the same hierarchy of intellectual value. Thus landscape was elevated by its association with human creations and simultaneously interwoven with an ordering of the arts based upon this evaluative structure.[133]

Finally, the mentality governing landscape discussion helps to explain the ease of transition between discourses connected to landscape. The interconnected intellectual hierarchy provided a clear rationale to link landscape both 'upwards' to morality and Christianity, and at the moral level 'horizontally' across the several moral discourses, such as patriotism, politics and agriculture. Connections which seem from the standpoint of a different division of knowledge arbitrary were clearly ordered: where landscape could, to the eighteenth-century observer, validate the value system of the *ancien régime*, revealing in Swift's phrase 'order from confusion sprung', the twentieth-century reader must be reassured that the apparent chaos of connections in landscape discussions is confusion from order sprung.

The result of a twentieth-century investigation of landscape as its own discourse is perhaps at first sight paradoxical: that landscape was not an independent discourse. Precisely because it held a subordinate discursive position, yet could shed light on more important issues, landscape exemplified an overarching system of values in eighteenth-century English society. Yet this system of value allowed for far broader discursive connections than most modern critics have admitted (as shown above). By a contextual recovery of these connections, we have come to see how the 'politicization' of landscape, so oft-discussed at present, was justified and that it was but one part of a far broader system of debate.

Debate within the *ancien régime* hierarchy (divorcing nature)

Given the lowly status of nature and landscape, one option available was to discuss nature in its own right, a move discussed earlier. This in no way had to question the hierarchy of value of the *ancien régime*, often only being expressed in larger works which also connect to the 'higher' matters.

The more serious question was as to the aspirations a writer could have in discussing landscape and nature. To what extent could moral and religious lessons be gleaned from the landscape? Were landscape descriptions suitable vehicles for instruction, or did they degrade that which they sought to incorporate? Was it better to aspire to a linkage to superior discourses or to remain content with accurate description? Within the landscape discourse, these questions came to a head in the debate over the 'pure view' of landscape.

The debate over description and the engrafting of ideas in topographical poetry was a question of the aspirations of that genre. Were the ideas natural reflections on the prospect or were they merely being juxtaposed? If the latter, topographical poetry was aspiring beyond its scope. This debate parallelled that in landscape painting where Reynolds assessed the merits of Claude's use of historical subjects in landscape. The superior worth of moral and historical subjects was not questioned, but the legitimacy of their engraftment onto landscape was:

> That the practice of Claude Lorrain, in respect to this choice, is to be adopted by Landschape Painters, in opposition to that of the Flemish and Dutch schools, there can be no doubt, as its truth is founded upon the same principle as that by which the Historical Painter acquires perfect form. But whether landschape painting has a right to aspire so far as to reject what the painters call the Accidents of Nature, is not easy to determine.[134]

This debate as to the linkage between landscape and nature to other levels of the hierarchy was complemented by two connected forms of scepticism. First, there was a widespread scepticism about the linkage of nature to morality and devotion expressed in the form of satire. The Horatian theme of retirement and rural simplicity was contrasted with the fashionable and ultimately superficial concern for prospects which went under the name of a love of nature and the countryside. Such attacks can be found throughout the period, becoming more frequent after the 1750s when *The World* and *Connoisseur* ran a series of articles satirizing taste in gardening.[135]

The second form of scepticism posed the same questions about the plausibility of linking landscape and nature to religion and morality in a more earnest manner. Given that nature was at the base of the hierarchy of intellectual inquiry, many felt it more important to focus upon moral and religious issues directly. The two scepticisms were linked, satire easily becoming more serious as in Graves's *Columella*:

> "Columella has a delightful place here," says Hortensius, by way of introducing conversation. "Yes, Sir," says the Rector, "if happiness depended

upon any particular place or situation, I know no one who has a better chance for happiness...".[136]

Graves's rector redirected attention from place to religion as the route to happiness. Johnson made a similar plea in favour of looking at morality directly rather than foppish concerns of landscape:

> Never heed such nonsense, a blade of grass is always a blade of grass, whether in one country or another: let us if we do talk, talk about something; men and women are my subjects of inquiry.[137]

This critique can be seen as that of the 'reactionaries' in the landscape discourse. An important strand of this came from those sympathetic to the High-Church position and who were formative of the right wing *mentalité*.[138] Graves had been a fellow of All Souls, Oxford,[139] Tucker a student at St John's and Johnson at Pembroke, the university having 'for the most part...remained firmly entrenched behind its ramparts of conventional religious observance and traditional divinity'.[140] As suggested earlier, Reynolds, Goldsmith and Baretti, all friends of Johnson, expressed scepticism in various forms about an unalloyed concern for nature and questioned whether discussions of nature had any right to connect themselves with higher concerns. Similarly, of the great satirists of landscape appreciation, Garrick was Johnson's friend as, in later life, was George Colman (both being sometime members of the Literary Club), the latter being a schoolfellow and friend of Robert Lloyd at Westminster.[141] Craddock was at Emmanuel College, Cambridge, which was unusual in Cambridge in that it 'resisted the winds of change and continued to remain true to the [High-Church] ideals which Sancroft had so firmly implanted'.[142] He was tutored by Richard Farmer, friend of Johnson and 'another old-school Tory'.[143] It is also noticeable that Johnson had connections with Patten, Horne and Wetherell, all of whom were Oxford Hutchinsonians sceptical of the nexus of Newtonian natural religion and physico-theology so important for the elevation of landscape discussion into higher intellectual realms.[144] Perhaps, then, we can see a consistent position emerging, coupling a Tory disposition and High-Church education, which developed a scepticism about the worth of the landscape discourse within the *ancien régime* scale of values, preferring to focus directly on morality and a more fideistic view of religion. This position emerged in the 1750s and 1760s, a period which also saw the collapse of the old Tory party after Culloden and 'marked the dawn of something of a High-Church revival'.[145] It is within this discursive position that Johnson's contributions to debates about landscape and the natural world can best be understood. Until recently, the views of this group have not been taken seriously: as they are being reinvestigated in English history, it is the aim of Part III of this book to investigate their ideas on landscape, rescuing them from the oblivion to which they have been consigned by their inconvenience to prevalent historiographical and contextual assumptions.

Debating the ancien régime *hierarchy*

It is noticeable how rarely the entire *mentalité* relating to the appreciation of landscape and nature was questioned. This may suggest the extent to which presuppositions of argument are not themselves open to debate within any discursive framework, but it also could indicate its strength and continued appeal.

The main sources of potential questioning of the hierarchy were the pure view of nature and the associationist argument in philosophy and aesthetics. The pure view of nature could suggest that not only was looking at nature in its own right valid, but also that it was as worthy as the concern for morals and religion. In fact, this possibility was not acted upon. The associationist argument did proceed further in questioning the presuppositions of intellectual inquiry. It suggested that the linkage of discourses to aesthetics in general and landscape in particular was based upon resemblance, habit and custom,[146] rather than due to their actual connection by analogy. Yet the extent to which this undermined the belief in the hierarchy and thus the understanding of landscape and nature outlined above was minimal.

This scale of values began to break down around the time of the French Revolution. Both Gilpin and Price still used an ethical justification for their approach to the picturesque, yet their work acted to create a separate aesthetic realm. The postscript to Knight's second edition of *The Landscape* amounted to a renunciation of the *ancien régime* mentality as applied to landscape:

> I assure Mr Repton, however, that I will never follow the example he has set, in his Letter to Mr Price, of endeavouring to involve speculative differences of opinion, upon subjects of elegant amusement, with the nearest and dearest interests of humanity...To say that his own system of rural embellishment resembles the British constitution, and that Mr Price's and mine resemble the Democratic tyranny of France, is a species of argument which any person may employ on any occasion, without being at the expense either of sense or science.

Knight's words act in several ways which betoken the end of the mode of thought landscape had previously worked in: the debate was over specific and factional politics rather than broader politico–ethical notions of governance; and, having rendered the analogical reasoning explicit, Knight rejected it, ridiculing the

> endeavour to find analogies between picturesque composition and political confusion; or suppose that the preservation of trees and terraces has any connection with the destruction of states and kingdoms.[147]

Landscape aesthetics at this time were dehistoricized, such that attempts like the Chartists' after the late eighteenth century to reconnect the political and the natural were 'counter-hegemonic':

> the specificity and *peculiarity* of the Chartist poetic being its yoking of explicitly political claims and aims to a landscape poetic that had, by the end of the 1830s, come to be habitually identified with concepts of aesthetic autonomy and a poetic of private meditation.[148]

To fuse nature and politics was now a self-conscious act of retrieval, rather than the product of a set of hegemonic cultural presuppositions as it had been for much of the eighteenth century.

The breakdown of the assumptions framing landscape debates was part of a larger pattern of discursive change. Pocock claims that the unifying theme in political discourse from 1500 to 1790 was 'the construction, crisis and survival of the unified sovereignty in church and state,' but that events 'occurring about the outbreak of war with revolutionary France, increasingly belong to a history shaped by forces other than those which have been invoked'.[149] This change broke the connection of religious, moral and political discourse so important to the *ancien régime* and so amply reflected in the discursive connections established in landscape debates. The factitious nature of politics and politics of nature became increasingly apparent:

> speculations concerning the nature of humankind, of society and even of Nature now [in the 1790s] came under ever closer scrutiny for their implications in the battle to maintain the ideological support of the established order in Church and State.[150]

When the structure of eighteenth-century landscape discussions was laid bare and its politicization became self-conscious, the response was to relinquish analogical reasoning. The natural world came to be apprehended scientifically or aesthetically, both approaches sharing a commitment to viewing nature in its own right rather than as a cipher for other issues. Scientific inquiry saw the traditional areas of 'natural philosophy and natural history ... made redundant by the new vocabulary of specialized subjects',[151] such that there was an increased distance between scientific inquiry and the language of moral and religious debate.[152] Aesthetically, Wordsworth's romanticism, for example, denied the legitimacy of associative reasoning, arguing that landscape

> — had not need of a remoter charm,
> By thought supplied, nor any interest
> Unborrow'd from the eye—[153]

John Ruskin, in the generation after the early romantics, could move between 'High Tory' and 'radical, [the] reddest of the red', but still uphold the aesthetic autonomy of landscape. Clearly, the analogical mode of linking landscape and politics of the eighteenth century had disappeared: where Johnson's aesthetic would not heed such nonsense as a blade of grass, preferring more important themes in his hierarchy of values, Ruskin's aesthetic urged that he begin with the close study of a single blade of grass.

Elements of the old structure of argument survived into the early nineteenth century: Aubin, for example, notes of that *locus classicus* of reflection on the landscape, the topographical poem, that 'the latest...by an author of any consequence is Bowles's *Banwell Hill* (1828)'.[154] Yet these were the vestigial remains of a once all-encompassing structure of thought. The natural world became something to be appreciated for its own sake after the 1790s, such that to both the arts after romanticism and the sciences after specialization, it is the 'ideology' of analogical approaches to landscape and nature in the eighteenth century which stands out, not how this debate was framed in the period. Yet it was only after the collapse of this framework that the presuppositions by which it has been criticized came to be intelligible.[155]

4
Latitudinarianism and Landscape: Low-Church Attitudes to Nature, 1660–1800

Having established that religion was the key factor structuring the diverse contexts in which landscape description was invoked in the eighteenth century, I wish to elaborate on the question of the positions adopted with respect to landscape by suggesting further that we can break down responses to landscape by theological orientation. In short, differing theological attitudes to the natural world and its role in the proof of God's existence and nature led different authors to deploy landscape imagery in different ways. As we saw at the beginning of Chapter 3, Latitudinarian theology was central to the 'long' eighteenth century's interest in landscape, and it was the dominant form of Anglicanism in the period. This chapter investigates that theology and shows that it resulted in the establishment of dominant conventions for the description of landscape by a genealogy of canonical authors. As such, this chapter is an essential context and benchmark against which Samuel Johnson's deployment of landscape imagery can be compared in Part III of the book.

Latitudinarian theology and nature

The theological and ecclesiological position called 'Latitudinarianism' is notoriously hard to define, and its nature varies according to whether it is taken as primarily a movement in church politics; a form of 'liberal' theology; as a precursor to eighteenth-century theology; or an historically specific creation of the Restoration.[1] Yet some general traits of low churchmanship in the era after the Restoration can be detected, and the leading lights behind the position were a close-knit group, normally having been educated together in Cambridge, before migrating to London.[2] Above all, the aim of Latitudinarianism was to ensure that the sort of theological factionalism which they saw as leading to the Civil War did not recur. This would be achieved by developing a more comprehensive and moderate approach within the Church of England, which would have the 'latitude' (hence the name, which was initially a pejorative one, given by High-Church opponents) to include all types of Protestant within the fold of the established church. The retrospective and the prospective sides of this creed, together with its rhetoric of inclusion, were nicely summarized by John Tillotson, the most important Latitudinarian spokesman:

> The manners of men have been almost universally corrupted by a Civil War. We should therefore all jointly endeavour to retrieve the ancient virtue of the Nation, and to bring into fashion again that solid and substantial that plain and unaffected piety, (free from the extremes both of superstition and enthusiasm) which flourished in the age of our immediate Forefathers.[3]

Tillotson's words also exemplify the repeated claim of the Latitudinarians that their position amounted to the revival of the Anglican *via media* which had been developed by Hooker and the Elizabethan settlement.[4] This sense of the importance of a moderate Protestantism was reinforced by the events of 1688, which the Latitudinarians supported wholeheartedly. Tillotson interpreted the Glorious Revolution (and there is no doubt that the term is appropriate in his case) as the last in a series of providential deliverances of the British Isles, whereby God 'hath been pleas'd to work for this *Nation* against all the remarkable attempts of Popery, from the beginning of our *Reformation*'.[5]

Most of the details of Latitudinarian theology and church politics have received extensive scholarly treatment elsewhere, and need not concern us here.[6] What is of great relevance to the present inquiry is the way in which Latitudinarian theology incorporated the natural world and, by extension, landscape into its modes of argumentation. As we will see, the Latitudinarians were distinguished in theological terms by their repeated recourse to the natural world as a mode of evidence which fitted their need for uncontroversial proofs of God and Christianity, to draw those of diverse religious beliefs into concord. Given that latitude became the 'mainstream' theological position for the upper echelons of the Church of England for at least the period from the accession of William and Mary to the later years of George III, it is important to understand how it treated the natural world, as this was to have a great effect on the cultural elites of England, who were also the main producers and consumers of literary depictions of landscape.

The Latitudinarian fascination with nature was not simply a belief in 'natural religion'. Natural religion argued that justifiable belief in God could be founded upon the exercise of reason alone. In this view, the operation of reason would lead the inquirer to ' "laws of nature" [i.e.] intuitively certain or self-evident moral principles, which are eternal and unchanging'.[7] This need have nothing to do with the observation of the natural world at all, but could be achieved by self-examination. There is no doubt that the Latitudinarians believed in the light of nature in this sense, but they also believed in the significance of the natural world and its landscapes as a proof of both the being and the attributes of God. In a revealing phrase, John Tillotson argued that

> In this visible frame of the world, which we behold with our eyes, which soever way we look, we are encountered with ocular demonstrations of the wisdom of God.[8]

The phrase is revealing for two reasons. First, it suggests that merely looking at the landscape before us is an inducement to devotion. The reasoning inquirer will find in the observation of nature evidence for the existence of God. Secondly, Tillotson adopts the rhetoric of empirical scientific inquiry in his reference to 'ocular demonstrations'. One of the most important ways in which Latitudinarian interest in nature as a proof of God was fuelled was by the findings of the 'New Science' of the Royal Society. The lines of influence also operated in the other direction, with scientific practice gaining in legitimacy by its connection with theology, which remained the lynchpin of scholarly inquiry and education. This was not simply an intellectual interaction, but was also personified by the significant number of Latitudinarian clergymen who were members of the Royal Society in its early years,[9] and, as we shall see, in the theological activities of scientists such as Robert Boyle. The close ties to the new science emphasized Latitudinarianism's claims to be a rational religion.

The Latitudinarians were distinguished from other theologies in the Anglican Church not by their belief in the natural world as evidence for the existence of God, but by the emphasis they placed upon this form of evidence. Both Tillotson and Samuel Clarke argued that natural religion was the foundation for revealed religion, a proposition with which the High-Church Anglicans disagreed.[10] In context, it is clear that both were referring to natural religion as opposed to the natural world, but their fellow Latitudinarian Isaac Barrow went further and suggested that the natural world was the best evidence of God's being and attributes:

> The best (no less convincing than obvious) arguments, asserting the existence of a Deity, are deduced from the manifold and manifest footsteps of admirable wisdom, skill and design, apparent in the general order, and in the peculiar frame of creatures; the beautiful harmony of the Whole, and the artificial contrivance if each part of the world.[11]

Barrow rendered explicit a truth that applies to Tillotson, Clarke and the other Latitudinarians: as a group, their theology gave the natural world and the appearance of the natural world seen as a landscape, an unprecedented role in the operation of Christian apologetics. On this basis was founded their self-proclaimed religion of reason.

The 'Latitudinarian moment' emphasized the natural world as a proof of God to an extent that was not matched by scholastic predecessors, evangelical successors or other theologies at the time. For the rhetoric of moderation which, as we have already seen, was so central to latitude narrowed its options in terms of routes to faith. On the one hand, the wranglings of Scholasticism were too abstruse to persuade people into faith and too contentious to generate a broad consensus. On the other hand, the flight from mathematical and philosophical demonstrations of God to a belief in faith divorced from proof was denounced as enthusiasm of the sort that had led to social breakdown in the Civil War. The rhetoric of moderation, then, homed in on the proof of God via the observation of nature. God's existence could no more be proved *directly* by the evidence of the senses observing nature than by mathematical demonstrations, but the evidence offered by observation of the natural world led from the visible towards the invisible, from secondary causes to the First Cause. Samuel Clarke sketched this indirect route to a rational faith:

That many *invisible* things are real, is evident from the continual Effects of Nature, which are all of them produced by invisible Powers; And from thence the *Being of God*, is strictly *demonstrable*. But they who have *not* capacities to apprehend the *Demonstration*, have yet sufficient *Reason*, from what they *are able* to observe and understand, to be *fully persuaded* of the *Truth* of God's Being, and his Government of the World.[12]

The increased theological sensitivity of nature amongst the Latitudinarians was also apparent in the increased concern that the term was used in ways which did not lead to idolatry. Once more, Samuel Clarke's *Sermons* led the way:

in Truth, inanimate *Nature* is nothing but an *empty Sound*; Unintelligent *Agents* and *Powers*, (as we improperly call them,) are nothing but *mere Instruments*; and the Whole Effect is *really* the *Operation* of *Him*, who is the *Author and God of Nature*.[13]

This caveat was designed to prevent the deification of nature, a departure from the moderate *via media* which became a possibility because of the unprecedented role which nature was given by Latitudinarian theology.

Robert Boyle and three Latitudinarian readings of the Book of Nature

We have already seen that the Latitudinarians' moderation led them to advocate a heavy reliance on observation of the natural world as a rational, uncontroversial and widely comprehensible proof of God. Given this role for the observation of nature, what remains to be shown is how this general focus of interest was rendered operational. The Latitudinarians developed

three main 'ways of seeing' nature, which, given their penchant for the
metaphor of 'the book of nature', might better be called three readings of
that book. These readings were all developed at the interface between theology,
science and nature, where so much Latitudinarian thought was generated,
and their form can be shown by recreating that interface, joining the arguments of Robert Boyle, himself being the personification of the Latitudinarian
Christian scientist,[14] with the arguments of the Latitudinarian theologians.

The argument from design

The best known Latitudinarian reading of the book of nature was the argument from design. This reading argued that the order, harmony and structure
of the visible world, at every level from the atomic to the universal, showed
the operations of an infinitely wise creator. This argument was not, of
course, new to the period, having been a commonplace in both Classical
and Medieval thought.[15] But the Latitudinarian generations could take the
argument up with renewed vigour, as scientific inquiry led to the proliferation
of facts and observations of the operation of the natural world. The increase
in information, coupled with the Latitudinarians' previously discussed desire
to emphasize natural proofs of God, led to a reliance on design in the 'long'
eighteenth century which allowed Newman to dub this the 'age of evidences'.

Boyle's construction of the argument from design shows how it was
modified in the light of the rise of scientific inquiry. For Boyle, 'the two chief
advantages, which a real acquaintance with nature brings to our minds, are,
first, by instructing our understandings, and gratifying our curiosities; and
next, by exciting and cherishing our devotion'.[16] In context, Boyle's notion
of a 'real acquaintance' is clearly a reference to the findings generated
by experimental method, which he contrasts with the *a priori* approach of
Aristotelian natural philosophy. Boyle's discussion in 'Of the usefulness of
Natural Philosophy' continues by suggesting that this real acquaintance has
led to an entirely new form of design argument:

> for the book of nature is to an ordinary gazer, and a naturalist [i.e. an
> experimental observer of nature], like a rare book of hieroglyphicks to
> a child, and a philosopher; the one is sufficiently delighted with the odd-
> ness and variety of the curious pictures that adorn it; whereas the other,
> is not only delighted with these outward objects, that gratify his sense,
> but receives a much higher satisfaction, in admiring the knowledge of the
> author.[17]

So, the scientific apprehension of the natural world adds a rational to the
aesthetic argument from design. The utility of this to Latitudinarianism as a
self-confessedly rational form of Christian religion is apparent.

Boyle's own deployment of the argument from design centred on the
animal creation: he argued that the heavens only showed the 'general

intendments of God in the universe', where from the 'bodies of animals it is oftentimes allowable for a naturalist, from the manifold and apposite uses of parts, to collect some of the particular ends' of God's design.[18] As such, the animal creation could reveal in more detail the intricacy and scope of the natural world's construction, thereby amounting to a more convincing and detailed argument from design. A number of later Latitudinarians followed Boyle's lead in this respect, notably William Derham in *Physico-Theology* and, at the end of the eighteenth century, William Paley in *Natural Theology* (1803). Equally, others preferred to follow the lead given by Isaac Newton, and therefore saw the most convincing arguments from design to rest in the 'system of the world', or operation of the universe. An obvious example was Richard Bentley, who was asked to give the first Boyle lectures in 1692. Boyle's commitment to the design argument as an inducement to piety had led him to found a series of lectures to prove God's existence by natural means. Bentley engaged in a correspondence with Newton on the theological implications of his system, and was persuaded that the Newtonian scale of analysis was more convincing as a proof of God:

These Reasons for God's Existence, from the Frame and System of the World, as they are equally true with the Former [i.e. those from the structure of animals], so they have always been more popular and plausible to the illiterate part of Mankind.[19]

To show that both types of design argument were equally overwhelming, Derham wrote a companion to his *Physico-Theology*, called *Astro-Theology*.[20]

It should be emphasized that the design argument, in whichever form, had a built-in restraint on the limits of science and reason in the proof of God. As God was omnipotent and omniscient, his full design of the universe was beyond our comprehension, as indeed was the design of any individual part of it since each part was linked to the whole system. A truly scientific design argument led simultaneously to piety and to humility: 'it assures us that some effects are possible, but cannot help us to determine what is impossible ... *his works (as Lactantius speaks) are seen with eyes, but how he made them, the mind itself cannot see*'.[21] Boyle from a scientific perspective also argued for the limits of reason. This argument was used to suggest that in the afterlife our understanding of the operation of the system of nature would be expanded, giving us expansive prospects of a sort unimaginable in this life. This emphasis on limits helped to forestall the dangers of hubris, and checked the tendency to conflate what we could understand of nature with God, something which the linguistic policing of the use of the term 'nature' (discussed above) also sought to prevent. This move kept Latitudinarianism within the fold of Christian religion, at one remove from the deist and other heterodox approaches to nature (see Chapter 5).

The aesthetics of design

As was seen above, science was seen by Boyle as allowing the proofs of God's existence derived from nature to pass from the aesthetic realm of the imagination to the rational realm of the understanding. This did not mean, however, that the aesthetic argument was to be jettisoned. As Boyle had said in the passage quoted above, the naturalist might be able to decipher the hieroglyphics of nature, but would also remain 'delighted with those outward objects, that gratify his sense'. Boyle continued to argue that the beauty of the face of the earth and the firmament could lead the mind to Christian piety.

A number of Latitudinarian preachers appealed to the aesthetic form of the design argument, being led into long landscape descriptions of the sort which were soon to be transferred to 'literature'. A short example came in John Ray's *Wisdom of God Manifested in the Works of Creation*:

> How variously is the Surface of it [the earth] distinguished into Hills, and Valleys, and Plains, and high Mountains affording pleasant Prospects? How curiously cloathed and adorned with the grateful verdure of Herbs and Stately Trees, either dispersed or scattered singly, or as it were assembled in Woods and Groves, and all these beautified and illustrated with elegant Flowers and Fruits.[22]

Particularly important to the aesthetic defences of the design of the earth in the scientific homiletics of Latitudinarian divines was the stimulus provided by Thomas Burnet's *Sacred Theory of the Earth*.[23] Burnet was himself on the Latitudinarian wing of Anglicanism (with a tendency towards rationalist heterodoxy), but argued that the mountains and irregular coastlines of the planet showed not harmony but fragmentation, not design but disorder. For Burnet, the earth showed the shattered ruins left by the Deluge, whose effects were superimposed on the wreck already caused by the Fall. In their Boyle lectures, both Bentley and Derham were led into long defences of not only the scientific importance of mountains and other signs of irregularity in the operation of the earth, but also the beauty of those features. Both concluded on the aesthetic point in similar terms, their position being summarized by comments Bentley made prefatory to a citation from *Paradise Lost* on the 'irregular' topography of paradise:

> we appeal to the sentence of Mankind, if a Land of Hills and Valleys has not more Pleasure too and Beauty than an uniform Flat? Which Flat, if ever it may be said to be delightful, is then only, when 'tis viewed from the top of a Hill... They [the poets] cannot imagine even Paradise to be a place of Pleasure, nor Heaven itself to be Heaven without them.[24]

It should be added that Latitudinarian preachers also had recourse to the aesthetic argument from design in 'standard' sermons delivered outside the context of the Boyle lectures. Isaac Barrow's sermons in particular, which had more rhetorical flourish than the plain style of most Latitudinarians, developed aesthetic arguments at some length, appealing to all the senses:

> [We are] invited to open all the avenues of our soul, for the admission of the kind entertainments nature sets before us ... doth she not everywhere present spectacles of delight ... to our eyes, however seldom any thing appears horrid or ugly to them? where is it that we meet with noises, so violent, or so jarring, as to offend our ears? All the air about us, is it not (not only not noisome to our smell, but) very comfortable and refreshing?[25]

This form of argument is especially important to the present study, as it not only brought the design argument within the purview of literature and *belles lettres*, but also focused on the appearance of the earth's surface, in other words on the landscape. As we will see, it was within this nexus that a number of canonical writers developed their landscape descriptions, although they have frequently been decoupled from their theological context in subsequent critical analysis.

Design and meletetics

The third reading of the book of nature encouraged by the Latitudinarian position was the rhapsody on some part of the creation, a procedure of meditative perception of nature Boyle labelled 'meletetics'. While latitude has always been defined by its rationality, this can be overemphasized by taking the criticisms of Low-Church Anglicanism made by its opponents at face value. Rationality, as far as the Latitudinarians themselves were concerned, was not designed to rule out spirituality; indeed, the two were supposed to be inseparable.

Boyle's meletetics, as set out in his 'Occasional reflections on several subjects', were a good example of the interweaving of rationality and spirituality. Based on the close (scientific) observation of small elements of the natural world, Boyle advocated devout contemplation of their wonders, which would render the world a series of 'lectures of ethicks or divinity'. He explicitly said this procedure could be applied to the observation of 'landskip'. As an example of Boyle's procedure, he argued that observing the moon could, by a meletetic parallel, lead the observer to think of the Christian dispensation:

> as the moon communicates to the earth the light, and that only, which she receives from the sun; so the Apostles, and first preachers of Christianity ... communicate to mankind the light, which themselves have received from the bright sun of righteousness.

Boyle believed a mind habituated to such a way of reading nature's book would 'live almost surrounded either with instructors or remembrancers': nature would be 'spiritualized'.[26]

While not explicitly derived from Boyle, Latitudinarian preachers did develop this third, more rhapsodic approach to nature and landscape in some of their work. Here, for example, is Derham's rhapsodic flight in praise of the design of the eye, which, as in Boyle, moved from close observation to high-blown praise. Thanks to the eye;

> We can, if need be, ransack the whole Globe, penetrate into the Bowels of the Earth, descend to the bottom of the Deep, travel to the farthest Regions of the World, to acquire Wealth, to encrease our Knowledge, or even only to please our Eye and Fancy.[27]

While all this is clearly true, it is not based on scientific argument, and Derham's aim was clearly to hit a higher rhetorical and emotional register, leading the reader to awe and wonder in the face of the creation. By rendering the familiar and everyday remarkable, Derham and Boyle were both trying to revive a sense of how extraordinary the ordinary in fact was, thereby 'spiritualizing' it.

Landscape was also drawn into the rhetorical side of Latitudinarian argument by means of extended metaphors. Perhaps the most accomplished practitioner of this mode of preaching was Isaac Barrow, and the following feast of landscape imagery is amongst his finest examples:

> He that is shut up in a close place, and can only peep through chinks, who standeth in a valley, and hath his prospect intercepted, who is encompassed with fogs, who hath but a dusky light to view things by, whose eyes are weak or foul, how can he see much or far, how can he discern things remote, minute, subtile, clearly or distinctly? Such is our case; our mind is pent up in the body, and looketh only through those clefts by which objects strike our sense; its intuition is limited within a very small compass; it resideth in an atmosphere of fancy, stuft with exhalations from temper, appetite, passion, interest...[28]

Barrow, in a different manner from Boyle, achieves the same effect of spiritualizing nature in general and the activity of viewing a landscape in particular. He was by no means alone, traditional Christian imagery of light and darkness, of sight and blindness, lending itself to such extended prospect metaphors, particularly to a group of divines so preoccupied with the natural world in any case.

The three Latitudinarian readings in the eighteenth century

Latitudinarianism, then, developed three ways of reading the book of nature which would lead the inquiring mind into a knowledge of God. The knowledge gained was not simply rational, but also appealed to the emotions and aesthetic sensibilities. My treatment of these matters has focused on the writings of the first generations of Latitudinarians, in the era from the Restoration to the turn of the century, but parallel arguments can be found echoing down through eighteenth-century devotional and homiletic literature. There is no need to go into detail here, but the lectures founded by Robert Boyle had a profound impact. While the Boyle lectures gradually lost their cultural prestige as the eighteenth century wore on, the published versions of the early lectures by Bentley, Derham and others remained standard devotional literature throughout the century. The design argument remained of great importance, as can be seen by William Paley's mammoth exposition of it in *Natural Theology* (1803). Paley came from a Low-Church Anglican context, and was associated with a later generation of 'Latitudinarian' (or simply liberal) theologians in Cambridge in the 1760s and 1770s. Over a century after Boyle had sought to 'spiritualize' nature both by science and rhetoric, Paley's position was still essentially the same:

> if one train of thinking be more desirable than another, it is that which regards the phenomena of nature with a constant reference to a supreme intelligent Author. To have made this the ruling, the habitual sentiment of our minds, is to have laid the foundation of every thing which is religious.[29]

It must be emphasized that I have only distinguished the three ways of reading the book of nature for heuristic purposes. In practice, all three modes of argument were closely interwoven in the design arguments developed by the Latitudinarians, as they were in large part simply different notes on the emotional register, to be hit in order to appeal to readers of different temperaments. As an example, we can look at William Paley's arguments concerning the design of the human eye by God in *Natural Theology*. Paley started with the complex structure of the eye, which he believed could not be formed by chance co-operating with the passage of time, but displayed the hand of an intelligent creator. This exposition demanded detailed discussion of the elements of the eye (the humours, the retina and so on), in the conclusion to which Paley agreed with Sturmius 'that the examination of the eye was a cure for atheism'. During the course of this scientific argument, Paley also hit a more rhapsodic note, notably describing the operation of the eye brow:

> an arch of hair, which, like a thatched penthouse, prevents the sweat and moisture of the forehead from running down into it.

This is an example of Paley's 'flights into hyperbole [which] strike the modern reader as ludicrous' but clearly link him to Boyle's meletetics. Interestingly, this interweaving of types of reading of the book of nature reached its apogee where Paley fully meshed the structure of the eye, the beauty of its objects of vision, and a meditative reflection in a discussion of the eye's perception of landscape:

> In considering vision as achieved by the means of an image formed at the bottom of the eye, we can never reflect without wonder upon the smallness, yet correctness, of the picture, the subtilty of the touch, the fineness of the lines. A landscape of five or six square leagues is brought into a space of half an inch diameter; yet the multitude of objects which it contains, are all preserved, are all discriminated in their magnitudes, positions, and colours. The prospect from Hampstead-hill is compressed into the compass of a six-pence, yet circumstantially represented.... If anything can abate our admiration of the smallness of the visual tablet compared with the extent of vision, it is a reflection which the view of nature leads us every hour to make, viz. that, in the hands of the Creator, great and little are nothing.[30]

Over time these ways of reading nature's book became disassociated from a specific faction in theology and church politics, being instead the general property of Low-Church modes of argumentation, which emphasized reasoned approaches to the inculcation of Christian belief. As such, an increased emphasis on the book of nature could also be found outside the Anglican context in which it had first been developed. The moderate Presbyterians of the Church of Scotland were drawn to such arguments,[31] one of the earliest examples being the poetry of James Thomson. As Thomson's most recent biographer has pointed out, Thomson was 'theologically liberal', having been educated by William Hamilton, who was 'as close as one could come to being a Latitudinarian in the Scottish church'.[32] Throughout *The Seasons*, Thomson argued that God's goodness could be seen in nature and, more specifically, in the differing opportunities the succession of the seasons provided:

> These, as they change, Almighty Father! These
> Are but the varied God. The rolling year
> Is full of thee. —

Just as with the Latitudinarians, this praise of nature was not allowed to slip into the deification of nature, which would move the argument from Christianity to heterodoxy:

> Nature, attend! join every living soul
> Beneath the spacious temple of the sky,

> In adoration join; and ardent raise
> One general song! To him —

nature could be called on to praise God, but was not itself God. The bulk of
the poem, of course, contained aesthetic descriptions of the landscapes
formed by the revolution of the seasons, interspersed with rhapsodies of the
sort encouraged by Boyle, such as Thomson's parallel of the return of Spring
to the Resurrection, and finally capped with the prolonged religious reflection
of the 'Hymn to Nature', from which both of the above quotations are
taken.[33] *The Seasons*, then, appealed at length to the aesthetic and meletetic
readings of the book of nature in ways Thomson obviously felt were consonant with his lax Presbyterianism.

It is the contention of the remainder of this chapter that the modes
of reading nature which the Latitudinarians pioneered in the Restoration
context, and which remained vital from Boyle to Paley, were of great significance to the ways in which the natural world in general and landscape in
particular were represented in eighteenth-century literature. The Low-Church
piety which characterized a number of the important writers traditionally
associated with the depiction of landscape in literature was no mere
accident, but significantly influenced both why and how they developed
those depictions. This genealogy developed the dominant approach to the
literary depiction of landscape in the long eighteenth century against which
Samuel Johnson's High-Church approach operated (see Part III)

The theological pleasures of the imagination: Joseph Addison and landscape

Addison as a Latitudinarian

Joseph Addison's Latitudinarianism is clear from the likes and dislikes in
both theology and church politics which he expressed throughout his written
oeuvre, and which he actively promoted in his career as a Whig politician.
This need to be shown, however, in the light of the comments of Addison's
modern biographer that 'his Christianity had always been of so doubtful an
orthodoxy that entry into Orders would not have been easy for him'.[34]

Although Joseph Addison came from the generation after the age of
Tillotson and the pioneer Latitudinarians, we can start with him as we did
with Tillotson, in a shared belief that the religious life of the nation had been
thrown out of kilter by the 'Swarms of Sectaries that over-ran the Nation in
the time of the great Rebellion', perverting religion by their 'enthusiasm'.[35]
Addison by no means saw this enthusiasm as a merely historical phenomenon, arguing in *Spectators* 185 and 201 against contemporary enthusiasts.
On the other extreme from enthusiasm, Addison also stigmatized Roman
Catholicism, most notably in his travel book, *Remarks on Several Parts of*

Italy, which was placed on the *Index Librorum Prohibitorum* by the Papacy.[36] Addison's anti-Catholicism recapitulated a trait of all the 'tolerant' Latitudinarians,[37] and extended into a hostility towards High-Church Anglicans, who were seen as leaning too far in the direction of Rome. In the fraught context of the 1715 Jacobite rebellion, Addison's essay series *The Freeholder* attacked the High-Church cry of the 'church in danger', which was directed at the Latitudinarians for their tolerant attitude towards Protestant Dissenters.[38] Equally, Addison's attack on an excess preoccupation with ceremony in faith in *Spectator* 213, although directed against the Jews and Catholics, could also clearly be applied to the High Churchmen, so his hostility to this deviation from the Latitudinarian *via media* was not just the product of the Jacobite threat. Finally, Addison also attacked those who went beyond either the enthusiastic or Catholic extreme to the outright rejection of Christianity in *Freethinker* 51 and in his play, *The Drummer*, in the character of Tinsel, a coffee house freethinker who has no intellectual basis for his ridicule of Christianity.[39]

Addison, then, defined his position as a Low-Church Anglican by negation, by his attacks on the positions which deviated from this standard. Importantly for the present study, all these deviations were associated by Addison with a superstitious, and therefore erroneous, reading of the book of nature which fell outside the parameters of the three Latitudinarian readings. This point is made repeatedly in the *Freeholder* essays, where Catholic Jacobitism is seen as fostering an irrational belief in natural omens (which is, given gender stereotyping of the period, predictably connected with femininity):

> The Party, indeed, that is opposite to our present happy Settlement, seem to be driven out of the Hopes of all human Methods for carrying on their Cause, and are therefore reduced to the poor Comfort of Prodigies and Old Women's Fables. They begin to see Armies in the Clouds, when all upon the Earth have forsaken them.[40]

The error of Jacobites and Catholics springs from an ignorance of science and the procedures of rational thought, which perverts their approaches to nature. For this reason, they make the same superstitious readings of the book of nature as uneducated rustics, of the sort Addison more gently satirized in *Spectator* 7. For Addison, the same ignorance of the natural world is shown by freethinkers, although they make more show of scientific rationality. He shows this in the character of Tinsel in *The Drummer*. Tinsel has learnt the jargon of Epicurean atheism as an explanation of the form of the landscape:

> I shall have time to read you such Lectures of Motions, Atoms, and Nature – that you shall learn to think as Freely as the best of us, and be convinced in less than a Month, that all about us is Chance-work.[41]

But because Tinsel has never studied nature by empirical and rational methods, he is later frightened into believing in a ghost, such that the errors of
the atheist and the enthusiast, of the rationalist and the rustic, are closely
allied. In Tinsel's character, Addison personified the Latitudinarian commonplace that without a rational approach to religion, people were liable to
be blown between the extremes of scepticism and credulity.[42]

Addison, then, had a well defined sense of the theological errors which
extremism could generate, and how this mapped onto a pre-rational comprehension of the natural world. He had an equally well-defined and- expressed
sense of the *via media* which the rational Anglican would tread. This was
literally made graphic in an essay he wrote in the *Tatler* about the 'ecclesiastical thermometer' (number 220), where he defined a scale ranging from the
ignorance of excess enthusiasm to the ignorance of atheism, with the true
church placed precisely at the mid-point. Addison was unstinting in his
praise of established Anglicanism, which was dominated by Latitudinarians
by the time he was writing. Once more, he was most explicit about this in
the *Freeholder*, his most politicized work:

> my Readers as they are *Englishmen* . . . by that Means they enjoy a purer
> Religion, and a more excellent Form of Government, than any other
> Nation under Heaven.[43]

Analysis of Addison's *oeuvre* also shows that his pantheon of heroes was
strongly Whiggish and Latitudinarian. In theological and church history, he
praised Tillotson (*Spectator* 557 and *Freeholder* 39) and Gilbert Burnet (*Spectator*
331). Similarly, in science he praised those who spiritualized nature: Thomas
Burnet (in his poem, *Ad insignissimum virum, Tho. Burnettum*),[44] Newton
(*Spectators* 543 and 565) and, above all, Robert Boyle 'who was an Honour to
his Country, and a more diligent and successful Enquirer into the Works of
Nature, than any other our Nation has ever produced'.[45]

There can be little doubt that Addison was strongly Latitudinarian in his
thought. Having mapped out the way Addison conceived of those who deviated from this *via media*, how these deviations led them into errors in the
reading of the book of nature and how Addison aligned himself with the
leading individuals who had established the nexus of theology, science and
reason which defined Latitudinarianism, it remains to show how his *oeuvre*
established in a literary context ways of describing nature and landscape in
accord with the idea of complex Latitudinarian theology.

'To moralize this natural pleasure': landscape in the Spectator

Addison's project of presenting landscape and the natural world in a literary
context in ways consonant with Latitudinarian theology shows continuities
with the aims of Robert Boyle, especially as set out in Boyle's meletetic treatise.
Both adopted the subject position of the Christian layman trying to inculcate

piety outside the traditional literary contexts of theological discourse.[46] Both also centred this lay piety on nature and landscape. As we have seen, Boyle described this as 'spiritualising nature', and Addison, in a self-evidently parallel formulation for a more literary realm, was to describe himself as 'moralizing natural pleasure' in *Spectator* 393.

Spectator 393 (iii. 473–76) is a useful epitome of the Addisonian presentation of the natural world to be discovered throughout the *Spectator*. Addison starts with an aesthetic description of the effects of an English summer and of the beauties of the springtime, including a passage from *Paradise Lost*. But he does not stop here, going on to the more abstract thought that the landscape is a continual stimulus to the imaginative powers of mankind: 'the Creation is a perpetual Feast to the Mind of a good Man', a message also reinforced by the Psalms. From the imagination, Addison's discussion then transfers to another mental faculty which can be activated by the face of nature, namely reason: 'Natural Philosophy quickens this Taste of the Creation, and renders it not only pleasing to the Imagination, but to the Understanding.' Finally, it is argued that this understanding, this 'rational Admiration' of nature, leaves the soul in a state 'little inferiour to Devotion'. Importantly, Addison clearly intimates that this progress from the appeal to the senses through aesthetic imagination and rational comprehension to spiritual improvement is a *method* of observing nature which can be cultivated, just as was Boyle's meletetics. This method is what he calls the endeavour 'to moralize this natural Pleasure of the Soul, and to improve this vernal Delight...into a Christian Virtue'. Once this method has become fully ingrained, the result is parallel to Boyle's description of the pious man being surrounded by instructors in nature: 'Such an habitual Disposition of Mind consecrates every Field and Wood, turns an ordinary Walk into a Morning or Evening Sacrifice.'

Addison's approach to landscape centres on this progress from imagination to understanding, and the fact therefore that the senses and reason, in their approach to nature, can be part of a larger spiritual activity. This can be seen by the fact that two major types of deployment of landscape imagery occur in the *Spectator* papers, which are aimed at these two faculties of imagination and reason, and which are broadly equivalent to the aesthetic and design readings of the book of nature discussed previously. Addison is explicit that these two faculties are in a hierarchy: 'The Pleasures of the Imagination, taken in their full Extent, are not so gross as those of Sense, nor so refined as those of the Understanding' (*Spectator* 411; see also *Spectator* 420). Further, this conception of twin ways to use landscape imagery to appeal to the faculties in a devotional project is itself derived from the traditional Christian idea of the position of mankind in the Chain of Being, which Addison set out in *Spectator* 519: 'Man...fills up the middle Space between the Animal and Intellectual Nature, the visible and invisible World.'

The pleasures of the imagination

Many historians of aesthetics have analyzed the series of *Spectator* essays on the pleasures of the imagination (numbers 411–21). It is quite clear that Addison's 'landscape aesthetic' is more accurately described as a theological landscape aesthetic: 'A beautiful Prospect delights the Soul, as much as a Demonstration' (*Spectator* 411). At other points, Addison made it clear that he conceived of aesthetic pleasure in general as an excess over mere survival given to man by God as a token of his benevolence (see *Spectators* 387 and 465). This excess also shows God's benevolence in another manner, in that the aesthetic sensation requires none of the intellectual capacities which the design argument directed at the understanding demands. As such, the aesthetic appeal of nature can lead more people to a state of devotion.

Analyzing this aesthetic pleasure in more detail, Addison divided the pleasures of looking at the face of nature into three kinds, the beautiful, the uncommon and the great (*Spectator* 413, in 540–44). In each case, he argued that this pleasure was a product of God, being designed to act on the soul. The longest explanation of this process came in his discussion of the effect of the great. The great is undoubtedly for Addison a category of landscapes: 'Such are the Prospects of an open Champian Country, a vast uncultivated Desart, of huge Heaps of Mountains, high Rocks and Precipices, or a wide Expanse of Waters.' But his main concern is not the description of such landscapes, but rather the effect of such landscapes on the imagination: 'We are flung into a pleasing Astonishment at such unbounded Views, and feel a delightful Stillness and Amazement of the Soul.' It is clear from *Spectator* 412 that such a process has a devotional aspect, especially as it is described as the imaginative equivalent of the mind grappling with eternity and infinity. But in *Spectator* 413 this is made explicit in a very Boylean discussion of the final causes of aesthetic pleasure:

> One of the Final Causes of our Delight, in any thing that is *great*, may be this. The Supreme Author of our Being has so formed the Soul of Man, that nothing but himself can be its last, adequate, and proper Happiness. Because, therefore, a great Part of our Happiness, must arise from the Contemplation of his Being, that he might give our Souls a just Relish of such a Contemplation, he has made them naturally delight in the Apprehension of what is Great or Unlimited.

This discussion of the great clearly suggests the reason for Addison's interest in aesthetic descriptions of landscape was that they provide a route to faith, as the Latitudinarians had claimed.

To analyze Addison's 'aesthetic' independent of the function of that project is to compartmentalize it according to subsequent divisions of knowledge to which he did not conform. Indeed, Addison's 'aesthetic' of the great has more in common with Boyle's arguments about how our limited powers of sensation and reason could lead us to an intimation of God's infinity in 'A discourse of things above reason.' Boyle argued that 'consideration of his [God's] works' was too great for the human intellect, and, in the recognition of this, we come to a knowledge that there must be a Creator beyond the reach of our minds. True to his empiricism, Boyle related this mental limitation back to the limits imposed by our senses: 'these instruments may be too disproportionate to some objects to be securely employed... we cannot by that [the eye] safely take the breadth of the ocean, because our sight cannot reach far enough to discover, how far so vast an object extends itself.'[47] Addison's and Boyle's arguments followed the same trajectory, starting with observation and being led from the great to a realization that there must be something beyond the great. Of course, their accounts differed in that Boyle's sense of being overmatched was empirical/scientific, where Addison's was aesthetic/literary, but both saw this excess as a way to spiritualize nature.

Reason and design

Accessible to fewer, but more satisfactory to Addison was the design argument, moving from nature and landscape to God via reason rather than the imagination. Here Addison shifted his focus to the other half of man's nature as conceived in the Christian version of the great chain of being. This argument was put by Addison in a linked series of more serious 'Saturday papers' in the *Spectator*.[48]

The series starts in *Spectator* 489 (iv. 233–36) where the pleasures of the imagination series left the argument, with the experience of the great in nature. The essay starts with an aesthetic description of a tempestuous sea scene: 'when it is worked up into a Tempest, so that the Horison on every side is nothing but foaming Billows and floating Mountains, it is impossible to describe the agreeable Horrour that rise from such a Prospect'. The paper then goes beyond the pleasures of the imagination faced with this seascape since 'the Imagination prompts the Understanding, and by the Greatness of the sensible Object, produces in it the Idea of a Being who is neither circumscribed by Time nor Space'. By turning attention from the landscape to the operation of the reason, Addison is closely recapitulating Boyle's line of thought discussed above.

Spectator 489 initiated a series of papers making this connection between the natural world and landscape and a reasoned approach to faith. In other words, *Spectator* 489 initiated a series of papers deploying the design argument. That Addison planned the essays as a series is made clear in the text. The next paper is *Spectator* 519, which opens by arguing that the pleasure of contemplating the material world is less than that in 'Contemplations on the World of Life'. The rest of the paper is a conventional argument to design looking at animals, and is complemented by the discussion of the wisdom of God which anatomical investigations have revealed and which forms the subject matter of *Spectator* 543. In both these papers, further, Addison recognizes, as did most discussions of design, that the proofs of contrivance to be found in the human body are still more striking than those in animals, largely because mankind has been given the reasoning faculty with which to discover all this skilful contrivance.

Following on from these papers, Addison wrote a series of more metaphysical Saturday essays concerning the proof of God through the consideration of infinity and eternity. As with the series which started with the contemplation of the ocean in number 487, this series originates in *Spectator* 565 with the way in which the rational faculty apprehends the face of nature. Looking at the firmament, Addison is led to consider his own insignificance in the words of Psalm 8: 'What is man that thou art mindful of him, and the son of man that thou regardest him?' In conventional fashion, *Spectator* 565 answers that an omnipotent and omniscient God can care for all beings, and that in these circumstances it is only rational that the immensity of the universe as we now understand it, thanks to Newton, leads us to praise God, a line of argument followed up in the complementary papers, *Spectators* 571 and 580.

These two sets of Saturday papers showed Addison giving a literary version of both Boyle's argument *to* design from the structure of the inanimate and animate world, and Newton's argument *from* the design and functioning of the universe. As we have seen, both were standard Latitudinarian arguments designed to appeal to our reason. We can further show that Addison added a caveat on the limits of reason, to check the possibility of hubris. Closing his series of essays on infinity and eternity, in *Spectator* 580 Addison had made it clear that our prospects and understanding of the natural system in the afterlife will surpass anything we can at present conceptualize. This argument was made still more clearly in *Tatler* 119, where Addison's 'Good Genius' tells him about our relish of the system of nature in the afterlife:

> We who are embodyed Spirits can sharpen our Sight to what Degree we think fit, and make the least Work of the Creation distinct and visible. This gives us such Idea's [*sic*] as cannot possibly enter into your present Conceptions.[49]

Addison's movement from the landscape and nature via reason to rational faith as set out in the *Spectator*, then, corresponded closely in all respects to the ways in which the Latitudinarian scientists and divines developed the design argument as a reading of the book of nature.

'Transcribing Ideas out of the Intellectual World into the Material'

There was a third way in which landscape imagery and the natural world was incorporated into Addison's essays. We have already seen in *Spectators*

489 and 565 how Addison could be led from observing the face of nature in the form of a tempestuous seascape and of the firmament to the realms of the understanding. He also allowed for a movement in the reverse direction, from the understanding to the imagination, from reason to aesthetics. This movement took two forms, which are set out in the last pleasures of the imagination papers, *Spectators* 420 and 421.

Spectator 420 (iii. 574–77) argues that 'there are none who more gratifie and enlarge the Imagination, than the Authors of the new Philosophy'. Addison's discussion of this movement of ideas from reason to imagination suggests the aesthetic benefit of this is that it allows us to be awed by the complexities of nature. At the small scale, microscopes leave us 'not a little pleased to find every green Leaf swarm[s] with Millions of Animals', while at the other end of the spectrum, Newtonian cosmography leaves us 'lost in such a Labyrinth of Suns and Worlds, and confounded with the Immensity and Magnificence of Nature'. In other words, natural philosophy creates new avenues leading the imagination to the experience of the great, overmatching our powers of comprehension. The paper closes by suggesting that in the afterlife we may be able to grasp more of this system.

Spectator 420, by its linkage of the findings of science with an aesthetic awe to induce piety created a pattern of argument similar to Boyle's meletetics. That this was the case can be seen from a *Guardian* essay where Addison used just such a movement of ideas. In *Guardian* 103, Addison started with a description of the fire work display which had put on to celebrate the peace of Utrecht in 1713. Addison uses this starting point in true meletetic style, blending science and rapture to inculcate a Christian message of humility:

> I could not forbear reflecting on the Insignificancy of Human Art, when set in Comparison with the Designs of Providence. In the Pursuit of this Thought I considered a Comet . . . as a Sky-Rocket discharged by an Hand that is Almighty. Many of my Readers saw that in the Year 1680, and if they are not Mathematicians will be amazed to hear that it travelled in a much greater Degree of Swiftness than a Cannon Ball, and drew after it a Tail of Fire that was Fourscore Millions of Miles in length. What an amazing Thought is it to consider this stupendous Body traversing the Immensity of the Creation with such a Rapidity, and at the same time Wheeling about in that Line which the Almighty has prescribed for it?[50]

Spectator 421 (iii. 577–82) opened another route by which a movement could occur from the findings of reason to the imagination. Here Addison argued that there is a pleasure of the imagination to be had from the

> Polite Masters of Morality, Criticism, and other Speculations abstracted from Matter; who, though they do not directly treat of the visible Parts of Nature, often draw from them their Similitudes, Metaphors, and Allegories.

The pleasure here was that the imagination and the understanding were entertained simultaneously, and, moreover, that the process of understanding was made easier as the ideas had been given 'Colour and Shape' by this transcription of ideas from the intellectual to the natural world. There is no need for any extended citation to show the extent to which Addison himself deployed this strategy, his numerous essays of visions and allegories being well known. Whether transcribing his Whiggish political ideals of liberty into landscape allegories, as in *Tatler* 161, or more religious messages as in the Vision of Mirzah (*Spectator* 159), Addison was consistently following his own precept of transcribing ideas from the realm of the understanding into the realm of nature and landscape to ease understanding. When done with a religious aim in mind, as in the Vision of Mirzah, the parallel with the Latitudinarian preachers is unmistakable.

I see, said I, a huge Valley and a prodigious Tide of Water rolling through it. The Valley thou seest...is the Vale of Misery, and the Tide of Water that thou seest is Part of the Great Tide of Eternity. What is the Reason, said I, that the Tide I see rises out of a thick Mist at one End, and again loses it self in a thick Mist at the other? What thou seest, said he, is that Portion of Eternity which is called Time, measured out by the Sun, and reaching from the Beginning of the World to its Consummation.

The parallel with Isaac Barrow's long allegory on the passions of mankind (cited earlier) should be apparent: both naturalize Christian arguments about human vision and its limits through the language of landscape description.

In the two ways of drawing the imagination into the propagation of the findings of the understanding which Addison set out in *Spectators* 420 and 421, we see him drawing on the third Latitudinarian reading of the book of nature outlined previously. Allegory and, to a lesser extent, meletetics, were an integral part of Addison's repertoire of uses of landscape imagery.

Conclusion: Addison's literary Latitudinarianism

The contexts in which Addison discussed the natural world and described landscapes were normally religious, and the different approaches he adopted closely map onto the readings of the book of nature developed by the Latitudinarians of his father's generation. As such, it seems an unavoidable conclusion that Addison's interest in and description of landscapes was Latitudinarian.

This conclusion might appear to be a recrudescence to Mandeville's line of criticism of Addison as a 'parson in a tie-wig'. If so, it is because such an approach to Addison has considerable merit in literary historical terms. Most approaches to Addison's concern with landscape have modernized him to an unwarranted extent, largely because they fail to take note of the way in which his pleasures of the imagination essays and his landscape descriptions are framed by a theological argument. Addison was in no way attempting to develop an independent aesthetic, nor simply to reflect his own love of landscape. The former view has been scotched in the course of my argument, and makes more sense retrospectively, after the development of a considerable body of aesthetic writings, which Addison helped to develop in the English language tradition, but which he could in no way foresee. The latter line of reasoning, most prevalent in Smithers' biography of Addison, casts him in the role of a proto-Romantic, 'ahead of his time' in terms of taste. Whatever Addison's private enthusiasm for his gardening pursuits, in his public role *qua* author, he justified landscape description and the observation of nature in more intellectual terms.

Seeing Addison as 'a parson in a tie-wig' also captures the truth that Addison is not simply a Latitudinarian theologian, his tie-wig makes him a *literary* author. While my main concern has been to show the theological bases of Addison's approach to landscape, there is no intention to suggest that the literary genres in which Addison wrote did not significantly influence the ways in which landscape was presented. For whilst, as has been shown, Addison deployed the three modes of reading the book of nature which the Latitudinarian theologians had developed, the *balance* between these readings was completely different in Addison than either Tillotson or Boyle. Addison's pleasures of the imagination essays and the landscape descriptions scattered throughout the *Spectator* meant that the aesthetic argument took on a significance it did not have to the contemporaneous Boyle lecturers, who indeed increasingly subordinated the aesthetic line to the argument from design. Addison's context in *belles lettres* led him to emphasize the aesthetic approach, a move which was to be repeated by other Low-Church literary authors through the century in their treatment of landscape. Indeed, as we shall see, the growth of aesthetic treatments of landscape in a literary context had an influence on the presentation of landscapes in devotional works later in the century.

So Addison's presentation of landscapes in his *oeuvre* amounts to a *literary Latitudinarianism*. Agreeing with Latitudinarian divines as to the importance of a rational approach to faith in the face of religious (and atheistic) extremism with its superstitious readings of nature, Addison took their basic modes of presenting landscapes and reworked them in ways appropriate to the genres in which the literary author operated. Given the reverence with which Addison's achievement was treated by subsequent generations of authors, and the popularity of his works with the reading public, his literary Latitudinarianism as applied to landscape marks the foundation of a significant tradition in the literary and intellectual history of eighteenth-century England.

Landscape, Latitudinarianism and literature at mid-century

Mid-century Latitudinarian novelists

One line of influence in eighteenth-century literary history leads from Addisonian *belles lettres* to the novels of the mid-eighteenth century. There is widespread agreement that Henry Fielding, Samuel Richardson and Laurence Sterne all speak from a Latitudinarian perspective, and that this influences their literature. As we will see by a brief excursus, none of these authors, however, expresses this through descriptions of landscape and the natural world to any great extent.

Martin Battestin points out that by the time Fielding wrote his novels, he had shifted from his youthful deism or freethinking to a rationalist Latitudinarian Anglicanism.[51] This shift was reflected in his novels, as Battestin has shown in his article, 'Tom Jones: the argument of design'[52] Fielding's deployment of the design argument in a literary context, however, was very different from that we have analyzed for Addison. Fielding incorporated the design argument at the level of narrative structure, such that 'form *is* meaning'. Further, this focus on plot means that the design argument manifests itself with respect to human nature rather than the natural world. As Battestin points out, the result is that characters are carriers of an abstract argument about the operation of providence which unfolds over the course of the whole novel. This status as carriers for larger ideas leaves many elements of the construction of *Tom Jones* poised between realism and allegory:

> Characters, scenes, the action itself – while maintaining an autonomous "reality" within the world of the novel – may owe their conception to some ulterior, abstract intention of the author. When the abstraction becomes so obtrusive that it dispels the illusion of reality in a fiction, we no longer have a novel, but an allegory or parable.[53]

This sense of shuttling between realism and allegory is clearly present on the one occasion on which the landscape is discussed at length, the description of the landscape of Allworthy's Paradise Hall. Fielding gave a long, realist description of the situation of the house, the lake beyond the groves and the valleys which opened from the house. But the realism of this description is undercut by its conclusion: after enumerating the beauties of this 'lovely prospect', Fielding peopled it with the

> one object alone in this lower world [which] could be more glorious, and that Mr Allworthy himself presented; a human being replete with benevolence, meditating in what manner he might render himself most acceptable to his Creator, by doing most good to his creatures.[54]

The passage converts a landscape description into a theological argument. Further, later reference to the landscape of Paradise Hall suggests it was described to create an emblem of Allworthy's good sense. Where under Allworthy the prospect owes 'less to art than to nature', the malevolent Captain Blifil is shown eagerly anticipating Allworthy's death, his head filled with 'projecting many other schemes, as well for the improvement of the estate as of the grandeur of the place'. The feeling of allegory is completed by Blifil's demise:

> while the captain was one day busied in deep contemplations of this kind, one of the most unlucky, as well as unseasonable, accidents happened to him. The utmost malice of Fortune could indeed have contrived nothing so cruel…In short…just at the very instant when his heart was exulting in meditations on the happiness which would accrue to him by Mr Allworthy's death, he himself – died of an apoplexy.[55]

Allworthy's natural benevolence is symbolized by the landscape owing less to art than nature, just as Blifil's malevolent scheming is symbolized by his schemes to override nature. The natural world in this case is an emblem of natural reason and its subversion. It is precisely because of the type of design argument centring on human nature adopted by Fielding, then, that landscape and the argument from the design of the natural world are not given a prominent role in Fielding's fiction.

For all the differences between Fielding and Samuel Richardson as novelists which have been discussed in criticism from Johnson to the present day, their treatment of landscape is similar. Richardson also has little use for descriptions of landscape and nature in the construction of his novels, which are also driven by a Latitudinarian piety, albeit in a far more direct fashion than Fielding's. As Richardson's fiction is primarily a fiction of interiors, indeed of enforced domesticity, there is little scope for the description of landscapes. Furthermore, perhaps the most interesting discussion of landscape, Clarissa's strategic description of the garden at Harlowe Place in Letter 86 recreates the sense of an interior, with the problems of surveillance, solitude and deception to the fore as they had been indoors. The fact that Clarissa's fall begins in a garden adds an obvious level of Christian symbolism to this landscape.[56] Yet *Clarissa* does make the same equation between virtue and the love of unspoilt landscape that *Tom Jones* made. As Lovelace admits, Clarissa 'always gloried in accustoming herself to behold the sun-rise; one of God's natural wonders, as she once called it'. Lovelace himself uses the argument from design when observing the animal creation, but his reflections, whilst in character, are less than Christian:

> there is nothing nobler, nothing more delightful, than for lovers to be conferring and receiving obligations from one another….A strutting

rascal of a cock have I beheld chuck, chuck, chucking his mistress to him, when he has found a single barley-corn . . . and when two or three of his feathered ladies strive who shall be the first for't (Oh Jack! a cock is a Grand Signor of a bird!) . . . [the successful hen] lets one see she knows the barley-corn was not all he called her for.[57]

Richardson does not describe landscapes or invoke the design argument as anything other than emblems of the characters with which they are connected.

Laurence Sterne was clearly a Latitudinarian in his theology, as is revealed by his *Sermons'* reliance on Tillotson and Samuel Clarke.[58] In this context Sterne advocated looking at landscape as an innocent pleasure, and contrasted this rational attitude to piety with the gloomy and superstitious rejection of the beauties of nature by Roman Catholics and Methodists.[59] At the other extreme, the *Sermons* also checked excess trust in the findings of science, arguing that secondary causes could never explain the operation of the natural world without reference to the divine design.[60] Sterne, therefore, established his Latitudinarian credentials as a rational and, because rational, a Christian observer of the natural world and its pleasing prospects.

Sterne's Latitudinarian attitude to nature and landscape did not, however, translate itself into a preoccupation with these topics in his literary *oeuvre*, as had been the case with Addison. This can be related to Sterne's literary positioning as a Latitudinarian *satirist* in contrast to Addison's as a Latitudinarian expositor.[61] Sterne's literature satirized deviations from the Latitudinarian *via media*, rather than defining that path. Where the *Sermons* were concerned with the 'solid purposes' of looking at nature, *Tristram Shandy* and *A Sentimental Journey* focused on solipsistic, sentimental and frivolous ways of viewing the landscape. Thus Yorick opines:

I pity the man who can travel from *Dan* to *Beersheba*, and cry, 'Tis all barren . . . I declare, said I, clapping my hands together, that was I in a desert, I would find out wherewith in it to call forth my affections.[62]

Yorick's brief mention of landscape here is simply a function of his desire to act in all matters sentimentally. The same can be said of Tristram Shandy's very similar statements in his travels in Book VII of *Tristram Shandy* (Chapter XLII). In both cases, the landscape is an emblem of sentimentalism and its excesses, and in both cases it is only one of the whole parade of such exemplars of that sentimentalism.

The great Latitudinarian novelists of the mid-eighteenth century, then, did not engage in any protracted literary representations of landscape and the natural world. In each case, their belief in rational religion and the argument from design was channelled into other aspects of their literature.

Meletetics in the mid-eighteenth century

To find the continuation of Addison's project of a literary Latitudinarianism centred on the design of the natural world and the appearance of landscape, we need to look in a different direction from that offered by the novelists. This continuation can be found in the work of Edward Young and James Hervey, although it took a different form because they developed a different balance between the various strategies for reading the book of nature.

Edward Young:

Young's background was a Latitudinarian one, his father being a friend of Tillotson's.[63] Young's admiration for Addison, whom he knew personally, as a Christian and a writer was clearly unbounded as can be seen from his praise of Addison in the *Conjectures on Original Composition*. Given this, it is unsurprising that Young tried to recreate elements of the Addisonian project in his own writings, rendering Christian instruction in palatable literary form.

> Since Verse you think from Priestcraft somewhat free,
> Thus, in an Age so gay, the Muse plain Truths
> (Truths, which, at Church, you might have heard in Prose)
> Has ventur'd into Light; well-pleas'd the Verse
> Should be forgot, if you the Truths retain[64]

Further, Young's Christian instruction, as Addison's and Boyle's, centred on landscape and nature, which led to a certain similarity in the ways they described their goals. Where Addison had wished to 'moralise the natural pleasure of the soul', and Boyle to 'spiritualise nature', for Young 'Nature is Christian' (NT, iv. 704), and 'The World's a System of Theology' (NT, vii. 1138). Man's position in this world, just as for Addison in *Spectator* 519, is midway in the Chain of Being:

> Who center'd in our make such strange Extremes?
> From different Natures, marvelously mixt,
> *Connection* exquisite of distant Worlds!
> Distinguisht *Link* in Being's endless Chain!
> *Midway* from *Nothing* to the *Deity*! (NT, i. 70–74)

For all these parallels with Addison, however, no one would say that the ways in which landscape is presented by the two are closely comparable. The origins of this difference can be traced to the different ways in which Addison and Young conceptualize mankind's position in the chain of being. For Addison, as we have seen, we are midway because our nature partakes of both the sensual and the rational. In Young's version, by contrast, we are midway between nothing and God. Vitally, Young linked God to the imagination, where Addison had linked imagination to the senses and reason to the divine. This was clear when Young wrote of the death of Grace Cole:

> She is dead to us; she is in another state of existence; we are in the world of reason; she is in the kingdom of imagination; nor can we more judge of her happiness or misery, than we can judge of the joy or sorrow of a person that is asleep. The persons that sleep are (for the time) in the kingdom of imagination too.[65]

Young, then, had to exalt the imagination as the route to faith to an extent Addison had never considered. Furthermore, Young saw the literary imagination in denominational terms, as something particularly encouraged by libertarian Protestantism, where Pope's Catholicism encouraged an adherence to literary tradition and imitation which stifled originality and imagination:

> His [Pope's] taste partook of the error of his religion; it denied not worship to saints and angels, that is, to writers, who, canonized for ages, have received their apotheosis from established and universal fame. True poesy, like true religion, abhors idolatry.[66]

As form is meaning in *Tom Jones*, so for Young the imaginative faculty has both a denominational meaning and generates literary imperatives of form.

Young's advocacy of the imagination, while distinguishing his theological-literary project from Addison's, did not mean that reason should be jettisoned, a move which would have placed him beyond the pale of Latitudinarian rationalism. He argued that 'Passion is Reason, Transport Temper' (NT, iv. 640). He expanded on this latter in *Night IV*, arguing that freethinkers deified reason, where true reason was inseparable from imaginative faith (iv. 743–47). This was a line Young also took in response to the growth of natural philosophy, which he saw as another way in which reason had been perverted from its natural alliance with religion:

> Born in an Age more Curious, than Devout;
> More fond to fix the *Place* of Heaven and Hell,
> Than studious *this* to shun, or *that* secure.
> 'Tis not the *curious*, but the *pious* Path,
> That leads me to my Point — (NT, ix. 1852–56)

Young's criticisms of reason fit into the pattern of acceptance of the limits of reason which all Latitudinarians had adopted. Where he differs from most of his predecessors, is in the extent to which, for Young, the religion of reason is an imaginative one.

If 'nature is Christian' and the realm of the Deity is the 'kingdom of imagination', it is not surprising that Young's literary readings of the book of nature differ from Addison's or Boyle's in the extent to which they are drawn to flights of imagination, starting from the face of nature. In other words, given the way in which Young conceptualized the natural world and the chain of being, it was to be expected he would be drawn towards the meletetic approach to an unprecedented extent, which is exactly what we find in *Night Thoughts*.

Young does still acknowledge the traditional argument from design, notably in a long passage towards the conclusion of *Night Thoughts*:

> GOD is a *Spirit*; *Spirit* cannot strike
> These gross, material, Organs; GOD by Man
> As much is seen, as *Man* a GOD can see,
> In these astonishing Exploits of Power:
> What Order, Beauty, Motion, Distance, Size!
> Concertion of Design, how exquisite! (NT, ix. 1419–24)

This passage is a conventional argument from design, with both God's existence and attributes (power and, by implication, wisdom) being visible in the Creation.[67] That this design was to be apprehended scientifically is clear from Young's praise of Newton in his correspondence, which was matched by a bust of Newton which he erected in his garden at Welwyn.[68] But Young was far more taken by the argument from design in its aesthetic form. This is most readily apparent, as is his position in the tradition of Addison, in his recapitulation of the theological aesthetic of the great from the *Spectator*. Looking at the heavens, Young apostrophizes:

> The Soul of Man, HIS Face design'd to see,
> *Who* gave these Wonders to be seen by Man
> Has *here* a previous Scene of Objects great,
> On which to dwell; to stretch to that Expanse
> Of Thought; to rise to that exalted Height
> Of Admiration; to contract that Awe,
> And give her whole Capacities that Strength,
> Which best may qualify for final Joy:
> The more our Spirits are inlarg'd on *Earth*,
> The deeper Draught shall they receive of *Heav'n*. (NT, ix. 70–79)

It is in his reliance on the meditative reading of the book of nature, however, that Young's Latitudinarian landscape aesthetic is distinctive. Throughout his *oeuvre*, Young recurs to this approach, and only two examples will be given here of his search into the natural world for 'emblems just' (NT, vi. 690) of moral and religious themes. First, we have seen that in Boyle's pioneering work on meletetics, he had exemplified his method in a discussion of the moon. Young takes up this theme in *Night Thoughts*, creating a parallel between the light of the moon and the light of God, as had Boyle:

> Vain is the World, but only to the Vain.
> To what compare we then this varying Scene,
> Whose Worth ambiguous rises, and declines?
> Waxes, and wanes? (In all propitious, *Night*
> Assists me Here) Compare it to the Moon;
> Dark in herself, and Indigent; but Rich
> In borrow'd Lustre from a higher Sphere (NT, iii. 420–26).

The other example is Young's approach to moralizing the fireworks of 1749, which celebrated the peace of Aix-la-Chapelle. This was a situation parallel to that Addison had used for his meletetic reflections in *Guardian* 103, and Young made exactly the same rhetorical move from fireworks to God's fireworks in the firmament:

> Instead of Squibbs, & Crackers, I shall humbly content myself with Sun, Moon, & Stars. These glorious Fireworks of that Great King who in ye noblest sense is ye *Author of Peace*.[69]

If the basic idea of moving from the human to the natural world to inspire awe was common to both Addison and Young, their different modes of wonder show their divergence. Where Addison had used this move to introduce a scientifically grounded sense of wonder in the size and speed of comets, Young's reflections were more rhetorically high-strung and rhapsodic:

> Hast thou ne'er seen the Comet's flaming Flight? . . .
> Thro' Depths of Ether; coasts unnumber'd Worlds,
> Of more than solar Glory; doubles wide
> Heaven's mighty Cape, and then revisits Earth,
> From the long Travel of a thousand Years.
> Thus, at the destin'd Period, shall return
> *He*, once on Earth, who bids the Comet blaze;
> And with Him all our Triumph o'er the Tomb (NT, iv. 705, 710–15).

Young pioneered a new approach within the Latitudinarian tradition, reassessing the respective weights to be given to the various reading of the book of nature on the basis of both his pessimistic assessment of the piety of the readership of literary genres and of his prioritization of imagination rather than reason as an intimation of immortality.

James Hervey: 'baiting the gospel-hook'

James Hervey is not a figure who can be described as central to any literary or theological canon for eighteenth-century England, but he is an appropriate figure for analysis in the present context for two reasons. First, Hervey shows that the arguments pioneered by the Latitudinarians branched out to Low-Church divines more generally as the eighteenth century progressed. Hervey's allegiance to the Church of England and its articles is not open to doubt,[70] but he was doctrinally far from being Latitudinarian in the traditional sense. Hervey was both heavily influenced by Methodism, being chaplain at Lincoln College, Oxford when John Wesley's Holy Club had just been founded, and he was a believer in the Calvinist doctrine of election which had been so vigorously rejected by the Latitudinarians. Secondly, Hervey shows that where Addison had been heavily influenced in his literary presentation of landscape and nature by devotional writings, his success led to a reverse flow of ideas and techniques from literature to devotional writings. If Addison was a parson in a tie-wig, Hervey was a belletrist in a pulpit.

Hervey was quite explicit about the debt his devotional writings owed to the literary realm, with citations of Milton, Addison, Thomson and Young peppering his hugely successful devotional manual, *Meditations and Contemplations*. Indeed, citing Addison's *Spectator* 393, Hervey went so far as to suggest that 'Upon the Plan of these Observations the preceding and following Reflections are formed'.[71] The debt to Young was just as great if not greater, as Hervey also shifted the Addisonian approach in the meletetic direction.[72] As with Young, this shift was justified by giving the imagination a greater role in the persuasion to piety: 'Allegory *taught many of the Objects to speak the Language of Virtue; while* Imagination *lent her Colouring to give the Lessons an engaging Air*' (*Meditations*, ii. p. xiii).

The presentation of landscape and the natural world in Hervey's *Meditations and Contemplations* (1746–7) was dominated to an unprecedented extent by meletetic reflections. Furthermore, the descent from awed rhapsody into mere bombast reached depths even Young's most stringent critics would have admitted he did not plumb. This can be seen in Hervey's religious parallel generated by the moon's control over the tides:

> O! Ye Mansions of Blessedness; ye Beauties of my Father's Kingdom; that far outshine these Lamps of the visible Heaven, transmit your sweet and winning Invitations to my Heart. *Attract* and *refine* all my Affections. With-hold them from *stagnating* on the sordid Shores of Flesh; never suffer them to *settle* upon the *Lees* of Sense (Meditations, ii. 98).

In fact, Young had made exactly the same parallel in *Night Thoughts*, but the effect does not strike the modern reader as quite as absurd:

Can yonder *Moon* turn the Ocean in his Bed,
From Side to Side, in constant Ebb and Flow,
And purify from Stench his watry Realms?
And fails her *moral* Influence? Wants she Power
To turn LORENZO's stubborn Tide of Thought
From stagnating on *Earth's* infected Shore
And purge from Nuisance his corrupted Heart? (NT, ix. 1201–207)

Clearly, Hervey's reflections on the moon fit into the tradition established by Boyle, continued by Addison and emphasized by Young, but he gives it a still higher emotional charge. Hervey also extended these meletetic reflections on nature from the short sallies of the imagination envisaged by Boyle to extended parallels. As an example appropriate to this study, Hervey's second meditation, 'Reflections on a Flower-Garden', had a set piece moralizing the prospect from the flower garden (*Meditations*, i. 137–52). Hervey opens with a predictable apostrophe:

O! What a Prospect rushes upon my Sight! How vast; how various; how "full and plenteous with all manner of Store!"...Methinks, I read, in these spacious Volumes, a most lively Comment (*Meditations*, i. 137).

He then goes into a long description and reflection on the various elements of the prospect – the fields, meadows, groves and so on – before adding a Christian refrain:

Only let me remind you of one very important Truth.... that you are *obliged* to CHRIST JESUS for every one of these Accommodations, which spring from the teeming Earth, and smiling Skies (*Meditations*, i. 146).

The refrain with which Hervey closed his moralized landscape description is important, as it emphasized the difference between his own approach and the rhapsodic enthusiasm for nature advocated by freethinkers inspired by Shaftesbury.[73] Hervey always made the orthodox Christian transition from nature to nature's creator when describing landscapes. The dangers of failing to do this were personified by Hervey in his other major work, *Theron and Aspasio* (1755). Theron was portrayed as liable to let his enthusiasm for prospects seduce him from the correct interpretation of those prospects as the works of God designed for our instruction, a refrain continually enforced by Aspasio. This had already happened several times in their dialogues before the best example, Dialogue 17. Theron is enraptured by a forest scene, and cries:

Give me the scenes, which disdain the puny assistance of art, and are infinitely superior to the low toils of man.

Theron does ascribe this to *ye greatness of our GOD*, thus recreating Addison's theological aesthetic of the great, but Aspasio recalls him to a still more orthodox, because Biblical, approach to nature:

> *Isaiah's* divine imagination was charmed with the same grand spectacle [cites Isaiah x. 33–4] . . . Then he passes by a most beautiful transition, to his darling topic, the redemption of sinners.

Theron is grateful for this check on his 'roving thoughts' which recalls him to 'a more excellent subject'.[74] Unlike Shaftesbury, then, Hervey's ultimate aim was to inculcate the Christian message.

Hervey, then, strenuously differentiated his meletetics from heterodox enthusiasm for nature. He also saw his emblematic method as rational, drawing a sharp distinction between meditating on the natural scene and being prey to superstitions about nature. In his prose reworking of *Night Thoughts*, 'Contemplations on the Night', Hervey dismissed superstitions such as the raven's cry being an omen of death:

> One cannot but wonder, that People should suffer themselves to be affrighted at such *fantastical*, and yet be quite unaffected with *real*, Presages of their Dissolution. Real Presages of this awful Event, address us from every Quarter. What are these *incumbent Glooms* [i.e. Night time], that overwhelm the World, but a *Kind* of *Pall* provided for Nature; and an Image of that long Night, which will quickly cover the Inhabitants of the whole Earth? (*Meditations*, ii. 60–61).

Hervey also highlighted his rational approach when he discussed comets in the *Meditations*, citing Newton and Derham to refute the 'Pretenders to *Judicial Astrology*' (*Meditations*, ii. 126–28, 119). Here Hervey showed that, for all his preference for the aesthetic and meditative approaches to the book of nature, he could draw on the scientific version of the argument from design in order to differentiate himself as a rational Christian from the excesses of irrational superstition.

However, meditative Hervey's approach to landscapes and the natural world, he still constructed his approach within the parameters of the Anglican *via media*, steering between the deification of reason and nature in Shaftesbury and the irrationalism of the superstitious. That he emphasized the aesthetic and meditative approach so strongly was due to a pessimism akin to Young's about the piety of his contemporaries in mid-eighteenth-century England. Similarly, the emphasis on landscape and the natural world was because they provided a bridge between the literary/aesthetic and the spiritual. In a long comment which also looks forward to Gilpin's picturesque, Hervey summarized his justification for a theological landscape aesthetic:

I wish you had taken minutes of what you saw most remarkable, in your tour through *Westmoreland* and *Cumberland*. A description of those counties would be very acceptable to us, who inhabit a more regular and better cultivated spot. Described in your language, and embellished with your imagination, such an account may be highly pleasing to all; and grafted with religious improvements, might be equally edifying. – Such kind of writings suit the present taste. We don't love close thinking. That is most likely to win our approbation, which extenuates the fancy, without fatiguing the attention. Since this is the disposition of the age, let us endeavour to catch them by guile; turn even a foible to their advantage, and bait the gospel-hook agreeably to the prevailing taste.[75]

Conclusion: The Janus face of mid-century literary history

Many critics have co-opted the writings of Young (and to a lesser extent Hervey) into a trajectory of literary history as 'pre-' or 'proto-Romantics'.[76] There has also been an argument that the mid-eighteenth century is to be conceptualized as an independent literary 'moment', sharply differentiated from the 'Age of Pope and Swift' and the 'Age of Johnson' which surround it.[77] The reading developed here does not challenge the findings developed by these approaches, but it does suggest that further light can be shed on mid-century literary history by looking to the lines of influence which connect it back to the intellectual history of the Restoration and early eighteenth century, as well as looking forward or seeing it as a self-defining moment. The themes of sentimentality, literary loneliness and melancholy which permeate the middle of the century have links back to a meditative theological tradition which have been ignored due to unexamined assumptions about the nature of rational religion and Latitudinarianism.

What is new in the presentation of literary landscapes in mid-century literature is the balance they strike between science, aesthetics and meditation. Young and Hervey both emphasized that they were rational Christians, adapting themselves to a rising taste for literature. This led them both to emphasis the intertwining of reason and imagination in a way Addison and his generation had not. The increased role for aesthetic and meditative approaches to the description of landscapes certainly meant they were presented in a new and heightened emotion tone, but the basic Low-Church project connecting literature, landscape and theology was unchanged. In historical context, the literary landscapes of Young and Hervey were both new and deeply rooted in traditional Anglicanism.

William Gilpin and the Latitudinarian picturesque

Gilpin and posterity

William Gilpin is remembered as the pioneer of an aesthetic approach to landscape, 'the picturesque'.[78] Indeed, the standard biography of Gilpin is exclusively concerned with his picturesque drawings and tours.[79] If that is Gilpin's legacy, it is certainly not the one he hoped to leave. Writing an autobiographical 'Memoir' for future generations of his family, Gilpin's own vision of his life was rather different:

> Thus Mr G[ilpin]. has given an account of the only two transactions of his life, which make it worth the attention of his posterity – his mode of managing his school at Cheam, which was uncommon – & his mode of endowing his parish school at Boldre, from the *profits of his amusements*.[80]

The profits Gilpin refers to were the result of publishing his picturesque tours, and therefore, as literature, they clearly occupy a marginal position in Gilpin's own estimation of his life's work. Gilpin was clearly most proud of the two schools he was connected with at Cheam and Boldre. From his own account of his 'uncommon' management of those schools, we learn much of his scale of values. The 'Memoir' suggests that Gilpin tried to inculcate both political and religious values of liberty and Protestantism in his pupils. Politically, the boys were taught the laws of the school, and, with literal Lockeanism, signed contracts promising to obey those laws, 'impressing young minds with an early love of order, law and liberty'. Equally, Gilpin's schools were uncommon in the emphasis they placed on moral and religious instruction, at the expense of a classical education because of Gilpin's belief that 'where one boy miscarries for want of classical knowledge, hundreds are ruined for want of religious principles'.[81]

In one sense, Gilpin's self-assessment of his achievements as a religious and political educator has started to attract attention through studies of the 'politics of the picturesque', which move the ground of an assessment of his life's work away from pure aesthetics.[82] But the role of religion in Gilpin's achievement as a writer has not attracted attention: while Barbier accepts that 'in the last analysis' nature for Gilpin is 'a divine work of art',[83] there has been no attempt to draw out the lines of influence between Gilpin's twin religious roles of Anglican clergyman and religious educator and his picturesque *oeuvre*. Following the contours established by the hierarchy of values in Gilpin's 'Memoirs', I will move from Gilpin's neglected moral and religious writings, in particular their construction of nature and landscape, to the picturesque tours, showing how an attenuated and aestheticized form of the Latitudinarian approach to the face of nature influenced Gilpin's picturesque.

Nature and landscape in Gilpin's didactic works

'Mr Gilpin's doctrinal views coincided, in many points, with those known by the denomination of "low church".'[84] This contemporary biographical opinion is vindicated by an analysis of Gilpin's religious and moral writings. Gilpin displayed all the traits of a Low-Church Anglicanism that have recurred in the course of this chapter: an opposition to excess church ornament;[85]

advocacy of toleration of Methodists and Quakers;[86] and an attack on both the excess rationalism of deists and Socinians, and on 'solfideism', the unqualified belief in grace over reason (*Dialogues*, pp. 131–33). The result is a defence of a rationalist Anglican *via media*. But while the rhetoric of the middle way is the same as it was a century previously in the Restoration founders of Latitudinarianism, the nature of that middle way had changed for Gilpin, becoming less doctrinal and dogmatic:

> we have the works of many excellent divines of the last century – Barrow, Mede, Sanderson, Tillotson, and many others; all of them full of matter, but formally digested, dry in their manner, and often, perhaps, intermixed with Popish controversies, and other points, which relate more to the times they lived in, than to ours (*Dialogues*, p. 283).

Gilpin here was part of the general movement away from historical and doctrinal modes of religious scholarship which occurred in eighteenth-century churchmanship, an approach he saw as 'no better than solemn trifling'.[87] A theology like Gilpin's, based on Latitudinarian rationalism, updated to strip away doctrinal and denominational conflicts, was left heavily reliant on the evidence of nature and, by extension, landscape.

In a phrase reminiscent of Edward Young, Gilpin stated that 'Nature never produced an atheist.'[88] From this starting point, he deployed natural knowledge and landscape imagery in a number of ways to make it positively inculcate Christianity. First, while Gilpin's picturesque is known for eschewing science in favour of aesthetics in the apprehension of nature, his didactic works did deploy the more scientific design argument. This is most fully expressed in one of Gilpin's *Dialogues*, 'The advantages of a town life, and a country life, compared.' Here two discussants, Sir Charles and Willis, accept that the works of nature lead us to 'that bounteous benefactor', but further that 'the works of nature...furnish many employments to the mind, more solid than looking at a prospect' (*Dialogues*, p. 181). The 'solid employments' in question are the two varieties of the design argument, the arguments *from* and *to* design. Sir Charles prefers Boyle's argument to design:

> as to the *starry heavens*...A gnat, or a beetle, which I understand better, is more the object of my attention; and, of course, a stronger argument to me of the Almighty power, than they are in all their vastness and magnificence (*Dialogues*, pp. 183–84).

By contrast, Willis prefers the Newtonian argument from design, and in his version shows it has close affinities with Addison's argument that the great leads us to a sense of the design by overwhelming our imagination:

> in all the works of God, there is something beyond human comprehension, which seems intended to teach us, at the same time, the omnipotence of

God, and the weakness of man.... If we could comprehend all the works of God, our minds, like the great Creator's, must be infinite. If the ocean could be fathomed, our ideas of its grandeur would in a degree subside (*Dialogues*, pp. 184–85).

Willis's acceptance of the scientific design argument, balanced by an emphasis on the limits of human knowledge, was a more general refrain in Gilpin's didactic works. As he preached, 'we cannot comprehend with our confined understandings, the whole mode of God's government'.[89]

Gilpin's second common position is a version of the aesthetic design argument. He starts with the argument that the beauties of landscape are at least innocent, or 'eligible' (*Dialogues*, p. 142), in that they do not encourage the spectator to vice. It was for this reason Gilpin could recommend gardening and looking at landscape as a pursuit appropriate for a clergyman.[90] But beyond this innocent pleasure, the aesthetic appeal draws the thoughtful person to devotion, an argument for which Gilpin cited Biblical support:

I have often admired that beautiful picture of patriarchal innocence and simplicity, in the history of Isaac...Nobody can read, that *Isaac went out to meditate in the fields at even-tide*, without conceiving him to be a man strongly impressed with a sense *of piety and devotion* (*Dialogues*, p. 173).

This trajectory, from innocent pleasure to Christian piety thanks to the beauties of the landscape, is repeated in Gilpin's 'Defence of the Polite Arts', an imagined dialogue between Sir Philip Sidney and Lord Burleigh:

it appears to me, that an admiration of the beauties of nature may be ranked in a still higher form, than that of administering merely to pleasure. Perhaps a person of your Lordship's serious disposition will not accuse me of enthusiasm, when I speak of these sublime appendages of landscape, as leading the mind to the great author of them (*Dialogues*, p. 401).

As we will see, the first part of this argument, conceptualizing a love of landscapes as an innocent pleasure, was part of Gilpin's explicit defence of the picturesque, while the second part of the argument, leading to the Creator, was implicit in Gilpin's understanding of the picturesque way of seeing the landscape.

The third way in which landscape and nature were incorporated into Gilpin's religious and moral argumentation was as what he variously described as emblems, parables and analogies. Gilpin's method here was akin to that of Robert Boyle's meditative *Occasional Reflections*, especially in what he called his 'Hints' for sermons, which were brief reflections jotted down as he walked or rode in his parish. Both Boyle and Gilpin, then, built a meletetic method where observations on the natural world were given spiritual significance by the creation of parallels between the appearance and operation of nature and Christian doctrine. Further, Gilpin thought these moral messages are actually implanted by God in nature, not merely the product of religiously motivated associative reasoning:

> religious uses...we draw from contemplating the works of God; particularly various analogies, corresponding with the truths of religion; and, indeed, as the same God is the author both of nature, and of religion, the analogy serves, I think, not only to *illustrate* the religion, but in a degree, to *confirm the truth* of it. This analogical method was taught us by our blessed Saviour, whose beautiful allusions are chiefly confined to the productions of nature (*Dialogues*, p. 201).

The parallels, allusions and analogies that Gilpin draws are of a type already familiar from Hervey and Edward Young: the argument from design is seen as parallel to the doctrine of the redemption, both showing grandeur, contrivance, utility and simplicity;[91] nutrient cycling in nature becomes an emblem of Christ's speech after the feeding of the five thousand (*Dialogues*, pp. 194–95); and 'the growth of corn' is found to give a 'strong representation...of the resurrection of the dead'.[92] As well as deploying this meditative technique himself, Gilpin encouraged others, especially the country congregation of the title of his *Sermons*, to give a spiritual reading of the landscape which surrounded them in their husbandry. This was the theme of his sermon, 'The husbandman', which opened with the predictable trope of nature as 'books of instruction wherever we throw our eyes', going on to exhort his audience to take the 'mixed instruction with the things of this world'.[93]

The final religious use of landscape and the natural world in Gilpin's didactic works is one also found in Fielding's *Tom Jones*. For if the landscape is full of analogies and lessons, so the cultivated or improved landscape can become an emblem of its owner. Gilpin drew on this approach in his *Moral Contrasts*. *Moral Contrasts* has two main characters, who inherit considerable estates at the same time. One, Sir James Leigh, is profligate, never having been given the sort of religious instruction which Gilpin practised as an educator, and which would curb his desires. Sir James's attitudes to the landscape are a mirror of his mind:

> Sir James Leigh was carrying on his improvements, as he called them, with a profusion of expence, that astonished every body. If you walked near his house, you saw groups of labourers, here, and there, and every where – removing ground – widening rivers – building bridges – or employed in other expensive operations; none of which was well considered, or was conducted with the least taste, or judgment; for he had

too high an opinion of himself to follow the advice of any one. His projects were all in opposition to nature.

Willoughby, Gilpin's pious contrasting character, is quite the reverse: a Christian education has led him to respect God's landscape, and therefore 'never to fight with nature'. As in *Tom Jones*, the 'most essential difference' between Willoughby/Allworthy and Sir James Leigh/Captain Blifil lay, for Gilpin, 'in the article of religion'.[94]

Gilpin's didactic works, then, developed a number of ways in which to co-opt the natural world and landscape into Christian and moral arguments. Nature became even more important than it had been for earlier generations of Latitudinarian divines, given Gilpin's distaste for any form of doctrinal or dogmatic theology.

The picturesque view of landscape: 'like the fly on the column'

Given Gilpin's pride in his role as a didactic and religious educator, and given that his didactic works were so concerned with nature and landscape, it would be surprising if Gilpin's picturesque tours contained no traces of the approaches to nature he developed in his clerical and pedagogic life. Analysis of the picturesque tours with Gilpin's didactic deployments of landscape in mind reveals a substantial overlap of arguments.

Gilpin's programmatic discussion of the picturesque, in his *Three Essays*, may appear to discountenance the religious and didactic potential of the tours in favour of a purely aesthetic reading. In the *Three Essays*, the possibility that observing nature will lead the picturesque tourist to nature's creator is hedged around with subjunctives:

we might begin in moral stile; and consider the objects of nature in a higher light... We might observe, that a search after beauty should naturally lead the mind to the great origin of all beauty... But tho' in theory this seems a natural climax, we insist the less upon it, as in fact we have scarce ground to hope, that every admirer of *picturesque beauty*, is an admirer also of the *beauty of virtue*; and that every lover of nature reflects, that

> Nature is but a name for an *effect*,
> Whose *cause* is God.—[95]

Gilpin seems doubtful that the trajectory from innocent to religious pleasures which his didactic argument (and his quotation from Young) described can be maintained in the context of literature and *belles lettres*. As such, he concentrates on the first stage of his argument, suggesting that the picturesque is at least a 'rational, and agreeable amusement', adding that even this may be of some use in an age teeming with licentious pleasure; and may even in this light at least be considered as having a moral tendency.[96]

Yet this discussion must be considered carefully. Gilpin does not deny the religious potential of picturesque tours if conducted by those who do tend to move from nature to nature's God. We know from his didactic works that Gilpin was habituated to just such a mental and meditative process. Therefore, whatever Gilpin's scepticism about the spirituality of picturesque touring in some hands, there is no need to doubt the possibility in Gilpin's own practice of the picturesque, as he was an individual for whom we know a group of indicatives can be substituted for his maze of subjunctives. Gilpin did, in fact, incorporate his theological and didactic approaches to landscape into his picturesque tours in two main types of argument, consonant with his own view of the tour genre as not simply didactic nor descriptive, but 'a species between both'.[97]

The first way in which Gilpin's religious approaches to nature carried over from his didactic works into the picturesque tours was in the occasional eruption through the otherwise aesthetic surface of the texts of meditative or meletetic reflections on the scenes before him. A good example of this comes in Gilpin's tour of the highlands, where the landscape leads to a reflection on the rise and fall of civilizations, this reflection being justified as

> high places, and extended views have ever been propitious to the excursions of imagination. As we surveyed the scene before us, which was an amusing, but unpeopled surface, it was natural to consider it under the idea of population.[98]

Similarly, in his tour to Cumberland and Westmoreland, Gilpin was led into a brief digression since 'rivers often present us with very moral analogies; their characters greatly resembling those of men'.[99] This analogical method of reasoning seems entirely out of place if Gilpin is taken as the pioneer of a purely aesthetic movement which refused all varieties of associative reasoning when confronted with a landscape. It is, however, entirely consonant with the methods developed in the didactic element of Gilpin's *oeuvre*. These meletetic reflections are not common, but their presence in the picturesque tours shows Gilpin transferring ways of seeing nature from his role as an educator to his aesthetics.

If the two examples cited so far suggest Gilpin inserted moral reflections but shied away from the religious, perhaps the greatest such reflection, in his *Remarks on Forest Scenery*, is explicitly Christian. Discussing the appearance of individual trees, Gilpin crosses from the descriptive to the didactic since:

> If a man were disposed to moralize, the ramification of a thriving tree affords a good theme.

The remainder of this moralization is contained in a long footnote. Gilpin introduces what he calls a 'short allegory', based on Horace's comparison of falling leaves and the mortality of mankind in the *Ars Poetica*. Gilpin compares the acacia he sees from his window with 'a country divided into provinces, towns, and families', and sees the leaves falling, as Horace had, as 'the most obvious appearance of mortality'. Having run through this parallel at some length, Gilpin gives a Christian turn to this moral theme: 'How does every thing around us bring it's [sic] lesson to our minds! Nature is the great book of God.' But, of course, the Christian revelation puts us in a better position to comprehend the spiritual lessons of nature than Horace:

Morality must claim it's [sic] due. Death in various shapes hovers round us. — Thus far went the heathen moralist. He had no other knowledge from these perishing forms of nature, but that men, like trees, are subject to death. . . . Better instructed, learn thou a nobler lesson. Learn then that God, who with the blast of winter shrivels the tree, and with the breezes of spring restores it, offers it to thee as an emblem of thy hopes.[100]

Trees in this meditation become a symbol of the Christian doctrine of the resurrection, just as corn had been in one of Gilpin's sermons. Where an aesthetic reading of Gilpin must view such moments as mere aberrations in his picturesque, seen in the context of his complete *oeuvre*, they show there was continuity in Gilpin's output as a writer. Moreover, seen in the context of the Latitudinarian literary tradition outlined in this chapter, Gilpin's moralizing is clearly part of the approach to nature which Addison had described in the same terms in *Spectator* 393.

It could still be said that such moments of moralizing in the picturesque tours, while fitting into Gilpin's approach to landscape and the natural world more closely than has hitherto been imagined, were still just that, moments unrepresentative of the descriptive approach which governed most of the text. This is undeniable, but it mistakenly assumes that there was no religious element to the actual process of viewing, describing and drawing the landscape according to picturesque conventions. In fact, the picturesque is a highly attenuated form of the aesthetic version of the design argument, coupled with the standard refrain of rationalist Christians about our limited views and comprehension of God's plan in the universe.

The indebtedness of the picturesque way of seeing to this combination of design arguments was gestured at in Gilpin's first published tour (along the Wye Valley), where he wrote:

Nature is always great in design; but unequal in composition. She is an admirable colourist; and can harmonize her tints with infinite variety, and inimitable beauty; but is seldom so correct in composition, as to produce a harmonious whole. . . . The case is, the immensity of nature is beyond human comprehension. She works on a vast scale; and, no doubt, harmoniously, if her schemes could be comprehended.[101]

In other words, the errors which the picturesque traveller discovers in nature's composition do not suggest the world is deformed or imperfect, as Thomas Burnet and David Hume had suggested, but that human understanding was imperfect, only comprehending parts of God's design. The picturesque perspective, then, was that of a being only midway in the chain of being, just as it had been for Addison and Young.

This argument was also mentioned in the Western tour,[102] but received its fullest exposition, together with its most explicit linkage to traditional Anglican design arguments, in Gilpin's North Wales tour. In a long passage, Gilpin laid out how the picturesque vision of mankind related to God's vision:

> In composition alone – I mean picturesque composition – nature yields to art. Nature is full of fire, wildness, and imagination. She touches every object with spirit. Her general colouring, and her local hues, are exquisite. In composition only she fails. We speak however in this manner like the fly on the column. Her plans are too immense for our confined optics. They include kingdoms, continents, and hemispheres; and may be as elegant, as they are incomprehensible. Could we take in the whole of her landscapes at one cast; could we view the Hyrcanian forest as a grove; the kingdom of Poland as a lawn; the coast of Norway as a piece of rocky scenery; and the Mediterranean as a lake; we might then discover a plan justly composed, and *perhaps* beautiful even in a painter's eye.[103]

The possibility is, then, that God's vision of the universe makes all appear picturesque, but that we cannot see at the requisite scale to appreciate this fact. The 'fly on a column' refers to George Berkeley's essay on freethinkers in *Guardian* 70. In that essay, Berkeley wrote of visiting St Paul's, and there beholding 'a Fly upon one of the Pillars'. He then compared the fly to a freethinker:

> For it required some Comprehension in the Eye of the Spectator, to take in at one view the various Parts of the Building, in order to observe their Symmetry and Design. But to the Fly, whose Prospect was confined to a little part of one of the Stones of a single Pillar, the joint Beauty of the whole, or the distinct Use of its Parts, were inconspicuous, and nothing cou'd appear but small inequalities in the Surface of the hewn Stone, which in the view of that Insect seemed so many deformed Rocks and Precipices. The thoughts of a *Free-Thinker* are employed on certain minute Particularities of Religion, the Difficulty of a single Text...without comprehending the Scope and Design of Christianity.[104]

Gilpin 'naturalised' the argument, shifting the design in question from a textual to a natural one, as might be expected, given his sceptical view of the worth of 'pedantic' doctrinal and dogmatic disputes.

Two important consequences follow from this. First, if the entire universe appeared picturesque to God's eye, those scenes which appear picturesque to what Gilpin called man's 'microscopic eye'[105] amounted to an intimation of God's existence and of the pleasures of futurity. The North Wales discussion of God's view and the limited comprehension of humans went on to recognize this implication in the vale of the Abbey Crucis, which was picturesque without the intervention of the picturesque eye to remodel it:

> as we can view only detached parts, we must not wonder, if we seldom find in any of them *our confined ideas* of a whole. Sometimes, however, we do; as in the valley we are now admiring.[106]

Secondly, the picturesque procedures of rearranging scenes pictorially and complaining of their defects in describing them, far from being a complaint as to the design of nature, should be viewed as showing humans how God sees the whole, and rearranging the landscapes before us to suit our microscopic eye. The picturesque, then, was justifying God's views to man by reducing God's aesthetic design of the whole to a scale at which humans could appreciate it, and could *see* that design, rather than simply assuming the beauty of the whole, as in the more traditional versions of the aesthetic design argument. As the fly should have been awed by St Paul's, so humans should humbly recognize their insignificance in the face of landscapes which dwarfed St Paul's:

> As we approached Cockermouth, the mountains, which occupy the middle of Cumberland, begin to make a formidable appearance. One of them, in particular, enlightened by an evening sun, seemed supported by vast buttresses, like some mighty rampart, in the times of the giant wars. Each buttress, I suppose, might be three or four times the height of St Paul's church. When nature in any of her frolic-scenes takes the semblance of art, how paltry in the comparison appear the labours of men![107]

While the theological grounding of both Gilpin's theory and practice of the picturesque cannot be doubted, it would also be absurd to deny that the picturesque amounted to an unprecedented 'aestheticisation' of Latitudinarian arguments. This was a consequence of Gilpin's assumptions that his audience only wanted a form of literature which was an innocent pleasure, and that only aesthetic justifications of religion were acceptable in an era wearied by doctrine and dogma. The result was that standard design arguments from nature to God appear in Gilpin's writings, shorn of explicit reference to their religious origins. A good example can be found in Gilpin's treatment of the sublimity of the ocean, which from Addison on had been one of the most popular natural-aesthetic experiences leading to God in literary descriptions of landscape. Where in his *Dialogues* (p. 402), Gilpin had deployed this argument with an explicitly religious conclusion, in his *Observations on coasts*, this was excised. Here two types of sublimity in the appearance of the ocean were discussed, those of the still and the tempestuous ocean. In both cases, Gilpin accepts that the mind is stretched by the immensity of the object before it, but does not then go on to view this as an intimation of God's grandeur.[108] Another sign of this aestheticization is the way in which scriptural citations in the picturesque tours are solely for their descriptive virtues. James Hervey had said the Bible had the best exemplifications of Addison's three species of beauty,[109] but Gilpin's tours took this thought further, only citing the Bible in aesthetic contexts. Examples of this can be seen in the highland tour, where Isaiah's description of ruins was cited for its 'beautiful *images of desolation*' and where the prospect of Loch Lomond reminded Gilpin of a description by the Prophet Joel.[110] This use of Biblical quotations fits with Gilpin's reluctance to be seen moralizing in a popular genre, as he felt this would offend his readers.

Given the ways in which Gilpin reworked the arguments from our limited views of the creation and the harmony of design, it is unsurprising that posterity has viewed his work as an exclusively aesthetic project. Most of the overt signs of didactic intent appear to be brief and uncharacteristic deviations from the main aims of his tours, while most of the references to scripture appear more aesthetic than didactic. Yet there can be no doubt that the way Gilpin constructed the picturesque as a category was theological: it was the aesthetics of a limited being, reconstructing the immense scenes around them as God might view them, thus giving the microscopic eye of man, a mere fly in the eyes of God, some idea of the beauty and harmony of the design of the whole fabric of the universe.

Ann Radcliffe and Latitudinarian gothic

Gilpin and Radcliffe: the landscape of romance

One element of William Gilpin's approach to landscape and the picturesque which has yet to be addressed is his fascination with imaginative landscapes. For the picturesque was not simply a way of seeing and drawing actual landscapes, but a formula which allowed for the construction of fictional landscapes. The type of fictional landscapes Gilpin admired were simultaneously fanciful and rational:

Some artists, when they give their imagination play, let it loose among uncommon scenes – such as perhaps never existed: whereas the nearer they approach the simple standard of nature, in it's [*sic*] most beautiful forms, the more admirable their fictions will appear. It is thus in writing romances. The correct taste cannot bear those unnatural situations, in which heroes, and heroines are often placed: whereas a story, naturally, and of course affectingly told, either with a pen, or a pencil, tho known to be a fiction, is considered as a transcript from nature; and takes possession of the heart.[111]

Gilpin's idea of a reasonable form of romance or novel sounds rather like the achievement of Ann Radcliffe's Gothic novels. Gilpin praised Radcliffe as a travel writer in his *Western Tour*,[112] and the sense of kinship between the two is enhanced by Gilpin's personal practice of creating fictional landscapes. In Gilpin's imaginative 'Fragment', one of a number of imaginative tours comprising landscape painting and description which he constructed, Gilpin's landscape contained a 'strange building' which could come out of Radcliffe:

Its form was that of the body of an immense church; but if any light was introduced, it must have been introduced from the top; for it seems never to have had any windows. For what use this strange edifice was originally constructed, we could not learn. The common opinion is, that it was built for a prison; & we heard a romantic story of a prince, who had been confined there, 30 years, with his daughter, a beautiful princess.

Even more reminiscent of Radcliffe, is the rational deflation of these romantic speculations by the cold light of the fictive traveller's reason:

I rather believe this strange edifice, instead of a prison, was built as a repository for the goods which were brought from all parts of the lake to the great fair at Vazner.[113]

Other portions of the 'Fragment', with their references to Italian monks and extraordinary abbeys only add to the sense of a link between Radcliffe's Gothic and Gilpin's picturesque.

This connection might appear only to contribute to the traditional interpretation of Gilpin as predominately an imaginative aesthetician, simply adding a Gothic side to his pre-Romantic sensibilities. Yet this ignores the fact that Radcliffe's practice of Gothic fiction is itself within the Latitudinarian tradition of literary representations of landscape and the natural world. It is to the demonstration of this claim that I now turn.

Radcliffe as a Latitudinarian

Criticism of the novels of Ann Radcliffe has traditionally focused on their relationship with the generic norms of Gothic fiction. This has been the case since the earliest biographical writings, namely Sir Walter Scott's assessment and the memoir attached to Radcliffe's posthumously published novel, *Gaston de Blondeville*.[114] Criticism developing this interpretation has seen Radcliffe in the forward-looking context of subsequent developments in the format of the Gothic novel. Radcliffe is accepted to have been an important innovator in the Gothic genre, but is also seen as falling short of its full development for several reasons, notably her penchant for explaining away supernatural events by naturalistic means. All such criticism was written a long time after the novels themselves, and thus has the benefit of seeing the generic trajectory of Gothic fiction in a way Radcliffe herself could not at the time of writing.

More recently, critics have sought to place Radcliffe's novels in their contemporary context in the 1790s. Here the argument has been over the ideologies Radcliffe was utilizing rather than the genre in which she operated, or rather, over the ideological resonances of the Gothic genre itself. It has been suggested that Radcliffe's work has radical strains in its presentation of women and in the portrayal of patterns of feminine sensibility more generally. Radcliffe is seen as representing a middle class, radical Dissenting tradition opposed to a Burkeian reassertion of aristocratic values. Equally, the limitations of Radcliffe's radicalism are also discussed, her providential conclusions of marital bliss proving as problematic today as they were to early nineteenth-century critics. Radcliffe emerges from this criticism, as she did from generic criticism, as a liminal figure. Her novels point towards radicalism, whilst she herself pulls back from endorsing such a position.[115]

It is suggested here that further light can be shed on the aims of Radcliffe's novels by treating them not in a subsequently crystallized generic context, nor in the contemporary context of the 1790s, but by placing them in the context of the intellectual milieu from which Radcliffe came. Such an approach reactivates a line of inquiry signalled but not taken by Radcliffe's early biographers. All these memoirs emphasized that Radcliffe's education inculcated values from a generation previous to her birth, thus placing her ideas as deriving from the early- to mid-eighteenth century intellectual arena. Discussing Radcliffe's reclusivity, Elwood argued she had acquired her ideas from 'the early impressions of education, and…the somewhat primitive and old-fashioned society with which she associated'.[116] I suggest that these values informed her presentation of the issues of nature, landscape, the supernatural and the providential in her novels.

The element of Radcliffe's old-fashioned society which is the key to her writing lies in her religious beliefs. Radcliffe was imbued with the tenets of the Latitudinarians. Whilst some modern critics have tried to connect Radcliffe with Dissent and radical religion, contemporary biographers were in no doubt of her orthodoxy:

> She was educated in the principles of the Church of England; and through life, unless prevented by serious indisposition, regularly attended its services. Her piety, though cheerful, was deep and sincere.[117]

What makes many modern commentators see Radcliffe's religious beliefs as more radical is their rationalism, but clearly this was compatible with the Anglicanism of the Latitudinarians. A cheerful but sincere piety was exactly the variety Latitudinarians had always encouraged. Radcliffe's London middle-class background was the traditional stronghold of Latitudinarian preaching. Latitude had established itself in London in the late seventeenth century, and its moral frameworks might well provide the old-fashioned education to which Radcliffe's biographers alluded.[118]

Radcliffe's novels developed in ways accordant with Latitudinarian theology and this substantiates the claim that her old-fashioned education in Anglican principles influenced her thematic interests and narrative strategies. First, attention will be paid to Radcliffe's presentation of landscape imagery, which follows the lines prescribed by the Latitudinarian doctrine of nature and the natural evidence of God. Then the broader framework of the 'explained supernatural' which Radcliffe deployed and the narrative closure she sought will be discussed in relation to Latitudinarian doctrines concerning the supernatural and the providential.

Gothic content: landscape, nature and religion

Radcliffe's novels are renowned for their abundance of landscape imagery, to the point that some contemporary reviewers found it excessive:

> We trust...we shall not be thought unkind or severe if we object to the too great frequency of landscape-painting; which, though it shews the extensiveness of her observation and invention, wearies the reader with repetitions.[119]

It is clear that Radcliffe found her landscape descriptions meaningful to an extent that many of her contemporaries did not. One reason for this lies in the theological purpose which many of those descriptions demonstrate.

Looking at picturesque and pastoral landscapes, Radcliffe's characters were continually led to an appreciation of God's benevolence. Thus in *A Sicilian Romance*, Julia's spirits are 'insensibly tranquilized' by a moonlight scene, as Young's had been in *Night Thoughts*:

> The night was still, and not a breath disturbed the surface of the waters. The moon shed a mild radiance over the waves...A chorus of voices now swelled upon the air, and died away at a distance. In the strain Julia recollected the midnight hymn to the virgin, and holy enthusiasm filled her heart.[120]

The civilized landscapes of human cultivation and activity, then, could lead the mind towards God. The contemplative mind does not only mirror the tranquillity of the scene, but is led beyond this to the Creator. It is in sublime and rugged landscapes, however, that the linkage of the Creator to his creation was most frequently made. As Ellena says in *The Italian*, such landscapes lead to a direct appreciation of God's grandeur:

> If I am condemned to misery, surely I could endure it with more fortitude in scenes like these, than amidst the tamer landscapes of nature! Here, the objects seem to impart somewhat of their own force, their own sublimity, to the soul. It is scarcely possible to yield to the pressure of misfortune while we walk, as with the Deity, amidst his most stupendous works![121]

Once more, looking at the landscape is the strongest proof of God's existence and one which can be turned to in any misfortune.

Radcliffe also showed her belief in the design argument *in propria persona* in her travel account, *A Journey made in the Summer of 1794*. This work is by no means as theologically laden as her novels, but Radcliffe was led to the argument from design by the sublimity of the Lake District. At Ullswater, the surrounding mountains were described as 'huge, bold, and awful; over-spread with a blue mysterious tint, that seemed almost supernatural'. She argued, in terms reminiscent of Young, and interesting for their apparent linkage of the writer's purpose with religion, that the overall experience of Ullswater 'inspires that "fine phrensy" descriptive of the poet's eye, which not only bodies forth unreal forms, but imparts to substantial objects a character higher than their own'.[122]

The sheer frequency with which Radcliffe links viewing the landscape with ascending to a mood of devotion is consonant with a Latitudinarian position, given the unusual emphasis it placed on naturalistic proofs of God. The argument from design was in no sense the preserve of the Latitudinarians, and Radcliffe's emphasis on the natural as proof of the divine could be part of a deist strategy conflating the two, but for the fact that the novels specifically work against such a position. It is only characters with evil intentions who conflate nature with the divine in Radcliffe's *oeuvre*. The first instance comes in *The Romance of the Forest*, where the Marquis of Montalt urges La Motte to kill Adeline:

> Truth is often perverted by education. While the refined Europeans boast a standard of honour, and a sublimity of virtue, which often leads them from pleasure to misery, and from nature to error, the simple uninformed American follows the impulse of his heart, and obeys the inspiration of wisdom.

Montalt goes on to list those peoples who sanction murder, arguing that they follow 'Nature, uncontaminated by false refinement' (*RF*, p. 222). The other important instance of nature being taken as divine is in similar circumstances, where Schedoni in *The Italian* encourages the Marchesa de Vivaldi to agree to the murder of Ellena di Rosabla (*I*, pp. 177–78). Radcliffe used the argument that nature is God to repulse her readers by its possible consequences.

If that were not enough, Radcliffe's opposition to such materialism is rendered explicit by La Luc, her portrayal of an ideal vicar. His speech deploys the Newtonian argument from design, coupled with a rhapsodic tone:

> When the imagination launches into the regions of space, and contemplates the innumerable worlds which are scattered through it, we are lost in astonishment and awe.... O! how expressively does this prove the spirituality of our Being! Let the materialist consider it, and blush that he has ever doubted (*RF*, pp. 275–76).

As opposed to conflating God with nature, the heroes and heroines of Radcliffe's novels ascended from nature to its Creator as the Latitudinarians envisaged. The difference can be seen in *The Romance of the Forest*. Where Montalt used nature to justify murder, Adeline, his proposed victim, responds differently:

> The scene before her soothed her mind, and exalted her thoughts to the great Author of Nature; she uttered an involuntary prayer: "Father of good, who made this glorious scene! I resign myself to thy hands" (*RF*, p. 22).

Nature has an author who can be perceived through it, but is not synonymous with it.

If Radcliffe's invocations of nature are to be distinguished from those of deists, it is also noticeable that it was the everyday course of nature which led to reflections on God, rather than extreme natural events. The novels contain a number of storm scenes and shipwrecks, but these are not interpreted as evidence of God's wrath. Julia and Ferdinand are shipwrecked in their attempt to escape from Sicily, but there is no suggestion that this is a divine judgement (*SR*, pp. 152–54), and the same holds for the Mediterranean storm in *The Mysteries of Udolpho* (*U*, pp. 484–85). This sets Radcliffe's interweaving of religion and the operation of the natural world apart from many in the Calvinist Dissenting tradition such as Daniel Defoe or, in her own era, William Cowper, who interpreted the Sicilian Earthquake as evidence of God's wrath at human sinfulness.[123] Again, Radcliffe had expressed this view of extreme natural events in her own travel writings. Travelling in a storm, she was impressed by the rational and measured effects of God's omnipotence, rather than being terrified by it as either capricious or wrathful:

> This display of the elements was the grandest scene I ever beheld; a token of GOD directing his world. What particularly struck me was the appearance

of irresistible power, which the deep monotonous sound conveyed. Nothing sudden; nothing laboured; all a continuance of sure power, without effort.[124]

If we look at the portrayal of true and false spirituality in Radcliffe's novels, this reinforces the contention that they are Latitudinarian in intent. True spirituality builds from the appreciation of the natural world. Just as nature was the foundation of a religious attitude in Paley's *Natural Theology* (see p. xxx), so for Blanche in *Udolpho*:

> God is best pleased with the homage of a grateful heart, and, when we view his glories, we feel most grateful. I never felt so much devotion, during the many dull years I was in the convent, as I have done in the few hours, that I have been here, where I need only look on all around me to adore God in my inmost heart! (*U*, pp. 475–76).

But Radcliffe was not engaged here in either radical anti-clericalism or in chauvinistic anti-Catholicism. Not all convents were dens of false spirituality because some Catholic characters build on a rational Christianity. The portrait of the superior of Santa della Piéta in *The Italian* is important in this respect:

> Her religion was neither gloomy, nor bigotted; it was the sentiment of a grateful heart offering itself up to a Deity, who delights in the happiness of his creatures; and she conformed to the customs of the Roman church, without supposing a faith in all of them to be necessary to salvation (*I*, p. 300).

The convent is set in a beautiful landscape which harmonizes with the reign of the superior and there is no sign of the gloomy seclusion suffered by so many of Radcliffe's heroines in convents. The superior's religion is cheerful yet deep, just as Radcliffe's was according to her early biographers. Radcliffe's attitude towards Catholics here was much more tolerant than that of the founders of Latitudinarianism, but this is unsurprising as English attitudes to Catholics became more tolerant throughout the century, notably in the 1790s when the defence of established order in Church and State became more important than denominational infighting.[125]

Equally, Radcliffe only portrays a Latitudinarian form of Catholicism in this positive light. As Saglia says, some form of cultural-geographic imperialism is at work here, where 'the narrator organizes the discourse on Italy so as to allow the Protestant, Northern, English characters to emerge as familiar figures in the cultural arena of fiction, while native presences are paradoxically objectified as distant, strange, and alien'.[126] Radcliffe's imagery of bogus spirituality draws on traditional Latitudinarian images of Catholicism as superstitious and hostile to science to construct it as 'distant, alien, and strange'. Radcliffe alluded to two such beliefs which had amused generations of Protestant travellers to Italy. Most notable is the tradition which peopled Mount Etna with devils, which Radcliffe mentions in *A Sicilian Romance* (*SR*, p. 28). There is also Paulo's appeal in *The Italian* (*I*, p. 78) to St Januarius, the Neapolitan patron saint who was alleged to have the power to stop Vesuvius erupting. Radcliffe probably got the details about both traditions from Patrick Brydone's popular travel narrative, *A Tour Thro' Sicily and Malta* (1773), which treated both beliefs with scorn.

Like Gilpin, Radcliffe contrasted scholastic pedantry with rational and vital modern approaches to faith in *A Sicilian Romance*, which was set in the sixteenth century:

> The dark clouds of prejudice break away before the sun of science, and gradually dissolving, leave the brightening hemisphere to the influence of his beams. But through the present scene [in the sixteenth century] appeared only a few scattered rays...Here prejudice, not reason, suspended the influence of the passions; and scholastic learning, mysterious philosophy, and crafty sanctity supplied the place of wisdom, simplicity, and pure devotion (*SR*, pp. 116–17).

Similar vignettes can be found in *Udolpho*, where reaching an appreciation of God through nature is compared favourably with 'all the distinctions of human system' (*U*, p. 48) and in the posthumously published *Gaston de Blondeville*. Set in the time of Henry III, *Gaston de Blondeville* contrasts the heroic Archbishop of York who is sceptical of sorcery and the supernatural with the Machiavellian prior, 'no true son of the church', who promotes sorcery.[127] Radcliffe has never had a reputation for historical accuracy, and whatever period she discussed, her pattern of ideal devotion looks decidedly Latitudinarian.

At times, the two visions of spirituality – scholastic/pedantic and rational/natural – are directly juxtaposed, as in Blanche's comments about convents cited above and Ellena's experiences in San Stefano. Whilst Ellena can be moved by a solemn church service (*I*, p. 85), her attention tends to stray. This occurs when she sees Olivia for the first time during a service, the remainder of which she spends trying to catch her attention (*I*, pp. 86–87). This inattention is in stark contrast to Ellena's response to the sublime landscapes around San Stefano, which rivet her attention to God as a service cannot:

> Hither she could come, and her soul, refreshed by the views it afforded, would acquire strength...Here, gazing upon the stupendous imagery around her, looking, as it were, beyond the awful veil which obscures the features of the Deity, and conceals Him from the eyes of his creatures,

dwelling with a mind thus elevated, how insignificant would appear to her the transactions, and the sufferings of this world!' (*I*, pp. 90–91).

Clearly, Radcliffe's novels relate to the general contours of Latitudinarian thought about the interconnection between the natural world and rational faith. The Latitudinarians portrayed themselves as treading the *via media* between the excesses of superstitious worship and scientific igno-rance found in Roman Catholicism, and the opposite extreme of irreligious scientism, represented by the deists who replaced the Christian God with a self-sufficient Nature. Radcliffe's heroes and heroines are clearly Latitudi-narian figures surrounded by religious extremists. Further, the way that the most Catholic characters such as Schedoni can invoke the deistic argument that nature is God is entirely traditional within Anglican apolo-getics, which argued that the two extremes on either side of the *via media* were interchangeable. The theology approved in Radcliffe's novels leads from nature to its creator. As with Gilpin, the progress from nature to faith is a more purely aesthetic one in Radcliffe than it had been in the early Latitudinarians.

Gothic form: explaining the 'explained supernatural'

Critical discussion of Radcliffe's novels has responded unfavourably to two of their recurrent narrative patterns. First, Radcliffe's technique of the 'explained supernatural', where apparently supernatural events are later explained by naturalistic means, has always been seen as a failure of nerve. Radcliffe is said to have taken the Gothic so far, but not been prepared to take it to its logical conclusion, reverting to an enlightenment rationalism in the last instance. By Radcliffe's death in 1823 the development of Gothic novels could be seen in perspective, and her transitional status was clear:

It is extraordinary, that a writer thus gifted should, in all her works inten-ded for publication, studiously resolve the circumstances, by which she has excited superstitious apprehensions, into mere physical causes. She seems to have acted on a notion, that some established canon of romance obliged her to reject real supernatural agency.[128]

The second characteristic, which has received less but equally unfavourable attention, is Radcliffe's way of resolving all her novels by conclusions in which virtue is rewarded and vice punished. Once again, this has been seen as a failure of nerve:

And did Mrs Radcliffe really write to enforce truths so excellent, but so commonplace? It is hard to believe it. But a certain formality, a love of trite and too evident conclusions, always were her errors.[129]

Kavanagh goes on to link this penchant for simplistic conclusions with the naturalistic explanation of the supernatural, both demonstrating that Radcliffe 'had a fine, but not a free imagination'.

While it is undeniable that Radcliffe's conclusions and explanations of the supernatural are somewhat mechanical in their effect, it would appear that both elements of her narrative patterning can be explained by reference to Latitudinarianism. As we have seen, Radcliffe follows a Latitudinarian form of argument with respect to the interrelation of nature and religion at a thematic level. She also, I suggest, structures her narratives in ways which reflect Latitudinarian arguments about the supernatural and the providential. Radcliffe repeatedly showed how chains of natural events could lead to providential outcomes, just as the Latitudinarians did. In her novels rational explanation is not antithetical to divine purpose. By the same token, what most see as providential intervention is simply superstition designed to mislead the credulous, just as the pomp and ceremony of Catholicism were to the Latitudinarians.

Looking first at the 'explained supernatural', Radcliffe rationally explains what appear to be omens or interventions from a supernatural world of ghosts and spirits. Normally, the resolution of supernatural appearances into their natural causes is delayed until the end of the novel, but in *Udolpho* there is one potted version when Emily sees a light glittering at the end of a lance:

> resembling what see had observed on the lance of the sentinel, the night Madame Montoni died ... She thought it was an omen of her own fate.

Her escort gives the light a natural explanation, based upon the electrical effects of thunder storms which Radcliffe had read of in Bertholon's writings on electricity. Emily's descent from Latitudinarian rationalism is reproached:

> you are not one of those, that believe in omens: we have left cowards at the castle, who would turn pale at the sight. I have often seen it before a thunder storm, it is an omen of that, and one is coming now, sure enough. The clouds flash fast already (*U*, p. 408).

Ugo, who utters this reproach, has acted as a good empirical rationalist, looking to correlate an apparently unusual effect with the natural conditions in which it occurs.

In all of her novels, Radcliffe develops the same pattern which this incident demonstrates on a smaller canvas. The supernatural is continually deflated, most obviously in *The Italian*. One of the central points of this novel is the education of Vivaldi, whose benevolent character is not supported by a rational approach to the natural world. Schedoni plays on Vivaldi's superstition to mislead him into thinking that the mysterious Nicola di Zampari is a ghost, a plan which eventually backfires. Descanting on Vivaldi's 'prevailing weakness' on his deathbed, Schedoni delivers what could be the text for the education of all Radcliffe's characters:

> the ardour of your imagination was apparent, and what ardent imagination ever was contented to trust to plain reasoning, or to the evidence of the senses? It may not willingly confine itself to the dull truths of this earth, but, eager to expand its faculties, to fill its capacity, and to experience its own peculiar delights, soars after new wonders into a world of its own! (*I*, pp. 397–98).

Schedoni advocates rational religion patterned, as Latitudinarianism was, on a solid foundation in truths of nature. Where a Catholic education encourages superstition, a rational faith tempers imagination by reason, a lesson the dying St Aubert had also imparted to Emily in *Udolpho* (*U*, pp. 79–80).

If, however, supernatural interventions do not occur to show the workings of God, this does not rule out providence: 'the most obvious residue of the spiritual after the supposed apparitions have been cleared away is Providence'. Whilst I agree with Clery that this means 'material existence is suffused by religion' in Radcliffe's novels, I would argue that the type of theology the novels develop can be specified more precisely.[130] Just as Latitudinarians believed natural laws created an ideal state of probation for moral actors, so the operation of Radcliffe's novels show virtue passing through rationally explicable affliction to reach a life of happiness. The moral of each is explicitly Christian: Paulo in *The Italian* likens Vivaldi's path to happiness to passing through purgatory to reach heaven (*I*, p. 413), but more redolent of the Latitudinarian view is the conclusion to *A Sicilian Romance*:

> In reviewing this story, we perceive a singular instance of moral retribution. We learn, also, that only those who do only THAT WHICH IS RIGHT, endure nothing in misfortune but a trial of their virtue, and from trials well endured derive the surest claim to the protection of heaven. (*SR*, p. 199).

The natural world is a state of probation, the main purpose of which is to act as a test of moral worth. Moral probity is available to all who follow the promptings of natural religion which allows all rational actors to distinguish right from wrong.

Natural laws produce providential outcomes; reason and faith are complementary, not contradictory. Within the narrative economy of Radcliffe's novels, this has to take a rather different form from that within Latitudinarian argument more generally. Where in Latitudinarianism the just dispensation of rewards can be postponed to the afterlife, as a novelist Radcliffe has to draw this forward to the present. The *terminus ad quem* moves from the afterlife to a narrative closure, but this still gives her works a strongly teleo-logical view of history akin to that of Latitudinarian thought. This can be seen most clearly in *A Sicilian Romance*'s discussion of spirits. Early in the novel, Madame de Menon argues that spirits may exist for 'Who shall say that any thing is impossible to God?' She argues that the restricted capabili-ties of human reason mean that much about the creation is mysterious to us:

> If we cannot understand how such spirits exist, we should consider the limited powers of our minds, and that we cannot understand many things which are indisputably true. No one yet knows why the magnetic needle points to the north (*SR*, p. 36).

This is the standard argument about the limits of human knowledge which Latitudinarians frequently used, arguing that in the afterlife we would gain a broader prospect of God's governance of the creation. In Radcliffe's novel, however, just as the rewards of providence are brought into the present day, so apparent apparitions from the spirit world are explained in this life. The education of the characters gives them the broader prospect which theology could postpone. In this case, secret cells in Mazzini's castle explain the sounds which had been interpreted as interventions from the spirit world (*SR*, p. 195).

While the form of the novel demanded that the working of natural laws for providential ends be explained within the lifetime of the characters, Radcliffe did still gesture to the broader prospect the afterlife would give us in terms familiar from Anglican apologetics. La Luc, who is more of a Latitu-dinarian than a Rousseauesque figure, puts the argument in its traditional form:

> [In] a future state...We shall then be enabled to comprehend subjects too vast for human conception; to comprehend, perhaps, the sublimity of the Deity who first called us into being. These views of futurity...elevate us above the evils of this world, and seem to communicate to us a portion of the nature we contemplate (*RF*, p. 274).[131]

Whilst virtue is rewarded within the human time frame of the novels, the ultimate reward will be in the afterlife. Radcliffe herself expressed such views in her private journals, drawing on the same theme and also empha-sizing that natural laws led to belief in a Christian God:

> saw the sun set behind one of the vast hills. The silent course over this great scene awful – the departure melancholy. Oh GOD! thy great laws will one day be more fully known by thy creatures; we shall more fully understand thee and ourselves. The GOD of order and of all this and of

far greater grandeur, the Creator of that glorious sun, which never fails in its course, will not neglect us, His intelligent, though frail creatures.[132]

The narrative elements of Radcliffe's novels which critics past and present have found least satisfactory relate back to a Latitudinarian position. Latitudinarianism created a distinctive interrelationship of nature, reason and religion, which supported a providentialism stripped of the supernatural and the superstitious. Just as Radcliffe articulated a rational approach to religion via an understanding of the natural world, so the overall patterns developed in her novels reflected a belief that the operation of natural laws was simultaneously providential. Natural laws create a moral state of probation for characters to endure, but these laws also ensure that virtue is rewarded both with happiness in this life and with a prospect of futurity. In this, Radcliffe's work has much in common with Battestin's analysis of Fielding's use of the design argument in *Tom Jones*. The major difference is that in Radcliffe's novels nature and landscape have a far more integral role in the exposition of the argument from design. Where Fielding was concerned with human nature, Radcliffe was interested in human nature in certain landscapes and with the human response to landscape as a key to their character.

Conclusion

If Gilpin was closely comparable to Radcliffe in his imaginative landscapes, Radcliffe follows Gilpin's pattern in terms of reputation. Both Gilpin and Radcliffe have been interpreted in almost exclusively aesthetic terms from soon after they wrote. In both cases, the way in which they aestheticized Latitudinarian approaches to landscape and the natural world to fit the tastes of their literary audience has led subsequent generations to construct them in exclusively aesthetic terms. Just as Gilpin's moralizing passages have been seen as aberrations from the picturesque ideal, so Radcliffe's explained supernatural has been viewed as a deviation from the canons of Gothic fiction, both aberrations being seen in terms of the inadequate realization of an ideal by its pioneer.

Most of the criticism of Radcliffe's novels was not strictly contemporary with their production. Beyond book reviews, criticism began with biographical assessments after Radcliffe's death in 1823. The gap between the writing of the novels and the onset of critical reflection is important. As intimated previously, this gap allowed Radcliffe to be assessed in terms of by-then established generic norms which she herself had helped to forge. Many of the values Gothic novels such as *Frankenstein* or *The Monk* upheld were polar opposites to those Radcliffe had demonstrated in her writings, notably the materialism of the former and the atheism of the latter. The genre of the Gothic novel developed in ways Radcliffe was not responsible for and showed no sign of concurring with. To recover the project with which Radcliffe herself was engaged in her novels requires looking back to the intellectual milieu she came from, rather than fitting her work within a generic framework defined during and by the seismic shifts occurring in English society and thought in the period from 1789 to 1832.[133] One of those shifts, and the one which obscured Radcliffe's own aims even from many of her earliest critics, was the collapse of traditional Latitudinarianism in the 1790s. As Fitzpatrick puts it, the 1790s saw

> the loss of coherence of traditional Latitudinarianism. A movement which had emerged as a moderate response to the factionalism and fanaticism of the early modern period, could no longer cope with the new rifts emerging in government and society as they entered the age of revolutions.[134]

The Latitudinarian position itself became dragged towards extremes which it had previously held in productive tension:

> Its combination of moderate scepticism and rationality was too easily separated out into conservatism and radicalism.[135]

Ann Radcliffe's position as a Latitudinarian novelist, then, was awkward, and the problems of articulating her position paralleled those which led to the collapse of the Latitudinarian position. Contemporary critics could easily see the supernatural framework of romances as too radical for comfort, and her use of it as too conservative. Modern critics have done much the same, seeing signs of radicalism and feminism in her works, coupled with egregious conservatism in her conclusions. Whilst criticism was dragged towards the extremes, there is no sign that Radcliffe herself was; her cheerful piety remained, whilst storms of controversy tried to pull her works towards the extremes her literary *via media* steered between so carefully.

Conclusion

Latitudinarian approaches to the natural world were clearly vital to the development of eighteenth-century literary depictions of landscape and the natural world. Many of the canonical authors traditionally associated with landscape – Addison, Young, Gilpin, Radcliffe – developed justifications of the activities of looking at and describing landscapes which were theologically informed by Latitudinarianism, and further developed descriptive conventions which reinforced the interconnection between rationality, religion and the apprehension of the natural world. Of course, the three readings of the book of nature discussed at the beginning of this chapter were neither new to the Restoration world nor exclusively linked to Low-Church Anglicanism, but their role was unprecedented. First, the *nexus* of science, rationalism, religion, nature and landscape was developed with a new intensity. Secondly, evidences from the natural world were linked more strongly than before to a Christian argument, as opposed to a defence of natural religion, or some form of heterodox response to established Christianity. Newman's characterization of the eighteenth century as the 'age of evidences', like most truisms, contains a substantial amount of truth. This nexus was taken up outside Anglican theology, notably by the moderate Presbyterians, such as James Thomson. As such, I suggest the approach to nature pioneered by the Latitudinarians amounted to a position in a *theological* discourse, rather than operating as a *denominational* discourse, in the way Jonathan Clark has described the debates surrounding the American War of Independence.[136] The theological position which drew people together and led them to represent landscape in similar ways involved rationalist proofs (as opposed to scriptural or historical/patristic) leading from nature and landscapes to Christianity (as opposed to leading to the deification of nature). It is against this context, which dominated the eighteenth-century nexus of landscape and literature that Samuel Johnson's approach can be understood in counterpoint.

In literary contexts, the theological arguments were unsurprisingly recast. While authors drew on the three readings of the book of nature in various ways, underlying this was a gradually increasing aestheticization of depictions of landscape, whether they were scientific (as in some of Addison's essays), meditative (as in Young and Hervey) or aesthetic (as in Gilpin and Radcliffe). This was particularly noticeable in the late eighteenth-century works of Radcliffe and Gilpin, both of whom had moved so far in the aesthetic direction that the theological element of their approaches to landscape has been almost completely ignored, in Gilpin's case despite his clerical status. As should be clear from all the authors this chapter has discussed, *aestheticisation* is not tantamount to *secularisation*. The interactions between these two processes were complex, especially as religious authors tried to harness the increasing didactic power of literature, to 'bait the gospel hook'.

Even taking into account this process of aestheticization, however, it is the rational approach to nature and landscape in literature which distinguishes all the authors this chapter has addressed from their predecessors and successors. If conventional literary history sees an era of Augustan rationalism being overtaken by sentimental and melancholic proto-Romanticism and then Gothic and Romantic literature, another trajectory defining rationalism in literature in theological rather than Augustan terms might discern longer continuities echoing through eighteenth-century literature, and moulding literary approaches to landscape. Subsequent to the fragmentation of Latitudinarianism, this tradition has itself been fragmented, elements being appropriated and appreciated under the rubrics of aesthetics (Addison, Gilpin), pre-Romanticism (Young, Hervey) and the Gothic (Radcliffe). But as this chapter has tried to show, there is much that draws these approaches to landscape and literature together. Above all, their total achievement involved an interaction of the rational, the beautiful and the Christian that defined and distinguished the Latitudinarian approach to landscape. As we shall see in Part III, Samuel Johnson, coming from a High-Church position which was very nervous about too great a reliance on nature and landscape as a proof of God, was to develop a very different approach to landscape which has led to his unwarranted marginalization from debates about what landscape tells us about the culture of eighteenth-century England.

Part III

Samuel Johnson, High Churchmanship and Landscape

Using the contextual method advocated in Part I of the book, Part II set up two historical contexts through which we can gain a deeper understanding of Samuel Johnson's uses of landscape imagery. As we have seen, there was an enormous diversity in the literary use of landscape imagery in the eighteenth century, but this diversity was structured by a religious mentality. Further, specific theological positions led to different uses of landscape themes as we have seen at length for the genealogy of Latitudinarian authors. Samuel Johnson was a High-Church Anglican at some distance (although not a total one) from Latitudinarianism. Part III of this book makes good the contention that Johnson's High Churchmanship led him to very different usages of landscape imagery, usages which have led to his neglect in recent scholarship concerning landscape. Our study of Johnson will draw landscape studies into contact with diverse and previously neglected contexts. Johnson, then, opens up in landscape studies a 'dark side' of a different type from that Barrell has investigated.

5
The Lexicon of Landscape: Johnson's *Dictionary* and the Language of Natural Description

Introduction: the *Dictionary* as an historical survival

The functions of the dictionary in eighteenth-century England differed from those associated with it today. The dictionary remained a repository of instruction, being a descendant of the Renaissance aspiration of creating a book of books.[1] Given a theory of language where words were of divine origin, a book of books also amounted to a redescription of God's book of nature. Whilst the origin of words in Adam's revelatory naming was no longer the reigning theory of language by the time of Johnson's *Dictionary*,[2] the encyclopaedic element in lexicography remained. Johnson was strongly influenced by the Lockean theory of language, in which names were arbitrarily attached to ideas, a theory far more sensitive to the dynamic nature of language use and the resultant problems of changing word meaning.[3] As such, his *Dictionary* had a more historicized view of language than its predecessors. Johnson in the 'Preface' recognized he was often giving the 'intellectual history' of words, a 'genealogy of sentiments'.[4] Combining the Renaissance demand that a dictionary be instructive with the Lockean theory of language left the dictionary as a book of books in a new sense; it became a record of the way previous books had used certain terms. Similarly, the connection of the Lockean dictionary with the book of nature was placed at a further remove; the dictionary becoming a record of the way a language of natural description had been deployed, and of the evolution of that usage.

Published in 1755, Johnson's *A Dictionary of the English Language* was not the first dictionary of the English language, nor was it technically original.[5] Yet it did advance the division and ordering of definitions,[6] and was the first English dictionary to use illustrative quotations.[7] That Johnson intended these quotations to instruct in the encyclopaedic tradition of dictionary making he himself made clear:

> When I first collected these authorities, I was desirous that every quotation should be useful to some other end than the illustration of a word.

Although he went on to admit that, having to edit his illustrative quotations, they 'are no longer to be considered as conveying the sentiments or doctrine of their authours', we can accept DeMaria's summary:

> the majority of Johnson's bibliographical decisions were, broadly speaking, linguistic ... But Johnson also made bibliographic choices that promoted his moral and educational purposes.[8]

These words point to an ambiguity: the extent to which the *Dictionary* is to be seen as a social document, the intellectual history of words; and the degree to which it is a more personal document, reflecting Johnson's own desire to instruct after a specific manner. On the one hand, it seems reasonable to accept the contention that 'from its quotations one gains some sense of the history of English culture',[9] but equally, 'dictionary-making involves thousands upon thousands of small decisions, many of which reveal the character behind them'.[10] In truth, the 40,000 words coupled with 116,000 illustrative quotations provides ample scope for both elements.[11]

Given that the *Dictionary* is an important document in the history of English culture, reflective of both its author and his authorities, what it suggests about the language used to describe landscape and the natural world is worthy of consideration, especially given the linguistic-contextual approach adopted here. The *Dictionary* shows how the language of natural description had been used, and with what other discourses it was connected. Yet the ambiguity of the *Dictionary*'s message must be accepted:

> context ... both prior and present, is all; this is the primary reason that it is very difficult to assess what Johnson 'says' in his *Dictionary* by extracting quotations from it and tying them together into themes.[12]

In recognition of this, I have worked back from Johnson's definitions and illustrations of selected terms important to the language of natural description to the sources of his illustrations, and studied previous dictionaries to find where Johnson's definitions mark significant departures from a tradition where 'lexicography progressed by plagiarism'.[13] This contextualization seeks to elucidate the central themes in the appreciation of landscape and nature as presented in the *Dictionary*, to show what discourses the *Dictionary* connected with landscape, and to use the provenance of illustrative quotations to give some insight into Johnson's position in relation to the approaches to landscape and nature being adopted in the mid-eighteenth century.

Framing assumptions for natural description

All discussions of landscape and nature in the *Dictionary* took place within parameters set by two groups of ideas: a Mosaic chronology of the Creation, and a strict hierarchy of intellectual value. These two sets of framing ideas acted to bound the language of natural description and to set its position in the division of knowledge established by the *Dictionary*.

The chronology of creation

In the *Dictionary*, the accuracy of the chronology of the Earth's creation as presented in *Genesis* was unquestioned. One of Johnson's aims was to let us

> Know how this *world*
> Of heav'n and earth conspicuous first began (Milton).[14]

Under the verb 'to create' Johnson placed at the head of his illustrations: 'In the beginning God *created* the heaven and the earth' (*Genesis* 1:1). The Nicene Creed was also cited to remind us in consequence of this that Christ must have been 'Begotten before all *worlds.*' The Earth, then, is not eternal, a point Johnson also made in one of his rare annotations to a book he was marking up for illustrative quotations for the *Dictionary*. On his copy of Hale's *Primitive Origination of Mankind*, Johnson rebutted an argument for the eternity of the earth, which Hale himself only put forward to refute it.[15]

The meaning of the creation that God is good and displays his benevolence by this act was frequently reiterated:

> Consider the immensity of the Divine Love, expressed in all the emanation of his providence; in his *creation*, in his conservation of us (Jeremy Taylor).

These themes were by no means restricted to illustrative material. Johnson's definition of 'paradise' as 'the blissful regions, in which the first pair was placed' left no doubt as to the status of the *Bible's* word: it was, quite literally, definitive.

The entry under 'paradise' compressed the providential scheme of history as mapped out in *Genesis*: having given his first definition as cited above, he illustrated this with Milton's

> Longer in that *paradise* to dwell,
> The law I gave to nature him forbids.

This reminded the *Dictionary*'s readers of the first pair's expulsion from the blissful regions as the middle of the drama, the state in which they still lived. The final quotation, under the only other definition of 'paradise', closed the drama, recalling the finitude of the created earth:

> The earth
> Shall all be *paradise*, far happier place,
> Than this of Eden, and far happier days (Milton).

The theme of the Happy Fall – *foelix culpa* – and final salvation were thus evoked to close the sequence the first definition had begun.

This emphasis on the Mosaic chronology reflected the degree to which, by drawing upon the physico-theologists, the poets they inspired, the *Bible* and *Paradise Lost*, Johnson was making such connotations unavoidable.[16] Yet the depth of Johnson's personal belief in the Creation story cannot be doubted, given his attack on Brydone's *A Tour through Sicily and Malta* which had questioned the account in *Genesis*:

> Shall all the accumulated evidence of the history of the world; – shall the authority of what is unquestionably the most ancient writing, be overturned by an uncertain remark such as this?[17]

Johnson himself had done much to marshal this accumulated evidence by placing it so frequently and so visibly under key definitions in his *Dictionary*.

One point suggests the permeation of Johnson's own beliefs into the texture of the *Dictionary* in the use of physico-theology. Whilst Johnson frequently cited the physico-theologists, he rarely cited arguments that earthquakes and the like reflected God's wrath. Grew, for example, argued

> if Earthquakes do oftener happen in Cities than in the Fields, Whatsoever Natural Cause hereof may be given; the same also shows, they were design'd to do it', his point being 'nor are we to look upon Providence, any other way by the Halves; but to own it, either in none, or in all Effects and Consequences.[18]

The *Dictionary* maintained a silence over this, focusing upon the more fundamental and less controversial expressions of the Divine Wrath in the Deluge and Conflagration. This may reflect Johnson's 'unwillingness to take sides in intramural religious disputes'[19] in the *Dictionary*, the theodicy question being so debatable. It may also reflect Johnson's personal scepticism that earthquakes and the like were signs of divine wrath against specific groups, a scepticism clearly expressed in his review of Jenyns's *Free Enquiry* in 1757.[20]

Intellectual and natural hierarchy in the *Dictionary*

The other overarching framework within which the language of natural description moved in the *Dictionary* was a strongly hierarchical one, which created a clear scale in the natural world, connected to an intellectual hierarchy.

In the natural world, Johnson's definitions of vegetable, animal and man relied upon an implied hierarchy:

> VEGETABLE...Any thing that has growth without sensation, as plants.
> ANIMAL...1. A living creature corporeal, distinct, on the one side, from pure spirit, on the other, from mere matter

and

> MAN [sense 10] not a beast' illustrated by [sense 1]:
> A creature of a more exalted kind
> Was wanted yet, and then was *man* design'd,
> Conscious of thought (Dryden).

Three of Johnson's quotations fill in further elements of this hierarchy:

> It is the saying of divine Plato, that man is nature's horizon, dividing betwixt the upper hemisphere of immaterial intellects and this lower of *corporeity* (Glanville).

> Angels, in their several degrees of *elevation* above us, may be endowed with more comprehensive faculties (Locke)."

> God being supposed to be pure spirit, cannot be the object of any *corporeal* sense (Tillotson).

There is, in other words, continued reference to a chain of being in the *Dictionary* even if not as a formal concept capable of being understood as Jenyns had believed.[22] This was apparent not only in Johnson's dislike of blurring the distinctions in the hierarchy,[23] but more directly in the *Dictionary* in the various illustrations of 'link':

> The moral of that poetical fiction, that the uppermost *link* of all the series of subordinate causes, is fastened to Jupiter's chair, signifies an useful truth (Hale).

> > While she does her upward flight sustain,
> > Touching each *link* of the continued chain,
> > At length she is oblig'd and forc'd to see
> > A first, a source, a life, a deity (Prior).

> > So from the first eternal order ran,
> > And creature *link'd* to creature, man to man (Pope).

> These things are *linked*, and, as it were, chained one to another: we
> labour to eat, and we eat to live, and we live to do good; and the good
> which we do is as seed sown, with reference unto a future harvest
> (Hooker).

The last illustration from Hooker suggests the ease with which this chain
was linked to a parallel hierarchy of values pertaining to human learning.
An important image in the *Dictionary* (and one which connected knowledge
with the order found in nature) is

> of knowledge as a chain, a ladder, or a pyramid that leads to the Most
> High...The final goal of this old noetic chain is God, but a complete
> expression of the old order seems to be just outside the bounds of the
> *Dictionary*'s treatment of knowledge...[24]

The chain of being was interconnected with the noetic chain of knowledge,
one being analogous to the other. In the chain of knowledge, the analysis of
nature (and landscape) was an activity of lowly significance, but because it was
more accessible to man's limited reason than the higher levels of morality
and divinity, it became valuable to the extent that, by analogy, it gave insight
into higher levels. This was the rationale behind the physico-theology which
Johnson relied on for illustrative quotations. But it also helps to explain the
discursive connections Johnson made in the *Dictionary* between the language
of landscape and other discourses. Johnson continually connected mundane
terms such as those used in the discussion of landscape with the more important realms of debate higher up the chain of learning. The *Dictionary* achieved
this connection of levels of discourse by the juxtaposition of illustrative quotations relating to the natural, moral and religious meanings of a word.[25]
   An example of this procedure can be found in Johnson's first sense of the
noun 'prospect', as a 'view of something distant', which is followed by three
illustrative quotations:

> Eden and all the coast in *prospect* lay (Milton)

> The Jews being under the oeconomy of immediate revelation, might be
> supposed to have had a freer *prospect* into that heaven, whence their law
> descended ([Allestree] Decay of Piety).

> It is better to marry than to burn, says St. Paul; a little burning felt pushes
> us more powerfully, than greater pleasures in *prospect* allure (Locke).

Even the strictly physical illustration from Milton is strongly religious, and
the next quotation from Allestree recalled a more important meaning of distant than the physical, this being reinforced by the fifth sense of 'prospect'
as 'view into futurity'. It is only in the third sense of a 'series of objects open

to the eye' that a purely physical sense of the word is given from Addison's *Letter from Italy*:

> There is a very noble *prospect* from this place: on the one side lies a vast extent of seas, that runs abroad further than the eye can reach: just opposite stands the green promontory of Surrentum, and on the other side the whole circuit of the bay of Naples.

But in the context of the surrounding illustrations –

> Him God beholding from his *prospect* high,
> Wherein past, present, future he beholds,
> Thus spake (Milton)

and

> Man to himself
> Is a large *prospect*, rais'd above the level
> Of his low creeping thoughts (Denham).

– the strictly physical meaning of 'prospect' in the language of natural description is unavoidably (perhaps designedly) conflated with moral and religious prospects. What Wimsatt said of Johnson's prose applies to the effect of his *Dictionary*:

> we have the universe of analogy within which Johnson moves. Johnson matches the physical scale of analogy, but especially the lower, mechanical, and elemental end of it, against the realm of mind.[26]

Yet the process was not one of matching two equal sides, for the language of natural description was repeatedly overwhelmed (as in 'prospect') by more important knowledge; the connection of the language of natural description to other knowledge is more important in Johnson's scheme of instruction than that language itself. This chapter seeks to follow the division of knowledge Johnson imposed, by looking to his deployment of the language of natural description via its discursive juxtapositions.

The discursive languages of landscape in the *Dictionary*

The connection of religion and nature in the *Dictionary*

Atop the pyramid of knowledge was knowledge of the divine:

> One science is incomparably above all the rest, where it is not by corruption *narrowed* into a trade, for mean or ill ends, and secular interests;

I mean, theology, which contains the knowledge of God and his creatures (Locke).

That religious knowledge must be the goal was reflected in the preponderance of religious associations juxtaposed with nature in the *Dictionary*, either natural evidences for the existence of God or a metaphorical conflation of the language of physical landscapes with divinity.

Natural evidences of God

The design argument. Given Johnson's reliance upon physico-theologists for illustrative quotations, the presence of the design argument using nature as a proof of the existence and attributes of God is not surprising.[27] Johnson's quotations reflect an orthodox deployment of the design argument, which also amounts to Johnson's longest and least problematic engagement with the Latitudinarian approach to landscape depicted in Chapter 4. The order of the earth was seen as proof of the existence of a God, and a refutation of the Epicurean argument that the universe is a random creation of the collision of atoms.

> The atoms which now constitute heaven and earth, being once separate in the *mundane* space, could never without God, by their mechanical affections, have convened into this present frame of things (Bentley).

Bentley also discussed the effect of this discovery:

> An astrologer may be no Christian; he may be an *idolator* or a pagan; but I would hardly think astrology to be compatible with rank atheism.[28]

Given the *Dictionary's* educative aims, this was undoubtedly the rationale behind the inclusion of so much material relating to this argument.

If the naturalist was forced to admit there is a Creator, the cure for atheism was apparent:

> *Atheist*, use thine eyes,
> And having view'd the order of the skies,
> Think, if thou canst, that matter blindly hurl'd,
> Without a guide, should frame this wond'rous world (Creech).[29]

The Boyle lecturers Johnson cited stressed that the evidence of nature was accessible to those not learned as to proofs of God and that, unlike revelatory evidences of God, it was common to all people and times. Ray urged that these [i.e. natural] Proofs taken from the Effects and Operations exposed to every Mans view, not to be denied or questioned by any, are the most effectual to convince all that deny or doubt of it. Neither are they only convictive of the greatest and subtlest Adversaries, but intelligible also to the meanest Capacities. For you may hear illiterate Persons of the lowest Rank of the Commonality affirming, that they need no Proof of the being of God, for that every Pile of Grass, or Ear of Corn, sufficiently proves that.[30]

The same argument was taken up by the philosophical divines of the Restoration.[31]
This group was also represented in the *Dictionary*:

The far greater part of men are no otherwise moved than by sense, and have neither leisure nor ability so far to improve their powers of reflection, as to be capable of conceiving the divine perfections, without the aid of *sensible* objects (Rogers).

This could be called a negative argument for the existence of God from the natural world, asserting that the world could not be as ordered as it is on any other hypothesis. This was summarized, appropriately, under 'nothing':

It is most certain, that there never could be *nothing*. For, if there could have been an instant, wherein there was *nothing*, then either *nothing* made something, or something made itself; and so was, and acted, before it was. But if there could never be *nothing*; then there is, and was, a being of necessity, without any beginning (Grew).

We do not create the world from *nothing* and by *nothing*; we assert an eternal God to have been the efficient cause of it (Bentley).

The design argument also had a positive side: by specifying how the earth was contrived for the convenience of man, it revealed God's goodness. Woodward's *An Essay toward a Natural History of the Earth* (1695) bulked large in the *Dictionary*'s arguments for the benevolent disposition of features on the earth's surface:

The scorched earth, were it not for this remarkably *providential* contrivance of things, would have been uninhabitable.

Such a mediocrity of heat would be so far from exalting the earth to a more happy and *paradisiacal* state, that it would turn it to a barren wilderness.

The sea was very necessary to the ends of providence, and would have been a very *wild* world had it been without.[32]

The divinity of our senses and mind. The design argument went back a stage: not only is that which we sense in the natural world evidence of God's existence and attributes, but so are the means by which we are capable of sensation.

> Having surveyed the image of God in the soul, we are not to omit those characters that God imprinted upon the body, as much as a spiritual substance could be pictured upon a *corporeal* (South).

Sensible proofs of God, then, were themselves based upon this further proof of God's goodness, for

> How could Man, particularly, view the Glories of the Heavens, survey the Beauties of the Fields, and enjoy the Pleasure of beholding the noble variety of diverting Objects, that do, above us in the Heavens, and here in this lower World, present themselves to our View every where; how enjoy this, I say, without that admirable Sense of *Sight*![33]

The entire relationship between landscape and the spectator was one interwoven with God in the *Dictionary*'s field of knowledge. This is a good example of Johnson the lexicographer also acting as a moralist: 'much of the vocabulary of natural science in Johnson's *Dictionary* is accompanied by reminders of God's benevolent omnipotence'.[34]

The limits of natural knowledge. Running alongside the argument from design was a cautionary recognition of the limitations of human reason's ability to understand the Creator. Johnson highlighted this theme under a word important in the language of the physico-theologists:

> In things the *fitness* whereof is not of itself apparent, nor easy to be made sufficiently manifest unto all, yet the judgment of antiquity, concurring with that which is received, may induce them to think it is not unfit (Hooker).

Hooker's words suggest a more traditionary and scriptural view of the evidences for God's existence. Within the Mosaic chronology which framed the interpretation of the *Dictionary*, the Fall prevented a complete understanding of God via nature and natural religion:

> Our sovereign good is desired *naturally*; God, the author of that natural desire, hath appointed natural means whereby to fulfil it; but man having utterly disabled his nature unto these means, hath had other revealed,

and hath received from heaven a law to teach him, how that which is desired *naturally*, must now supernaturally be attained (Hooker).[35]

One of Johnson's favourite theologians and a source for the *Dictionary*, Baxter, took this further, arguing that 'nature', since the Fall,

is now a very hard book...The common people have not leisure for so deep and long a search into nature, as a few Philosophers have made.[36]

This reversed the position of the Boyle lecturers, who had argued that the book of nature was the most legible postlapsarian source of evidences for God.

Johnson provided space for this view, juxtaposing it with physico-theological arguments. The principal source for reminders of the limitations of human knowledge of God via natural evidences was Hooker.[37]

No man can attain *belief* by the bare contemplation of heaven and earth; for that they neither are sufficient to give us as much as the least spark of light concerning the very principal mysteries of our faith.

There is a knowledge which God hath always revealed unto them in the works of nature: this they honour and esteem highly as profound wisdom, *howbeit* this wisdom saveth them not.

Hooker's point here was a commonplace in homiletical literature: that the 'mysteries of our faith' – meaning principally Christ's expiation of our sins – cannot be known from the face of nature or from natural reason. As such, 'Scripture must be sufficient to imprint in us the character of all things necessary for the attainment of eternal life.'[38]

This connects to the persistent concern in the *Dictionary* with human ignorance. Johnson chose to highlight this by his frequent use of Baker's *Reflections upon Learning*, Glanville's *Scepsis Scientifica* and Browne's *Pseudodoxia Epidemica* as sources.[39]

It is *evident*, in the general frame of nature, that things most manifest unto sense have proved obscure unto the understanding (Browne).

Furthermore, reasoning itself is not without presuppositions which cannot themselves be justified by reasoning. As such, to arrive at an understanding of God independent of traditional structures of knowledge – one of the possible misconceptions of natural evidences of God – is impossible. This point was forcefully put by Johnson's citations under 'science':

No *science* doth make known the first principles, whereon it buildeth; but they are always taken as plain and manifest in themselves, or as proved and granted already, some former knowledge having made them evident (Hooker).

The systems of natural philosophy that have obtained, are to be read more to know the hypotheses, than with hopes to gain there a comprehensive, *scientifical*, and satisfactory knowledge of the works of nature (Locke).

The delusion of our senses. Just as the physico-theological argument from design moved back from the sensed to the senses, so scepticism about natural evidences was coupled with caution about the accuracy of our senses and our understanding of what they revealed about the natural world.

A vast variety of phænomena, and those many of them so *delusive*, that it is very hard to escape imposition and mistake (Woodward).

This argument was carried forward by Glanville speaking of 'the *Imposture* and *fallacy of our Senses*', going on

and yet to speak properly, and to do our *senses* right, simply they are not deceived, but only administer and occasion to our forward *understandings* to deceive themselves.[40]

Even the man who did not deceive himself by his senses or understanding would only understand a small part of God's nature. The Infinite cannot be understood by the operation of a finite understanding working on a small part of God's creation:

it is impossible for a man of the greatest parts to consider any thing in its whole extent, and in all its variety of *lights* (Addison).

Clarke, a preacher Johnson admired but did not cite in the *Dictionary* due to his heterodoxy on the Trinity, summarized the relation of the great chain of being to the senses:

we [humans] can fix on but one side of a thing; and consider it only in one view at once; But [God's] 'tis a *perfect* comprehension of everything, in all possible respects at a time, and in all possible circumstances *together*. Again; it is not, as Ours, and possibly that of higher Beings than we [i.e. angels], only a *superficial and external* knowledge of things, but an *intimate and thorough prospect of their very inmost nature and essence*.[41]

Reason and scepticism, idolatry and orthodoxy. In the field of knowledge created by the *Dictionary*, these two approaches to the use of nature and landscape to prove the existence of God were not conflictual. Hooker, as well as persistently asserting the inferiority of natural evidences to scriptural, could also say:

Nature and scripture, both jointly and not *severally*, either of them, be so compleat, that unto everlasting felicity we need not the knowledge of any thing more than these two may furnish our minds with.

As far as evidences of God were concerned, the *Dictionary*, like Johnson, 'liked to have more',[42] provided that their limitations were respected.

The relationship between reason and revelation was fully articulated in Johnson's choice of illustrative quotations under 'evidence' and 'evident':

There are books extant, which they must needs allow of as proper evidence; even the mighty volumes of visible nature, and the everlasting tables of right reason (Bentley).

However mighty these volumes, their status in relation to fallen man was made clear by the qualifying term 'even'. We can see, then, that even when citing Latitudinarian theologians, Johnson had an impulse to downplay the role of landscape and natural proofs of God.

The position adopted by the *Dictionary*, on what was one of the central concerns about the status of landscape and nature in the eighteenth-century field of knowledge, amounted to something approaching an Anglican orthodoxy. The writers Johnson cited on this subject were explicit about the *via media* they trod, and the extremes on either side.[43] On the one hand, they sought to avoid the deification of Nature. Those who stressed the limitations of human reason did so to oppose 'men [who] are fond of Learning almost to the loss of Religion', arguing that 'Religion suffers by their Contentions about it, and we are in danger of running into Natural Religion'.[44] The *Dictionary* actively contributed to avoiding this extreme, not only by citing orthodox authors, but also by Johnson's first definition of 'nature' as 'an *imaginary* being *supposed* to preside over the material and animal world'.[45] This definition precluded the deification of nature far more categorically than, for example, Bailey's definition:

the system of the world, the Machine of the Universe, or the Assemblance of all created Being; the universal Disposition of all Bodies; also the Government of divine Providence, directing all Things by certain Rules and Laws.[46]

The *Dictionary* consistently attacked idolatry,[47] and idolatry of nature was checked at all points. Clarke provides an eloquent summary of the argument behind this principle of selection employed in the *Dictionary*:

Men, therefore who in Christian countries, where the Gospel is preached, pretend to believe in the *God of Nature*, and yet at the same time reject the revelation of the *Gospel*, which is so *agreeable to* and *perfective of* the Law of Nature; do, generally speaking, in *pretense* only, and not in reality, show any more regard to *natural* than to *revealed* Religion, falling for the most part into absolute Atheism.[48]

The other extreme Anglicanism and the *Dictionary* sought to avoid was that of an enthusiasm which denied any place for reason (and therefore for evidence from nature) in religion. The denial of reason was seen as a feature both of Roman Catholicism and of extreme (or uncatholic) Protestantism.[49] Baker linked the two errors of enthusiasm and naturalism, arguing that

after Men have try'd the force of natural Reason in matters of Religion, they will soon be sensible of its weakness, and after they have run them-selves out of breath and can centre no where they think they can find it...they will take up with an *Infallible Guide* [i.e. Rome].[50]

· Within the Anglican Church from which most of his quotations came, Johnson was able to reflect the consensus various factions displayed on this issue. Partly because of the frequency of quotation, but also due to the pattern of argument established, the *Dictionary* reflected most closely Hooker's position in the *Ecclesiastical Polity*. Johnson cited heavily from Book I, Section 3, 'The law which natural agents observe, and their necessary manner of keeping it', which sustained a precise awareness both of nature as evidence of God,[51] and of the relation of nature to God.[52]

Religion and the metaphorical language of landscape

In addition to this debate over natural evidences, the *Dictionary* persistently connected landscape to religion by the metaphorical use of physical termin-ology in theological contexts. In doing this, it reflected the strength of pic-torialist theories of language in England in the lateseventeenth century on which it relied so heavily for illustrative material.[53] It also reflects language usage in eighteenth-century England where

writers, aware of the older, more pictorial senses of words of Latin origin, used them as the starting point of metaphors, sometimes extended so far as to become little allegories,[54]

a practice Johnson frequently used in his own writings.[55] This itself reflected the 'universe of analogy' where natural, moral and spiritual were interpene-trative and mutually informing, which received its greatest expression in Butler's *Analogy of Religion* (1736).

The view from above. God was portrayed as on high, his view being akin to that of the viewer of a prospect, but still more comprehensive:

> Jehova, from the summit of the sky,
> Environ'd with his winged *hierarchy*,
> The world survey'd (Sandys).[56]

God's view from this eminence was comprehensive in several ways. In terms of spatial extent:

> He that laid the foundations of the earth cannot be excluded the secrecy of the mountains; nor can there any thing escape the *perspicacity* of those eyes, which were before light, and in whose opticks there is no opacity (Brown).

God's prospect also extended through all the time he had created:

> All *futurities* are naked before that All-seeing Eye, the sight of which is no more hindred by distance of time than the sight of an angel can be determined by distance of place (South).

Wilkins summarized the prospect of God:

> He hath a perfect Comprehension of all things, that have been, that are, or shall be...So that this Attribute of his must be infinite and unbounded, both *extensive*, with respect to the several kinds of Objects which it comprehends; and likewise *intensive*, as it sees every single Object with a most perfect infallible view.[57]

One of the most important metaphors of truth in the *Dictionary* (as in Johnson's *Vision of Theodore*) is as height:[58] the view from that height is another significant theme and one which connected the language of natural description with that of religion.

The mists below. If God, at the head of the chain, has perfect vision in time and space, the view of man will be correspondingly limited. This allowed the use of a physical language of clarity and obscurity of views to be deployed with reference to things spiritual, and the *Dictionary* reflects the prevalence of such an elision.[59]

> Our understandings lie grovelling in this lower region, *muffled* up in mists and darkness (Glanville).

Similarly, Milton reminds us, 'God, to remove his ways from human *sense*, / Plac'd heav'n from earth so far.' Just as those who stressed the limitations of natural evidences pointed to the ease with which the senses are deceived, so this was extended to mental sight. This point is made clear by definitions in the *Dictionary*, where the physical and mental sense of a term were placed alongside one another: thus 'shortsighted' was both 'unable by the convexity of the eye to see far', and 'unable by intellectual sight to see far'. Similarly, and remembering Glanville's 'mists and darkness', 'foggy' was both 'misty; cloudy; dank; full of moist vapours', and 'cloudy in understanding; dull'.

Man's cloudy prospect in the present related not only to his position in the chain of being, but also to his postlapsarian position in the Mosaic chronology. South says of man's understanding before the Fall:

> it was vegete, quick, and lively; open as the Day, untainted as the Morning, full of the Innocence and Spriteliness of Youth; it gave the Soul a bright and a full View into all Things; and was not only a Window, but itself the Prospect.

> Adam, then, 'had no Catechism but the Creation, need no Study but Reflection, read no Book but the Volume of the World.'[60]

In a postlapsarian state, man must take more conscious prospects, both inward and forward. Inwardly, man must take a moral prospect of his conduct:

> Man to himself
> Is a large *prospect*, rais'd above the level
> Of his low creeping thoughts (Denham).

> That pow'r which gave me eyes the world to view,
> To view myself infus'd an inward light,
> Whereby my soul, as by a *mirror* true,
> Of her own form may take a perfect sight (Davies).

South also said, in a passage strikingly reminiscent of the allegorical topography of Johnson's *Theodore*,

> the Change and Passage from a State of Nature to a State of Virtue, is laborious ... The Ascent up the Hill is hard and tedious, but the Serenity and fair Prospect at the Top, is sufficient.[61]

We make this inward prospect because of our consciousness of futurity, of a future prospect. The *Dictionary* insistently reminds of the urgency of salvation, often in a metaphorical language of landscape:

> To him, who hath a *prospect* of the different state of perfect happiness or misery, that attends all men after this life, the measures of good and evil are mightily changed (Locke).

Salvation, the *terminus ad quem* of the cosmological drama, closed the analogy with viewing the landscape:

The Spheres of Mens Understandings are as different, as Prospects upon the Earth. Some stand upon a Rock or a Mountain, and see far round about; Others are in an hollow, or in a Cave, and have no prospect at all...And yet the fairest Prospect in this life is not to be compar'd to the least we shall have in another. Our clearest Day here is misty and hazy! We see not far, and what we do see is in a bad Light. But when we have got better Bodies in the first Resurrection...better Senses and a better Understanding, a clearer Light and an higher Station, our Horizon will be enlarged every way, both as to the Natural World, and as to the Intellectual.[62]

Metaphorical vision. Johnson undoubtedly allowed political concerns to be interwoven with the language of landscape. Just as man's view when compared to God's was cloudy, a similar loss of vision could occur in the political sphere:

My people's eyes were once blinded with such *mists* of suspicion, they are soon misled into the most desperate actions (King Charles).

The resultant loss of political order could be modelled on the situation at the Deluge, for just as

the whole universe would have been a confused *chaos*, without beauty or order (Bentley) so,
had I followed the worst, I could not have brought church and state into such a *chaos* of confusions, as some have done (King Charles).

Yet the *Dictionary* highlighted the religious counterpoint to this topographical imagery: the overseeing prospect remained that of God, and it was Man in general, not classes of men, who was condemned to the imperfect vision of a position down on the land. For as long as the language of landscape retained the theological resonance it had in the *Dictionary*, the overview of the ruling elites, itself justified in a religio-political language, had to recognize its limitations within the same framework of Anglican orthodoxy: in the *Dictionary*, 'the politics of the soul are generally more important than the politics of states', with 'politics' being shown as a 'branch of ethics'.[63]

Johnson's *Dictionary* 'is in itself an important incident in the history of philosophic words and in that of the interaction between natural philosophy and the rest of life and literature',[64] a finding which can be extended to the overlapping language of landscape and natural description discussed here.

The existence, prevalence and legitimacy of this strategy was recognized by contemporaries:

> the true meaning therefore, when God is said to *Be in Heaven*, is to express his *Height and Dignity*; not in *place*, but in *Dominion* and *Power*: It being only a *similitude* drawn into common Speech from the situation of Things in Nature . . . by an easy *figure* of Speech, whatsoever is above us *in Power*, we are from hence used to represent as being above us *in Place*.[65]

The *Dictionary* acted to affirm this strategy as one which both rendered the truths of religion more readily comprehensible and dignified the language of nature. It imbued natural description with a theological resonance. As such, it was not only at the level of discursive connections that a religious apprehension of the natural world remained vital, nor even at the level of the structure of argument;[66] the language in which the natural world was described made conflations of Christianity and landscape unavoidable.

Aesthetics and natural description

Johnson's *Dictionary* has generally been cited in the context of landscape studies for its failure to define the term 'picturesque'[67] and for its lack of an explicitly natural-aesthetic definition of 'sublime' even in the 1773 edition, this despite Johnson's admiration of Burke's *Philosophical Enquiry*.[68] This has often been seen as proof of Johnson's scepticism over landscape aesthetics.[69]

This image, however, is modified by looking at the illustrative material Johnson cited in the *Dictionary*, which shows a considerable awareness of the aesthetic language of natural description. Wimsatt's point that 'almost a complete grammar of neo-classic aesthetic may be illustrated from Johnson's quotations'[70] can be extended to include the description of the natural world in aesthetic terms. Under 'beauty', Johnson cited Pope:

> He view'd their twining branches with delight,
> And prais'd the *beauty* of the pleasing sight.

Addison's *Spectator* papers, 'The pleasures of the imagination' (numbers 409 and 411–21), were the most frequently mined source of quotations on natural beauty, again allowing Johnson to draw on Latitudinarianism:

> If the natural *embroidery* of the meadow were helpt and improved by art, a man might make a pretty landskip of his own possessions.

> *Hilly* countries afford the most entertaining prospects, tho' a man would chuse to travel through a plain one.

And under the key term:

We are like men entertained with the view of a spacious *landscape*, where the eye passes over one pleasing prospect into another.

Clearly, Johnson recognized that the beauty of the landscape was a linguistic category independent of its utility.[71]

Similarly, and despite the lack of definition, Johnson did recognize the landscape-aesthetic sensation of the sublime. Again, Addison's *Spectator* served as the source of this idea:

When the sea is worked up in a tempest, so that the *horizon* on every side is nothing but roaming billows and floating mountains, it is impossible to describe the agreeable horrour that rises from such a prospect.[72]

The same grandeur of landscape was evoked by Pope:

Sudden the thunder blackens all the skies,
And the winds whistle, and the surges roll
Mountains on mountains, and *obscure* the pole.

It is apparent, then, that Johnson was conversant with the sublime in landscape and thought it worthy of inclusion in the *Dictionary*. Johnson's sensitivity to the idea of the sublime did not develop only during his travels in the 1770s, even if this is when he first experienced it.

Returning to Johnson's definitions of aesthetic terms relating to landscape, these appear less reactionary when viewed in the context of the tradition of English lexicography. The term 'picturesque', for example, had never been defined in any English dictionary from Cawdrey's on, and so its omission from Johnson's wordlist can be seen as the continuation of a tradition rather than a sign of scepticism. The *OED* itself only cites four uses of the term picturesque prior to 1755, the first in 1703. Similarly, 'sublime' in previous dictionaries had been used to mean grandeur, but only as related to style in writing, not awe in the face of nature. Under 'sublime' (sense 7, 'of things in nature and art'), the *OED*'s first example is from Evelyn's *Diary* (published, 1700), the next from Kames's *Elements of Criticism*, published seven years after the *Dictionary*. Again, we cannot be surprised that this sense of 'sublime' did not appear in Johnson's *Dictionary*.

Furthermore, comparing Johnson's *Dictionary* with its predecessors, we find in the terminology of landscape as elsewhere, that the careful discrimination of shades of meaning gave a far richer lexicon of landscape than had previously existed. Johnson, like Martin, promoted the physical sense of the term landscape above the painterly.[73] Johnson's *Dictionary* was the first to recognize a non-theatrical meaning of the term 'scene' as 'the general appearance of any action; the whole contexture of objects; a display; a series; a regular disposition',[74] and thus to recognize 'scenary' as 'the appearance of place or things'. Similarly, under 'prospect' Bailey's only definition is 'a view or sight afar off; an Aim or Design'.[75] Johnson extended the meaning of 'prospect', his second and third senses for the first time pointing to the use of the term to describe landscapes: '2. Place which affords an extended view' and '3. Series of objects open to the eye.'

Yet it would not be in keeping with Johnson's own beliefs nor the educative function he saw the *Dictionary* as having if the aesthetics of landscape had not been tempered by an affirmation of its subordination to issues of utility. The *Dictionary* was insistent on this with respect to gardens:

> I am more pleased to survey my rows of colewarts and cabbages springing up in their full *fragrancy* and verdure, than to see the tender plants of foreign countries kept alive by artificial heats (Addison).

Bacon added his weight:

> Fine devices of arching water without spilling, be pretty things to *look* on, but nothing to health,

before extending the idea to architecture:

> *houses* are built to live in, not to look on; therefore let use be preferred before uniformity, except where both may be had.

This prioritization of use was firmly embedded within the neo-classic aesthetic the *Dictionary* expresses; Johnson used Pope's *Epistle to Burlington* to summarize the argument:

> There's something previous ev'n to taste; 'tis *sense*,
> Good *sense*, which only is the gift of heav'n,
> And, though no science, fairly worth the sev'n:
> A light within yourself you must perceive;
> Jones and Le Nôtre have it not to give.

Johnson's position in the *Dictionary* was not, therefore, a simple scepticism, but a qualified approval similar to the 'philosophical view of landscape' he was to adopt as a traveller. He refused to substitute 'a purely aesthetic for a human and philosophical view of the landscape' because of his adherence to the scales of value which frame the interpretation of the language of landscape in the *Dictionary*.[76]

Summary: the language of landscape in the first edition of the dictionary

'The science of Johnson's Dictionary is that of a generation or of a century before the Dictionary appeared.'[77] It is the same generation of writers which provides the instruction on how to view nature and landscape, and their inclusion amounts to a significant statement in mid-eighteenth century England. For whilst many continued to seek evidence of God in the land, the discourse of natural aesthetics was also showing increasing independence from any instructive purpose.[78] By refusing to give much ground from the position of the late seventeenth- and early eighteenth-century physico-theologists, Johnson in the *Dictionary* was sounding a considerable note of scepticism about the worth of such an independent and aesthetic inquiry into landscape and nature. In this field, the *Dictionary* does appear to be a 'conservative project'.[79]

In another sense, however, there simply was no field of landscape and nature in the *Dictionary*. As DeMaria has suggested,

> the field of human knowledge, is only partly divisible into recognisable academic categories, such as politics, natural history or literary criticism. The ethical part of knowledge is so pervasive that the whole book is more easily broken into topics like Death, Judgement, Happiness, Freedom...[80]

Landscape was at best a useful emblem and evidence for concerns higher up the noetic chain.

Reinforcing the orthodoxy of nature: the fourth edition of the *Dictionary* (1773)

The fourth edition of the *Dictionary*, published in 1773, was the only one to be significantly altered in Johnson's lifetime, some 15,000 changes being made: 'the fourth edition is now accepted as the best, representing most fully what Johnson wanted the *Dictionary* to be'.[81] These alterations do modify the patterns of the language of landscape, although the broader scales of value remain unaltered.

By the inclusion of material from Reynolds's *Discourses*, Johnson recognized the growing independence of aesthetic discussion about nature. The most important addition of this kind was the new sense of 'prospect': 'View delineated; a picturesque representation of a landscape', a definition illustrated by Reynolds's words:

> Claude Lorrain, on the contrary, was convinced, that taking nature as he found it seldom produced beauty; his pictures are a composition of the various draughts which he has previously made from various beautiful scenes and *prospects*.

Three things are worth noting about this. First, Johnson's definition included the word 'picturesque' which was omitted from the first and fourth editions' wordlist: given its use here, this must be seen as an oversight rather than a display of recalcitrance. Secondly, the Reynolds quotation came from a passage in which the legitimacy of landscape subjects aspiring to moral and historical themes was questioned.[82] Thirdly, Reynolds (unsurprisingly) was used to bolster the neo-classical aesthetic which in the first edition had primarily been articulated by Dryden's DuFresnoy, a point further exemplified under the new seventh meaning of 'nature' in the 1773 edition:

> The works, whether of poets, painters, moralists, or historians, which are built upon general *nature*, live for ever; while those which depend for their existence on particular customs and habits, a partial view of nature, or the fluctuation of fashion, can only be coeval with that which first raised them from obscurity.

Juxtaposing the two Reynolds quotations in their contexts suggests that whilst Johnson was prepared to incorporate a new meaning of prospect, he wished, as Reynolds had, to assert its subordinate status, suggesting doubts that the depiction of the natural world could legitimately engraft the themes of general nature.[83]

The most substantive of Johnson's alterations took place under the words 'nature' and 'natural', the aim being to clarify some of the implicit argument of the first edition. Under 'natural', Johnson cited Wilkins's definition:

> I call that *natural* religion, which men might know, and should be obliged unto, by the meer principles of reason, improved by consideration and experience, without the help of revelation.

This was not, however, a defence of the sufficiency of natural religion, either for Wilkins in his *Principles and Duties of Natural Religion* or as he was presented in the *Dictionary*. The quotation under 'natural' must be seen in the context of the criticism of it added under 'supernatural':

> No man can give any rational account how it is possible that such a general flood should come, by any natural means. And if it be *supernatural*, that grants the thing I am proving, namely, such a supreme being as can alter the course of nature (Wilkins).

Despite this 'at the heart of the work lies a tension between its implicit claims to a unified authority and the presence of other diffuse and disparate – and sometimes competing – authorities'.[84] The fragmentary format of the *Dictionary* threatened to obscure the orthodoxy of its attitude to nature. Johnson acted to combat this in 1773 by an unusual means: he added a long epitome of Boyle's *A Free Enquiry into the Vulgarly Receiv'd Notion of Nature* (1685) to make the *Dictionary*'s position clear. The purpose of Boyle's *Free Enquiry* was to clarify the meaning of the term 'nature' and to remove the possibility of equivocal or heretical uses. The opening and closing paragraphs of the *Dictionary*'s epitome are of particular importance:

Nature sometimes means the Authour of Nature, or natura naturans; as, nature hath made man partly corporeal and partly immaterial. For nature in this sense may be used the word creator.

Nature is sometimes indeed commonly taken for a kind of semideity. In this sense it is best not to use it at all.

Boyle adopted a nominalist ontology of nature, such that the particulars in the natural world 'are denied the power to cause change in and of themselves. God's Will, therefore, is the only causally efficacious agency in nature.'[85] Boyle stressed the omnipotence of God's Will, which could change even the laws of nature. As such, natural reason could not, without the aid of revelation, understand God. If 'a spiritually imbued material world provided a usable vision of a self-moving and self ordering system, independent of superintendence by spiritual intermediaries',[86] the vision Boyle put forward and Johnson concurred with suggested the reverse; that nature could not be self-ordering, and that God was its ordering principle.[87]

This was not an alteration of the orthodox doctrine put forward by the *Dictionary*, but a more didactic presentation, akin to many of Bailey's encyclopaedic definitions in the *Universal Dictionary*, which achieved a unified voice by allowing only one speaker – the lexicographer.[88] It was the need to clarify orthodoxy which also perhaps motivated the significant reduction in quotations from Thomson's *Seasons* in the fourth edition. Thomson's language was 'unorthodox and peculiar, often concerned with expressing the higher truths of God's presence in nature'.[89] Thomson's vocabulary sprang from the physico-theology Johnson relied upon so heavily, but as the century progressed and that approach was increasingly overtaken, it could be seen as rendering nature as the semideity Boyle and Johnson opposed.[90]

The fourth edition is more strident in its political–theological argument, this being a response to the challenges to the established status of Anglicanism in the 1770s.[91] This challenge, expressed in the Feathers' Tavern Petition of 1771, was widely identified with Arianism and Socinianism which emphasized the sufficiency of natural religion.[92] The first edition of the *Dictionary* had countered this approach by attacking the vanity of overlooking the limitations of human reason. The fourth edition was more anxious to bring its argument about nature and God to the reader's attention without ambiguity, and thus more insistent on its Anglican orthodoxy.[93] In the face of these challenges to church and state, Johnson penned some of his most well-known political tracts, contributing to the emergent right-wing *mentalité*.[94] The language of landscape in the fourth edition partook of this more self-conscious defence of orthodoxy, precisely because its function in the *Dictionary* was as a carrier of more important issues in the hierarchy of knowledge. Johnson's orthodox language of landscape was increasingly the denominational discourse of a High-Church Tory.

Concluding comments: Johnson's *Dictionary* and modern debates on the language of landscape in eighteenth-century England

The language of landscape has received considerable attention from modern scholars.[95] That this lexicon had a religious element was accepted by Barrell:

> [a] frequent eighteenth-century image of social organisation – also originally employed in attempts to describe the order of the universe – [was] of society as a landscape, as a painting, or as a landscape painting, in which the various objects in the view, in which light and shade, may appear in one perspective to be in no relation or even to be in conflict with one another, but can, from the correct viewpoint, be seen in 'just harmony and proportion.'[96]

Barrell's words point to two widely accepted points about the language of landscape: first, that religion was not the origin of this language, and any religious connotation had rapidly been jettisoned.[97] Secondly, it is argued that the language of landscape became almost exclusively tied with a political language of class consciousness. This is the view of the outsider and

> the crucial phrase here is perhaps 'commanding height', a phrase borrowed of course from the language of military tactics, and by no means used, by eighteenth-century poets, without a sense of embattled hostility to what is being commanded, the landscape below.[98]

Fabricant has gone on to stress that the class-based language of landscape is supplemented by its gendered desire to order and control Nature, the 'coy or seductive maiden'.[99]

The evidence from Johnson's *Dictionary* suggests that some modification of these points is needed. First, Johnson stressed the continued vitality of the religious in landscape terminology. The sheer density of quotations built upon this connection overwhelms any other in the *Dictionary*. Secondly, the political language of landscape in eighteenth-century England could not escape from the political–theological nexus: the commanding prospect inevitably recalled the prospect of futurity and a superior Surveyor. Putting these two qualifications together gives a somewhat different perspective on the field of knowledge in which landscape participated in eighteenth-century England, one which, by the detailed study of one book, supports the findings of the broader survey in Chapter 3. If landscape continued to be strongly connected with religion, and the language of landscape was political in an ethical sense, the language of landscape appears religious in far more than the vestigial sense currently accepted. Moreover, the debates over the language of landscape, as both the first and fourth editions of the *Dictionary* show, were part of the denominational discourse of *ancien régime* England. The universe of analogy into which the *Dictionary* wove the language of landscape was clearly one appropriate to an intellectual *milieu* in which religious and hierarchical thinking remained of central importance. However different the scales or value established in eighteenth-century England and in the *Dictionary*, the contextual historian must accept that the contemporary meanings of the language of landscape differed from what appears most significant to our own times.

6
The Moral Landscape: Johnson's Doctrine of Landscape, 1738–59

Johnson's *Dictionary* is a liminal document with a complex interaction of the compiler and his authorities which makes the post-structural problem about the existence of authorial intention particularly apparent. This chapter moves on to discuss Johnson's own writings as a poet, novelist, essayist and homilist. Here we see Johnson manipulating the language of natural description and landscape appreciation for his own ends and drawing on specific sources to do this. This gives a greater insight into which of the numerous sources Johnson cited in his *Dictionary* his own writing derived from, and thus moves us towards his authorial motivations and his location in eighteenth-century debates on landscape and nature. As we shall see, this also moves us from the points where Johnson overlapped with Latitudinarian uses of landscape imagery to the ways in which he was at odds with that dominant discourse.

From politics to morality, 1738–49

Country rhetoric and landscape binaries

Johnson's first major work after moving to London was a poem entitled *London* (1738). This poem, together with a number of political pamphlets from the same period, places Johnson in the tradition of 'country' rhetoric adopted in opposition to Walpole's government.[1] In *London* Johnson established a political geography of corruption built around the opposition of town and country: 'the contrast between country and city became, so to speak, an extended metaphor containing his essentially political theme'.[2] This contrast also became an historical geography, past and present landscapes being evoked for political purposes. I will suggest that, whatever the adequacy of the political categories Johnson evoked, he was unable to fix them convincingly to his political geography and consequently to the 'look' of the landscape. It is the failure of this country politics of landscape in Johnson's earliest writings which led him to different uses of landscape themes and imagery, emergent in the late 1740s.

London opens with the poet accompanying Thales as he 'bids the town farewell'.[3] The opposition of town and country is established at the outset with apparent certitude: Thales is

> Resolved at length, from vice and London far,
> To breathe in distant fields a purer air,
> And, fix'd on Cambria's solitary shore,
> Give to St. David one true Briton more (ll. 5–8).

Vice and London, and the purity of distance form the basic geographical framework within which the poem operates. That this is also an historical geography is suggested by the reference to 'true Britons'; Thales is re-enacting the flight from the corrupt/conquered centre to the pure and defiant margins, preserving the true spirit of English/British liberty.[4] Purity lies at a distance from London in space and in time, and the two are connected, historical sites of virtue offering an escape from present-day corruption for the same geographical reason of remoteness.[5]

Where in later work Johnson argued that happiness derived from our mental state, in *London* it was a matter of geography. Thus Thales seeks 'to find some happier place' (l. 43) and can visualize the pastoral landscape which will satisfy his requirements:

> Some pleasing bank where verdant osiers play,
> Some peaceful vale with nature's paintings gay (ll. 45–46).

The received image of Johnson hardly suggests that Thales's eulogy of a pastoral retirement can be sincere: yet

> when Johnson wrote *London* he had been in the metropolis hardly more than a year...and even after going to London and throwing in his lot with Grub Street he still nourished the ideal of a quiet life in the country.[6]

For Johnson to write within the patriot idiom, he had to establish a political geography consonant with it; for these purposes the notion of a corrupt metropolis from which the virtuous would retire to an innocent countryside was essential, and fitted with the established interpretation of Juvenal's third satire which Johnson was imitating.[7]

The countryside as innocent is opposed in *London*'s political geography to the corrupt metropolis. London was the spatial centre of vice; it 'sucks in the dregs of each corrupted state' (l. 96). This corruption could not come from rural Britain, so Johnson's political geography had to widen its scope, London being

> – the needy villain's gen'ral home,
> The common shore of Paris and of Rome (ll. 93–94).

Geography and political values interpenetrated: Britain could not be dissociated from liberty; London, therefore, which did not bear these political values, could not geographically be associated with Britain, and is thus a 'French metropolis' (l. 98). Britain is not envisaged as creating its own corruption, there being no source from which this could come within the nation; the change comes from Paris and Rome which, influenced by Catholicism, do not embody the same fusion of geography and liberty.[8]

Given the purity of Britain as a land and people, should not London be a metropolis of liberty? The implication is that it once was; as Johnson's presentation of the country was tied with the historical geography of ancient Britons, so London's present corruption is contrasted with its previous status. Johnson 'realised early in life the power of place to evoke history',[9] and used the connection at Greenwich:

> On Thames's banks, in silent thought we stood,
> Where Greenwich smiles upon the silver flood:
> Struck with the seat that gave Eliza birth,
> We kneel, and kiss the consecrated earth;
> In pleasing dreams the blissful age renew,
> And call Britannia's glories back to view (ll. 21–26).

Tudor Greenwich must have been a reflection of 'Britannia's glories' (now found exclusively in the country) to give birth to an uncorrupted monarch. This is in stark contrast to George II, who, born outside the influence of British values, living in corrupt (foreign-invaded) London, and continually returning to Hanover, is bitterly attacked towards the end of the poem (ll. 242–47). How Greenwich has declined since Elizabeth's time is suggested in *Marmor Norfolciense*: Johnson's Walpolian commentator, Probus Britanicus, urges that Greenwich hospital, an asylum for precisely those naval officers who made England 'The guard of commerce, and the dread of Spain' (l. 28), be converted into apartments for a society of (pro-Walpole) commentators. Thales could agree with Probus that 'the situation of Greenwich will naturally dispose them to reflection and study',[10] but where he is led to a despairing reflection on the gap between Elizabeth's birthplace and modern London, the Society of Commentators will in this situation merely mirror the corruption in the landscape around them.[11]

This is not simply a geography, the implication being that it can be seen; that there are landscapes of virtue and vice. Thus Thales is led into his attack on London, which forms the bulk of the poem, as he '*eyes* the neighb'ring town' (l. 34, emphasis added).[12] The historical geography of *London* is also visualized in changes in the landscape. For this reason Edward III is invited to 'survey' Britain to see how this proud nation has capitulated to foreign influence:

> Illustrious Edward! from the realms of day,
> The land of heroes and of saints survey;
> Nor hope the British lineaments to trace,
> The rustic grandeur, or the surly grace (ll. 99–102).

If Edward could 'survey' the decline of British landscape due to his historical position, Thales can 'eye' it by virtue of his geographical perspective. He was born

> – from slav'ry far,
> I drew the breath of life in English air (ll. 117–18).

Thales, coming effectively from a different country, can see the disparity between London and Britain. Johnson adopted the same strategy in his political essay, 'Eubulus on Chinese and English Manners.' Eubulus argues that

> the satisfaction found in reading descriptions of distant countries arises from a comparison which every reader naturally makes, between the ideas which he receives from the relation, and those which were familiar to him before.[13]

Eubulus is amazed at the disparity between Chinese political probity and the (unstated, but English) political knavery he is accustomed to.[14]

There is, however, something of a contradiction in asserting the visibility of urban corruption: Thales wants to say both that he can see corruption in the landscape and that what distinguishes townspeople is their ability to dissimulate and thus deceive the eye. That the appearance of commerce and urban life was shifting and unreliable was a commonplace in political language:[15] yet if it is 'the mimick's art' (l. 134) which achieves success in London, does not the landscape, given the homology of place with political personality the poem asserts, then become a stage which is hard to read as direct visual evidence of corruption? Moreover, if Thales can see the corruption behind London's dissimulation, he must be more knowing than his birth 'from slav'ry far' would suggest. The rigid division of town and country suggested by the opening lines is called into question: if knowledge is based on place, and Thales claims to understand both the town and the country, the two systems cannot be mutually exclusive.

The poem itself blurred the town and country division by occasional suggestions that the countryside could not be sealed from the corrupting influence of the town. When Edward III surveys, he cannot expect to find 'the rustick grandeur' (l. 102) of the fourteenth-century countryside. However pure the country may be in comparison with London, it is not what it once was. This is because Walpole's pensioners 'raise palaces, and manors buy' (l. 57) such that the 'nation', not just London, is left 'groaning' (l. 65).[16] This argument is also expressed in *Marmor Norfolciense*, where the (patriot) prophecy suggests:

> Then thro' thy fields shall scarlet reptiles stray,
> And rapine and pollution mark their way...
> The teeming year's whole product shall devour,
> Insatiate pluck the fruit, and crop the flow'r:
> Shall glutton on the industrious peasants spoil,
> Rob without fear, and fatten without toil.[17]

However much it suited the political geography of *London* to see the country as untouched, the sources of urban corruption had rural power bases. *London* has to admit this: to argue that the countryside was innocent despite corrupt local landlords would be to divide (political) personality from the character of a place. Yet this would render the standard against which London is measured and the source of Thales's vision of metropolitan degeneracy radically incomplete.

Despite admissions that the country has been affected by urban vice, Thales expands on retirement to the 'fair banks of Severn or of Trent' (l. 211):

> There might'st thou find some elegant retreat,
> Some hireling senator's deserted seat;
> And stretch thy prospects o'er the smiling land,
> For less than rent the dungeons of the Strand;
> There prune thy walks, support thy drooping flow'rs,
> Direct thy rivulets, and twine thy bow'rs;
> And, while thy grounds a cheap repast afford,
> Despise the dainties of a venal lord (ll. 212–19)

Weinbrot has argued that 'Thales describes an eighteenth-century country estate.'[18] In fact, the imagery is curiously split between agriculture and gardening, subsistence and aesthetics. The ground affords a 'cheap repast' which is preferred to dainties, yet the 'labour' undertaken – pruning walks and twining bowers – is hardly redolent of georgic values. Moreover, the fact that the land continues to 'smile' despite a period of control by a hireling senator raises the prospect (literally) that the landscape of the countryside is as deceptive *vis-à-vis* political practice as the townscape with its 'mimick's art'.

Thales's reassertion, then, of the virtue of retreat to the country is undercut by its own inconsistencies. This becomes still more apparent towards the end of the poem:

Scarce can our fields, such crowds at Tyburn die,
With hemp the gallows and the fleet supply.
Propose your schemes, ye Senatorian band,
Whose Ways and Means support the sinking land;
Lest ropes be wanting in the tempting spring,
To rig another convoy for the k—g (ll. 242–47).

The fields of England here have a direct connection to politics in the city and thus to corruption. If the land appears to 'smile', this is deceptive given its purpose; supporting the (foreign) king and executing those (patriots) who oppose his corrupt government.[19] The claim that Thales has a true view of both town and country has collapsed, both sites being deceptive. Thales's parting suggestion that the poet 'fly'st for refuge to the wilds of Kent' (l. 257) sounds increasingly arbitrary.

In *London* and other patriot writings, Johnson attempted to establish a series of interrelated binaries of town and country, present and past, deceit and honesty. They are overdrawn: 'the reader will doubt whether London is entirely evil, that all men in all places are evil, that total seclusion will bring happiness to the good man'.[20] The simple political and historical geography Johnson's patriot stance wished to create, together with its visualization in the landscape, was undermined by the transgression of categories, leaving landscape as a more opaque revelation of political health. What distinguishes these works of the 1730s in terms of landscape ideas, is the attempt to link the concept of place to politico-moral ideas of vice and virtue. Johnson's interest in landscape and place was to continue to build around its connection with moral issues, but never again in so direct or causal a manner.

New uses and altered doctrines of landscape in the 1740s

Walpole's successors proved patriot rhetoric to have been as incoherent as Johnson's efforts in *London* to inscribe it in space. Johnson's disillusionment with patriot politics was reflected in a shift in the centre of gravity in his ideas, with an increasing emphasis on unalloyed moral issues and a marginalization of direct political discussion.[21] The change also brought the Christian element of Johnson's doctrine to the fore. The changing contours of his thought led to an alteration in the use of landscape and nature as concepts, and their connection to a different set of discursive concerns. With *The Vanity of Human Wishes*, *The Vision of Theodore* and *Irene*, we can see the emergence of the approach to landscape the mature Johnson was to adopt in subsequent didactic works.

Mind and Place

At first sight, it could appear that little had changed from *London* in Johnson's picture of the innocent countryside. *The Vanity of Human Wishes* (1749) speaks of 'the hind'

> Untouch'd his cottage, and his slumbers sound,
> Tho' confiscation's vulturs hover round.
> The needy traveller, secure and gay,
> Walks the wild heath, and sings his toil away (ll. 35–38).[22]

Similarly, in his only play, *Irene* (1749, written 1737–49), Aspasia, who 'carries the burden of right thinking and right acting'[23] would 'chearful . . . follow to the rural cell' (IV. i. 110).[24] The success of such retirement, is also pointed to at the outset of *The Vision of Theodore* (1748): Theodore's cell is 'a place where all real wants might be easily supplied' (p. 195).[25]

It would be mistaken, however, to think that Johnson was still trying to invest happiness in place. In *Vanity*, the happiness of the hind and needy traveller is based upon their poverty; the wealthy in rural areas are prone to all the problems 'confiscation's vultures' pose in the town. Aspasia provides the key: she is cheerful in a rural cell, for 'Love [will] be my wealth, and my distinction virtue' (IV. i. 111). Place cannot sanctify a person; it is the virtue of the person which makes them comfortable in a place. As such, a mental-moral geography, with contentment at a site relating to the morality of those who live there, replaces the political geography of *London*, where removal to a pure place created a tranquil mind. The binary of town and country was thus discarded in favour of the active struggle of individual minds between vice and virtue, regardless of location. By moving to an unclouded assessment of virtue decoupled from landscape, Johnson was signalling his distance from Latitudinarianism, his refusal to see landscape as a bearer of real (as opposed to allegorical) moral meaning.

This can be seen in the *Vanity* and the *Vision*. The old man in the *Vanity*, fearful of death and surrounded by the vultures who do not perplex the hind, finds:

> In vain their gifts the bounteous seasons pour,
> The fruit autumnal, and the vernal flow'r,
> With listless eyes the dotard views the store,
> He views, and wonders that they please no more (ll. 261–64).

The joys of nature become invisible to the troubled mind; they cannot act as a recuperative agent. Even Theodore, whose real wants are supplied, says, 'I stood one day beholding the rock that overhangs my cell, [and] I found in myself a desire to climb it' (p. 196). The virtuous mind can become troubled for no reason, at which point the pleasures of situation are not enough to prevent the need for change. Johnson is moving towards a picture of the restlessness of the human mind central to the *Rambler* ('this invisible riot of the mind' – *Rambler* 89), which puts the static geography of *London* into motion: by reversing the causal connection of mind to place and then recognizing the instability of mental states, the binaries which organized *London* are broken down.

The emergent analogy of physical and moral vision of landscape

With Johnson's detachment of mental well-being from physical landscapes, landscape concepts and imagery became available for alternative uses. Unlike for Latitudinarian authors, landscape began to play a merely metaphorical role in Johnson's writings of the 1740s which it retained in the essays of the 1750s. The design argument was not deployed, nor is meletetics or aesthetics, producing a distinct approach to landscape from the three readings of the book of nature favoured by Latitudinarian authors.

The opening lines of the *Vanity* are suggestive of this:

> Let observation with extensive view,
> Survey mankind, from China to Peru (ll. 1–2).

In the prose equivalent Theodore is told to 'survey...and be wise' (p. 198). This survey is not of a physical landscape: where in *London* Edward III and Thales could see corruption in the actual landscape, in the *Vanity* and *Vision* this is not the case. The *Vanity*'s survey is of various moral exemplars and the impossibility of earthly happiness, not the connection of happiness, virtue and place synthesized in a visible prospect.[26] Similarly, Theodore's is not a physical survey: his vision is on a 'small plain' when he 'had not advanced far' up Mount Teneriffe (p. 197).

> Johnson uses the convention of the mountain-which-affords-an-elevated-view-of-humanity in a delightfully unconventional way...[It] tells us clearly and forcefully that if we wish to survey human existence, all we need to do is stand still and look about us. Johnson uses Mount Teneriffe only to show us it is of no special use.[27]

Johnson's focus remained on man's mental states (with a shift away from political ones), but a diminished belief in their capacity for visualization marginalized landscape and sublimated its terminology, giving it a more insubstantial existence.[28]

Johnson extended the imagery of landscape into a considerable moral and allegorical structure in his writings. The foundation is laid in the three works under review. The opening lines of the *Vanity* are followed by an extension of the moral prospect:

> And watch the busy scenes of crouded life;
> Then say how hope and fear, desire and hate,
> O'erspread with snares the clouded maze of fate,
> Where wav'ring man, betray'd by vent'rous pride,
> To tread the dreary paths without a guide,
> As treach'rous phantoms in the mist delude,
> Shuns fancied ills, or chases airy good (ll. 4–10).

The language of deluding mists became a commonplace with Johnson. In the *Vision* the top of the Mountain of Existence is shrouded by a mist (pp. 204, 209). By contrast, and within limits set by Christianity, reason provides a clear view: man chasing airy good in the *Vanity* proves, 'How rarely reason guides the stubborn choice' (l. 11). Metaphorical landscape imagery was connected in a consistent manner to mental states, reason equating to clarity, and passions to the obscuring of our view:

> The gleams of reason, and the clouds of passion,
> Irradiate and obscure my breast by turns (*Irene* V. v. 44–45).

As well as making the ability to survey nature a metaphor for moral perspicuity, Johnson also created a moral topography, of particular importance in *Vanity*. To rise in this moral landscape is to ensure subsequent fall.[29] Wolsey, whose 'restless wishes' tower to 'new heights' (l. 105) must subsequently

> — sink beneath misfortune's blow,
> With louder ruin to the gulphs below (ll. 127–28).

The same fate awaits the scholar who seeks the 'glitt'ring eminence' (l. 166) of knowledge, for, in an image combining Johnson's metaphors of moral vision and topography, few see 'unclouded... the gulphs of fate' (l. 312). This language was replicated in *Irene*, where Cali, having deposed Mahomet will look 'down, contemptuous, from his fancy'd height' (IV. iv. 2). Irene speaks of 'the precipice of pow'r' (III. viii. 113) from which she will eventually fall and which Aspasia (again reverting to the language of moral vision) refers to as 'the glitt'ring fallacy to view' (III. viii. 128). Theodore, upon experiencing a desire to climb a mountain, examines whether

> my heart was deceiving me... and that my ardour to survey the works of nature, was only a hidden longing to mingle once again in the scenes of life (p. 196).

Only once he is assured that he is not seeking social eminence does Theodore begin to climb a physical eminence.

With Theodore's fear that the desire to survey nature related to the lure of earthly pleasures, Johnson began to develop a notion that nature is a seductive distraction from more important issues.[30] Theodore is seduced by the 'small plain' on which he rests on his ascent of Teneriffe: 'once I had tasted ease, I found many reasons against disturbing it. The branches spread a shade over my head, and the gales of spring wafted odours to my bosom' (p. 197). Whilst this leads to Theodore's instructive vision, the dangers of ease are apparent, and contribute to Irene's downfall as she is courted by Mahomet:

> If greatness please thee, mount th' imperial seat;
> If pleasure charm thee, view this soft retreat;
> Here ev'ry warbler of the sky shall sing;
> Here ev'ry fragrance breathe of ev'ry spring.
> To deck these bow'rs each region shall combine,
> And ev'n our Prophet's gardens envy thine (II. vii. 84–89).[31]

If reason gives a clear prospect of life's snares, the indolence encouraged by soft retreat clouds vision, and leads away from the morally acceptable path. The imagery of a path is closely related to the moral topography Johnson established. This is most apparent in *Vision*: education confines those who ascend the mountain of existence 'to certain paths too narrow and too rough' (p. 199). It is notable, given the sensual danger of landscape imagery, that successful climbers did not 'regard the prospects which at every step courted their attention' (p. 205).

Landscape analogy and theology

Johnson's moral language of landscape related back to his 'master image of the voyage or journey of life'.[32] This image was a Christian commonplace, derived from classical rhetoric, and it was persuasion to a Christian lifestyle which determined Johnson's deployment of landscape ideas and imagery.

Reason may give a clear view, but is still limited: in the *Vision* Reason cannot see through 'a mist before you settled upon the highest visible part of the mountain' (p. 204). This mist is 'pierced only by the eyes of Religion', an answer all three works point to. In response to the famous question posed at the end of the *Vanity* – 'Must helpless man, in ignorance sedate, / Roll darkling down the torrent of his fate?' – Johnson argues that we must not 'deem religion vain' as God's 'eyes discern afar' (ll. pp. 345–46, 350, 353). The grand claims for human vision gradually decline in 'the verbal descent from the empyrean – from "survey" to "remark" to "watch" and finally to the unpretentious "say"'.[33] The conclusion invokes God's vision of the moral prospect and inverts the human position in Johnson's moral topography: 'in the concluding paragraph we are looking up toward the *skies* . . . Our wishes *rise*, we *raise* our prayers, our devotion *aspires*'.[34]

Johnson implicitly rebuked Thales's impious plea: 'Grant me, kind Heaven, to find some happier place' (*London*, 1. 43) on Earth. He exchanges such an earthbound vision:

> For faith, that panting for a happier seat,
> Counts death kind Nature's signal of retreat (*Vanity* ll. 363–64).

In the *Vision* also, only by following Religion can the summit of the Mountain of Existence be reached, atop which are the 'temples of happiness'.[35] As Johnson's language of landscape became more ethereal, so the notion of happiness was released from locative associations: '*London* points towards Gehenna, the *Vanity* towards the city of God and the paradise within.'[36]

Given that nature is only fully legible to God, Johnson could attack the delusions of men wishing to control nature or believing they could read the future in nature. Johnson's megalomaniacal characters speak of dominating nature, and this is a sign of their mental imbalance. Charles XII wishes 'all be mine beneath the polar sky' (*Vanity* 1. 204), and Xerxes's situation is still more futile: 'The waves he lashes, and enchains the wind' (1. 232).[37] In his inevitable fall from the precarious eminence he aspired to, Xerxes's dreams of dominion are shattered by reassertion of the dominion of nature over man:

> Th' insulted sea with humbler thoughts he gains,
> A single skiff to speed his flight remains;
> Th' incumber'd oar scarce leaves the dreaded coast
> Through purple billows and a floating host (ll. 237–40).

Similarly, Mahomet in *Irene* expounds 'His vast designs, his plans of boundless pow'r':

> When ev'ry storm in my domain shall roar,
> When ev'ry wave shall beat a Turkish shore (I. v. 41–43).

As there is no location of happiness without faith, so no extent of dominion can result in happiness, contrary to these deluded fancies. After Irene's death Mahomet realizes this: he no more burns 'for fame or for dominion' in 'the tasteless world' (V. xii. 43, 48).[38]

Irene also undermines the impiety of those who would read nature prophetically.[39] At the outset, Demetrius (together with Aspasia the moral arbiter[40]) has to correct Leontius who argues:

> — That power that kindly spreads
> The clouds, a signal of impending show'rs,

> To warn the wand'ring linnet to the shade,
> Beheld without concern, exposing Greece,
> And not one prodigy foretold our fate (I. i. 31–35).

Leontius is looking to nature to foretell the collapse of Greece, when the decline of civic virtue is a less opaque cause (if less visible in the landscape). Demetrius responds:

> Can brave Leontius call for airy wonders,
> Which cheats interpret, and which fools regard? (I. i. 42–43).

It is the corrupt characters in the play who look to prognostics, leaving their Christian rivals to reprove them for their idolatrous readings of nature, which are false in all cases. Abdalla, plotting against Mahomet believes.

> The fav'ring winds assist the great design,
> Sport in our sails, and murmur o'er the deep. (IV. iii. 6–7).

In fact, Mahomet has known of the conspiracy since Mustapha informed him in Act II, scene vi.

It is the apostate Irene who is most mistaken in reading nature. Like Thales she believes landscape reflects political health:

> O! did Irene shine the Queen of Turkey,
> No more should Greece lament those prayers rejected.
> Again should golden splendour grace her cities,
> Again her prostrate palaces should rise (III. viii. 51–54).

But the ruins of Greece, as Demetrius (V. v. 17) points out, are irreparable, and the recovery of virtue is not related to the appearance of Constantinople. Irene makes the same error in act III, scene viii, likening those who 'mount the precipices of pow'r' (113) to an archangel who 'Directs the planets with a careless nod, / Conducts the sun, and regulates the spheres' (123–24). The comparison forgets the limited view of man, and Aspasia chastises her:

> Stoop from thy flight, trace back th'entangled thought,
> And set the glitt'ring fallacy to view (127–28).

The exchange between Aspasia's Christian views and Irene's impious readings of nature reaches a peak in the final act. In scene i, Aspasia beseaches:

> Attention rise, survey the fair creation,
> Till conscious of th'incircling Deity,
> Beyond the mists of care thy pinion tow'rs (V. i. 5–7).

The creation can show us the existence of God, and points our attention to a world beyond the present. In contrast is Irene's superficial reading of nature as revealing the fortunes of individuals in scene ii:

> See how the moon through all th'unclouded sky
> Spreads her mild radiance, and descending dews
> Revive the languid flow'rs; thus Nature shone
> New from the Maker's hand, and fair array'd
> In bright colours of primaeval spring;
> When Purity, while fraud was yet unknown,
> Play'd fearless in th'inviolated shades.
> This elemental joy, this gen'ral calm,
> Is sure the smile of unoffended Heav'n (ii. 1–9).

Irene's evocation of Edenic language is ironic given her apostacy, and her reading of nature is entirely erroneous: Cali has already been condemned to the torture (IV. viii. 21–25) which will implicate Irene in conspiracy, and lead to her death. Aspasia again corrects her, urging that vice will not be rewarded, but that Irene may be cast down by a flaming bolt. Given her own recurrent belief in prognostics, Irene's response applies more to herself:

> Forbear thy threats, proud prophetess of ill,
> Vers'd in the secret counsels of the sky (ll. 26–27).

Irene's confidence is based on her superstitious reading of nature, which ignores the fact that nature is deceitful in its appearances; Aspasia's submissive view, which sees God in nature but does not presume to arrogate divine knowledge of the workings of nature to frail human beings is supported by the play's outcome.[41]

Conclusion: revisioning nature and landscape

Landscape emerges as far less concrete in Johnson's writings of the 1740s. This lack of concern for the physical character of the scene was noted by an anonymous contemporary in his *A criticism of Mahomet and Irene*.[42] Nature is still instructive, something surveyed to gain knowledge, but it is not the transparent revelation of individual and social virtue suggested by *London*. The instruction we receive is tempered by the limits of the human mind. It is this vision forged in the 1740s which is elaborated in the essays of the 1750s.

Johnson's moral doctrine of landscape in the periodical essays, 1750–1759

Johnson's periodical essays, the *Rambler* (1750–2), *Adventurer* (1753–4) and *Idler* (1758–9), have always been seen as central to his work as a moral writer.[43]

They are a continuation of the programme established in the 1740s. The sheer volume of work contained in the essays meant that landscape imagery figured in an increasingly broad range of situations, and also that Johnson articulated at more length the role of landscape and the natural world in a moral life.

The uses of landscape imagery

Landscape participated in the instruction delivered by the essays in a number of ways which can initially be analyzed independent from what Johnson wanted to say about the natural world. The imagery of landscape appeared far more frequently than any actual discussion of landscape. It is important to reflect on this, as it is one of the dominant impressions the student of landscape receives from the essays, and suggests ways in which landscape concepts and imagery were understood and used in the eighteenth century which are different from those discussed in modern landscape studies.

Johnson's use of landscape imagery can be analyzed in terms of four spectra of strategies.

From Oriental tale to allegory

Johnson's creation of Oriental tales required some exotic landscape background. But the points Johnson was trying to make in the Oriental essays were universal:[44] the dangers of excessive desire (*Rambler* 38), the uncertainty of happiness (*Ramblers* 204 and 205) or frustrated wishes (*Idler* 101). The level of exotic landscape 'colour' in these essays varied. Johnson could, with his extensive geographical reading,[45] establish extended exotic landscape descriptions, as in *Rambler* 204:

> [the palace of Dambea] stood in an island cultivated only for pleasure, planted with every flower that spreads its colours to the sun, and every shrub that sheds fragrance in the air. In one part of this extensive garden, were open walks for excurions in the morning; in another, thick groves, and silent arbours, and bubbling fountains for repose at noon.

In short, the garden contained 'all that could solace the sense[s]',[46] and Johnson's interest was to show the insufficiency of this to the achievement of (even earthly) happiness. 'We shall comb the *Rambler* in vain for images and descriptions calculated to evoke specific scenes, sounds, and smells from Arabia or Persia – perhaps because Near East literature was most eloquently represented for Johnson not by the *Arabian Nights* or its jaded continuations but by the *Bible*.'[47]

Johnson's focus on the instructive point behind the exotic setting also explains the ease with which he could shift locations, as in his 'Greenland history' (*Ramblers* 186 and 187).[48] Here Johnson burlesqued primitivism and pointed out that all civilizations can develop elaborate literary symbolism based on the natural world as they experience it. Anningait courts Ajut with a poem saying 'her fingers were as white as the teeth of the morse, and her smile grateful as the dissolution of the ice' (v. 213). Just as the Oriental appears exotic and improbable to the European, so Anningait warns,

> we live not, my fair, in those fabled countries, which lying strangers so wantonly describe; where the whole year is divided into short days and nights ... with flocks of tame animals grazing in the fields about them (v. 214).

Anningait can also convert a basic Christian message – the night cometh when no man can work – into the natural imagery of the Arctic: 'a few summer days, and a few winter nights, and the life of man is at an end' (v. 215).

In most cases, however, Johnson was far less careful to establish his exotic setting. In his tale of Hamet and Raschid (*Rambler* 38) on 'the plains of India', no local colour is given at all (with the exception of the character names and the name of the Ganges): the plains are suffering a drought, and the Genius of Distribution comes to them offering water. Raschid, ignoring the essay's message that 'mediocrity is best', asks for the Ganges to water his grounds. The result is disastrous, but non-localized: 'the flood [of the Ganges] rolled forward into the lands of Raschid, his plantations were torn up, his flocks overwhelmed, he was swept before it', until Johnson remembers his setting with a somewhat arbitrary piece of local colour as 'a crocodile devoured him' (iii. 206–10).[49]

The non-local nature of Johnson's Oriental tales is suggested by the ease with which they develop into allegory. Where Hamet and Raschid nominally 'localise' a general message, the apparently local tale of Obidah (*Rambler* 65) is generalized by its last paragraph. Obidah, journeying on 'the plains of Indostan', passes through a valley, as in the garden at Dambea 'all his senses ... gratified'. In the midday heat he choses a parallel path, loses his way and, as storm clouds gather, he fears for his life until he stumbles on a hermit's cottage. The hermit allegorizes all that has gone before: 'remember, my son, that human life is the journey of a day', where it is dangerous to 'enter the bowers of ease'. Where Bunyan's By Path Meadow[50] is clearly allegorical from the outset, no generalized name reveals Obidah's by-path to be moral until the hermit saves him. At this point the application of the motto of the paper, from Horace's *Satires*, becomes apparent: 'The chearful sage, when solemn dictates fail, / Conceals the moral counsel in a tale.' Yet the reality of Obidah's journey, as well as its allegorical status, is reasserted by the hermit's last words: 'Go now, my son, to thy repose, commit thyself to the care of omnipotence, and when the morning calls again to toil, begin anew thy journey and thy life' (iii. 344–49).[51]

Imagery of the natural world is vital to Johnson's allegorical essays. In his 'Allegory of Rest and Labour' (*Rambler* 33, iii. 179–84), the landscape of Rest in the world's earliest stage is destroyed by violence. This causes a change in the landscape: 'amidst the prevalence of this corruption, the state of the earth was changed; the year was divided into seasons; part of the ground became barren, and the rest yielded only berries, acorns and herbs'. Labour rescues the people from Famine, and this again can be seen in the prospect: 'the face of things was immediately transformed; the land was covered with towns and villages, encompassed with fields of corn, and plantations of fruit-trees'. The conclusion, that a balance of Rest and Labour is needed, means the prospect should contain not only the 'crouded storehouses' of Labour, but the 'artificial grottos with cascades' of Rest. *Rambler* 67, 'The garden of Hope', is more concerned for the vertical view of the 'craggy, slippery, and winding path' up to Hope, a view reminiscent of the *Vision*. From the gate of Reason, a path – the 'Streight of Difficulty' – can be climbed to the throne of Hope, but the other gate, that of Fancy, has no access, the 'mountain' being 'inaccessibly steep, but so channelled and shaded, that none perceived the impossibility of ascending it'.[52]

The Oriental tales and allegories draw, then, transparently unreal landscapes for the purpose of clear moral instruction. But landscape imagery does fulfil in these cases an important role, and therefore cannot be dismissed. Johnson's allegories were among the most popular of his essays;[53] the prioritization of didactic discourse in the period meant that landscape, to be discussed at all, had to visualize these concerns, and allegory and the Oriental tale were two of the most well-established strategies for achieving this.

Moral to theological language of landscape

The moral language of landscape which Johnson developed in the 1740s was extended in the essays.[54] That such imagery was of great importance to Johnson's didactic purposes cannot be doubted: 'images drawn from nature are more abundant in the essays than any other kind'.[55] As with the allegories, this could visualize Johnson's message, the brevity of references to 'eminences' and the like being compensated for by their frequency and the moral expectations Johnson's consistent usage of topographical language established for the reader.

Moral topography. The Johnsonian language of 'slopes', 'eminences' and 'paths' established in the 1740s was used repeatedly in the 'moral geography'[56] of the essays. As in the Vanity, Johnson analyzed human hope in terms of those who 'have panted for a height of eminence denied to humanity' (R[ambler]. 17, iii. 96). The 'highest eminences of greatness' (R. 29, iii. 160) are subject to sudden subversions by the slights of those who 'look up with envy to the eminences before them' (R. 58, iii. 311). Not only worldly greatness, but also virtue is a 'summit' reached by steps as treacherous as those in the Garden of Hope (R. 70, iv. 4).

The paths of 'truth' are 'narrow' (R. 121, iv. 282) tempting man to seek 'some by-path, or easier acclivity' which 'cannot bring him to the summit, [but] will yet enable him to overlook those with whom he is now contending for eminence' (R. 164, v. 107).[57] Any rise in moral or social eminence is allied with an increased potential to fall: 'how swiftly it may fall down the precipice of falshood' (R. 104, iv. 193).[58] This will even impinge on those who 'walk the road of life with more circumspection, and make no step till they think themselves secure from the hazard of a precipice' (R. 184, v. 203).

Moral hydrology. *Irene* had begun the use of fluvial and oceanic imagery, Abdalla speaking of Reason as 'the tim'rous pilot, that to shun / The rock of life, for ever flies the port' (III, I, 46–47).[59] But this only became 'a recurring image'[60] in the *Rambler* essays.

Johnson elaborated the image of being 'afloat in the stream of time'. This must not be used as an excuse for indolence:

> he that floats lazily down the stream, in persuit of something borne along by the same current, will find himself indeed move forward; but unless he lay his hand to the oar, and increases his speed by his own labour, must be always at the same distance from that which he is following.[61]

The image of the stream of life is used, then, to argue the need for an active moral life in which we avoid being 'the slaves of external circumstances' who 'roll down any torrent of custom' (R. 70, iv. 6).

Riely has argued that Johnson's use of fluvial imagery depicts 'man at the mercy of the elements'.[62] Yet Johnson's imagery encourages resolve, not fatalism; 'to faint or loiter, when only the last efforts are required, is to steer the ship through tempests, and abandon it to the winds in sight of land' (R. 207, v. 313). To give oneself up to chance, in another oceanic image, is 'to put to sea in a storm because some have been driven from a wreck upon the coast to which they were bound' (A[dventurer]. 69, ii. 392). The ultimate answer can only be an active one coupled with humility: 'when he has contended with the tempests of life till his strength fails him, he flies at last to the shelter of religion' (I[dler]. 89, ii. 278).[63]

The flowery landscape of danger. In *Rambler* 2, Johnson spoke of those 'more inclined to pursue a track so smooth and so flowery, than attentively to consider whether it leads to truth'. This established a connection between the flowery and the false, whose 'unaccustomed lustre dazzles' (R. 172, v. 148), which persisted in Johnson's essays. This is connected with the steep path to virtue and truth as its easy alternative – the flowery plain at the bottom of the Mountain of Existence in the *Vision*. The opposition is fully expressed in *Rambler* 151: as we pass into intellectual maturity,

the painted vales of imagination are deserted, and our intellectual activity
is exercised in winding through the labyrinths of fallacy, and toiling with
firm and cautious steps up the narrow tracks of demonstration (v. 40).

Yet many abrogate their intellectual responsibility: 'few that wander in the
wrong way mistake it for the right; they only find it more smooth and
flowery, and indulge their own choice' (R. 155, v. 62).[64]

Riely has argued that 'the large proportion of images expressing the
beauty of growing things (trees, flowers, plants) is evidence that Johnson was
a fond and careful observer of natural scenery'.[65] The images under discussion
are, in fact, so conventional as to reveal nothing of Johnson's observational
powers,[66] so embedded within a transparently moral language as to have
little connection with natural scenery, and so tinged with connotations of
turpitude as to render Johnson's attitude anything but fond.

Moral and theological vision of man. Johnson frequently speaks of the need
to 'survey the moral world' (R. 70, iv. 1), starting with 'a survey of ourselves'
(R. 28, iii. 156). He is more elaborate on the limitations of such moral surveys: 'Johnson's ... periodical essays ... make much of seeing as perception;
but they also, as the aptly-named *Spectator* does not, make much of the difficulties of seeing accurately.'[67] Human eyes are easily deceived by appearances because of the limitations of human (moral) vision. This can be
voluntary: 'as a glass which magnifies objects by the approach of one end to
the eye, lessens them by the application of the other, so vices are extenu-
ated' (R. 28, iii. 153). In this case it can be coped with: 'it is the business of
moralists to detect the frauds of fortune, and to show that she imposes upon
the careless eye ... it shakes off those distinctions which dazzle the gazer'
(R. 58, iii. 311). Mental vision, however, is still limited: 'the eye of the mind,
like that of the body, can only extend its view to new objects, by losing sight
of those which are now before it.' (R. 203, v. 295). Too minute an attention
can lead to an imbalance in our moral life: 'an object, however small in
itself, if placed near to the eye will engross all the rays of light; and a trans-
action, however trivial, swells into importance, when it presses immediately
on our attention' (R. 106, iv. 202).

This limitation is established in contradistinction to another eye and prospect, the *deceptio visus* being a traditional Christian topos.[68] The most important prospect is the 'prospect into futurity' (R. 28, iii. 159), for this 'spreads by
degrees into the boundless regions of eternity' (R. 167, v. 125).[69] Yet our vision
is at its weakest here: our position is better than the 'darkness and uncertainty
through which the heathens were compelled to wander' (R. 29, iii. 158), but,
'to take a view at once distinct and comprehensive of human life ... is beyond
the power of mortal intelligences' (R. 63, iii. 336).[70] We are directed, then, to
remember a 'more extensive comprehension' (A. 128, ii. 481) 'whose eye takes
in the whole of things' (A. 107, ii. 445).[71]

Nature pointing the moral and making moral Points

The natural world was also used in the moral structure of many essays. Usage ranged from acting as a simile for a moral stricture to providing evidence for a moral position being established.

The prospect of nature, at its simplest, could provide an opening into a reflection: 'the tendency to begin with the natural world and suddenly turn to the moral is common'.[72] Thus *Rambler* 169 begins with the reflection that 'natural historians assert, that whatever is formed for long duration arrives slowly to its maturity', and in the second paragraph suggests 'the same observation may be extended to the offspring of the mind'[73] (v. 130). The converse could also occur, Johnson initially giving the moral, and then providing a lengthy description of nature: the best example is his Greenland history (*Rambler* 186 and 187), where the first three paragraphs point to the importance of comparing our situation (in this case in terms of climate) with that of others. Complaints that we do not live 'in the vales of Asia' will give way if we reflect how much we 'owe to providence' that we do not live in Greenland (v. 211–12).[74] The remainder of these two *Ramblers* is the Greenland romance of Ajut and Anningait.

Johnson could also create a more complicated interweaving of moral point and natural parallel. *Rambler* 5, 'A meditation on Spring', opens with a character who could suppress his discontent by postponing his hopes to the spring each year, before moving on (paragraph 6) to the 'immoderate pleasure...of this delightful season'. Paragraph 9 refers back to the human realm, suggesting 'there are men to whom these scenes are able to give no delight', and who seek company at all times in preference to rural beauty. Johnson then reverts to the 'volume of nature', and the instruction we can receive from the landscape. The concluding paragraph, like *Rambler* 111, returns attention to his younger readers by a shift akin to the hermit's in *Rambler* 65: 'the younger part of my readers, to whom I dedicate this vernal speculation, must excuse me for calling upon them, to make use at once of the spring of the year, and the spring of life' (iii. 25–30).[75]

At the other extreme, nature does not merely provide a parallel to man's activity, but is causally connected. In *Rambler* 80, Johnson reflects that 'providence has made the human soul an active being', adding, 'the world seems to have been eminently *adapted* to this disposition of the mind' (iv. 55–56, emphasis added). In this stronger argument, nature does not just allow for *reflection upon* the human moral state, but *reflects* that state. We can gain fresh insight into the human condition by looking to nature.[76] This relates back once more to Johnson's Christianity: 'the depravation of human will was followed by a disorder of the harmony of nature' (I. 89, ii. 275).

The essays, then, see Johnson deploying nature in various ways to make moral points, his range being from piquant parallel to the natural evidences argument of physico-theology. In Johnson's hands the look of the land became a flexible tool: it always served a higher moral purpose, but was not always simply a vehicle for a message.

Portraits, satire and landscape

Johnson's characters in *Vanity* and *Irene* displayed their mental imbalances via some false reading of nature. This general strategy is continued in the essays by Johnson's portraits. The tone, in line with generic norms, is usually lighter,[77] and the follies revealed more trifling, though they occasionally display a vicious streak worthy of Mahomet.

At the most frivolous level, gardens and travel become a site for the display of absurd sensibility. In *Rambler* 34, Anthea is part of 'a small party...viewing a seat and gardens.' Arriving at the gardens, 'Anthea declared that she could not imagine what pleasure we expected from the sight of a few green trees and a little gravel.' She then progresses around the garden in fear of a frog hopping towards her (iii. 184–9). Anthea's character flaw, that she 'mistakes cowardice for elegance, and imagines all delicacy to consist in refusing to be pleased', is a minor one, and yet it is illustrated by reference to landscape.[78]

A similar folly is that of false sensibility, as portrayed in the cits Mercator (*Adventurer* 102) and Drugget (*Idler* 16). Drugget is a trader who 'thought himself grown rich enough to have a lodging in the country, like the mercers on Ludgate-Hill'. The Idler

> found him at Islington, in a room which overlooked the high road, amusing himself with looking through the window, which the clouds of dust would not suffer him to open. He embraced me, told me I was welcome into the country (ii. 50–53).

Mercator, a far wealthier cit, buys a country house and engages in extensive landscaping, though he does not care for the results:

> I ride out to a neighbouring hill in the centre of my estate, from whence all my lands lie in prospect round me; I see nothing that I have not seen before, and return home disappointed.

Mercator looks back to his 'happy days of business' and realizes that he is 'condemned by a foolish endeavour to be happy by imitation' of concerns, aesthetic, social and intellectual, which he does not possess (ii. 435–40).[79]

Excess credulity is part of Drugget's folly. Similar credulity leads Shifter (*Idler* 71, ii. 220–24) 'born in Cheapside', to seek the pastoral delights he has read of. The essay is a series of scenes in which this is exposed:

he saw some reapers and harvest-women at dinner. Here, said he, are the true Arcadians, and advanced courteously towards them, as afraid of confusing them by the dignity of his presence. They acknowledged his superiority by no other token than that of asking him for something to drink.[80]

Rural characters also betray a range of follies: the obsessive housewife, Lady Bustle's garden reflects her ruling error of economizing at the expense of giving her relative a moral education, being 'nothing either more great or elegant, than...the same number of acres cultivated for the market' (R. 51, iii. 273–79).[81] Quisquilius the rural virtuoso goes to the other extreme, allowing 'tenants to pay their rents in butterflies', but betrays a similar mental imbalance encouraged by his rural situation (R. 82, iv. 64–70).

Tom Tranquil, in *Idler* 73 (ii. 227–29), is credulous in a different manner, being surrounded by flatterers and too weak to dismiss them. This again is displayed in a landscape context:

[one] fills his garden with statues which Tranquil wishes away, but dares not remove... another has been for three years digging canals and raising mounts, cutting trees down in one place, and planting them in another, on which Tranquil looks with serene indifference... Another projector tells him that a water-work, like that of Versailles, will complete the beauties of his seat.

The conclusion, Tranquil's bankruptcy, shows how folly can lead to personal ruin, and thus opens a darker side to these landscape-related portraits.[82] The malevolence of Squire Bluster in *Rambler* 142 is another example. He tyrannizes the neighbouring labourers, his enclosed garden symbolizing a sheltered upbringing by his grandmother who taught him (like Lady Bluster) 'all the lower arts of domestick policy', but no compassion (iv. 388–93).[83] Given the Christian framework of the essays, the folly of Prospero is still more troubling:

at the age of fifty-five, has bought an estate, and is now contriving to dispose and cultivate it with uncommon elegance. His great pleasure is to walk among stately trees, and lye musing in the heat of noon under their shade; he is therefore maturely considering how he shall dispose his walks and his groves, and has at last determined to send for the best plans from Italy, and forbear planting till the next season. Thus is life trifled away in preparations to do what can never be done (R. 71, iv. 8–9).

Johnson's purpose is not to satirize landscape taste as such, but to use landscape as a site of human folly and malevolence.[84] Johnson repeatedly turns the satire back onto the satirist and the reader.[85] Amid the everyday follies of credulity, sensibility and vice displayed in the sphere of landscape, Johnson returns us to the question posed by his motto to *Idler* 88 and *Adventurer* 137: 'What have I done today?'

Conclusions

All these strategies by which landscape imagery and language were deployed in the essays were interwoven. The allegories and Oriental tales are distinguished by their length and consistency from the more fragmented use of the moral language of landscape, with paragraph length metaphors straddling the two categories. Equally, the use of the moral language of 'slopes' and 'gulphs' shades into the role of pointing the moral message, just as Johnson's character sketches are simply a different vehicle for general moral points which can use landscape as part of their apparatus.

The result is that the essays are full of references to landscape, but very little is about landscape *per se*. Just as a debate has emerged over whether Johnson's diction generated concrete images or tended towards stately generalization,[86] so Johnson's use of landscape imagery is hard to categorize as relating to actual landscapes or simply metaphorical ones. This in its own right is suggestive of the position of landscape and the natural world for Johnson, and broadens our understanding of what 'talking about' landscape could mean in eighteenth-century England.

The doctrine of landscape

In addition to providing Johnson with an array of terms and images he could use in establishing his views on moral life, landscape and the natural world had to fit into that analysis of human conduct in its failings and ideals. However much, then, the essays may appear to be full of references to, but not about, landscape, they do contain the elements of a consistent approach to landscape, placing it in a moral framework. Landscape, nature and rurality participate in Johnson's view of both moral error and moral striving.

Human error: the delusions of place

One of the recurrent themes in Johnson's essays is that happiness is not to be found in any particular place, and, by extension, that to seek happiness in place is to distort the relationship between mind and place, and undervalue the free will the human condition displays. The theme is prosecuted by deflating the joys of various country idylls.

The pastoral.   Johnson is well known in this context for his attacks on the pastoral.[87] In *Ramblers* 36 and 37 Johnson does not simply denigrate the pastoral, arguing that its images 'have always the power of exciting delight, because the works of nature, from which they are drawn, have always the same order and beauty' (R. 36, iii. 196). The objection is to the delusion which the pastoral produces. Thousands of pastoral pieces are not 'enlarged with a single view of nature not produced before', nor 'any new application of those views to moral purposes' (R. 36, iii. 197). Johnson implies that we voluntarily participate in the delusion: 'we readily set open the heart, for the admission of its images' (R. 36, iii. 195), not because we believe them to be true, but because we wish them to be.[88] It is the dream of a place where the struggle to forge a moral and (thus) happy life is no longer needed which Johnson opposes.

The pastoral has some specious confirmation in the rural prospect. The urbanite Euphelia (R. 42, iii. 227–31), who 'never went to bed without dreaming of groves, and meadows, and frisking lambs', at first sees the pastoral she has heard of inscribed in the landscape. 'A few days brought me to a large old house, encompassed on three sides with woody hills, and looking from the front on a gentle river, the sight of which renewed all my expectations of pleasure.'[89] Going down into the landscape, she cannot sustain the delusion, confessing with absurd artlessness, 'I have tried to sleep by a brook, but find its murmurs ineffectual.' No matter how much we wish the pastoral image to be true, it no more captures human than landscape reality.

Fashionable retirement.   Euphelia also belongs to another category Johnson repeatedly ridiculed, those who retire in the summer to the countryside.[90] Euphelia found that when she had exhausted the novelty of the countryside, she 'had not in myself any fund of satisfaction'. Johnson generalized the message in *Rambler* 124. The fashionable in solitude 'must learn, however unwillingly, to endure themselves' (iv. 296).

It is not retirement *per se* which Johnson undermined, but fashionable retirement by those who could gain nothing from rural retreat. The wealthy, as in the pastoral, 'are seduced into solitude merely by the authority of great names', historical examples of 'statesmen and conquerors' substituting for the literary ideal (A. 126, ii. 471–76). Yet modish retirement does 'not appear either to find or seek any thing which is not equally afforded by the town and country'. The delusion here tackled is one which substitutes place for mind, denying the centrality of rational life to human happiness. 'They [who] purpose nothing more than to quit one scene of idleness for another' can only recreate 'the tediousness of unideal vacancy' in another location (R. 135, iv. 349–54).[91]

Gardens.   Another error was that of viewing the Edenic garden of felicity as an attainable earthly ideal. Johnson's picture of gardens was designed to undercut such an image, portraying them in three ways.

First, the garden is a site of affectation. Tetrica, like Anthea (R. 34) endeavours 'to force respect by haughtiness'. She can display this by excess sensibility – 'if she takes the air, she is offended with the heat or cold, the glare of the sun, or the gloom of the clouds' – or by affected aesthetics – 'she quarrelled with one family, because she had an unpleasant view from their windows' (R. 74, iv. 26–27).[92] Eriphile, whose solitude has developed into a 'spiteful superintendance of domestic trifles', keeps her garden beautiful to the eye, but this betrays a folly far from innocent or Edenic.

> She lives for no other purpose but to preserve the neatness of a house and gardens, and feels neither inclination to pleasure, nor aspiration after virtue, while she is engrossed by the great employment of keeping gravel from grass (R. 112, iv. 235).

Second, the garden is a display of wealth. In *Adventurer* 119 (ii. 464), Johnson speaks of the restlessness of the prosperous:

> one man is beggering his posterity to build a house, which when furnished he never will inhabit; another is levelling mountains to open a prospect, which, when he has once enjoyed it, he can enjoy no more.

The reason why such 'splendor and elegance' are esteemed is 'principally as evidences of wealth'.[93] Those like Drugget who seek wealth also dream of displaying it in the purchase and improvement of a garden.[94]

As *Idler* 30 states, it is not only money but time which is a burden. The third image of the garden, then, is as a novelty to relieve the tedium of existence. As such, landscaping figures in the quests of a number of Johnson's characters to avoid boredom. In an oriental setting, Almamoulin at one time 'laid out gardens', and 'opened prospects', but whilst 'these amusements pleased him for a time ... langour and weariness soon invaded him' (R. 120, iv. 278).[95]

This desire to display wealth can lead to disaster: 'such as ruin their fortunes by expensive schemes of buildings and gardens' do so because 'they carry on with the same vanity that prompted them to begin' (R. 53, iii. 285). This point is personified by Bob Cornice's route to the Fleet prison. Cornice became reputed for his house and gardens, alterations being made every day 'without any other motive than the charms of novelty'. He was building a grotto 'when two gentlemen who had asked permission to see his garden, presented him a writ and led him off to less elegant apartments' (A. 53, ii. 370).[96]

Travel. As with retirement, Johnson argued that travel is pointless if the mind is not ready: in *Idler* 97 Johnson castigates those content to 'gratify ... [the] eye with variety of landscapes'. The vacuous traveller may 'conduct his readers thro' wet and dry, over rough and smooth', but will do so 'without incidents, without reflection' (ii. 298–99). The result for the traveller is not a broadening of the mind, but at best no change, and at worst the boorishness of the seaman in *Rambler* 197 and 198, who is contemptuous of those who have not travelled, but has not extended his view of humanity by his travels.

The description and practice of travel repeat the follies exhibited in gardening. Travel accounts are exaggerated to display the delicacy of sensibility and to impress others: Will Marvel is the most famous example of this 'ambition of superior sensibility' (*Idler* 50, ii. 157).[97] Johnson explicitly links this exaggeration to vanity in *Adventurer* 50, 'On Lying' (ii. 364).

Travel can also relate, as gardening does, to the vanity of conspicuous expenditure. In *Rambler* 115, Hymeneaus is courting the aptly named Charybdis,[98] who proceeds to suck all his money from him as they travel:

> she soon after hinted her intention to take a ramble for a fortnight, into a part of the kingdom which she had never seen . . . She had no other curiosity in her journey, than after all possible means of expence (iv. 251).

Johnson's rural prospect.   The final deflation of errors about place is of Johnson's own image in the late 1730s of an innocent countryside. The leaders of rural society are not pleasant characters in the essays. The general point is that 'money, in whatever hands, will confer power', and rural areas are no exception. Although Johnson does attack the invasion of the countryside by nabobs in Mercator's character,[99] he is not arguing that there has been a decline in rural values due to this. 'There is no suggestion in Johnson's fiction that British society used to be better than it now is.'[100] In stark contrast to Thales's historical geography of rural purity is *Rambler* 161's defence of the history of a garret:

> the power or wealth of the present inhabitants of a country cannot be much increased by an enquiry after the names of those barbarians, who destroyed one another twenty centuries ago, in contests for the shelter of woods or convenience of pasturage (v. 90).

Blame for rural malevolence is frequently ascribed to the misunderstanding of economics,[101] but Johnson is not advocating a romantic ignorance of financial affairs: he is no more flattering to profligate heirs who bankrupt their estates.[102]

The pastimes of the rural wealthy are also attacked: Johnson's response to gardening and travel has already been discussed, and to this can be added his view of hunting. Mercator speaks for Johnson: 'I could discover no music in the cry of the dogs, nor could divest myself of pity for the animal whose peaceful and inoffensive life was sacrificed to our sport' (A. 102, ii. 439).[103]

Yet Johnson's is not an attack on the squirarchy. His picture of the lower ranks in rural society is equally damaging. The credulity of Shifter in *Idler* 71 is matched only by the malevolence of the rural labourers he encounters: the innkeeper sells goods at a higher price than in London, the messenger is dilatory and avaricious. One farmer threatens to indict him for trespass, another sells him a blind horse.[104]

Johnson's position has all but reversed from *London*. Where urban emulation was deceit, it is now a way in which 'the peculiarities of temper and opinion are gradually worn away'. Where rural simplicity was honesty, it is now pernicious:

> in the country every man is a separate and independent being...Every one indulges the full enjoyment of his own choice...without...considering others as entitled to any account of his sentiments or actions (R. 138, iv. 365 66).

Conclusion: Johnson's scepticism. The analysis presented suggests that Macaulay's famous judgement that Johnson 'had studied, not the genus man, but the species Londoner...[H]is philosophy stopped at the first turn-pike-gate'[105] is inaccurate, but reflects a certain truth. Johnson was sceptical about the pleasures of the landscape, but this was because he did study the motivations of the genus man, and realized they only altered their expression, not their existence, beyond the turnpike-gate. Johnson most certainly was, however, sceptical of the pleasures of landscape, such that to reverse Macaulay's position on Johnson's aesthetics[106] does not improve the situation. Johnsonian scepticism is anything but blind: it is integrated into two of his great moral themes, the vanity of human wishes and the freedom of the mind.

Human striving: place and the moral life

Johnson did not stop at scepticism. Once the delusions of place had been removed, landscape and the natural world could be reintegrated into the moral lives of free actors. Some introduction to Johnson's view of place in the moral life can be found in *Adventurer* 126: while attacking the follies of fashionable retirement, acceptable situations for retreat are also suggested. Solitude can be the parent of philosophy, for 'some studies require a continued prosecution of the same train of thought, such as is too often interrupted by the petty avocations of common life'.[107] Another group whose retirement 'intitles them to higher respect' is those who seek to 'employ more time in the duties of religion'. Johnson's view of place in the moral life revolves around three categories suggested by this positive assessment: religious knowledge, knowledge and self-knowledge.

Landscape, nature and religious knowledge. The appearance and structure of nature had a role to play in the proof of God's existence in the essays. In *Rambler* 83:

of all natural bodies it must be generally confessed, that they exhibit evidences of infinite wisdom, bear their testimony to the supreme reason, and excite in the mind new raptures of gratitude, and new incentives to piety (iv. 73).[108]

The natural world also provided evidence for the chain of being,

which might extend the sight of the philosopher to new ranges of existence, and charm him at one time with the unbounded extent of the material creation, and at another with the endless subordination of animal life (R. 9, iii. 49).

That Johnson's view of the natural world was both Christian and hierarchical cannot be doubted: it is a 'pyramid of subordination' (R. 145, v. 9). *Idler* 52 argues that man's superior intellectual faculties allow him to 'forget the wants and desires of animal life for rational disquisitions or pious contemplations' (ii. 161).[109] Evidence that the look of nature can provide proof of the Mosaic chronology is less obvious, but Johnson does argue that 'the depravation of human will was followed by a disorder of the harmony of nature' (I. 89, ii. 275).

The essays regulate the use of nature as a proof of God's existence. Faith is essentially non-sensual and above visualization. If in *Idler* 89 the disorder of nature reflects man's fall, it is also the case that 'the supreme being is invisible' and it is dangerous to 'chain down the mind...to the present scene'. The result is that proofs of God from nature, whilst deployed in the essays, are rare when compared with the practice of Latitudinarian writers such as Addison: Johnson's emphasis was on 'external' evidences of the truth of Scripture.[110]

Contradicting *Idler* 89, Johnson argues that

if the extent of the human view could comprehend the whole frame of the universe, I believe it would be found invariably true, that Providence has given that in the greatest plenty, which the condition of life makes of greatest use; and that nothing is penuriously imparted or placed far from the reach of man, of which a more liberal distribution, or more easy acquisition would increase real and rational felicity (I. 37, ii. 115).[111]

Yet we cannot see this harmony, only assert its existence as 'the works and operations of nature are too great in their extent...to be reduced to any determinate idea' (R. 125, iv. 300). Given the limited scope of our view, any design argument will be similarly limited: for Johnson, unlike Butler, this was not a comforting thought. Any argument for design from the appearance of nature could be as mistaken as that made by the vulture in Johnson's original *Idler* 22: the vulture says they would rarely eat the flesh of men, 'had not nature, that devoted him to our use, infused into him a strange ferocity'. The opinion of another vulture is that 'men had only the appearance of animal life, being really vegetables with a power of motion' (ii. 317–20). Man's misreading of the message of nature, which Johnson dealt with in *Irene*, again came in for scrutiny. In *Adventurer* 50, 'On Lying', Johnson referred to those 'to whom portents and prodigies are of daily occurrence; and for whom nature is hourly working wonders invisible to every other eye' (ii. 364).[112] Johnson opposed the notion that calamities were the result of God's providential intervention in the operation of nature: 'they are no particular marks of divine displeasure' (A. 120, ii. 470).[113] Even in the humourous *Idler* 11 on the weather, Johnson attacked those who believed in its causal effects as succumbing to 'the idolatry of folly'.

The appearance of nature can lead us to religious knowledge, but not a fully developed faith. The 'tendency among orthodox writers such as Johnson to withdraw God from any direct involvement in the world'[114] did not desanctify the appearance of nature, but gave it a more limited role than did the Latitudinarian practice of landscape description.

Progress, nature and knowledge. Johnson in the essays was in no doubt that geographical knowledge had increased, enhancing the capacity to control nature. Speaking of the cumulative effect of human effort, *Rambler* 43 refers admiringly to 'those petty operations, incessantly continued, [which] in time surmount the greatest difficulties...mountains are levelled, and oceans bounded, by the slender force of human beings' (iii. 235).[115] Control over nature was positively evaluated, as 'the natural progress of the works of men...from rudeness to convenience' (I. 63, ii. 196). Johnson's hostility to primitivism is well known,[116] and emerges in his receptivity to knowledge of nature's operations. The progress spoken of in *Idler* 63 is one in which we learn how to avoid being 'incommoded by heat and cold, by rain and wind'; by this 'the mind is set free from the importunities of natural want'.

This progress was driven by the mind's insatiable curiosity: 'science, though perhaps the nursling of interest, was the daughter of curiosity'. This 'desire of knowledge' means 'we do not see a thicket but with some temptation to enter it, nor remark an insect flying before us but with an inclination to persue it'.[117] However useful natural knowledge may become, it is curiosity which is the motive force behind its acquisition, 'for who can believe that they who first watched the course of the stars, foresaw the use of their discoveries to the facilitation of commerce, or the mensuration of time?' (R. 103, iv. 185–6)[118] Finally, this curiosity is itself related back to its source: 'I cannot but consider this necessity of searching on every side for matter on which attention may be employed, as a strong proof of the superior and celestial nature of the soul of man' (R. 41, iii. 221–22).

A similar nexus of ideas supports Johnson's views of the knowledge attainable through travel.[119] *Idler* 97, as we have seen, castigated those who travel 'without incidents, without reflection', but it also commends those who understand that 'the great object of remark is human life'. His notion of human life is a broad one, and includes the natural knowledge of foreign lands: 'every nation has something peculiar in its manufactures...its medicines, its agriculture', and the useful traveller brings home such knowledge (ii. 300). That the origin of instructive travel is curiosity is made clear by Johnson's interpretation of Acastus's reason for joining the Argonauts:

> Acastus was soon prevailed upon by his curiosity to set rocks and hardships at defiance, and commit his life to the winds; and the same motives have in all ages had the same effect upon those whom the desire of fame or wisdom has distinguished from the lower orders of mankind (R. 150, v. 35).

The serious traveller can also be spurred to moral reflections on human transience, as in *Rambler* 165 and *Idler* 43, where travellers return to their birthplaces to find their friends dead or forgetful of their existence.[120]

Johnson's view of the worth of natural knowledge is qualified by the role it plays in his view of moral life. Natural knowledge is better than idleness:[121] if man is to overcome his 'habitual drowsiness', 'he must, in opposition to the Stoick precept, teach his desires to fix upon external things' (R. 89, iv. 107).[122] Yet this must be subordinate to moral principles.[123] It is this which gives Johnson his sceptical tone on natural knowledge:

> if, instead of wandering after the meteors of philosophy...the candidates of learning fixed their eyes upon the permanent lustre of moral and religious truth, they would find a more certain direction to happiness (R. 180, v. 186).

These two points may appear to be in conflict, but are linked by the dynamic nature of Johnson's moral thought: 'innocent gratifications must be sometimes with-held; he that complies with all lawful desires will certainly lose his empire over himself' (I. 52, ii. 163). This general point is applied specifically to natural knowledge in *Rambler* 24: Socrates is praised for having turned the Greeks

> from the vain persuit of natural philosophy to moral inquiries, and turned their thoughts from the stars and tides, and matter and motion, upon the various modes of virtue, and relations of life (iii. 132).

This message is then personified in Gelidus, who, in the course of his studies, 'has totally divested himself of all human sensations', his focus on physical nature leaving him 'unmoved by the loudest call of social nature' (iii. 133). Johnson habitually illustrates the danger of focusing on trifles with the example of the pursuit of natural knowledge: 'when we examine a mite with a glass, we see nothing but a mite' (R. 112, iv. 236).[124] Even Quisquilius the virtuoso describes himself, in a comical reminder of his loss of perspective, as 'an enemy to trifles' (R. 82). The need to balance progess in natural knowledge with roundedness in moral life set the bounds of Johnson's view.

Nature, self-knowledge and society. *Rambler* 7 discussed the worth of retirement for religious purposes: 'it is necessary that we weaken the temptations of the world, by retiring at certain seasons from it' (iii. 40). Here is the third way in which place can contribute to the moral life: it can provide a changed environment to facilitate self-assessment. Man may

> hope, by retirement and prayer, the natural and religious means of strengthening his conviction, to impress upon his mind such a sense of the divine presence, as may overpower the blandishments of secular delights (R. 110, iv. 225).

That Johnson assigned such a position to retirement was unusual, and marks one of his points of departure from the period's views on the uses of place: William Law and Johnson 'stand alone ... among Protestant Englishmen of the eighteenth century, as even qualified endorsers of the monastic ideal, and of ascetic practices'.[125] The retirement Johnson advocated was connected with his distrust of the senses. The contemplation retreat was not based on natural evidences for God, but the reverse: it puts us 'in such a situation by retirement and abstraction, as may weaken the influence of external objects' (R. 28, iii. 156).

The natural and self-knowledge gained by rural retreat for Johnson must be reintegrated into society to be of value. The portrait of Vivaculus (*Rambler* 177) is important here: his retirement is exemplary, having a large collection of books and broad interests. Yet he finally goes back into society, finding his 'mind contracted and stiffened by solitude' (v. 169). Individually, we 'may corrupt our hearts in the most recluse solitude, with more pernicious and tyranical appetites and wishes, than the commerce of the world will generally produce' (R. 8, iii. 43), just as rural characters can develop greater absurdities than urbanites. Socially, knowledge must be reintegrated for the common good.[126] For this reason 'the solitary philosopher' becomes an icon of maladjustment to the realities of the world: 'he is at last called back to life by nature or by custom, and enters peevish into society, because he cannot model it to his own will' (R. 89, iv. 106). The scepticism Johnson displayed to monasticism[127] is based upon the same complaint:

> by debaring themselves from evil, they have rescinded many opportunities of good; they have too often sunk into inactivity and uselessness; and

though they have forborn to injure society, have not fully paid their contributions to its happiness (R. 131, iv. 335).

Conclusion: coelum, non animum mutant. Johnson's positive views on the role of natural knowledge, landscape and place combine with his deflation of myths of happiness residing in certain places to form a coherent effort to recentre the relationship between mind and place towards the former. Johnson cited Horace's *Epistles* in *Rambler* 135: *Coelum, non animum mutant*, which he translated as 'place may be chang'd, but who can change his mind?'[128] The mind overwhelms any concern for place, such that the effects of human passions – the malevolence of squires, the boorishness of 'swains', the tedium of a vacant mind in the country – cannot be overcome by a change of place, except for the briefest span of time. The inefficacy of place, landscape and idyllic dreams in the face of the passions of the mind is summarized in *Rambler* 6, on Cowley's dream of retiring to America.

The general remedy of those uneasy without knowing the cause, is change of place; they are willing to imagine that their pain is the consequence of some local inconvenience, and endeavour to fly from it, as children from their shadows (iii. 32).

Cowley 'never suspected that the cause of his unhappiness was within, that his own passions were not sufficiently regulated' (iii. 35). The essays, by contrast, never allow the reader to suspect otherwise.

'The foundation of content must spring up in the mind' (R. 6). It is for this reason, Johnson's stress on human free will, that any form of environmental determinism is attacked so strongly.[129]

They who believe that nature has so capriciously distributed understanding, have surely no claim to the honour of serious confutation. The inhabitants of the same country have opposite characters in different ages (R. 122, iv. 289).

The farcical environmental explanations of the rage for writing (*Adventurer* 115) and the theory of the garret (*Rambler* 117) show Johnson turning to satire to rebut what he considered a pernicious doctrine. Environmental determinism was the theorization of the errors of retirement and the like, all of which involved the resignation of our responsibility for our actions, and an imbalance in the relationship between mind and place.

Johnson's views on landscape and the natural world as exhibited in his essays do not perhaps appear to be what the twentieth century would understand as discussions of these topics at all.

The consistency with which Johnson maintains priorities, and subordinates lesser values to greater, is one of his distinctive strengths as a critic and moralist. To understand his opinions on any one issue, we must understand its relative importance within the largest context.[130]

To present a contextual reading of Johnson's attitudes to landscape and nature alone, then, would be a contradiction in terms, for his aim was always to demand that any interest we had in the natural world was itself contextualized in a moral and Christian framework. For Johnson, the importance of place, landscape and nature in our lives had been exaggerated and fantasized: as such, he had to explode the myths about place and reintegrate what remained into a broader vision. For the delusions of place, Johnson substituted a Christian moral doctrine of mind and place, which, ultimately, focused the reader's mind on 'a change not only of place, but the manner of his being' (R. 78, iv. 4/).

Contexts for Johnson's moral doctrine of landscape

Periodical essays and landscape 1710–60

Johnson's essays have been seen as less interested in issues pertaining to landscape and nature than either the *Spectator* and its early eighteenth century imitators, or the 'lighter' satirical periodicals of the middle decades of the century, such as the *World* and *Connoisseur*. Boswell's belief that the *Spectator* and the *Tatler* were the last of their kind,[131] coupled with Johnson's own argument that the *Connoisseur* and *World* 'wanted matter'[132] has led to a belief in the singularity of Johnson's essay papers in the mid-eighteenth century.[133] Certainly, Johnson's moral doctrine of landscape as outlined, appears far removed from most present-day discussions of what it meant to write about landscape in eighteenth-century England. How did Johnson's use of landscape and nature in his essays relate to the context of that 'rather strange literary form, peculiar to the eighteenth century' the periodical essay?[134]

Landscape and nature in early eighteenth-century periodicals

Nature and God. A vital distinction between the early periodicals associated with Addison and Steele, and those of Johnson is that (as we saw in Chapter 4) 'Addison's major proof for divine existence was based on natural data.'[135] A comparison of Johnson's 'vernal reflection', *Rambler* 5, with Addison's in *Spectator* 393 shows the differing attitudes to nature. Johnson repeatedly moved from the natural to the moral and back again, never allowing description of natural beauties to take centre stage, whereas Addison, whilst wishing to 'moralise this natural Pleasure of the Soul',[136] had longer descriptive passages, allowing the natural pleasures of spring landscapes a greater role. Moreover, where Johnson's main focus was on man's moral state, only once referring to the beauties of spring as a proof of God's existence (paragraph 15), Addison continually referred back to the Creator.[137]

The argument as deployed, particularly by Addison, differed from Johnson in its emphasis on the beauty of nature as an evidence of God. The natural-aesthetic pleasure of the soul in *Spectator* 393 feeds into the theological:

> the Chearfulness of Heart which springs up in us from the Survey of Nature's Works is an admirable Preparation for Gratitude...Such an habitual Disposition of Mind consecrates every Field and Wood, turns an ordinary Walk into a Morning or Evening Sacrifice.[138]

Spectators 387 and 393 were among Addison's religious Saturday papers, and in theme they link thematically with his series on 'The Pleasures of the Imagination' (numbers 411–21). These papers connect God, nature and aesthetics in the same manner, the argument being particularly apparent in number 413, on the final causes of aesthetic pleasure.

The fact that nature and landscape were aesthetically pleasing, and that this as well as their mere existence was a proof of God's goodness, allowed early eighteenth-century periodicals to include extended descriptions of nature of a sort which Johnson never did.[139]

Controlling the interpretation of nature.   Given that nature and landscape in early periodicals had a more theological role, their interpretation was a more sensitive issue than it was to be in Johnson's essays. Addison makes the assertion that nature is not politicized: 'there is no party concerned in speculation of this nature' (*Guardian* 160).[140] Yet as we have seen in Chapter 4, notably in Addison's *Freeholder* essays, nature and therefore landscape clearly participated in denominational and political debate.

Observing nature was seen as destroying the arguments of freethinkers. 'The Atheist has not found his Post tenable, and is therefore retired into Deism, and a Disbelief of revealed Religion only' (Sp. 186).[141] Berkeley in the *Guardian* repeatedly challenged (in a language familiar from Johnson's essays) the 'short views' of Freethinkers. This argument is elaborated using landscape imagery in numbers 39 and 62, reaching its apogee with the author's discussion of the fly on the pillar which was co-opted to Gilpin's picturesque (see above, Chapter 4).[142]

Akin to the myopia of those who could not see the harmony of nature was the error of those who saw omens in nature. Johnson did, in *Idler* 10, satirize the Jacobite Tom Tempest who 'can recount the prodigies that have appeared in the sky, and the calamities that have afflicted the nation every year from the Revolution' (ii. 34), but for periodicals earlier in the century, the problem was more serious. While 'the great and memorable Darkness which happened at the time of the Crucifixion of our Saviour was preternatural',[143] Catholics and Jacobites were seen as superstitiously invoking omens in nature in the present era. Both groups were particularly attacked for superstition by the Whig *Freethinker*.[144] *Freethinker* 16 argued that Protestants should avoid the 'Follies and Dotages' of 'the Papists',[145] by learning 'some of the most familiar and evident Truths in Natural Philosophy, particularly concerning Meteors, and the Sun, Moon, and Stars, which so much astonish and terrify the Vulgar...'[146] A similar '*Preparation* of Philosophy' is recommended in number 46 against the Jacobites:

> we have seen, within these three years [written in 1718; i.e. since the 1715 Jacobite uprising], several remarkable Lights in the Air; one very extraordinary and total Eclipse of the Sun; and are this Night to see the silver Light of the Moon stained as it were with Blood, and wholly overshadowed by the Earth; and yet the course of Nature and our political Concerns go on, as they went before; the Seasons of the Year continue their usual Vicissitudes; the Protestant Succession remains in full Force.[147]

There was no removal of the Providentialist mentality, but a demand that it be Protestant and Hanoverian:[148] thus the *Freethinker*, speaking of the English Reformation could opine that

> the Messenger, who carried the King's Letters of Submission to that *See* [ot Rome], was then actually upon the Road; and hindered by an accidental (I may say Providential) Severity in the Weather, from arriving at *Rome*.[149]

Periodicals at this time attacked superstition about the natural world, from the 'unintelligible Cant' of the Rosicrucian who 'jumbled natural and moral Ideas together' (*Spectator* 574) to the 'impudent Mountebank who sold Pills which (as he told the Country People) were very good against an Earthquake'.[150] Johnson's essays, by encouraging a withdrawal of attention from natural to scriptural evidence of God,[151] betray little of the factional tensions of their predecessors. In Johnson, the Anglican *via media* is assumed, and the politics of prognostics is replaced by a more general ridicule of omens, Tom Tempest's Jacobite being no worse than Jack Sneaker's Whig providentialism in *Idler* 10,[152] both serving as examples of moral myopia.

Mind and Place. The Addisonian view of the role of place in our mental and moral lives differs from the Johnsonian by according it greater significance. Where for Johnson retirement into the country, if successful, is built upon turning our view in upon ourselves, for Addison the mental eye is spiritually nourished by the inner *and* outer view:

> in Court and Cities we are entertained with the Works of Men, in the Country with those of God...Faith and Devotion naturally grow in the

Mind of every reasonable Man, who sees the Impressions of Divine Power and Wisdom in every Object on which he casts his Eye (*Spectator* 465).[153]

Addison's belief in such a retirement is personified in Sir Andrew Freeport's departure from the city in *Spectator* 549.

Of greater interest in the light of Johnson's doctrine relating mind to place is Richard Steele's position. For Steele, rural retirement is often a sham:

it has been from Age to Age an Affectation to love the Pleasure of Solitude, among those who cannot possibly be supposed qualified for passing Life in that Manner (*Spectator* 264).

He quotes (as Johnson would) Horace's *coelum, non animum mutant* in *Spectator* 80, and provides his own version of the thought in *Tatler* 93: 'Men may change their Climate, but they cannot their Nature.'[154] Addison opened his account of Freeport's retirement with the thought that 'most People begin the World with a Resolution to withdraw from it into a serious kind of Solitude or Retirement', and then gave an example of its success. Steele's parallel reflection in *Spectator* 27, 'there is scarce a thinking Man of the World... but lives under a secret Impatience... Retirement is what they want' leads to the less flattering but more Johnsonian conclusion:[155]

the same Passions will attend us where-ever we are, till they are Conquer'd, and we can never live to our Satisfaction in the deepest Retirement, unless we are capable of living so in some measure amidst the Noise and Business of the World.[156]

The two essayists cannot be wholly separated: Steele is more positive about rural retreat in *Spectator* 406, and Addison expresses scepticism about place at some points in *Spectator* 15. But, in general, Steele's position was far more sceptical, and established a precedent for the position Johnson's essays were to take.

The language of landscape, 1710–60. The language of landscape applied for moral purposes is found throughout the period, and connects the earlier essays with those written at mid-century. The imagery analyzed for Johnson's essays was common, with references to the journey[157] and voyage[158] of life legion.

Physical and moral vision were conflated: Berkeley's point about the short view of freethinkers could be generalized to include the short view of man compared with the whole system of nature, and with the view he will have in the afterlife.[159] The overall impression, however, is that this language is used less frequently than in Johnsonian essays, and tends to be used in extended analogies, rather than as an habitual way of characterizing moral issues.

Visions, Oriental tales and allegorical topographies codified and extended to full essay length the language of moral visions, surveys and prospects. Best known is Addison's 'Vision of Mirzah' (*Spectator* 159).[160] Such allegorical topographies continued to be a regular part of the periodical repertoire until at least 1760, Goldsmith's *Citizen of the World* 31 being a symbolic garden with two paths of virtue and vice.[161] All such allegories had classical sources, notably in Prodicus's *Choice of Hercules* and Cebes's *Picture of Human Life*.[162]

The lighter tone of landscape instruction, 1710–60

Satire in the early eighteenth-century periodical essay. The satirical attacks on affectation with respect to landscape taste which Johnson made formed part of a tradition in periodical essays. Pope's *Guardian* 173 is well known: comparing the taste of the ancients, as shown in Homer's Garden of Alcinous, with that of the present day, he makes a list of the absurd topiary creatures of 'a Virtuoso Gardener'. The paper also satirized the cit's taste, which was to become a standard theme: 'I know an eminent cook, who beautified his country seat with a coronation dinner in greens.'[163] Pope also included a humourous hint of a connection between impiety and false taste in gardening:

he [the virtuoso] is a Puritan Wag, and never fails, when he shows his Gardens, to repeat that Passage in the Psalms, *Thy Wife shall be as the fruitful Vine, and thy Children as Olive Branches round thy Table.*

Satire on travel as ostentation also began in the early eighteenth-century essays. Johnson's attack on false travel accounts follows a pattern established by Addison's *Tatler* 254, which, under the motto *Splendide Mendax* – Bravely False (Horace, *Odes* 3, 11, 35) – talked of travel in remote countries giving 'the Writer an Opportunity of showing his Parts without incurring any Danger of being examined or contradicted'. These attacks had their serious side, warning of the dangers of sending the young and impressionable to Catholic countries, whose rituals and pomp are *Splendide Mendax*,[164] a point made repeatedly in the *Freethinker*'s 'unconcealed tirades against alleged Catholic practices'.[165] The satire also counterpointed the belief Johnson displayed in the worth of instructive travel: Steele's satire in *Spectator* 364 then turned to exemplars of 'the true End of visiting forreign Parts', 'to look into their Customs and Policies'.[166]

Rural superstitions and the absurdities of rural life were satirized in a way akin to Johnson's.[167] The *Freethinker* attacked the Sir Roger DeCoverly type of rural tory as

a distinct Race of Animals, a Breed of Creatures resembling Men, not to be found (as I am told by Travellers) out of this Island; and, methinks, they are justly supposed, by our Philosophers, to have a nearer Affinity to their Dogs than to the human Species (number 134).[168]

Satire on virtuosi also saw them as a distinct race, whose 'Speculations do not so much tend to open and enlarge the Mind, as to contract and fix it upon Trifles' (Tat. 236). Addison, no more wishing to discourage the acquisition of knowledge than Johnson, however 'would not have a Scholar wholly unacquainted with these Secrets and Curiosities of Nature' (Tat. 216).[169]

Later periodicals. Periodicals of the mid-eighteenth century have been seen as 'adopting the role of witty and ironical commentator upon contemporary foibles and fashions'.[170] The *Connoisseur* and the *World* certainly do have a greater bulk of satire on gardens and the like than their predecessors. Best known is *The World*, number 15, which provides a political history of gardening, followed by 'proofs of the degeneracy of our national taste' in Squire Mushroom's villa. Number 59 plays on the folly of gentlemen building follies, and number 65 on wooden structures in gardens, which the writer mistakes for Diogenes's tub. Number 76 satirizes both the improver, who must change all to make money, and his critic: the affectation of both sides leads to the ironic suggestion that the improver

> upon the arrival of his VISITORS, take care to purge their visual nerves with a sufficient quantity of CHAMPAIGN; after which ... they never SEE a fault in his IMPROVEMENTS.

The *World* also attacks travel, or 'that migrating distemper' (number 18). A humourous portrait of travel in number 93 encapsulates the tone of the *World*: 'he never mentioned France, but to condemn the post horses; nor took notice of any circumstance in his passage over the Alps, except the loss of his hat and perriwig'.

This lighter tone, however, did not free issues of landscape and nature from the sphere of moral discourse. The attack on foppish travel was based on motives familiar from the earlier periodicals. Travel should be productive of knowledge of men and manners: indeed, the eidolons of the *World* (Adam Fitz-Adam) and the *Connoisseur* (Mr Town) are both well travelled, and it is this experience which allows them to satirize others.[171] The *World* also points to the religious dangers of travel. The advice on how to write a bogus travel account in number 107 demonstrates a Swiftian mode of defending Anglican orthodoxy:

> in treating of the Indian manners and customs, you may make a long chapter of their conjuring, their idolatrous ceremonies, and superstitions; which will give you a fair opportunity of saying something smart on the religion of your own country.[172]

The satire on gardening for which the *World* and *Connoisseur* have become best known has to be juxtaposed with the purpose of the satire, which was demonstrated by the moral ideal of gardens opposed to the fashionable reality. The Squire Mushroom portrait in *World* 15 was followed in the next number by a portrait of a blissfully married parson which recalls Alworthy in *Tom Jones*:

in the seven years of their retirement, they have so planted their little spot, that you can hardly conceive any thing more beautiful... The produce of these fields supplies them abundantly with the means of bread and beer, and with a surplus yearly for the poor, to whom they are the best benefactors of any in the neighbourhood.

Similarly, the moral of the satire of Diogenes's tub in number 65 is one Steele and Johnson made about the relation of mind and place:

he would hardly imagine that even the most elegant palaces could add any degree of worth to the possessor, whose character must be raised and sustained by his own dignity, wisdom and hospitality; remembering the maxim of Tully, *non domo dominus, sed domino domus honestanda est.* [The owner should not be embellished by the house, but the house by the owner.][173]

The affectation of modern gardens posed moral threats. Gardens were persistently seen as sites of sexual licentiousness by mid-century essayists (but not by Johnson). Goldsmith's Chinese eidolon, Altangi, found Vauxhall gardens 'falls [little] short of *Mahomet's Paradise!*', traditionally associated with heathen sexuality in eighteenth-century thought.[174] Vauxhall, for *World* 63, provided for the body not the mind, in contrast to the owner Tyers's private garden, the Denbies:

I have heard that the master of Vaux-hall, who so plentifully provides beef for BODILY refreshment, has, for the entertainment of those who visit him at his country-house, no less plentifully provided for the MIND; where the guest may call for a scull to chew upon the instability of human life... I wish that this grand purveyor of beef and pastry would transfer some of the latter to his gardens at Vaux-hall.[175]

The dangerous sexuality associated with London gardens[176] was related back to idolatry: 'these persons of Taste may be considered as a sort of learned idolators' who have 'introduced the Heathen Mythology into our gardens'.[177]

Both the *World* (number 1) and *Connoisseur* (number 71) justified their satirical approach in moral terms, defending Christian morality by ridiculing that which opposed it. Moreover, both could drift into more serious papers relating to design arguments,[178] the relation of natural evidences to other proofs of God,[179] and the providentialist interpretation of earthquakes.[180]

Conclusion: Johnson in the periodical essay context: changing strategies and Christian motives

What has been observed in the evolving discussion of landscape and nature in the periodical essays of the first half of the eighteenth century is a shift in emphasis from landscape as theological evidence towards landscape as a satirical tool, a move parallel to the 'aestheticisation' noted in Latitudinarian approaches to landscape description in Chapter 4. The underlying motive, however, remained. It was, as Johnson put it in his last *Rambler* paper, instruction 'conformable to the precepts of Christianity' (v. 320). When viewed in context, then, at least the motives for Johnson's moral approach to landscape appear less distant from those of the period than standard interpretations may suggest. Johnson's main distinction from the mid-century periodical, that he sought to inculcate precepts 'without any accommodation to the licentiousness and levity of the present age' (v. 320), amounts to a different strategy from both the early- and mid-century periodical. Johnson rarely returns to landscape as a natural evidence of God's existence as the *Spectator* had. Equally, he refused to let his satirical vehicle, be it landscape, gardens or rusticity, escape from the instruction he wished to convey, which frequently occurred in later periodical essays. As such, both the early- and mid-century periodicals had longer descriptions of landscapes, and are in that sense 'more interested' in the subject. Yet the reasons for these descriptions were in no way divorced from Johnson's reasons for lacking them.

Homiletics and landscape, 1660–1760

A question which emerges from the periodical essay context is as to the sources for the language of landscape which Johnson employed more frequently than others, and also the sources of the approach to mind and place in the essays.

Johnson's sermons and landscape

An entry into these questions is provided by an analysis of Johnson's sermons. The parallel between Johnson's essays and sermons, already well established,[181] can be shown to have meaning for a study of landscape: Johnson's language and doctrine of landscape forged in the essays can be found replicated in the sermons.[182]

Johnson's homiletical language of landscape. The use of a physical language of landscape imagery in moral contexts was easily transferable to Johnson's sermons: 'the natural world is . . . a favourite source of imagery'.[183] Johnson speaks of the need for 'a survey of the moral world' (p. 13) and of ourselves, which is the aim of his sermons as of his essays. This is most elaborately rehearsed in Sermon 19 (pp. 206–207) where Johnson speaks of three surveys, all of which confirm the necessity of charity. First, if we look up to heaven, which we have been taught to consider as the particular residence of the Supreme Being, we find there our Creatour... whose infinite power gave us our existence, and who has taught us, by that gift, that bounty is agreeable to his nature.

Secondly, 'if we cast our eyes over the earth, and extend our observations through the system of human beings, what shall we find but scenes of misery' which necessitate a benevolent response. Finally, 'if we turn from these melancholy prospects, and cast our eyes upon ourselves, what shall we find, but a precarious and frail being' who may as soon need to receive, as have the power to dispense, charity. This language reached its greatest intensity in the sermon Johnson wrote for his wife's funeral, where references to 'that dissolution which shall put an end to all the prospects of this world' (p. 263), death 'open[ing] prospects beyond the grave' (p. 263), 'a sudden abruption of all their prospects' (p. 267) and 'new prospects open[ing] before us' (p. 270) follow hard on one another.

Johnson's moral topography is also found in the sermons. Truth is an eminence, and 'he [that] already has a wide extent of science within his view' will rarely labour to climb higher (p. 90). It is easy to 'deviate from the paths of truth' (p. 76), preferring 'the crooked paths of fraud and strategem' (p. 243). As the rise to greatness leaves us close to a precipice in Johnson's essays, so 'we all stand upon the brink of the grave' (p. 161; see also p. 15).

The imagery of the journey and voyage of life is common. The journey of the traveller is, in a more Christian idiom, juxtaposed with the pilgrimage of life:

as it is the business of a traveller to view the way before him, whatever dangers may threaten, or difficulties obstruct him, and however void may be the prospect of elegance or pleasure; it is our duty, in the pilgrimage of life, to proceed with our eyes open, and to see our state; not as hope or fancy may delineate it, but as it has been in reality appointed by divine Providence (p. 160).

The 'stream of life' (p. 164) and the 'rocks on which conscience... is wrecked' (p. 87) recur at greatest length in Sermon 20 (p. 216), where 'a man ventures upon wickedness, as upon waters with which he is unacquainted', the parallel continuing for a whole paragraph on his 'quitting the shore', and his danger of being 'dashed against a rock, sucked in by a quick sand' until habit makes him lose his fear, such that 'he is driven onto the boundless ocean, tossed about by tempests, and at last swallowed by the waves'.[184]

Moral vision is discussed in the language of physical vision, and in this homiletical context the limits of human vision are emphasized: 'to judge only by the eye, is not the way to discover truth' (p. 85; see also pp. 44, 70, 90, 239). Our views are not only limited, but easily 'dazzled by specious appearances' (p. 150; see also pp. 53, 154, 176).

The sermons contain less imagery than the essays,[185] but landscape imagery, as demonstrated, is not uncommon, and the imagery of cultivation, which is rare in Johnson's essays (though not infrequent in periodical essays in general[186]), is used. Johnson's 'political' sermons (numbers 24 and 26) suggest the need to pull 'up the roots' rather than mow 'down the heads of noisome weeds...in political, as well as natural disorders' (p. 253, see also pp. 257, 285).[187] Johnson also used the organic natural imagery more traditionally associated with conservatism:

the laws of a civilised and flourishing people, like mature and vigorous fruit-trees, though they afford shade, ornament, shelter and sustenance, to their proprietors, are yet rooted in obscurity.[188]

The sermons participate in the same nexus of values as the essays, where natural imagery is used to emblematize the moral world: the transitions are signposted for the auditors of the more overtly didactic sermons with such phrases as 'to sow and to reap are figurative terms' (p. 115, see also pp. 13, 253). In Johnson's sermons 'transitions from topic to topic are impossible to miss',[189] as, reponding to the demands of the genre, Johnson revealed the workings of the 'universe of analogy'[190] by which he connected natural and moral languages in the *Dictionary* and his essays.

The doctrine of landscape in Johnson's sermons. The relationship between natural evidences and revelation implicit in Johnson's essays understandably becomes more important in the sermons. Natural evidence does point to a beneficent Creator: we can 'endeavour to deduce the will of God from the visible disposition of things' (p. 39).[191] The limits of our knowledge of nature, 'when the wisest, and most arrogant philosopher knows not how a grain of corn is generated, or why a stone falls to the ground' (p. 71),[192] however, mean that revelation is indispensable.[193] This influenced the infrequency of Johnson's unqualified acceptance of natural description, and his preference to use the language of landscape aesthetics in a moral context.

In Sermon 3 Johnson charts the progress from innocent pleasures to vice:

publick spectacles, convivial entertainments...sports of the field...all of them harmless, and some of them useful, while they are regulated by religious prudence, may yet become pernicious, when they pass their bounds, and usurp too much of that time which is given us, that we may work out our salvation (p. 37).[194]

Sermon 10 made the relevance of this to the natural world clear in a portrait reminiscent of Prospero in *Rambler* 71:

the man who died yesterday, had purchased an estate, to which he intended some time to retire; or built a house, which he was hereafter to inhabit; and planted gardens and groves, that, in a certain number of years, were to supply delicacies to his feasts, and shades to his meditations. He is snatched away (p. 112).

The difficulty of shutting off the senses made Johnson fear the lure of gardens, but also led to his defence of the external trappings of religion, as potentially harmless ways of leading from the senses to an habitual reverence for the religious establishment.[195]

Johnson was always aware of the ease with which reading nature for evidence of God's existence could become idolatry: this was the ultimate example of the slide from virtuous pleasure to vice, and the possibility cast its shadow over Johnson's entire attitude to landscape and nature. Johnson's late political writings responded to a parallel slide into superstition as he saw it, attacking Wilkite patriots as those who claim 'political prescience', who 'hear the thunder while the sky is clear'.[196] As with the *Freethinker*, then, Johnson poured scorn on the credulity of those who believed in a politicized nature; but the critique of prognostics had shifted from an attack on Tories to one on Wilkite patriots.[197] In *Taxation no Tyranny* (1775), Johnson attacked the belief that 'change of place' entitled Englishmen in America to different representation or taxation.[198] This was Johnson's political transmutation of the Horatian point about place and mind.[199]

The homiletical context of Johnson's views on landscape

The parallels between Johnson's essays and sermons reflect a more general convergence of the sermon and the essay which had occurred since the printing of sermons.[200] Downey argues that in the second quarter of the eighteenth century, 'politics ceased to be a subject of major importance in the pulpit ... Ethical preaching completely triumphed',[201] which brought the sermon still closer to the subject matter of Johnson's essays.

Given this convergence, the possibility emerges that the Johnsonian language and doctrine of landscape, as found in his essays, derived from the Restoration homiletical tradition, as Addison's style and content had.[202] Given the strong effect the *Spectator* had on the development of the periodical essay genre, and that Johnson drew on this group of preachers extensively for illustrative quotations for the *Dictionary*, there are two routes by which the seventeenth-century divines could have made their mark upon Johnson the essayist.

Linguistic influences. The conflation of physical and moral vision Johnson played on so extensively is frequently used in Restoration sermons. The limited view of man was emphasized, most elaborately by Isaac Barrow as we saw in Chapter 4.[203] Prospects could be dangerous, Rogers's words recalling

Johnson's views of the danger of painted vales: 'all the visionary Beauties or Glories of the Prospect, the Paint and Imagery that attracted our Senses, fade and disappear, and leave us disconsolate in the Midst of a frightful Scene of Guilt, Temptation, and Misery'.[204]

The range of Johnsonian imagery can be found in seventeenth-century sermons: Bentley spoke of men as 'spectators in this noble Theatre of the World',[205] but far more common is an image in which man is a moral actor *in* the landscape rather than overlooking it. Only in salvation is a clear prospect gained. Thus, 'he that judges amiss ... [is] like a traveller, who being uncertain whether he goes in the right way, wanders in a continual perplexity',[206] as did Obidah in *Rambler* 65. Tillotson used imagery which conflated the voyage and the journey of life:

> we are upon our journey travelling towards our heavenly Country where we shall meet with all the delights we can desire, [therefore] it ought not to trouble us much to endure storms and foul ways, and to want many of those accommodations we might expect at home.[207]

Taylor, in an image comparable to Johnson's in the tale of Hamet and Raschid (*Rambler* 38), compared avarice to the man

> that had rather lay his mouth to Euphrates than to a petty goblet, but if he had rather, it adds not so much to his content as to his danger and his vanity. For so I have heard of persons whom the river hath swept away.[208]

Both Johnson and the Restoration homilists had a common source for their use of landscape imagery in the rhetorical training which remained a standard element of education in England.[209] Cicero and Quintilian spoke of 'occular demonstration', where the theme was visualized in words, one branch of this being 'the clear and vivid description of places' or 'topography'. By using such topographical imagery in theological contexts, both Johnson and the homilists hoped to move their readers or listeners to embrace Christian truths.[210]

Conceptual Influences. The view of nature the seventeenth-century divines established can also be compared with Johnson's position. All accepted the importance of the design argument,[211] there being 'hardly any thing in Nature, from whence the certainty of the Being of God, may not justly and reasonably be deduced'.[212] Clarke made a telling statement in favour of natural evidence: 'to speak *otherwise* of Faith, and to represent it as of Evidence *Superiour* to Sense or Reason; is to open the Door to all the absurdities of Transubstantiation'.[213] The design argument was, as for Addison, an aesthetic one:

every thing contributes somewhat to the use and benefit, or to the beauty and ornament of the whole: No weed grows out of the earth, no insect creeps upon the ground, which hath not its elegancy and yields not its profit.[214]

More insistently than the Boyle lecturers, however, Restoration homiletics balanced the use of natural evidences with their limitations: 'natural light or evidence is so unsuccessful in the world, that it loudly telleth us something is yet wanting'.[215] For Barrow 'a pretence to natural knowledge, and acquaintance with these things hath been so much abused to the promoting of atheism and irreligion'.[216] Revelation had provided the key and only after this could natural evidences become intelligible: 'How hard it is for Natural Reason to discover a Creator before revealed'.[217] The Anglican tradition, then, sought a balanced view; to claim too great a role for nature was to move towards the covert atheism of feigned natural religion, but to deny nature was to move towards fideistic Catholicism.

The distrust of the senses is a standard element of Christianity, but what links Johnson and the seventeenth-century divines is the degree to which it is an everpresent thought. In a revealing comparison, suggesting how the interpretation of landscape was connected with the immateriality of God, and reminiscent of the opening of *Rambler* 125, South argued:

As it would be extremely sottish and irrational for a *blind Man* to conclude, and affirm positively, That *there neither are, nor can* be any such Things, as *Colours, Pictures,* or *Landskips*, because he finds, that he cannot form to himself any true *Notion, Idea* or *Mental Perception* of them; So it would be equally or rather superlatively more unreasonable for us to deny the great Articles of our Christianity because we cannot frame *in our Minds* any Clear, Explicit, and Exact Representation of them.[218]

Because we cannot comprehend God's infinity, there is a danger we will worship that which, from our limited view, appears to be infinite but is material: Nature. As even Clarke, a great defender of natural religion, put it, '*Nature* . . . is nothing but an empty word.'[219]

The tendency to grasp the sensual and visible led Restoration sermons into an analysis of man's curiosity as the source of unhappiness, as in Johnson. Interestingly, the examples used are drawn from the fields of landscape and natural knowledge, especially by Taylor:

it is a thousand pities to see a diligent and a hopeful person spend himself in gathering cockle-shells and little pebbles, in telling sands upon the shore, and making garlands of useless daisies.[220]

Gardens could be understood in a similar framework:

> oh remember when you are tempted to please your eyes, your tast[e] and
> sensual desires, that these are not *Eternal* pleasures!...Houses and Lands
> are not *Eternal!*[221]

The retirement of the cit and the philosopher was discussed. Taylor characterized the ambition of all 'usurers and merchants, all tradesmen and
labourers' as being to 'buy up three acres of ground'.

> And is this sum, that is such a trifle, such a poor limited heap of dirt, the
> reward of all the labour and of all the care...and can it be imaginable,
> that life itself, and a long life, an eternal and happy life...that such a
> kingdom should not be worth the praying for [?][222]

To retire was, even if philosophically motivated, 'too like the *hideing of our
talents*'.[223]

The strength of such sensual pleasures, in a passage reminiscent of Johnson's
explanation of the lure of the pastoral in *Rambler* 36, is that,

> the World has the unhappy Advantage of pre-ingaging our Passions; of
> furnishing us with the first Sentiments of Pleasure, at a Time when we
> have not Reflection enough to look beyond the Instrument to the Hand
> whose Direction it obeys.[224]

It is only by keeping our eyes on the eternal that we stop the slide from
innocent pleasures to vices Johnson was to discuss in Sermon 20:

> Vice first is pleasing, – then it grows easy, – then delightful, – then frequent –
> then habitual, – then confirmed; – then the man is impenitent, – then he is
> obstinate, – then he resolves never to repent, – and then he is damned.[225]

Conclusion: Johnson in context

Johnson's approach to the issues of landscape and nature is a Christian one.
His refusal in the essays and sermons to allow such issues to slip from their
context in a moral life allies him more closely with Restoration sermons
than with the periodical essay tradition. In weighing innocent pleasures
against their decay by habit into moral imbalance, Johnson was mirroring
the pattern of argument of preachers of the late seventeenth century. In
exemplifying this shift in the spheres of gardening, retirement and virtuosi,
he was employing the same illustrations as Taylor, South and Baxter. As
such, Johnson was attached to 'a more archaic manner' of sermon than that
of Clarke, Tillotson or Barrow.[226] Addison's approach was more closely
based upon the latter triumvirate, together with the Boyle lecturers, such

that he was more positive towards natural evidences of God, and therefore the moral tone of his essays was different, landscape rarely appearing simply as an emblem. Johnson and Addison's essays, then, derive their views of landscape and nature from the Restoration homiletical tradition, but from different parts of the (mainly) Anglican spectrum. Johnson's essays are not unusual in the use of landscape to point morals, the moral context of landscape issues remaining everpresent as the periodical essay developed between Addison and Johnson, but they do make landscape more transparently emblematic, largely by reverting to the language of landscape established by seventeenth-century sermons to a degree which no other eighteenth-century periodical did. Hudson's contention that 'Johnson was one of the last and most interesting spokesmen for a tradition which, at least as an important literary and theological force, was dying – but dying with absolute confidence'[227] is an important one, and helps us to understand his position in landscape debates. Johnson's characteristic preoccupations were those of a Christian moralist, drawing upon a language and doctrine of landscape derived from seventeenth-century sermons, and stretching back to the roots of Elizabethan Anglicanism.

Mind and Place in *Rasselas*: an epitome of the Johnsonian doctrine of landscape

Rasselas (1759) comes at the end of the most active decade in Johnson's writing career and closes a twenty-year phase in which his position as a moralist was forged. An element of this position was a response to the role of landscape, place and nature in human life, and Johnson concentrates his ideas on this in *Rasselas*. Seen in the context of the previous twenty years, it is not the Oriental background to *Rasselas*, intensively studied as it has been,[228] that illuminates Johnson's attitudes to landscape and nature, but the way he situates these issues in the 'choice of life'.

Many structural divisions of *Rasselas* have been suggested,[229] but the most important section from the present perspective is Chapters 19–22, which in a compressed space deflated a number of delusions about place and its relationship with human happiness. These chapters act as a unit (although I will not speculate on whether Johnson intended this), with the claims being made for the pleasures attached to place ascending from the aesthetic to the sensual, from the sensual to the mental, and from the mental to the divine. Johnson treats these delusions of place with increasing firmness and ridicule, accordant with the danger of the error being committed in the choice of life.[230] Johnson returned to delusions throughout *Rasselas*, but these four chapters concentrate the message he had been making since the *Vanity*, redirecting our hopes from happiness in place to virtue in the mind. They act as a chain of events, a structural characteristic of philosophical tales such as *Rasselas*. Keener argues the logic of the chain of events, which examined empirically, opposed the rationalism of the Great Chain of Being:[231] appropriately, Johnson's chain of events in Chapters 19–22 ends with the destruction of rationalistic conceptions of the Chain of Being, incorporating them as the highest and most profane delusion about man's understanding of nature, and completing a sequence which began with the farcical (but potentially pernicious) errors of the pastoral. The net effect of these chapters is to provide a masterful précis of the High-Church approach to the role of landscape in moral life, and its message can be directly counterpointed to a Latitudinarian reading of the same themes thanks to Ellis Cornelia Knight's *Dinarbas* (1790), which sought to rework *Rasselas* from a different theological perspective. It is with *Rasselas*, then, that the tension between Johnson's approach to landscape description and that of the dominant Latitudinarian genealogy of Chapter 4 was brought into sharp relief.

Place and literary aesthetics: 'A glimpse of pastoral life'

After escaping from the Happy Valley, and an initial period seeking a choice of life in Cairo, Rasselas and his group set out to find a hermit, to 'enquire whether that felicity, which publick life could not afford, was to be found in solitude'.[232]

In Chapter 19, 'A glimpse of pastoral life', the route to the hermit's cave 'lay through fields, where shepherds tended their flocks' (p. 76). Imlac points out that 'pastoral simplicity' has frequently been celebrated as the ideal situation. In the two remaining paragraphs of the chapter, a familiar Johnsonian pattern related the truth of the pastoral to the 'hunger of the imagination'[233] for a site of human felicity. They find the shepherds 'so rude and ignorant, so little able to compare the good with the evil of the occupation ... that very little could be learned from them' (p. 77). What 'opinion of their own state' (p. 77) the shepherds could convey showed

> their hearts were cankered with discontent; that they considered themselves as condemned to labour for the luxury of the rich, and looked up with stupid malevolence toward those that were placed above them.

Although not in public life, the shepherds show an awareness of their relation to that life; there can be no rural existence sealed off from urban realities, as Johnson had suggested in *London*. Johnson did not portray their anger as the righteous response of the oppressed, as Knight's sequel and response to *Rasselas*, *Dinarbas* (1790) did. In Chapter 8 of *Dinarbas*, 'Apology for rusticity', Rasselas finds 'the poets have written after nature', and that shepherds working in the lands amidst 'these images of the power and goodness of the Deity must expand their hearts'.[234] Knight's shepherds are only surly to those who are haughty in their rank, therefore adopting a 'slave's' response to 'despotism'. Knight's shepherds, then, are both innocent and politically informed. Johnson's (pre-Revolution) swains are neither. The final paragraph of Chapter 19, having shown rural reality, allows the drive to self-delusion to reappear. Whilst Rasselas's sister Nekayah 'would never suffer these envious savages to be her companions' (p. 77), she will not believe 'all the accounts of primeval pleasures...fabulous'. Her image of the pastoral blends the courtly with the simple:

> with a few virtuous and elegant companions, she should gather flowers planted by her own hand, fondle the lambs of her own ewe, and listen, without care, among brooks and breezes, to one of her maidens reading (p. 77).

Nekayah's delusion is to believe that for some time in the future ('she hoped that the time would come') or the past ('primeval pleasures') pastoral descriptions have plausibility.

The strength of such delusions, as in *Rambler* 36, is due to the impression nature makes on the young, when 'nature was yet fresh' (p. 16). *Rasselas* shows that the mature mind can derive pleasure from the beauties of nature, but only for a brief period. This is shown in examples of pleasurably instructive travel in the story: at the start of their journey to the pyramids (Chapter 31), the company 'stopped from time to time...and observed the various appearances of towns ruined and inhabited, of wild and cultivated nature' (p. 114).[235] Such responses can only be brief, while the concern of *Rasselas* is with longer term happiness.[236] The aesthetic moment passes into vacuous gazing: Pekuah, in the Arab's house, initially enjoyed wandering 'from one place to another as the course of the sun varied the splendour of the prospect' (p. 137), but found the limitations of such pleasures rapidly: 'I was weary of looking in the morning on things from which I had turned away weary in the evening' (p. 138). Imlac's youthful response to the sublime ocean shows a similar progress in a higher key: '[I] imagined that I could gaze round for ever without satiety' with 'soul enlarged by the boundless prospect'; but 'grew weary of looking on barren uniformity, where I could only see again what I had already seen' (p. 35).[237] A literary or aesthetic approach to scenery, then, is insufficient.

The recurrence of delusion is shown in *Rasselas*: after Chapter 19, all three young seekers after the choice of life at some point return to the pastoral ideal. Rasselas 'would willingly believe [happiness] to have fled from thrones and palaces to seats of humble privacy and placid obscurity' (p. 101) despite what he has already seen. Pekuah has dreamed, as the lottery-obsessed shopkeeper did in *Rambler* 181, that she 'built new palaces in more happy situations, planting groves upon the tops of mountains' (p. 153), as if this could provide more lasting pleasure than the prospects from the Arab's house. Finally, Nekayah confesses: 'I have often soothed my thoughts with the quiet and innocence of pastoral employments, till I have in my chamber heard the winds whistle and the sheep bleat' (p. 153).

In *Dinarbas*, Knight pictured the aesthetic pleasures of landscape as a far more permanent consolation for the human condition: Rasselas in confinement finds one of 'the resources of solitude' to be composing poems on the prospect, and enjoys such pleasures during 'several months of confinement' (Din. 149). For Johnson in *Rasselas* the literary and aesthetic response to nature may recur because we wish it to be true, but it cannot take a lasting hold when humans are faced with more pressing concerns.

Place and the sensual: 'the danger of prosperity'

Departing from the shepherds, Rasselas and company enter a thick wood. Here they find a cultivated area, where

> the shrubs were diligently cut away to open walks where the shades were darkest; the boughs of opposite trees were artificially interwoven; seats of flowery turf were raised in vacant spaces, and a rivulet, that wantoned along the side of a winding path, had its banks sometimes opened into small basons, and its stream sometimes obstructed by little mounds of stone heaped together to increase its murmurs (p. 78).

The garden *appears* to be a 'harmless luxury' (p. 78), leaving 'all the face of nature smiling around the place' (p. 79). The owner, surrounded by chearful servants, is both wealthy and popular, and Rasselas hopes 'that he should find here what he was seeking'. But the central point of the chapter is that, as the owner of these delights puts it, 'appearances are delusive' (p. 79). It is the prosperity which has allowed the construction of the garden which also puts the owner's life in danger. The Bassa of Egypt is his enemy, 'incensed only by my wealth and popularity'. When an attack comes, the deceptive appearance will finally be stripped away: 'then will my enemies riot in my mansion, and enjoy the gardens which I have planted'. The apparent freedom of the garden is deceptive, for the man is confined to 'exile' (p. 80).

That the travellers should be deceived by the beauty of the landscape shows again the ease with which an exploded delusion can return. For the deceptive seduction of the senses is the same as the 'tasteless tranquility' (p. 60) of the Happy Valley, whence the group had fled. The Happy Valley, Johnson's most famous and most studied landscape,[238] appears to be a paradise on earth. The Valley also has its own garden, where

> the air [was] always [kept] cool by artificial showers. One of the groves, appropriated to the ladies, was ventilated by fans, to which the rivulet that run through it gave a constant motion; and instruments of soft

musick were placed at proper distances, of which some played by the impulse of the wind, and some by the power of the stream (p. 23).

Tomarken argues that the 'gardening allusions in the happy valley refer...to actual horticultural practice'.[239] Yet it should be apparent that Johnson's concern is not for a realistic depiction of gardening practice in eighteenth-century England: the descriptions of both the Happy Valley and its garden are designed, with the greatest economy possible, to describe a landscape which appeals to all the senses.[240] Also, again as in Chapter 20, the apparent freedom of the valley is, in fact, a confinement, but where the wealthy man's exile is hard to see, the Happy Valley's is a physical one, 'surrounded on every side by mountains of which the summits overhang the middle part' (p. 8).

Johnson, in the Happy Valley chapters and in Chapter 20, established a pattern of argument: all that nature can offer to the senses is present, and yet the inhabitants are dissatisfied. The purpose of these descriptions, then, is to show that man is not simply sensate, but cognate, and cannot be satisfied with 'whatever the senses can enjoy'. Rasselas's speech to the animals in the Happy Valley (pp. 13–14) shows that it is cognition which is the cause of feelings of confinement:

every beast that strays beside me has the same corporal necessities with myself, he is hungry and crops the grass, he is thirsty and drinks the stream, his thirst and hunger are appeased, he is satisfied and sleeps... I am hungry and thirsty like him, but when hunger and thirst cease I am not at rest.[241]

The 'latent sense' for which the Happy Valley provides no stimulus is the mind. Rasselas, moreover, sees that the human burden is also a blessing, 'nor do I, ye gentle beings, envy your felicity; for it is not the felicity of man'. To find happiness in sensate pleasures is to reduce man to the level of the animals, and Johnson associates the ignorant with bestiality in *Rasselas*.[242] Pekuah describes the diversions of the women in the Arab's harem as 'only childish play': 'I could do all which they delighted in doing by powers merely sensitive, while my intellectual faculties were flown to Cairo.' Revealingly, Pekuah compares the women who could be thus satisfied to animals: 'they ran from room to room as a bird hops from wire to wire in his cage. They danced for the sake of motion, as lambs frisk in a meadow' (p. 138). These women also show their emptiness by gazing at the landscape: 'part of their time passed in watching the progress of light bodies that floated on the river, and part in marking the various forms into which clouds broke in the sky' (p. 139).[243] All the delights place, landscape and nature can offer to the senses are examined and found wanting in Chapter 20's compression of the sequence of gratification–satiation–confinement established in the Happy Valley.

Place and the mental: 'the happiness of solitude'

In Chapter 21, 'The happiness of solitude.The hermit's history', the group finally reach the destination of their rural ramble. The hermit's cave and the surrounding landscape blend the simplicity of the shepherds with the improvement of the wealthy man:

> it was a cavern in the side of a mountain, over-shadowed with palm-trees; at such a distance from the cataract, that nothing more was heard than a gentle uniform murmur, such as composed the mind to pensive meditation, especially when it was assisted by the wind whistling among the branches. The first rude essay of nature had been...much improved by human labour' (p. 80).

The hermit's conversation matches his environment in its moderation: 'his discourse was chearful without levity, and pious without enthusiasm' (p. 81). The hermit, using one of Johnson's oceanic images, says that his first pleasure upon retiring had been an aesthetic one: 'I rejoiced like a tempest-beaten sailor at his entrance into the harbour, being delighted with the sudden change of the noise and hurry of war, to stillness and repose' (p. 82). But 'when the pleasure of novelty went away, I employed my hours in examining the plants which grow in the valley, and the minerals which I collected from the rocks' (p. 82). Natural knowledge is the element added by Chapter 21 to the pleasures of place, obviating the criticism of 'merely sensual' pleasures in Chapter 20.

The hermit finds, however, that the mental pleasures of the natural world provide no more lasting fulfilment than the sensual; as the Happy Valley became 'tasteless tranquility', so the hermit confesses 'that enquiry is now grown tasteless and irksome'. The hermit's assessment of retirement includes most of Johnson's complaints against it. The hermit has only been secured from acting viciously by 'retiring from the exercise of virtue'. If he cannot act viciously, the hermit thinks viciously ('my fancy riots in scenes of folly', p. 83) because he lacks the society which will polish out the peculiarities of behaviour.

The views of Knight's Rasselas in *Dinarbas* are somewhat different. Rasselas reflects 'if I am incapacitated from doing good, I am at least prevented from committing ill' (Din. 131); the same equation is used by Johnson's hermit, but the latter is 'ashamed to think' (p. 82) this was the only way he could avoid vice. Knight's Rasselas not only gets protracted aesthetic (sensitive) pleasure from the landscape, but natural knowledge never grows irksome as it does for the hermit: 'How can a man think himself alone while surrounded with the noblest works of his Creator?' (Din. p. 132). For him it 'affords a *constant* field for meditation' (Din. p. 132 emphasis added). Johnson's hermit never links his examination of plants and minerals to his contemplative meditation: as in *Rambler* 135, natural evidences are unimportant in the discovery of God in rural retirement. It is 'stillness and repose', not God's handiwork which facilitates reflections on the divine. As Knight's priest figure Elphenor espouses Latitudinarian beliefs Johnson could not have agreed with,[244] so Knight's Rasselas in retirement links his contemplation to the landscape in which the contemplation occurs, prioritizing natural evidences in a way quite different from the High-Church Johnson.

Returning to *Rasselas*, at the beginning of Chapter 22, several responses on the hermit's story are given. The view of one 'more affected with the narrative than the rest' (p. 85) is that 'the hermit would, in a few years, go back to his retreat, and perhaps, if shame did not restrain, or death intercept him, return once more from his retreat into the world'. This view is one Johnson himself had adopted in *Rambler* 63. An unsettled mind will not be able to find happiness in the sensual, the mental or the social pleasure offered by town or country, sociable conversation or nature's handiwork. This is the opinion of the hermit as well, who puts the other side of the same argument about the relationship between mind and place: 'To him that lives well...every form of life is good' (p. 81). And yet, given the impossibility of living contently, the only solution is to focus our mind on a place beyond earthly places. As the commentator on the hermit says 'the time will surely come, when desire will be no longer our torment, and no man shall be wretched but by his own fault' (p. 85).[245]

Johnson's attitude in *Rasselas* to the relationship between place and mind is, however, a balanced one. If place cannot palliate mental unease, the curiosity represented by our wanderlust and interest in the natural world is a function of that mental capacity which distinguishes us from animals. Pekuah's lack of conversation in the Arab harem is because the women, true to their sensate lives, betray no curiosity in the natural world: 'they had seen nothing; for they had lived from early youth in that narrow spot' (p. 139). The intelligent characters in *Rasselas*, by contrast, display a healthy interest in geography and natural knowledge: Imlac, because of his travels has 'a mind replete with images' (p. 54),[246] and the astronomer is 'delighted' with Imlac's narrative of his journeys (p. 142).

Moving from the individual to the social, it is this curiosity about the natural and human world which distinguishes Europeans from 'Asiaticks and Africans' (p. 47). The relationship of Europe to other continents is like that of Pekuah to the Arab's seraglio: 'They are more powerful...because they are wiser; knowledge will always predominate over ignorance, as man governs the other animals' (p. 47). Europeans, through their interest in the natural world, have a greater power over it:

> we suffer inclemencies of weather which they [the Europeans] can obviate...they have roads cut through their mountains, and bridges laid upon their rivers (p. 50).

Johnson, then, partakes of the opinion of seventeenth-century geographers 'that the Orient was being outstripped and outdated by Western science'.[247] But he put this down to their curiosity (mind), not any advantages conferred on them by climate or landscape. Said has argued that landscapes were important to the visualization of European superiority over the Orient.[248] In *Rasselas* a landscape of bridges and roads may reflect superior knowledge, but only in Knight's *Dinarbas*, which comes in the late eighteenth century, the time of 'modern Orientalism's' genesis,[249] is European superiority directly reflected in the *natural* condition of the landscape. In *Dinarbas*, the titular hero from Abyssinia reflects in Europe that 'nature...ever remains beautiful in temperate climates'. Dinarbas has been 'led to adore the goodness of the Creator of the Universe', which has been lavished not 'in countries where the northern blasts deface the charms of fertility, or [in the south] where the too ardent rays of the sun dry up its sources', but in temperate Europe (Din. 174). By contrast, Johnson's only answer as to the cause of the superior knowledge of the Europeans is 'the unsearchable will of the Supreme Being' (p. 47). Johnson's central concern, then, was mind not place: the European had no providential dispensation, and God's hand was no more visible in the European landscape or climate than in any other. As Said says that 'when Pope proclaimed the proper study of mankind to be man, he meant all men',[250] such was Johnson's meaning in his famous lines 'Let observation with extensive view, / Survey mankind from China to Peru'. This is made explicit by Imlac's discussion of pilgrimage. Contrary to Knight, 'that the Supreme Being may be more easily propitiated in one place than in another, is the dream of idle superstition' (p. 48). Place may act as a mental stimulus, and it is only willed activity and curiosity which can lead towards a more reasoned choice of life; not a geographical determinism.[251]

   To complete the Johnsonian argument, whilst geographical curiosity may help to develop the mind, there is a final sense in which place is unimportant to the developed mind.[252] As the hermit's happiness will only occur in a final state, so the old man in Chapter 45 has 'ceased to take much delight in physical truth; for what have I to do with these things which I am soon to leave?' (p. 155). This expresses a wisdom which derives from the curiosity which drives man into the natural world and various locations being given a religious turn, which recognizes the vanity of attachment to place, at a sensual or mental level, and converts curiosity into hope. The hope is still, in a sense, locative, as the old man suggests: it is the 'hope to possess in a better state that happiness which *here* I could not find' (p. 156, emphasis added). In Rasselas's learning process, he has to convert a geographical complaint into an earthly one. Where in the Happy Valley he says of his discontent 'man has surely some latent sense for which *this place* affords no gratification' (p. 13, emphasis added), he has to learn from the words of Imlac that 'Human life is *every where* a state in which much is to be endured' (p. 50, emphasis added).[253]

Place and the divine: 'the happiness of a life led according to Nature'

The opening of Chapter 22, as already stated, sees a discussion of the hermit's history. After the religious conclusion of the most affected of the assembly, the philosopher of nature puts forward his view that 'the time is already come, when none are wretched but by their own fault' (p. 85). This is an impious interjection into a conversation which had been directed to the conclusion that the only consolation is religious.[254] The philosopher of nature represents the final form of delusion with respect to nature: accepting the need to look beyond place as a source of happiness, he believes this can be achieved by reason rather than faith. This is the most important error in the ascending scale set out in Chapters 19–22, as it is the most directly impious, arguing we can make heaven on earth.

The philosopher's route to happiness is 'to live according to nature, in obedience to that universal and unalterable law with which every heart is originally impressed' (p. 85–86). The language parodies that of the deists such as Tindal and the heterodox Clarke.[255] Johnson was pitting the Clarke of the Boyle lectures against the Clarke of the sermons who recognized that 'Nature is nothing but an empty word.' Because it is an empty word, the philosopher of nature suffers a delusion that neither the hermit nor the prosperous man does. When confronted with actual experience, both the hermit and the rich man were willing to confess the unhappiness of their lives in nature. The philosopher of nature is also more deluded than the shepherds, for whom ignorance does not prevent the recognition that the rich live happier lives (or at least lives worthy of envy). The philosopher, by contrast, cannot see the emptiness of his own words, as demonstrated by the fact that he is the only one in the four chapters on nature whose position does not develop in his encounter with Rasselas's party: he believes Rasselas satisfied with his explanation of what it means to live according to nature, and departs 'with the air of a man that had co-operated with the present system' (p. 89).

To live according to nature is said to be

> to act always with due regard to the fitness arising from the relations and qualities of causes and effects; to concur with the great and unchangeable scheme of universal felicity; to co-operate with the general disposition and tendency of the present system of things (p. 88).

Earlier, the philosopher of nature had given an apparently meaningful explanation of how to achieve this:

> let them observe the hind of the forest, and the linnet of the grove: let them consider the life of animals, whose motions are regulated by instinct; they obey their guide and are happy. Let us, therefore, at length, cease to dispute, and learn to live (p. 87).

Man, in a reverse of Johnson's own position, can grasp the whole of the system of nature without limitation, by rejecting reason in favour of the senses. This is very different from Imlac's way of learning from the conies in the Happy Valley. For Imlac, 'human reason borrowed many arts from the instinct of animals' (p. 57): by analyzing rationally what animals do instinctively, man is able to make the progress in controlling nature he ascribes to the Europeans. The philosopher of nature, by contrast and with considerable contradiction, reasons that we should follow our instincts.[256] The level of absurdity in the philosopher's gravitas, and the unquestionable error of his position, suggest that this was a stance Johnson wished to attack as strongly as possible. The exceptional vitriol for an eighteenth-century book review which Johnson used against Jenyns[257] is matched by the directness of the satire in this chapter, and both are Johnson's response to the impious hubris of those who ignore the limitations of man's mental vision. Just as Johnson's imagery was in accordance with the dictates of classical rhetoric, so was this strategy of defeating opponents by turning them into subjects of ridicule.[258]

Johnson returns to this error of believing a man can see the whole system of nature in three other portraits in *Rasselas*. In 'A dissertation on the art of flying' (Chapter 6),[259] a scientist builds wings to allow man to fly. From here the prospect would be one so far equated with the divine: 'to survey with equal security the marts of trade, and the fields of battle... [to] examine the face of nature from one extremity' to the other!' (p. 26). In Chapter 18, a similar error is made by a sage who proposes a stoic view of life. For Rasselas, the stoic 'looks down on the scenes of life changing beneath him' from 'the unshaken throne of rational fortitude'. There is a strong sense of impiety in describing a man as looking down from a throne above the world, which matches the fact that 'Rasselas listened to him with the veneration due to the instruction of a superiour being' (p. 74). Imlac suggests the stoic may be somewhat lower in the chain of being: 'the teachers of morality... discourse like angels, but they live like men' (p. 74). Imlac proves correct, and the fall of the sage is as crushing as that of the would-be flier: on the death of his daughter he declares, 'my views... are at an end' (p. 75). Johnson's point is that to survey life in a detached way is impossible for a human being. We are not viewing the theatre of life, but are actors on its stage; we cannot raise ourselves above the prospect because we are in it.[260]

The final portrait to deal with the desire for a Godlike perspective, that of the astronomer, is also the most extended and sympathetic. The astronomer indulges 'in imaginary dominion' (p. 146) over the course of the seasons. He confesses to Imlac that

in the hurry of my imagination I commanded rain to fall, and, by comparing the time of my command, with that of the inundation, I found that the clouds had listned to my lips (pp. 146–47).

The pattern is like the pastoral delusion: just as Nekayah, faced with the shepherds will not relinquish her belief, so the astronomer will not accept that he lacks dominion over the seasons, even though he 'cannot prove it by any external evidence' (p. 147). As the hermit had feared, so the astronomer is an example of the progress of innocent pleasures into vicious beliefs in the course of a solitary life. In a passage similar to that by Taylor on the progress to vice, Imlac says fancy

> grows first imperious, and in time despotick. Then fictions begin to operate as realities, false opinions fasten upon the mind, and life passes in dreams (p. 152).

The astronomer must recognize his position as a man who has not been 'singled out for supernatural favours or afflictions' (p. 163).

This last point returns to Johnson's attack on seeing providence or omens in nature. Rasselas, on digging the tunnel to escape from the Happy Valley found a fissure which he 'considered as a good omen'. Imlac's reply is crushing:

> do not disturb your mind...with other hopes or fears than reason may suggest: if you are pleased with prognostics of good, you will be...a prey to superstition (p. 59).

That Rasselas learns the lesson becomes apparent in Chapter 28, where he says,

> let us cease to consider what, perhaps, may never happen, and what, when it shall happen, will laugh at human speculation. We will not endeavour to modify the motions of the elements (p. 103).

This discussion makes it clear that 'all natural and almost all political evils, are incident alike to the bad and good' (p. 101).[261] The error of the believer in prognostics is akin to that of the philosopher of nature: both assume that within our earthly life a moral calculus exists, whereby the justice of God can be seen. Nature is useful not so much as evidence of God's existence, nor as a revelation of his will, both of which lead too easily to impiety and idolatry, but as a way of elaborating an already-known set of religious precepts. As Imlac puts it,

> the plants of the garden, the animals of the wood, the minerals of the earth, and the meteors of the sky, must all concur to store his mind with inexhaustable variety: for every idea is useful for the inforcement or decoration of moral or religious truth (pp. 42–43).

Dinarbas took a different line. The priest Elphenor attacked 'the strangest of all errors, judicial astrology' (Din. p. 145). This is, to say the least, odd, given his own history in which he was

> 'beloved; – but, at the moment in which I was to have been united to the object of my affection, an awful event separated us – a flash of lightning reduced my bride to ashes as she received my vows at the altar' (Din. p. 192).

In *Dinarbas*, Rasselas forms 'a system' of simplicity as the key of life. The result is a construction the philosopher of nature would have applauded: 'Simplicity is the child of nature: the love of it seems implanted in us by Providence' (Din. p. 203). His elaboration of this system, although seriously intended, is what Johnson would call philosophical cant:

> the variety of nature is infinite; but it is harmonized by general effect. The verdant leaves of the trees participate of the azure of the sky, and their trunks of the colouring of the earth . . . even in contrast there is an imperceptible connexion that unites the whole. Without one great plan, to which all is subservient, our general conduct in life, and our finest productions of art or genius, are like a republic without laws, or a monarchy without a king (Din. p. 204).

Knight through her Rasselas repeats the sentiments of the philosopher of nature, because she wishes to achieve a providential overview. Johnson's doctrine and the narrative position he adopts is that of a human being in rather than above the moral prospect.

> From a place where everyone seems required to know absolutely everything . . . it [*Rasselas*] changes to a place where everyone must learn to tolerate vast reaches of uncertainty.[262]

Thus Johnson has a 'conclusion in which nothing is concluded', whilst Knight seeks a narrative closure in which, on this earth, Providence has favoured the virtuous.

Johnson's views were made clear in his review of Jenyns's *Free Enquiry* (1757).[263] His main point with regard to cosmological speculation is made in the first paragraph: '*we see* but *in part*' (*Jenyns*, p. 171), and as such the structure of the universe lies 'out of the reach of human determination'. Jenyns's argument, that pain in one part of the universe cannot be eliminated without causing greater pain elsewhere is then nonsense, in that we have not the faculties with which to judge such a statement. To come up with such notions is, therefore, in Lockean terms, 'to impose words for ideas' (*Jenyns*, p. 301). When an author does speak of the whole, he applies to it (as Nekayah realized, *Rasselas*, p. 105) reasonings derived from a small part: 'Many words easily understood on common occasions, become uncertain and figurative when applied to the works of Omnipotence' (*Jenyns*, p. 252). The great chain of being, as living according to nature, is such a transference from part to whole, which reduces ideas to words. Johnson concurs with Jenyns that it is 'pompous nonsense' to speak of virtue as residing in 'conformity . . . to the fitness of things' (*Jenyns*, p. 303), but adds the chain of nature to the catalogue of absurdities. It is built to prove the existence of a benevolent Creator, but only has any force if we already accept this as a premise (*Jenyns*, pp. 172–73). Johnson is not seeking to deny a cosmic hierarchy, merely that we can understand any part of it beyond that available to empirical examination. Towards the end of the review, Johnson's position on religion is made clear: 'its evidences and sanctions are not irresistible, because it was intended to induce, not to compel, and that it is obscure, because we want faculties to comprehend it' (*Jenyns*, p. 306). It is this position which guided the exceptionally keen satire in Chapter 22 of *Rasselas* on the hubris of rationalism, and forms the background to the Johnsonian doctrine of landscape.

Conclusion: an Anglican Tory view of landscape and nature

Hudson has argued that *Rasselas* follows the 'three part division of human aspirations made by the Anglican theologian and philosopher Richard Hooker' into the sensual, the intellectual and the spiritual.[264] This argument relates to the structure of the book as a whole, but can be seen operating within Chapters 19–22 to regulate the 'choice of life' with respect to nature. The Hookerian triad is dealt with, the sensual in Chapters 19 and 20, the mental in Chapter 21 and the divine in Chapter 22.

This hierarchy in relation to the position of landscape and nature in Johnson's thought is particularly apparent in *Rasselas* due to its compression into a few chapters, but is also the principle behind the fragmented discussions in his essays, poems, political writings and sermons. Such a hierarchy, I have suggested in Chapter 3, guided discussions of landscape and nature in the 'long' eighteenth century. Johnson is distinguished by his refusal to let landscape or nature escape from such a framework. While Johnson used the natural as an analogy for the moral world, he was unwilling to let it do more than point a moral, for the use of natural evidences in more elaborate analogies led to the confusion of the figurative and the actual which he attacked in the Jenyns review, and which amounted to idolatry of nature. Everett has suggested the importance of Butler's *Analogy of Religion* to a 'Tory' view of landscape which fused the moral and natural worlds.[265] Johnson stressed rather the equally Butlerian point about the limitations of human reason as a restraint upon fanciful analogy. As such, Johnson's view of landscape emphasized a Hookerian *hierarchy* of concerns, in which landscape is less important than the moral and the religious, rather than Butlerian *analogy*, in which landscape reflects moral and religious matters. Johnson sought to check the fanciful inscription of values into physical landscapes, preferring them to be viewed empirically.[266] This is because the great duties of religion are revealed in scripture, and, true to Hooker's *Laws*, the laws which God has made man follow are different from those of nature.

Johnson's view of landscape is 'Tory' in the sense that he stresses the limits of human reason.[267] This comes in turn from an Anglican position which points to the limitations of natural evidences. The genealogy of this view in Johnson's thought is from Hooker through Taylor and South to William Law, whose formative influence Johnson himself admitted.[268] This can broadly be called a High-Church tradition,[269] and is the key to understanding Johnson's moral approach to landscape. It is the continual stress on the need to act in a Christian way in every action which Law points to in *A Serious Call* and that this lifestyle is an inner one, which is reflected in Johnson's vigorous refusal to let the aesthetic attraction of landscape escape a Christian hierarchy of values. If we term Johnson's position a High-Church Anglican view of landscape, it can be seen not as a blindness to the natural world, but as a principled scepticism which refused to allow landscape and nature to become more central in our lives than was rational for a voluntaristic agent whose main aim was salvation. It is such a view that *Rasselas* epitomized.

7
The Empirical Landscape: Johnson and Factual Description of the Natural World, 1735–75

It has been suggested that Johnson saw two ways to use language: a style appropriate for 'literature' and another for 'science'.[1] Related to this, Johnson's moral doctrine of landscape[2] runs parallel to a set of strictures on the description of actual landscapes and the environment. Both 'uses' of landscape, the didactic and the empirical, share a focus on truth and the central place of morality in a rational life, yet the 'universe of analogy' central to the *Dictionary* and essays,[3] which manifested itself in the use of landscape terminology in allegorical ways, is replaced in empirical contexts by the 'will to verisimilitude', which manifested itself in precise description.[4] For Johnson it was vital that the two types of knowledge, and therefore of landscape ideas, were not confused.

The position of the natural environment in Johnson's theory of factual description, 1735–75

Johnson's is not simply a theory of travel[5] or of natural description.[6] Though the approach to the natural environment is the main focus of the present study, this is best seen as the specific application of a more generalized theorization of factual description. As with the didactic use of landscape as a vehicle, the concern for the natural environment is subsumed in a broader understanding of the means to instruction in empirical work. Johnson does not have a 'theory of landscape': what I discuss are Johnson's opinions on these issues within the context of his approach to knowledge, seen in turn in the wider milieu of eighteenth-century English thought.

The preface to the *Voyage to Abyssinia*
Most of the elements of Johnson's approach to factual description are present in his first extant publication, the preface to his translation of Lobo's *Voyage to Abyssinia* (1735).

Lobo is praised for indulging in 'no romantick absurdities or incredible fictions'.[7] The next paragraph specifies how this is achieved and the style in which it should be conveyed to the reader. Lobo 'consulted his senses not his imagination', and related his findings by a 'modest and unaffected narration'. By implication, Johnson set out here the sources of error in empirical accounts which he would continue to criticize throughout his life. Any form of reasoning which did not rely on the senses, such as analogy or allegory, was inappropriate: by rendering information the platform for speculation, such approaches violated the demands of factual description. The other form of error criticized by implication was the affectation of a more 'elevated' style than the transmission of information necessitated: such a style could intervene in the direct relation of evidence and obscure its meaning.

The *Preface* made the presuppositions behind this position clear. To hold that facts sensed should be directly conveyed to the reader was not a simple position: at least in Johnson's thought, it was backed up by a view of mankind, which in turn was grounded in a view of God's role in the human and natural worlds. Johnson could not argue that Lobo had told the truth, as he has no way of verifying the account: his acceptance is based on the observation that

> whatever he [Lobo] relates, whether true or not, is at least probable, and he who tells nothing exceeding the bounds of probability, has a right to demand, that they should believe him, who cannot contradict him (p. 3).

This form of argument was one derived from seventeenth-century Anglican attacks on Pyrrhonian scepticism.[8]

Lobo's account was 'probable' because it reinforced a presupposition Johnson and the Augustan tradition[9] held: the uniformity of man. In Lobo's account, there

> are no Hottentots without religion, polity, or articulate language, no Chinese perfectly polite, and compleatly skill'd in all sciences: he [the reader] will discover, what will always be discover'd by a diligent and impartial enquirer, that wherever human nature is to be found, there is a mixture of vice and virtue, a contest of passion and reason (pp. 3–4).

This moral uniformitarianism was grounded on the belief that 'the Creator doth not appear partial in his distributions, but has balanced in most countries their particular inconveniences by particular favours' (p. 4). Such a view had as its corollary a view of the functioning of the natural world. In the *Preface*, Lobo's veracity is further demonstrated in that 'the reader will here find no regions cursed with irremediable barrenness, or bless'd with spontaneous fecundity, no perpetual gloom or unceasing sunshine' (p. 3). Such a belief was important not only at the regional level, but also in the encounter with each element of the natural world:

he [Lobo] meets with no basilisks that destroy with their eyes, his crocodiles devour their prey without tears, and his cataracts fall from the rock without deafening the neighbouring inhabitants (p. 3).

Johnson's translation further emphasized God's impartiality by removing references to special providence.[10] This affected the presentation of the natural world, forcing Lobo's account to fit more closely with the strictures of the preface: for example, 'Johnson omits Lobo's long justification of the plague of grasshoppers as a means of saving the souls of starving people.'[11] This is not a 'secular' view of the natural world, but one sceptical of man's ability to 'see' God in nature. Johnson wrote as one of 'those who believe the Holy Scriptures are sufficient to teach the way of salvation' (p. 5). The *Voyage*, then, strongly supports the subsequent attack on providential interpretations of the everyday operation of the natural world in *Irene*.[12]

Johnson's criticism of seeing God's hand in nature would become less partisan, but in 1735 it was anti-Catholic.[13] In the preface, the *prima facie* rationality of being sceptical of the accounts written by certain denominations and nationalities was not denied. Whilst Lobo's account was probable, Johnson could understand the reader that may 'not be satisfied with a popish account of a popish mission', suggesting such a reader 'may have recourse to the history of the Church of Abyssinia, written by Dr Geddes' (p. 4). This was part of a wider scepticism in England about the truthfulness of Jesuit travel accounts[14]: Johnson stressed not only the 'partial regard' paid 'by the Papists to their church', but also 'by the Jesuits to their society' (p. 4), as errors Lobo had avoided.

Johnson points to a third variety of 'partial regard' which could have vitiated Lobo's account: that 'paid by the Portuguese to their countrymen' (p. 4). That Lobo has transcended nationalism to give a truthful account is seen as the exception, being 'contrary to the general vein of his countrymen' (p. 3). Johnson's attack on the Portuguese relates to the belief that their colonists, along with the Spanish, were unchristian in a way the English were not, which originated with Elizabethan travel writers.[15] LeGrand, Lobo's French translator, is praised

for having dared so freely in the midst of France to declare his disapprobation of the Patriarch Oviedo's sanguinary zeal, who was continually importuning the Portuguese to…propagate by desolation and slaughter the true worship of the God of Peace (p. 4).

LeGrand himself, however, is later castigated for 'preferring the testimony of Father du Bernat, to the writings of all the Portuguese Jesuits…This is writing only to Frenchmen and to [anti-Jesuitical] Papists'(p. 5).

The church and state nexus for Catholic countries rendered their descriptions suspicious to the young Johnson: the burden of proof lay with the writer to demonstrate he was not 'byass'd by any particular views' (p. 5) by confirming the uniformity of man and God's impartiality. A set of binaries was established not dissimilar to those discussed for *London*: in both cases, certain positions – locative in the case of *London's* country/city division, national/denominational in the case of the Lobo preface – were seen *a priori* as leading towards falsehood. In both cases, truth is opposed to conscious falsehood rather than to simple error.[16] The origin of the viewer determines his ability to find the truth. This division related to origins in *London* and the Lobo preface may derive from a common source: as *London* was an attack on Walpole, so Johnson translated Lobo, a book derogatory towards French, Portuguese and Jesuit interests, at a time when Walpole was in alliance with Portugal and France.[17] Even here, however, the difference between the didactic and the empirical modes can be found. For whereas the city dweller in *London* would never achieve true vision, Lobo is proof that the background of a writer can be overcome to achieve factual description. Lobo, by 'consulting his senses' transcends his origins, such that his mistakes are errors not falsehoods: '[if] any argument shall appear unconvincing, or description obscure, they are defects incident to all mankind, which, however, are not too rashly to be imputed to the authors' (p. 5).

Equally distinctive of the preface to *Lobo* is the avowal that Johnson's approach to factual description is grounded in his Anglicanism. Where his subsequent writings eliminated their politicized element as his didactic writings had, Johnson's denominational position remained important to his approach to the structure[18] and content[19] of description, albeit in a more indirect manner.

Johnson's reviews and prefaces

Johnson continued to discuss the requirements of factual description in the forty years intervening between the *Voyage* and his *Journey to the Western Islands of Scotland* (1775) in the more fragmentary form of reviews and prefatory material.[20]

A shift away from the binaries of the Lobo preface can be seen when compared with the review of Du Halde's *Description of China*. Where Lobo's Catholicism and nationality had been *prima facie* grounds for scepticism, in the Du Halde review the burden of proof had shifted to those who doubted the accuracy of the account. Du Halde's narrative, constructed from the accounts of missionaries, is positively treated:

what may not be expected from the united Labours of Travellers like these, Men not intent, like Merchants, only on the Acts of Commerce . . . nor engaged, like Military Officers, in the Care of subsisting Armies . . . but vacant to every Object of Curiosity, and at Leisure for the most minute Remarks [?][21]

The translation was published by Cave, proprietor of the *Gentleman's Magazine*, but Johnson's position appears more than a 'puff' for the man who had rescued him from poverty. Johnson went on to deny that 'Want of Veracity' was due to falsehood:

> If we consider the Nature of the Contradictions discovered in Descriptions of remote Countries, we find them generally such as could not be produced by any apparent Influence, they do not often serve to confirm any Opinion favoured by the Authors, they can neither gratify a Party, nor promote any particular Views [this last phrase is used in the Lobo preface, p. 5, to make the point Johnson here criticises] and therefore must be reasonably considered, rather as Errors than Falshoods (p. 320).

Johnson concludes his argument on this point by suggesting the source of travellers' errors. They are

> Strangers to the Language of the Nations which they describe, suspected and insulted by the People, excluded from the View of those Places which must excite their Curiosity, and afraid of appearing too attentive and inquisitive, lest they should be seized as spies (ibid).

The review goes on to discuss 'the geographical Description' (p. 321) Du Halde provides, in the light of an approach to factual description which Johnson repeatedly elucidated in reviews and prefaces of this period. The central aim must be truth, which involved trusting the senses rather than books. Thus Johnson reviewed Browne's *Civil and Natural History of Jamaica* favourably:

> How much he has added to the history of Sir Hans Sloane we are not able to tell, having not compared them, but have reason to believe that he has generally trusted his own eyes, and then, though he should have discovered no new animals or vegetables, his book is still useful, as the accounts of the two observers necessarily illustrate one another.[22]

The ideal traveller combined, as in the Lobo review, a stylistic criterion with this observational one: 'the qualifications of an American traveller are knowledge of Nature, and copiousness of language, acuteness of observation, and facility of description'.[23] Johnson believed the search for truth in geographical descriptions had led to progress: as he put it in the preface to Macbean's *Dictionary of Ancient Geography*, 'there is no use in erring with the ancients, whose knowledge of the globe was very imperfect' (*Prefaces*, p. 135). In this respect, Johnson adopted the most common position in the aftermath of the 'Battle of the Books': 'all those activities that seemed to work by accumulation, such as the sciences and philosophy, were won for the moderns'.[24]

Geographical description needed another element beyond truth: it had to be useful. Such a linkage remained vital throughout the eighteenth century,[25] and Johnson often affirmed his support for useful knowledge. The Du Halde review enumerated the kinds of information he considered useful:

> the Situation and Extent of every Province is accurately laid, the Cities are enumerated and described, the different Manufactures and Commodities mention'd, and the various Products and Qualities of the Soil minutely specified (p. 321).

What is more, the utility of the knowledge conveyed by accurate description stretched across the ranks. In the dedication to Adams's *Treatise on the Globes* (1766), Johnson suggested

> geography is in a peculiar manner the science of Princes... Your MAJESTY must contemplate the scientifick picture... and consider, as oceans and continents are rolling before You, how large a part of mankind is now waiting on Your determinations' (*Prefaces*, pp. 3–4).

This is not the most reliable source, given Johnson's flippancy at having 'dedicated the Royal Family all round',[26] yet Johnson repeated this position in the *Preceptor*, a work he took more seriously, given its function of educating the young: 'If the Pupil is born to the Ease of a large Fortune, no Part of Learning is more necessary to him, than the Knowledge of the Situation of Nations' (*Prefaces*, p. 182). The *Preceptor* goes on to recommend geography as useful to the 'Learned Professions' (a point repeated in the preface to Macbean's *Ancient Geography*) and to those 'designed for the Arts of Commerce, or Agriculture' (ibid). By 1773, Johnson could have said of his own *oeuvre*, let alone eighteenth-century English writing in general, that 'the necessity of Geography to historical, political, and commercial knowledge, has been proved too often to be proved again' (*Prefaces*, p. 134).

The virtues of description lay in truth and utility, but this, in turn, was part of a hierarchy of concerns in Johnson's thought: morality could not be compromised by the dictates of utility. For this reason Johnson's opposition to slavery did not conflict with his desire for exploration and the expansion of geographical knowledge. The utility of knowledge was to be deployed to advance the physical condition of all: Johnson spoke of 'the time when science shall be advanced by the diffusion of happiness' (*Prefaces*, p. 4). In contrast to the Lobo preface, Johnson's preface to *The World Displayed* (1759) collapsed the notion of Portuguese colonialism being unchristian in a way English colonization was not: in attacking 'their treatment of the savage people', Johnson spoke of 'the practice of all the *European* nations, and among others of the *English* barbarians that cultivate the southern islands of *America*' (*Prefaces*, p. 227).[27] This focus on morality legislating the deployment of useful knowledge was part of a broader Christian position. The preface to the *World Displayed* and the dedication to Adams's *Treatise*, both of which attacked Europe's treatment of the people of other regions, also made clear Johnson's ideal of the usage of the fruits of geographical description in a Christian moral context, hoping that

> multitudes who now range the woods for prey, and live at the mercy of winds and seasons, shall by the paternal care of Your MAJESTY enjoy the plenty of cultivated lands, the pleasures of society, the security of law, and the light of Revelation (*Prefaces*, p. 4).[28]

The relationship between knowledge of the natural environment and Christianity was also made clear in Johnson's prefaces. As in the Lobo preface, Johnson held religions at a distance from natural knowledge. Thus, Johnson was enthusiastic in the concluding paragraph he contributed to Kennedy's *Astronomical Chronology* (1762), for this strengthened scriptural religion by establishing 'the truth of the Mosaical account': 'the universe bears witness to the inspiration of its *historian*', rather than directly to the Creator. This was not a simple argument from design. Rather, 'the validity of the *sacred writings* never can be denied, while the moon shall encrease and wane' (*Prefaces*, p. 77, emphasis added). In short, the argument from design is not an independent proof of God, but a supplemental proof of the truth of scriptural revelation. Johnson did, in fact, praise the design argument in Section VIII of his preface to the *Preceptor*, but this was qualified in Section IX, which points to the limitations of 'the Reason...strengthened by Logic, or the Conceptions of the Mind enlarged by the study of Nature' (*Prefaces*, p. 185).

The theory of description in the *Journey to the Western Islands*

Seen in this context, the theoretical *obiter dicta* in the *Journey* appear as a continuation of established themes.

To achieve truthful description, Johnson made several recommendations. First, detailed description was encouraged: defending his own 'diminutive observations' on Scottish windows, Johnson argued that 'the true state of every nation is the state of common life'.[29] By implication, the details of an individual's or a group's relation to the natural environment can reveal significant information. Johnson demonstrated this when landing at Raasay, where 'we had...some difficulty in landing'. He was led on to the reflection

> that rocks might, with no great labour, have been hewn almost into a regular flight of steps; and as there are no other landing places, I considered this rugged ascent as the consequence of a form of life inured to hardships, and therefore not studious of nice accommodations (*Journey*, p. 47).

This did not go unnoticed at the time, Johnson being attacked for his 'great *scrupulousity* of *minute investigation*'.[30] Johnson was part of the trend to close observation opposed to 'reverie's protracted but objectless moment',[31] of which Hanway, with her determination to travel sentimentally,[32] was an undistinguished representative.

Accuracy was also achieved by measurement and making notes *in situ*. Johnson measured an old fort on Inchkeith (*Letters*, ii. 54), and admitted on Iona that he 'brought away rude measures of the buildings, such as I cannot much trust myself, inaccurately taken, and obscurely noted' (*Journey*, p. 124). A similar pattern of stricture and self-deprecating deflation occurs in Johnson's thoughts on making notes *in situ*. Johnson expends three paragraphs on 'how much a few hours will take from certainty of knowledge', and in provocative proximity, at the end of a further three paragraphs, admits, 'I committed the fault which I have just been censuring, in neglecting, as we passed, to note the series [of islands] of this placid navigation' (*Journey*, p. 123).[33]

The descriptive pitfalls Johnson sought to avoid also followed from his earlier work. His only criticism of Browne had been 'not omission, but unnecessary diligence ... it should be considered that what has been already compleatly described, it is of no use to describe again'.[34] It followed from this that Johnson justified not describing Edinburgh as it was 'too well known to admit description' (*Journey*, p. 1).[35] Similarly, Johnson eschewed extensive description of places where he could not collect sufficient material to furnish a full discussion: 'To *Ulva* we came in the dark, and left it before noon the next day. A very exact description therefore will not be expected' (ibid., p. 118).

The aims of Johnson's travels were repeatedly described by him in the same terms. From Scotland he wrote to Mrs Thrale, 'I have many pictures in my mind, which I could not have had without this Journey' (*Letters*, ii. 94).[36] The *Journey* itself expressed the same notion in different terms. New ideas are the outcome of the mediation of received ideas and new sensations:

> ideas are always incomplete, and ... till we have compared them with realities, we do not know them to be just. As we see more, we become possessed of more certainties, and consequently gain more principles of reasoning, and found a wider basis of analogy (*Journey*, p. 31).

His discussion of Martin's tours of Scotland links this process of comparative reasoning with his strictures on Browne. Both Martin and Browne 'had not knowledge of the world sufficient to qualify him for judging what would deserve or gain the attention of mankind'. Where in Browne this led to a discussion of everything, in Martin it led to 'many uncouth customs that are now disused' being overlooked

> [as] the mode of life which was familiar to himself, he did not suppose unknown to others, nor imagined that he could give pleasure by telling

that of which it was, in his little country, impossible to be ignorant (*Journey*, p. 52).

As with the reviews and prefaces, there was a hierarchy of concerns which focused Johnson's attention in the *Journey*. First, the present was more important than the past for the traveller to acquire ideas about: 'events long past are barely known; they are not considered' (*Journey*, p. 6). This is the temporal equivalent of Johnson's (spatial) exhortation to make notes *in situ*, and moulded the foci of Johnson's attention when viewing 'some stones yet standing of a druidical circle, and what I began to think *more worthy of notice*, some forest trees of full growth' (*Journey*, p. 15 [emphasis added]).[37] The focus, then, was on the utility of the observed; a corollary was that it was more important to observe people, especially in their interaction to form a 'system of life',[38] than to observe nature. To be 'the mere lover of naked nature' (*Journey*, p. 130) was to forget the hierarchy of human significance as surely as (albeit less seriously than) the colonist had.

The origins of 'Johnson's' theory

It would appear, then, that Johnson's approach to description was fairly consistent throughout his life, the major shift being the extension of the criteria of assessment to all observers, replacing the assumption that some groups, because of their origins, could not obey these prescriptions. Even in Lobo, however, the Johnsonian criteria for positive evaluation of a description were fixed: truthful reporting of the evidence of the senses in a plain style, that evidence being gathered with its usefulness to man in mind, as well as the broader moral and religious truths of a Christian life.

Curley has suggested that 'the specific Renaissance source…the single most important influence on Johnson's philosophy of travel' was Howell's *Instructions* (1642).[39] Whilst Johnson did cite Howell, it seems mistaken to attribute to him the significance Curley does. To speak of writer A as 'influencing' writer B, it is necessary

(a) that there should be a genuine similarity between the doctrines of A and B; (b) that B could not have found the relevant doctrine in any writer other than A.[40]

There can be no doubt that (a) is partially satisfied in the case of Howell and Johnson,[41] but (b) most certainly is not. The ideas the two are said to share are 'a humanist study of men and manners, rather than nature or artifacts', an attention to 'the cultural diversity and moral uniformity of mankind' and a focus on 'moral and spiritual enrichment' in travelling.[42] Such ideas were common throughout the period.[43] For example, Bacon's essay 'Of Travel', which was written long before Howell's *Instructions*, and cited by Johnson under 'travel' in his *Dictionary*, focused on the importance of studying men and manners, and encouraged the keeping of a diary of remarks made *in situ*, as did both Howell[44] and Johnson. The uniformity of human nature was a widespread assumption,[45] which precludes it leading back from Johnson to Howell.

If we also look to Tucker's *Instructions for Travellers*, contemporaneous with Johnson, it contains parallels to Howell: both stress that the traveller must be unshakeable in his own religion and well versed in his national history and constitution before travelling;[46] both also emphasize that travel is best when it 'rubs off local Prejudices',[47] yet leads not to the affectation of foreign airs but a rational pride in England.[48]

This is not to suggest that Johnson's theory was 'Baconian' rather than 'Howellian', nor to convert Tucker into another follower of Howell, but to suggest that Johnson's approach to travel and factual description is a collection of post-Reformation commonplaces, and thus cannot be seen as specifically 'influenced' by anyone, or indeed as distinctively *Johnsonian* rather than just *Johnson's paraphrase*.

Johnson's remarks and their referents

Johnson passed judgement on a large number of travel accounts and natural histories. By referring back to these accounts, with particular reference to their remarks on landscape and the environment, I have sought tentatively to reconstruct the approaches to describing the natural world which Johnson liked and disliked.[49] The source of Johnson's judgements is not clear, and was unlikely to be the presentation of the environment, but consistent patterns of description in both approved and disapproved categories suggests that the authors' methods of describing landscape were both part of a more general approach to factual description and significantly correlated with the bases of Johnson's assessments.

The characteristics of positively assessed works

Looking at the prefatory material of the descriptive works, Johnson recorded his approbation of, the guidelines he advocated are much in evidence. Bell, whose work Johnson advised Boswell to read (*Life*, ii. 55), says he 'took notes...by way of diary, from time to time, during the course of my travels'.[50] Bell also recorded his observations 'without attempting to embellish them, by taking any of the liberties of exaggeration, or invention, frequently imputed to travellers'.[51]

The prefatory material also suggests a common focus on 'useful knowledge'. Browne's *Jamaica*, which Johnson reviewed favourably, was said to be

> interspersed with such remarks and observations as I could find well grounded or attested, and likely to prove of any service to mankind...
> and I hope by these means to render it an agreeable entertainment to the lovers of Natural History in general; profitable to such as live in those

parts in particular; and useful to such as may be induced to visit, or practice in, the like climates.[52]

The advance in such knowledge was also pointed out in a passage on Vesuvius similar to Johnson's preface to Macbean's *Ancient Geography*:

> 'Tis pretty surprizing that the Antients, whose Wisdom and Sagacity are cried up, as greatly superior to those of the Moderns, should have invented so many Fictions with regard to these burning Mountains.[53]

Twiss, about whom Johnson was more ambivalent, saw his *Travels through Portugal and Spain* as advancing geographical knowledge as he 'had never seen any satisfactory account of those two kingdoms'.[54]

Soyer's biography, prefixed to Blainville's *Travels*, a work Johnson recommended, summarized the ideal traveller. Soyer says "Twas Mr *de Blainville's* Custom to set down his hints daily.' Blainville's limits were those of a mind learned enough to extract useful reflections from travelling:

> next to Languages, the Knowledge of Geography and Chronology, of History Ancient and Modern, are highly necessary to a Traveller, and in all these Branches of Learning our Author excelled.

Blainville was also 'a true Citizen of the World', 'judging with Impartiality of all Nations'. Finally, he combined all this with religious orthodoxy:

> in his [Blainville's] Opinion...Superstition destroys the very Foundation of Religion; at the same time that 'tis the Bane of all useful Knowledge... [Yet] he never attacks...[ideas] which are commonly received among true Christian churches.[55]

Turning to the descriptions of landscape made by these authors, all were able to describe a scene clearly. Keysler's description of the Falls at Schaffhausen is a good example of straightforward description, wherein the elements of the prospect are described in terms of colour and mental response:

> the fall divides itself into three streams, of which, the green beds and silver vortices make an agreeable contrast to the beholder; but at the same time his mind cannot help being filled with a mixture of dread and amazement at the roar of the waters...[56]

Twiss and Brydone were more self-consciously artful in their relations, using literary references to build their word pictures.[57] In each case the writer described a landscape he has seen, with more general reflections emerging from close observation and clear description.

The works Johnson evaluated positively also made a connection between the beauty of a prospect and its utility. Indeed, an attractive landscape not converted to human use was seen as lacking a vital element, as Bell demonstrated at the Bakall [Baykal] Lake:

> in surveying these fertile plains and pleasant woods, I have often entertained myself with painting, in my own imagination, the neat villages, country-seats, and farm-houses, which, in the process of time, may be erected on the banks of the rivers, and brows of the hills.[58]

Given the values Johnson bore in mind when assessing the descriptions of others, the following description by Keysler, blending aesthetics with utility, and focusing on the condition of the poor, seems one Johnson could only have approved of:

> The *Lago Maggiore* is every way environed with hills covered with vineyards and summer-houses. Above the vineyards are plantations of chestnut-trees, the fruit of which, in the northern parts of *Italy* are consumed in such quantities, that when chestnuts are in great plenty, the price of corn falls, especially in *Genoa*. They continue fresh and green till *Christmas*; but the common people eat them till *Easter*... Along the banks of the lake are fine rows of trees, and walks arched with vine-branches; but it is on the left-hand of the lake... where these natural discoveries are seen in their greatest perfection; as this spot is exposed to the south, it produces a generous wine...[59]

Johnson's positive assessment seems to be linked with the scientific status of the author: Twiss, Brydone and Keysler were all Fellows of the Royal Society, and Bell, whilst on diplomatic service, engaged in scientific speculation.[60] The scientific approach was linked to an attack on superstition. Keysler speaks of Mount Solfatura, which was said by locals to be inhabited by spirits:

> Even the light of Christianity has not expelled these chimeras... the vulgar believe, that these apertures are spiracles, if not of hell, at least of purgatory; and these idle notions are carefully promoted by a *Capuchin* convent.[61]

Particular scorn was reserved for the Neapolitan patron saint, Januarius, whose relics were supposed to be able to stop the eruptions of Vesuvius.[62] Twiss was one of many to ridicule Januarius,[63] being part of an attack on superstition which is the keynote of his *Travels*.[64] Johnson's apparent approval (albeit ambivalent) of Twiss (see *Life*, i. 345) could hardly have been made if he disapproved of the position Twiss adopted on omens and miracles in the natural world throughout his narrative. This was not simply anti-Catholicism, Twiss also attacked the popularity of almanacs in England.[65]

The attack on superstition was not equivalent to these authors deploying a secular form of description. As with Kennedy's *Astronomical Chronology* (1762), the attack on superstitious readings of God's presence in natural events combined with a defence of the scriptural account of the natural world.[66] Keysler argued of fossils that 'many difficulties present themselves, unless recourse be had to the effects of the general deluge',[67] and in China, Bell related that 'as to Noah's flood, he [the Emperor] affirmed, that, at or near the same time, there was a great deluge in China'.[68] Blainville and Keysler further elaborate on the relationship between God, man and nature in their accounts. Blainville displays a scepticism about the design argument, where man's

> blind self-love makes him refer all to himself as the chief End; he imagines he can penetrate into the most hidden Secrets of Nature; yet in Truth he strictly speaking knows nothing at all.[69]

Similarly, Keysler's discussion of fossils (*lusus naturae*, the sports of nature) led him to reflect on the meaning of 'Nature':

> either it is a non-entity, to which no art or regularity of action can in anywise be attributed; or it is a being which, without a gross violation of the regard due to it, cannot be said to sport.[70]

This parallels Boyle's *Free Enquiry* which Johnson paraphrased in the *Dictionary*.[71] To observe nature factually without recourse to miracles and myths was not to secularize it, but, for Johnson and those travellers he approved of, to remove the danger of idolatry and thus establish a distinctively Christian interpretation of nature which recognized in contradistinction the scriptural origin of faith.

As the use of design arguments as inducements to faith was, for Johnson, a confusion of categories, so was the reverse notion that natural evidences could undermine scripture. It was on this basis that he attacked Brydone,[72] who criticized not only superstition, but Mosaic Earth history,[73] on the basis of Recupero's analysis of lavas at Vesuvius. Johnson's response is revealing:

> Shall all the accumulated evidence of the history of the world; – shall the authority of what is unquestionably the most ancient writing, be overturned by an uncertain remark such as this? (*Life*, ii. 468).

He points to the large number of travel accounts (such as Bell's and Keysler's) which supported the Mosaic account, but also to the age of scripture and the long-standing belief in the Bible as independent support of its truth.

Johnson, because of his High-Church sympathies, was more open to the 'appeal to antiquity'[74] than many contemporaries, thus playing down the role of science and nature in the proof of Christianity.[75]

It can be seen, then, that the works Johnson admired achieved a fidelity of description which encompassed useful knowledge turned to morally responsible ends, and encouraged speculation and science within the bounds of an orthodoxy which refused to idolize 'Nature'. These works, therefore, amply display the qualities Johnson consistently valued at the theoretical level.

The characteristics of negatively assessed works

Johnson's dislike of Pococke's *Description of the East* (1743) seems different from that he expressed for the travels of Wraxall, Chandler and Forster. Pococke was persuaded to 'give an account of his travels', rather than a simple antiquarian work, yet antiquarianism bulks large. Those descriptions of the face of modern Egypt which Pococke provides are simple enumerations of the elements in view, and link beauty with utility.[76] Johnson's objection, I infer, was not to Pococke's mode of description, but its infrequency, such present-day prospects being interspersed between long descriptions of temples, detailed discussions of the errors of ancient geographers, and the comparison of modern places in Egypt with their scriptural and classical equivalents.[77] In short, Pococke subordinated present utility to past conjecture in a book which invoked the conventions of factual description.

Chandler's *Travels in Asia Minor* made occasional reference to Pococke, but when Johnson wrote to Mrs Thrale 'do not buy Chandler's travels, they are duller than Twiss's' (*Letters*, ii. 209), he was not referring to the same dullness as the antiquarianism of Pococke. Chandler's travels aimed to record antiquities,[78] but the journal he published was not of his discoveries, but of his travels. With Chandler publishing his antiquarian findings elsewhere and failing to concentrate on useful knowledge, the resultant narrative sounded at times like the personification of the traveller of *Idler* 97 'without incidents, without reflection':

> we then had heavy showers and hard gales, by which we were drawn out of our way, and our masts endangered. Light airs and clear weather followed; the sky blue and spread with thin fleecy clouds... We stood for Corsica with a brisk gale and a great swell... We shipped several seas, and tossed prodigiously.[79]

If Pococke erred for Johnson by focusing on the past rather than the present, Chandler focused on a present as devoid of useful knowledge as the antiquarian past. Wraxall, whom Johnson described as 'too fond of words' (*Letters*, ii. 209), had as little useful content as Chandler, but was a more sentimental traveller, perhaps thereby provoking Johnson's comment.

Wraxall's approach to describing a landscape is one indicator of the grounds for Johnson's judgements. Wraxall's prospect on leaving Königsberg where he had met an attractive woman was at the opposite pole to the factual account:

I stopped the carriage on a rising ground at a little distance from the town, and looking back on its numerous spires, which were gilded by the eastern sun, a tear of vexation and despair stood in my eye, and diminished the prospect; but fancy penetrated the gloom, and saw her from the highest tower in Köningsberg, wave her handkerchief in the wind...'[80]

It might appear odd that Forster's *Voyage to the south seas*, the unofficial but scientific account of Cook's second voyage, was disliked by Johnson as betraying 'a great affectation of fine writing' (*Life*, iii. 180). Yet on closer analysis the judgement becomes comprehensible in the light of Johnson's presuppositions. In the preface, Forster expounded a descriptive relativism: 'two travellers seldom saw the same object in the same manner, and each reported the fact differently, according to his sensations, and his peculiar mode of thinking'.[81] While Johnson was prepared to impute differences in accounts to error, he did not go on to suggest conflicting accounts could all be correct. Moreover, Forster was prepared to engage in speculations such as the following:

When we saw the most beautiful fishes in the sea, the dolphin and bonito, in pursuit of the flying fish, and when these forsook their native element to seek for shelter in the air, the application to human nature was obvious. What empire is not like a tumultuous ocean, where the great in all the magnificence and pomp of power, continually persecute and contrive the destruction of the defenceless?[82]

Given Johnson's approach, this was certainly affectation, in that the shift from the natural to the human realm, which he made so frequently in didactic writings, was here being used in a descriptive genre where it was inappropriate. It is also possible that Forster violated Johnson's fundamental opinions by suggestions of heterodoxy.[83] Forster sought in the narrative 'to lift the soul into that exalted station, from whence the extensive view must *justify the ways of God to man*',[84] a presumptuous aim for a travel account.[85] Forster's willingness to invoke miraculous intervention in the *Voyage*, as in Plymouth where the Resolution slipped her moorings,[86] can also only have annoyed Johnson.

As was suggested earlier, Johnson saw the possibility of two varieties of error in factual description, the stylistic and the generic. Chandler and Wraxall were guilty of stylistic affectation, failing to describe the natural world closely, the result being devoid of content. Pococke's error was to invoke the

generic expectations of the descriptive travel account for a work of antiquarian interest. Forster, like Chandler and Wraxall, adopted too 'literary'
a style, but this was the product of a generic error, namely, the failure to
divide factual and didactic approaches, leading to analogical rather than
logical connections in his argument.

It would appear, then, that Johnson's theory of factual description resulted
in remarkably consistent assessments of various works over a prolonged
period. It has also been shown that these judgements resolved into approaches
to the description of the natural environment and of landscape which
Johnson favoured and disliked.

Comparison I: Johnson's assessment of Jonas Hanway as a traveller

Maxwell records Johnson as saying of Jonas Hanway that he 'acquired some
reputation by travelling abroad, but lost it all by travelling at home' (*Life*, ii. 122).
By comparing this assessment with Hanway's *Historical Account* (1753) and
his *Journal* (1756), a final approach may be made to Johnson's ideas about
factual description.

In the *Historical Account*, Hanway's aim was to boost British trade and
accurately delineate Persia.[87] Such aims, as we have seen, Johnson approved
in Bell's commercial embassies and Twiss's travels to Spain. The *Historical
Account* was highly factual, giving evidence on city populations, on the
latitudes of cities, and the products of rural areas. Hanway blends a concern
for utility and aesthetics, as Keysler and Blainville had.

Hanway did not avoid moralizing,[88] but the reflections did derive from
the landscapes he saw:

> Spring was already advanced... the variety of verdure on the different trees
> on the mountains, the lawns and cornfields, filled the imagination with
> the most pleasing ideas... The return of spring naturally delights the mind;
> and to me it afforded so much the more pleasure, as my past winter had
> been attended with many circumstances of distress. How happy thought
> I, might Persia be, if a general depravity of manners [were eliminated].[89]

Moreover, Hanway attacked superstitions about nature:

> the Persians demonstrate the highest superstition... they think meteors
> that resemble falling stars and which are vulgarly called so, are the blows
> of angels on the heads of the devils.[90]

He also drew attention to the fact that 'there is no word used in so loose and
indefinite a sense as that of nature', and opposed those who would make

nature an independent principle as forgetting 'him who is the first cause, and has the course and government of nature in his hands'.[91]

By contrast, the only 'business' of the *Journal* was 'the sight of agreeable objects'.[92] Rather than being an accurate journal made *in situ*, Hanway admits 'five of our eight days journey were elapsed, before a thought of my writing a journal was suggested'. What is more, Hanway's journal was converted into the form of letters for publication.[93] Given Johnson's ideal of factual description, a retrospectively written journal published in the form of bogus letters was likely to attract his scorn.

The *Journal* reversed the *Historical Account*'s ordering of observation and reflection in some cases, and presented the whole in an affected style:

> I remember the remarks you made on the wisdom of our forefathers, in chusing the most delightful situation for their convents and nunneries.... these pious mortals ought to be indulged in the enjoyment of beautiful situations. Their innocence cannot be endangered by it; it rather teaches them the more exalted love of him, by whose power the face of the earth is covered with so many objects to delight the heart. *Nettly-abby* by which we now passed, is most pleasantly situated.[94]

Such a passage, where the reflection precedes the observation of the object which is supposed to facilitate it, justified Johnson's comments in his review:

> We are told much that might have been as well told without the journey. Digression starts from digression, and one subject follows from another with or without connexion.[95]

Hanway himself had seen the journal as 'only a *vehicle*' for moral reflections, or, more whimsically, as a '*medley* of both worlds', present and future.[96] His aim, lamentably in Johnson's opinion, was to mix the didactic and the factual.

Hanway's medley also, unlike the *Historical Account*, confused the hierarchy of values Johnson believed factual accounts should bear in mind. As in the Netley Abbey reflection, aesthetics was seen as harmless, and it became more important than useful knowledge.[97] Johnson had little time for the notion of viewing nature as an innocent pleasure,[98] and must have been even more sceptical of Hanway's Shaftesburian rhapsodies on nature:

> True taste in the arrangement of material objects, such as delight the senses, or exalt the heart, seems to have a great analogy with the harmony, or order, which the love of virtue inspires. It now occur'd to my mind, of what little worth all these glories would be, to me who might be snatched away from them if I had no hopes beyond the grave, nor any

sure ground on which to build those hopes! Of what moment are all the works of art or nature, compared to the happiness of a future State?[99]

Hanway was 'a Latitudinarian' who 'moved in an Evangelical direction',[100] his tone in the *Journal* being akin to Hervey's *Meditations* (see Chapter 4), such that differing interpretations of the evidences of faith may have contributed to Johnson's hostile response to the *Journal*. But the *Journal* received 'nearly universal condemnation' for its failure to recognize the function of a travel journal.[101] Johnson's attitude to factual description, even if inflected by his denominational position, was widely held by his contemporaries.

Johnson and the structuring of empirical description

To what extent did Johnson's approach to writing up his observations in *Journey to the Western Islands of Scotland* reflect the theory of factual description we have outlined?

Structured description in the *Journey*

In the *Journey* Johnson displays an ability to build highly structured descriptions, by the organization of simple descriptive paragraphs into a coherent argument. He showed this initially in his description of Highland life (*Journey*, pp. 33–38) and then on Skye at Ostaig (pp. 63–99), on Raasay (pp. 48–51) and on Coll (pp. 103–13).

The Highland description will be discussed later, but all three island descriptions follow a similar pattern. In each case, Johnson started with physical geography. At Ostaig this entailed a description of the climate of an island 'in the fifty-seventh degree' (*Journey*, p. 63). On both Raasay and Coll, estimation of the size of the island started the discussion (ibid. p. 48, 103), followed by a brief comment on the face of the island:

> *Col* is not properly rocky; it is rather one continued rock, of a surface much diversified with protuberances, and covered with a thin layer of earth, which is often broken, and discovers the stone (p. 103).[102]

Climatic and physical-geographical openings led into a discussion of soil and vegetation. The Ostaig description began with the 'no great exuberance of vegetation' (p. 64) and the soil it was produced by, whilst on Coll the pattern was reversed, the sentence on soil leading to one on the island's vegetation.

These basic outlines of the physical environment led into the utility of that environment to man. Soil and vegetation led on to agriculture, which was dealt with at length in the discussion of island life at Ostaig, where the agricultural tools, productivity of the land, difficulty of transporting the results and its mode of storage were discussed (pp. 65–66). At Ostaig, Johnson then discussed garden cultivation, before a discussion of minerals (or the lack of them) on the islands and kelp burning. This passage has no equivalent in the shorter description of Coll, but both then discuss animals, largely with regard to their use to man as food or motive power (ibid., pp. 67–68, 103). Johnson's general point was that 'in the penury of these malignant regions nothing is left that can be converted to food' (p. 67).

Having begun with the environment and its use to human society, Johnson then transferred his attention to human life in this environment. In Ostaig, the transition follows smoothly from the discussion of animals:

> man is by the use of firearms made so much an overmatch for other animals, that in all countries, where they are in use, the wild part of the creation sensibly diminishes' (p. 68).

Johnson starts with the physical characteristics of the islanders, which are related to the environmental conditions with which the discussion began:

> the inhabitants of *Sky* ... are commonly of the middle stature ... In regions of barrenness and scarcity, the human race is hindered in its growth by the same causes as other animals (ibid., p. 68).

From the individual as affected by the environment, Johnson turned to the society the island environment sustained. This discussion opened with an enumeration of the ranks of island society, starting with the laird and progressing down the hierarchy. Johnson's narrative played off the old system in which 'the inhabitants are of different rank, and one does not encroach here upon another ... he that is born poor can scarcely become rich' with the dissolution of that system 'since money has been brought among them' (p. 70).[103]

After social structure, Johnson dealt with the lifestyle the islands sustained at Ostaig (p. 83 ff) and on Coll (p. 107 ff). This analysis began with the 'habitations of men'. The description on Coll added another dimension by viewing these buildings as part of a settlement pattern, this also serving a comparative purpose:

> as we travelled through *Sky*, we saw many cottages, but they very frequently stood single on the naked ground. In *Col*, where the hills opened a place convenient for habitation, we found a petty village, of which every hut had a little garden adjoining; thus they made an appearance of social commerce and mutual offices ... (ibid., p. 108).[104]

The discussion at Ostaig continued the analysis of the 'system of insular life' (*Tour*, p. 313) with 'their food [which] is not better than their lodging'

(*Journey*, p. 83), also pointing to the inconveniences of peat as a fuel. This can be seen as taking the utility of the environment discussed earlier, and considering it from the angle of sociable existence rather than mere survival.

On Skye, Johnson went on to less material elements of social life. Having admitted that 'the distance of one family from another, in a country where travelling has so much difficulty, makes frequent intercourse impracticable' (p. 85), he lists the 'few pleasures' island life does admit of. Musical life is discussed (p. 85), as is education (p. 86), before Johnson gets to 'the religion of the Islands'. Religion was not discussed on Coll, but was on both Skye and Raasay (pp. 51–53), where it was the concluding element in a discussion, like that on Coll, which began with the extent of the island. On Skye, religion came towards the end of the description, coupled with 'the various kinds of superstition which prevailed here' (p. 88). Concluding the structured discussion at Ostaig, 'a miniature gem' of travel writing,[105] is a discussion of the scanty evidence as to the history of Highland society and learning, and Johnson's scepticism over Ossian.[106]

To summarize, Johnson in his discussion at Ostaig and in abridged form on Coll and Raasay builds closely structured descriptions, which start from the physical environment, considered in terms of its utility to man, before looking to man's material existence in such an environment, and society's operation, both material and intellectual. As such, the hierarchy of concerns which constantly fed into Johnson's understanding of the theory of factual description, taken from various post-Reformation sources, manifested itself not only in his critical activities as a reviewer, but also in the construction of his published efforts at the fusion of truthful observation of useful information and its conveyance in a plain style.

The structured nature of Johnson's approach has been noted previously by O'Flaherty and Jemielity.[107] Curley suggests 'Johnson made use of the exhaustive topics of inquiry recommended by the Royal Society' in Boyle's *General Heads for the Natural History of a Country*.[108] This is plausible, the Royal Society's approach being influential, Johnson quoting Boyle on the correct understanding of 'nature' and reviewing the *Philosophical Transactions*.[109] But Boyle's *General Heads* did not discuss how to structure the published discussion of factual enquiries: the book only gave headings under which the traveller could profitably collect useful information. Furthermore, its only interest with humans was with 'their Stature, Shape, Features, Strength, Ingenuity, Dyet, Inclination, that seem not due to Education'.[110] Johnson's interest went far beyond the natural history of humans, this being only the point of departure for a discussion of island society as it had been influenced by education.

The patterning of Johnson's descriptions was not rigid. Once the pattern was established, it allowed for variations around the theme, the best example being the description of Iona. On arriving there, Johnson delivered one of his best-known passages:

to abstract the mind from all local emotion would be impossible, if it were endeavoured, and would be foolish, if it were possible. Whatever withdraws us from the power of our senses; whatever makes the past, the distant, or the future predominate over the present, advances us in the dignity of thinking beings (*Journey*, pp. 123–24).

This amounts to a rejection of Johnson's usual structure, and a modification of the priorities of description, justified by the 'local emotion'. As such, Johnson started (pp. 124–26) with the remains of the monastery and their history. The normal pattern is reversed, observations on the present day and the natural world coming afterwards, introduced by the connecting observation 'that ecclesiastical colleges are always in the most pleasant and fruitful places' (p. 126). The description then continues in a vein akin to that on Coll. The differences between Johnson's approach on Iona and Hanway's at Netley Abbey should be apparent. Although Johnson reversed his normal order, his treatment of the remains of the monastery remained factual, his observations being conveyed in a simple style. Johnson used the situation of 'ecclesiastical colleges' to introduce a brief discussion of the current state of Iona, where Hanway used the same idea to make the transition from moral reflection to the previously unintroduced object in the landscape. Johnson, then, varied the patterning of his description but never abandoned its goals.

The strong structural control demonstrated in Johnson's *Journey* seems to support the notion of 'a static, not a dynamic thinker'.[111] Johnson admitted that the topics of the *Journey* had been in his mind before he left London, adding 'and books of travels will be good in proportion to what a man has previously in his mind; his knowing what to observe' (*Life*, iii. 301). As the last clause makes clear, this does not mean Johnson did not observe, but that he knew how to observe. Even if Johnson exaggerated, the published *Journey* displayed retrospectively a structured control of its material.

A number of commentators have attempted to prove the opposite, adding to the attack on Macaulay's picture of Johnson as a bigoted Tory:

the observation of particulars . . . is a difficult process . . . Struggle is mirrored in style. A smooth and ordered presentation is sacrificed in the interest of factual completeness . . . Paragraphing is erratic. At many points the sentences and paragraphs end inconclusively.[112]

This seems an exaggeration: at Ostaig, the paragraphing shows clearly the progress of topics in the order previously discussed. Extra paragraphs of reflections are inserted, such as the discussions of emigration, but these reflections emerge from the structure at a logical point (the tensions in the island system of ranks) and are followed by a continuation of the structure of argument where reflection left off.[113] The dynamic view of Johnson as a traveller finds fullest articulation in Radner's 'The significance of Johnson's changing views of the Highlands'.[114] Radner points out that Johnson's famous remarks on the lack of trees in Scotland, in fact, show a shift in his opinion. At the beginning of the *Journey*, Johnson argued that the lack of trees reflected Scotland's improvidence, since 'to drop a seed into the ground can cost nothing' (*Journey*, p. 7). On Mull, however, Johnson stressed the difficulties of planting, and that only 'the speculatist hastily proceeds to censure' (ibid., p. 116). But this need not 'illustrate his evolving compassion for the Islanders as he wrote the *Journey*'[115] for several reasons. First, only one of these comments was made about islanders, the other referring to Lowland Scotland, of the differences between which and the 'insular system of life' Johnson was well aware. What amounted to improvidence in the Lowlands, where 'possession has long been secure, and inheritance regular' (ibid., p. 7) was not so on Mull where 'the consequence of a bad season is... not scarcity, but emptiness' (ibid., p. 115).[116] Given this geographical point, it could be argued that the consistent application of pre-existent principles led to Johnson's different (rather than changing) views. At the very least, the difference cannot be shown to mean change, and thus cannot arbitrate between the dynamic and static interpretations. The notion of an evolving compassion is further confused in this case by the fact that we have no evidence that Johnson's reflections were felt at the time. The passage on trees on Mull in the *Journey* has no equivalent in Johnson's letters (see *Letters*, ii. 103–104): if Johnson's compassion was evolving, it was certainly only 'as he wrote the *Journey*' in a far more literal sense than Radner meant. Given that this passage was a subsequent insertion, it is possibly a response to a similar reflection in Pennant's *Tour in Scotland* on Skye that 'poverty prevents him [the farmer] from making experiments in rural economy'.[117] More generally, such reflections on island life were not without precedent, Borlase's *Observations*, which Johnson reviewed, saying of the Scilly isles 'the true spirit of Planting has either never reached here, or has been forced to give way to more necessary calls'.[118] Finally, the classical origins of Johnson's remarks on trees render them 'participation in what was a commonplace of the western humanistic tradition'.[119] It was because the educated classes of eighteenth-century Britain shared this classical training that Johnson's remarks on trees generated a degree of controversy which no other comment he made on the physical environment did.[120]

The foci of attention in Johnson's subsequent travels to Wales (in 1774) and France (in 1775) suggest little change in attitude, although the lack of a finished work designed for publication prevents a definitive conclusion. But it would appear that factual description took place within a structure which in turn reflected a hierarchy of values, which it, by its position in that hierarchy, could not upset.

The construction of Johnson's structured descriptions of the natural world

Before discussing Johnson's actual approach to the natural world in the *Journey*, it is instructive to observe how it was constructed.[121] Edwards has spoken of 'the mutations of the written word as it moves into print... the ideal paradigm [being] of journal, manuscript draft, and printed version'.[122] In the case of Johnson in Scotland, the evidence surviving is his letters (*Letters*, ii. 46–119) and the final narrative. A book of remarks is missing,[123] which may well have provided factual remarks. Yet some details of the emergence of Johnson's public approach to landscape and the natural world can be gleaned by a comparison of the letters and the *Journey*.

The first point of note is the removal of personal references in the letters' response to landscape from the *Journey*. The general admission to Mrs Thrale that 'I travel with my mind too much at home, and perhaps miss many things observable, or pass them with transient notice' (*Letters*, ii. 95) was removed, as was its effect on the letters. Thus when 'offered' the isle of Isay, Johnson added to Mrs Thrale, 'my Island would be pleasanter than Brighthelmston, if You and Master [i.e. Mr Thrale] could come to it, but I cannot think it pleasant to live quite alone' (ibid., ii. 71). The whole incident and this reflection is removed from the published version.[124] In the transition from private to public, Johnson rendered his account 'at once more precise and impersonal'.[125] This brought his own presentation in line with his structures against sentimental travellers such as Wraxall, and followed the traditions of eighteenth-century travel accounts.[126]

Another mutation in the *Journey*, and one of the vital significance to the public presentation of Johnson's approach to the natural world, is the emergence of the structured descriptions. The descriptions of Ostaig, Raasay and Coll all have antecedents in the letters (*Letters*, ii. 97–98, 82–89, 100–101, respectively), but none have the clarity and order of the final description.[127] The material brought together in the set piece at Ostaig in the *Journey* is found in the letters under various entries for Dunvegan and Talisker.[128] At the close of his letter of 30 September 1773, Johnson attempts a more rounded summary of Skye, but the two paragraphs give little indication of the organization to be found in the published Ostaig passage, moving from climate to gardens to agriculture, before returning to Skye's climate and closing with a paragraph on the island's animals. The effect of the revisions for the published version, then, is to separate observations on the natural world from those on human society and to put the observations in the two realms into a recognizable order.[129]

If the island set-pieces were rendered more organized, so the earlier stages were rendered more connective by the addition in the *Journey* of more observations on incidents and prospects on the road than the letters contained. Thus the letters say 'we travelled towards Aberdeen, another university, and in the way dined at Lord Monbodo's' (*Letters*, ii. 57), where in the *Journey* mention of Monboddo is sandwiched between two inserted paragraphs. The first filled in the prospect between Montrose and Aberdeen, saying it

> exhibited a continuation of the same appearances. The country is still naked, the hedges are of stone, and the fields so generally plowed that it is hard to imagine where grass is found for the horses that till them. The harvest, which was almost ripe, appeared very plentiful (*Journey*, p. 9).

After Monboddo, Johnson added a reflection on his experiences in travelling:

> the roads beyond Edinburgh, as they are less frequented, must be expected to grow gradually rougher; but they were hitherto by no means incommodious (ibid.).[130]

Another example, different from such glimpses on the road, is Johnson's construction of the prospect of Loch Ness in the *Journey* (pp. 22–24) when compared with the letters (*Letters*, ii. 65). Hart summarizes the comparative qualities of the final version:

> It concentrates on 'stationing' the observer, offering dramatic plausibility for his acts of perception, and enabling him to perceive distinctly the separate elements of his total prospect...

Moreover, the appearance of Loch Ness is separated as a concern from more factual matters, with 'questions of size and depth, details of the road and its construction...kept for distinct paragraphs',[131] a presentation identical to the controlled paragraphing of the structured island descriptions. Presumably much of this observational material on roadside prospects must have been in the missing journal(s). Parallel material can be found in Johnson's French journal, with brief notes such as

> the appearance of the country pleasant. No hills, few streams, and one hedge. I remember no chapels nor crosses on the road. Pavement still, and rows of trees (*Diaries*, pp. 238–39),[132]

which could have been expanded into brief paragraphs of the sort found in the *Journey*. Johnson was not known for his observancy on the road: when in the Harwich stagecoach, rather than looking out at the geography of England, Johnson 'had...*Pomponius Mela de situ orbis*, in which he read occasionally, and seemed very intent upon ancient geography' (*Life*, i. 465),[133] but such brief notes could easily be made whilst reading.

The published version of Johnson's *Journey*, then, situated his comments on the natural world in a more rigorously structured account, organized the discussion of the physical and natural environment in its own right, and increased the attention paid to the landscape during the course of the tour, especially its early stages.

The conceptualization of nature in the *Journey*

The natural world is presented descriptively as the point of departure, being at the base of a hierarchy ascending to the non-material elements of human society. As such, the natural environment is described factually, as at the Buller of Buchan:

> it is a rock perpendicularly tubulated, united on one side with a high shore, and on the other rising steep to a great height, above the main sea. The top is open, from which may be seen a dark gulf of water which flows into the cavity, through a breach made in the lower part of the inclosing rock (*Journey*, p. 14).[134]

Johnson never hypostatizes 'Nature', but discusses it as an assemblage of climate, rocks, soil, vegetation and minerals. In this he was part of 'the nominalistic impulse',[135] most clearly expressed in Boyle's *Free Enquiry*, which Johnson epitomized for the 1773 edition of his *Dictionary*, the year of his tour in Scotland. Nature was not personified in the *Journey*: at Loch Ness, Johnson had described 'rocks which rise on three sides' as 'standing like barriers of nature placed to keep different orders of Being in perpetual separation' (*Letters*, ii. 66). This comment, with its undertone of personification, is not found in the published version.

Nature, the collection of physical attributes of an area, is connected to human society only by chains of causation, not by direct causation. The latter would lead to a form of environmental determinism, which Curley argues Johnson espoused: while

> theoretically, Johnson denied that climate controlled individual behaviour or totally accounted for national character...in practice, Johnson certainly stereotyped the French people in a prejudicial manner.[136]

Obviously, the one does not preclude the other, unless the national stereotype is built on climate, which does not hold in Johnson's case. To take Johnson's description of the Highlands (*Journey*, pp. 33–38), he seeks to explain the distinctive characteristics of its inhabitants by a chain of reasoning. The physical character of the Highlands makes them hard to invade: attackers are 'exposed to every power of mischief from those that occupy the heights; and every new ridge is a fortress', from which the locals can escape, knowing 'where the bog has firmness to sustain them' (p. 34). Invaders of barren areas are soon 'dislodged by hunger'. The difficulty of conquest leaves the inhabitants rough in their manners as 'men are softened by intercourse mutually profitable' (ibid.), a situation exacerbated by the fact that 'a tract intersected by many ridges of mountains, naturally divides its inhabitants into petty nations' (p. 35). This form of explanation can be distinguished from the determinism of Howell: 'the barrener a Countrey is, the more Masculine and Warlike the spirits of the Inhabitants are... and herein Nature may seeme to recompence the hard condition of a Countrey'.[137] The direct connection of climate and human 'spirit' as well as the personification of Nature were not part of Johnson's approach. The last paragraph of Johnson's description of the Highlands made his position clear: 'such *are* the effects of habitation, and such *were* the qualities of the Highlanders' (*Journey*, p. 38, emphasis added). The physical environment had not altered, but exposure to society meant the Highlanders were 'now losing their distinction'. The physical environment is vital to human society as material needs must be satisfied from the collection of possibilities the climate, soil and vegetation provide, but its effects are mediated through human society.

Johnson's 'philosophical' description of the natural world in comparative perspective

Boswell, comparing Pennant's travel books with Johnson's, argued the former 'shews no philosophical investigation of character and manners, such as Johnson has exhibited in his masterly *Journey*' (*Life*, iii. 274). This characterization of Johnson as a philosophical traveller has been taken up by many:[138] it gains further meaning when the structuring of his descriptions, with their ascent of a hierarchy of topics, is compared with other contemporary structures of description.

Johnson's approach differs from Pennant's chronological accounts, in which 'the actual time spent at a given place in travel corresponds closely with the proportions of the written account'.[139] As an example, Pennant's first tour included in one day the Fall of Foyers and Loch Ness.[140] It was only after this chronological account that Pennant gave a scientific account of the characteristics of Loch Ness.[141] Johnson also travelled this route in one day,[142] but his account is very differently arranged, the appearance and characteristics of the Loch being described together before a discussion of the waterfall. Pennant's approach, with its textual division of observation from reflection, also led him to insert long historical digressions from his actual travels,[143] where Johnson incorporated historical material into his structure of description. As such, the effect of Pennant's travels can be more fragmented, Johnson's ordered arrangement of all the facts he wishes to relate at a place justifying the label philosophical.

Johnson's approach is also philosophical in contrast to a chorographical approach. Martin's preface to his *Description of the Western Islands of Scotland* announces it to be a geographical work. This assertion is confirmed on Skye, where he observes that 'this Isle is naturally well provided with variety of excellent Bays and Harbours', before enumerating the various harbours on Skye.[144] It was the Elizabethan topographer Leland who 'established a pattern to be followed by future regional writers...describing in a consecutive manner various features'.[145] Johnson's account of Skye is divided by locational headings, but the geographical peculiarities of these four places play a small role in the material found under the headings, a point reinforced by the distillation of information the letters reveal Johnson had collected in various places into the description 'at' Ostaig.[146]

Despite the divergences between philosophical, chronological and chorographical approaches, they were all structured around the same values. In each case, travellers sought in justifying their accounts 'to advance reasons more consequential than those of enjoyment',[147] As such, all emphasized the production of useful knowledge, and dealt with the physical environment only insofar as it affected human society. Johnson's approach is distinguished by the clarity of its organization, but the focus on the utility of the physical environment was a venerable one, as Munro's *Description of the Western Isles* (1594) demonstrates. Munro always discussed the face of islands with an eye to utility: Mull, for example, was 'ane grate rough ile, noch the less it is fertile and fruitful'.[148]

Johnson was one of the last representatives of this tradition, focusing on the environment for its utility, and thus linking the natural and human worlds into a comprehensive descriptive account. Subsequent to Johnson, travel accounts in Scotland start to betray rather more specialized concerns. This may be a response to 'the reading public's insistence upon novelty', which left travellers in 'search of something new to say about previously described areas'.[149] It also perhaps reflects the increasing independence of a scientific and an aesthetic approach to the description of the natural world in the late eighteenth century. On the scientific side, Walker's reports (1764–71), whilst they follow a structured approach to the natural world, lack a concern for island social life. Similarly, Knox's *Tour* (1787) was sponsored by the British Society for Extending the Fisheries. Johnson would have had little disagreement with Knox's statement that 'in the selection of objects, I have had national utility principally in view', yet Knox's focus on fisheries made the utility attended to a narrower one.[150] On the aesthetic side, Gilpin made his picturesque tour in 1776, and was followed by others, Simond speaking of the 'beautiful nakedness' of Glen Croe.[151] This characterization was something of a contradiction in terms to Johnson who described the same place as 'a black and dreary region' (*Journey*, p. 132). Whilst the speed of this change should not be exaggerated, it is symbolic that Garnett, the frontispiece to whose *Observations on a Tour through the Highlands* (1800) informed the reader he was 'Member of the Royal Medical, Physical, and Natural History Society of Edinburgh; the Literary and Philosophical Society of Manchester; the Medical Society of London; the Royal Irish Academy; and Professor of Natural Philosophy and Chemistry in the Royal Institution of Great Britain' was accompanied by the painter Watts, both deploying their own skills to produce the illustrated account.

Comparison II: Boswell as a traveller: figures in the landscape

Boswell's *Tour* made two attempts at a fully structured descriptive argument. The 'survey of Rasay' (*Tour*, p. 168 ff) starts with the dimensions of the island (*Tour*, p. 168), as did Johnson (*Journey*, p. 48) but goes on to the laird's seat and the chapel (*Tour*, p. 169) before reverting to the vegetation, lakes, animals and climate (*Tour*, pp. 170–74). As such, Boswell does not follow Johnson's progression from the natural to the human via utility, but oscillates.[152]

To some extent, these differences reflect different principles of organization. Boswell's 'survey' concludes with 'some gold dust, – some fragments of Dr Johnson's conversation' (*Tour*, p. 174), which speaks of his concern for people in the landscape rather than the natural world as such. But Boswell also attempts to organize his survey in a chorographical manner starting with 'the south end of the island', then Dun Can mountain (both, *Tour*, p. 170) and finally 'the north end of Rasay' (*Tour*, p. 172). His survey of Rasay complete, Boswell the next day (11 September) adds a passage on the exercise of the law on the island (*Tour*, p. 177), which Johnson's approach would have incorporated into the survey in a published version.

Boswell's *Account of Corsica* (1768) also showed a very different structural approach from Johnson's, separating a discussion of the physical geography and history of the island from his journal, the latter being chronologically presented. Moreover, the natural history, described in Chapter 1 followed a chorographical approach, enumerating all the harbours as Martin had on Skye, then the towns and intervening countryside, starting at Capo Corso in the north and working clockwise, before discussing the interior.[153] Boswell went on to discuss the rivers, animals, vegetation and mines on the island[154] in a way akin to the approach fostered by the Royal Society.

That Boswell was less conscious than Johnson of the organization of factual description can be discovered by comparing the manuscript *Tour* with the published version.[155] At some points, the transition shows increasing organization. This occurs in the Raasay survey, for example, where the final version (*Tour*, pp. 173–74) separates off a paragraph on animals and one on climate. By contrast, the manuscript version[156] had a similar paragraph on animals confused by a Johnsonian anecdote (which was eventually omitted), then a paragraph on Boswell's wife and daughter (omitted from the final version as inappropriate),[157] and a paragraph on a barn in Raasay to which are adjoined the observations on climate (*Journal*, pp. 148–49). Equally, however, Boswell's alterations for the final version made the description less clear on Coll. Having discussed the physical geography of the island, the manuscript begins a new paragraph on the islanders, reflecting a transition to the material elements of human society, where the published version runs the two things together (cf. *Journal*, pp. 281–82 and *Tour*, pp. 300–301). As such, Boswell's approach did not consistently add polish and clarity to his descriptions.

Turning to content, Boswell saw his approach to the natural world as similar to Johnson's: 'I have a notion that he [Johnson] at no time has had much taste for rural beauties. I have myself very little' (*Tour*, p. 112). Boswell was aware of the limitations of his verbal landscape descriptions (see *Journal*, pp. 148, 180), but he rarely attempted them. His lack of interest in rural beauties is confirmed by his journals, where, even on the Grand Tour, he only recorded comments such as 'the prospect was horribly grand' of the Alps, but did not feel compelled to describe it.[158] Why was Boswell so lacking in concern for natural aesthetics? Part of the answer, at least during the early years, is: 'I have been a week in Siena and have not as yet seen any *maraviglia*... I have been so busy with women that I have felt no curiosity about inanimate objects.'[159] But a more convincing answer for the entirety of his life is

for Boswell, coming from a Calvinist background stressing the fallen nature of humanity and the ineffectiveness of the human will, nature is a realm providing little spiritual sustenance.[160]

Boswell, like Johnson, emphasized the visible church[161] and correspondingly downplayed the role of nature in proving the existence of God. This was probably an important reason for Johnson and Boswell's shared dislike of rural beauties.

Boswell was, however, responsive to landscape in one situation, which first manifested itself in his visit to Rousseau.[162] Having said little of nature, Boswell heard that Rousseau 'will walk in such wild places for an entire day' and begins to imitate this:

to prepare myself for the great interview [with Rousseau], I walked out alone. I strolled pensive by the side of the river Reuse, in a beautiful wild valley surrounded by immense mountains, some covered with frowning rocks, others with clustering pines, and others with glittering snow. The fresh, healthful air and the romantic prospect gave me a vigorous and solemn tone.[163]

It was not simply Rousseau's love for nature which led Boswell to landscape appreciation, but any famous figure he could connect to a landscape. Thus he commented on Iona 'what an addition was it to Icolmkill to have the Rambler upon the spot!' (*Journal*, p. 331).[164] Johnson was also incorporated into the fusion of figure and landscape at the family estate of Auchinleck:

it was my intention to erect a monument to him here, among scenes which, in my mind were all classical; for in my youth I had appropriated to them many of the descriptions of the Roman poets. (*Journal*, p. 374).

Boswell's iconography was not a retrospective construction, for he had written to Wilkes in 1765,

> when I am at Auchinleck in a sweet summer season, my imagination is fully persuaded that the rocks and woods of my ancestors abound in rural genii. There is hardly a classical spot which I have not upon our own estate.[165]

Yet the connection of classicism and the Auchinleck landscape was severed tragically for Boswell: 'after [his wife] Margaret's death he could hardly bear to go to Auchinleck even on a visit'.[166]

Boswell's connection of the figure to the landscape could lead him to a personification of nature which separated his approach to rural beauties from Johnson's factual description. Looking upon Loch Moidart, Boswell recalled that Prince Charles had landed there on the flight from Culloden:

> the hills around, or rather mountains, are black and wild in an uncommon degree. I gazed upon them with much feeling. There was a rude grandeur that seemed like a consciousness of the royal enterprise, and a solemn dreariness as if a melancholy remembrance of its events had remained (*Journal*, p. 247).

The landscape here became a vehicle for Boswell's emotions and started to act in its own right, a position far removed from Johnson's image of nature as a series of discrete physical conditions.[167] The personification of the landscape also led to a form of environmental determinism most notable in the *Account of Corsica*. Boswell argued in his introduction that 'liberty is... natural... to mankind'.[168] The physical geography of Chapter 1 implied that liberty was quite literally natural in the Corsican case:

> the interiour parts of the island... have a peculiar grand appearance, and inspire one with the genius of the place; with that undaunted and inflexible spirit, which will not bow to oppression.[169]

This thesis, not unlike Howell's, posited a quite different relationship between the environment and society from that which Johnson propounded.

Boswell, then, did not deploy the same model of structured description as Johnson, nor was he able consistently to apply any one approach. This reflected uncertainty as to whether the natural world was too trivial to be of concern given his overriding concern for role models, something to be connected with those individuals, or an important determinant of the character of a people. Boswell provided an admirable summary of the principle behind his shifting stance: 'A landscape or view of any kind is defective, in my opinion, without some human figures to give it animation' (*Journal*, p. 331).

The content of Johnson's landscapes of factual description

Within the structured approach Johnson adopted, and given the view of nature he articulated, the question remains of what he actually 'saw' (or, more accurately, saw and then wrote about) in the landscape. Out of the welter of material Johnson was confronted with, that which he chose to highlight when describing the landscape perhaps gives some insight into what he considered important intellectually and what he believed would elevate factual description. A point of entry into such an enquiry can be gained by considering Johnson's well-known passage about Slains Castle:

> We came in the afternoon to *Slanes Castle*, built upon the margin of the sea, so that the walls of the towers seem only a continuation of a perpendicular rock, the foot of which is beaten by the waves. From the windows the eye wanders over the sea that separates Scotland from Norway, and when the winds beat with violence must enjoy all the terrifick grandeur of the tempestuous ocean. I would not for my amusement wish for a storm; but as storms, whether wished or not, will sometimes happen, I may say, without violation of humanity, that I should willingly look out upon them from Slanes Castle (*Journey*, p. 13).

This passage sees Johnson's description slip into an aesthetic response to the landscape, which is then checked by a moral caveat, in its turn qualified by a religiously inspired view of the landscape, critical of notions of special providence outside the purview of natural laws. This passage, I suggest, epitomizes three central areas of Johnson's response to what he saw in the landscape and also articulates the interlinkages between them.

The aesthetics of landscape

Johnson's comment at Slains on 'the terrifick grandeur of the tempestuous ocean' clearly deploys the aesthetic language of the sublime. There has been some discussion of Johnson's position on landscape aesthetics, suggesting he was more sensitive than was once thought,[170] but also that this was restrained by his moral imagination.[171] I do not differ from this opinion, but would seek to analyze Johnson's responsiveness to various landscapes by putting it in the context of contemporary descriptions and descriptive practices, which suggest that if Johnson was not as insensitive to landscape as Macaulay suggested, he was less concerned with it than many of his contemporaries.

Johnson's visual landscapes in comparative perspective

Slains was not the only occasion on which description tipped over into aesthetic response: a few paragraphs later, Johnson's description of the Buller of Buchan, previously cited for its precision, also used the language of

Burke's sublime, being 'a dark gulf of water,' 'which no man can see with indifference, who has either sense of danger or delight in rarity' (*Journey*, p. 14). The published version here mingles novelty, danger, darkness and depth, all elements of Burke's sublime. Johnson's letters reveal a calmer response: the Buller was 'black at a great depth', but 'there was no danger' (*Letters*, ii. 62). In both cases, however, the language of discussion is the sublime.

Johnson was also able to combine the language of the beautiful with that of the sublime, as at Loch Ness:

> On the right the limpid waters of *Lough Ness* were beating their bank, and waving their surface by a gentle agitation. Beyond them were rocks sometimes covered with verdure, and sometimes towering in horrid nakedness (*Journey*, p. 22).

Even M'Nichol, in his attack on Johnson's *Journey* had to admit the aesthetic resonance of this passage in his own inimitable fashion:

> When riding along the side of Loch Ness, a ray of good humour seems to have stolen into the Doctor's mind...Even his own description of the scene through which he passed, in spite of all his endeavours to the contrary, conveys enough to the mind of the reader to make him regret that he has not a more perfect view.[172]

There is another sense of aesthetic pleasure in Johnson's account of the night voyage to Mull:

> The day soon failed us, and the moon presented a very solemn and pleasing scene. The sky was clear, so that the eye commanded a wide circle: the sea was neither still nor turbulent: the wind neither silent nor loud. We... therefore contemplated at ease the region through which we glided in the tranquillity of the night (*Journey*, p. 123).[173]

Johnson could also use that basic picturesque technique, the comparison of two prospects, as he did at Slains Castle:

> Dr Johnson observed, the situation here was the noblest he had ever seen, – better than Mount Edgecumbe, reckoned the first in England; because, at Mount Edgecumbe, the sea is bounded by land on the other side, and, though there is there the grandeur of a fleet, there is also the impression of there being a dock-yard, the circumstances of which are not agreeable (*Tour*, p. 102).

Having said this much for Johnson's aesthetic response to landscape it should be remembered that this does not amount to a 'vindication' of

Johnson, the factual description of landscape being no 'better' or 'worse' than its aesthetic description. By the same token, to suggest Johnson's response to landscape was guided by concerns of morality is not to show that 'the improper judge of landscape was not Samuel Johnson but William Gilpin',[174] for this is simply to reverse the prioritization of Romanticism's response to landscape. To refuse to see Johnson as equally responsive (in aesthetic terms) to landscape as his contemporaries is a factual judgement, which, while revealing something about Johnson's intellect, does not lead to an evaluation of praise or denigration. Such a judgement can be justified by a comparative approach since, for landscape aesthetics as for travel liars, 'the best clues...can be found by comparing the various accounts of one particular trip or different trips to the same place – in short, by setting one traveller against another'.[175]

Johnson's approach to landscape gardens was not suggestive of aesthetic sensitivity. Visiting Baronhill on Anglesey, Johnson's response was unusually long by the standards of his journals:

[the] garden is spacious and shady, with large trees and smaller interspersed. The walks are strait and cross each other with no variety of plan but they have a pleasing coolness and solemn gloom, and extend to a great length (*Diaries*, p. 202).

Johnson does not note the situation of Baronhill, which 'rises from a swelling lawn above Beaumaries in the midst of a thick grove',[176] nor the prospect from Baronhill, which was described by most as its greatest glory. Pennant was the most fulsome about the prospect:

The view is justly the boast of the island. The sea forms a most magnificent bay, with the *Menai* opening into it with the grandeur of an *American* river. The limit of the water in front is a semicircular range of rocks and mountains, the chief of *Snowdonia*, with tops spiring to the clouds, and their bottoms richly cultivated, sloping gently to the water edge. The great promontory *Penmaen Mawr*, and the enormous mass of *Llandudno*, are rude but striking features, and strong contrasts to the softer parts of the scenery.[177]

Johnson, then, is not enthusiastic by the standards of his generation, a view supported by his response to other gardens, including Hagley. Here Johnson noted that 'from the farthest hill there is a very wide prospect' (*Diaries*, p. 218), but did not record what was in that prospect, unlike Pococke and Young, neither of whom is remembered as an aesthete.[178] Byng was 'disappointed with Hagley', but went on to explain why in his private journal:[179] 'Hagley is deficient of water and gravel, two great charms.'[180] In contrast to both positive and negative responses to landscape gardens, Johnson appears simply indifferent.

Johnson omitted discussion of the prospect of towns as he approached them, which was one of the ways in which discussion of the land as an organized landscape came into the English tradition of topographical writing. The tradition stretches back to Leland,[181] who wrote of Johnson's birthplace – 'Lichfield is built on a low-lying flat site, and it is only the close and cathedral church, with a long street to the north of the town bridge, that occupy rather higher ground. There is no evidence that there was ever a town wall.'[182] While the militaristic element was diluted over time, the town prospect itself remained and became aesthetic. Johnson did not make such observations, which were recommended in many instructions for travellers, including Howell's.[183] This was most obvious at Edinburgh, 'a city too well known to admit description' (*Journey*, p. 1).[184] The experience was repeated later, with Johnson only saying 'we came to St Andrews' (*Journey*, p. 2), missing a chance to give a prospect of the city which for Macky 'at three Miles distance makes a very august Appearance'.[185] M'Nichol was mistaken, however, in his belief that Johnson's omission represented an anti-Scottish feeling,[186] for similar prospects are missing for Paris, where the aerial view of the city from Notre Dame was a travellers' commonplace,[187] and most of the English cities Johnson visited, even the magnificent situation of Durham eliciting only a brief comment, and that because it was 'a place of which Mr Thrale bad me take particular notice' (*Letters*, ii. 49).[188]

A third point of comparison, discussed by Brownell,[189] is Johnson's response to the situation of castles. As with gardens and towns, he did not see castles as part of an assemblage related to their situation. Johnson gleaned historical ideas from viewing castles, admitting Caernarvon 'surpassed my Ideas' (*Diaries*, p. 204), but by the time he reached Conway being 'afforded . . . nothing new' (ibid., p. 210). This can be compared with aesthetic types of response. Craddock wrote that

> the prospect from the castle [Denbigh] is most enchanting. Beneath, the vale of Clwyd displays her bosom, profusely gay to the admiring spectator. The banks of the river Clwyd decorated with seats, the towns of Rhythin and St Asaph, with the mountains rising at a distance, form a most delightfull view.

Pennant disagreed with some of Craddock's assessment, but his prospect was also aesthetic, allowing the castle to frame the view: 'the prospect through the broken arches is extremely fine. . . . a rich view, but deficient in water: the river *Clwyd* being too small to be seen'.[190] Simond's picturesque reversed Johnson's prioritization of the historical over the aesthetic: 'ruined castles, with the usual stories about Cromwell's cannon. He was a great master of the picturesque, and his ruins are always in the best taste.'[191]

Fourthly, Johnson's response to regions he visited was not dictated by scenic beauty. Johnson frequently visited the Derbyshire–Staffordshire area, staying at Ashbourne. 'The places Johnson visited' were not, to his mind, 'noteworthy for their pervasive uniqueness',[192] but this was not the response of others. Warner was rapturous at Matlock:

> Here a scene burst upon us at once, impossible to be described – too extensive to be called picturesque, too diversified to be sublime, and too stupendous to be beautiful[193]

Johnson's only comment was 'at Matlock'! (*Diaries*, p. 166) This could be put down to the brevity of Johnson's journal entries, but Walpole shows how the same format could reflect a much more aesthetic response: 'Matlocke. Most beautifull Scene, rocks, woods, cascades, mineoi and walks and seats.'[194]

It can be concluded, then, that in a variety of circumstances Johnson showed himself to be not simply a 'philosophical traveller', but actually uninterested in the relationship between a feature and its situation. Johnson looked aesthetically (as in his conceptualization of nature) to discrete entities rather than their assemblage.

Sensory response to landscape

If Johnson was not responsive to visible aspects of landscape, he was responsive to soundscapes. This can be seen at Slains Castle, where he spoke of 'the winds [which] beat with violence' (*Journey*, p. 13). At Dunvegan, his analysis was careful:

> we...suffered the severity of a tempest, without enjoying its magnificence. The sea being broken by the multitude of islands, does not roar with so much noise, nor beat the storm with such foamy violence, as I have remarked on the coast of *Sussex* (*Journey*, p. 56).

Here Johnson engaged in a comparative analysis of soundscapes akin to his visual comparison of Slains Castle and Mount Edgecumbe, the difference being that Johnson considered the soundscapes worthy of publication, where his visual comparison is only preserved by Boswell. The comparison continued on Coll where Johnson recorded, 'I know not that I ever heard the wind so loud in any other place' (*Journey*, p. 103).

Just as Johnson's most visually aesthetic responses were related to the sublime, so were these soundscapes:

> the eye is not the only organ of sensation, by which a sublime passion may be produced...The noise of vast cataracts, raging storms, thunder, or artillery, awakes a great and aweful sensation in the mind, though we

can observe no nicety or artifice in those sorts of music. The shouting of multitudes has a similar effect.[195]

A further soundscape in the *Journey* has affinities with this passage:

the cataracts which poured down the hills, on one side, and fell into one general channel that ran with great violence on the other. The wind was loud, the rain was heavy, and the whistling of the blast, the fall of the shower, the rush of the cataracts, and the roar of the torrent, made a nobler chorus of the rough musick of nature than it had ever been my chance to hear before. The streams . . . were so frequent, that after a while I began to count them; and in ten miles, reckoned fifty-five (*Journey*, p. 132).

As with Burke, it was cataracts which produced the sublime sensation, which both liken to music of 'no nicety'. The juxtaposition of this idea in the *Enquiry* with the shouting of multitudes may well have led Johnson to run together natural sounds, the multitude of 'voices' produced by the large number of streams, and the tradition of the rough music,[196] to come up with his phrase 'the rough musick of nature'. Regardless, the passage again records Johnson's far more imaginative participation in the audible than visual aesthetic of the natural world. For Hill 'the man who wrote this noble passage had not surely that insensibility to nature which is so often laid to his charge'.[197] In fact, the insensibility Johnson was charged with is a visual one, not impinged upon by this passage:

with respect to sight, it must be noted, that he was one of that class of men, who, from a defect in the visual organs, are termed myops, or near-sighted persons . . . the consequence whereof was, that in lieu of those various delightful prospects which the face of nature affords . . . his mind was presented with an universal blank.[198]

Moreover, Johnson's responsiveness to soundscapes was not merely aesthetic, but part of the close observation factual description demanded.[199]

Literature and landscape

At Slains Castle, 'there is a bow window in the drawing-room to the sea. Mr Johnson repeated the ode, *Jam satis terris*' (*Journal*, p. 73), this being the description of a flood in Horace's *Odes* (I. ii).

This incident is one of a series in which Johnson's response to an actual landscape was mediated by literature. He evoked a familiar topos connecting literature and landscape normally reserved for the Grand Tour in Italy: 'we went forwards the same day to Fores, the town to which Macbeth was travelling, when he met the weird sisters in his way. This to an Englishman is classic ground' (*Journey*, p. 18). Literature was not only used by Johnson in situations where an obvious application existed. His response could be similar on barren ground on Coll:

we also passed close by a large extent of sand-hills, near two miles square. Dr Johnson said, 'he never had the image before. It was horrible, if barrenness and danger could be so.' I heard him, after we were in the house of *Breacacha*, repeating to himself, as he walked about the room, 'And smother'd in the dusty whirlwind, dies.' Probably he had been thinking of the whole simile in *Cato* of which this is the concluding line; the sandy desert had struck him so strongly' (*Tour*, p. 291).[200]

Finally, Johnson also expressed his pleasure at being on hospitable ground through literary allusion. 'In *Ramsay*, If I could have found an Ulysses, I had fancied a *Phaecia*' (*Journey*, p. 54) and 'at *Dunvegan* I had tasted lotus', both references to the *Odyssey*.[201] These were further instances of the phenomenon (discussed above) of Johnson intermixing sublime and beautiful:

such a seat of hospitality, amidst the winds and waters, fills the imagination with a delightful contrareity of images. Without is the rough ocean and the rocky land, the beating billows and the howling storm: within is plenty and elegance, beauty and gaiety, the song and the dance (*Journey*, p. 54).

The beauty is provided by human civilization amidst barrenness, a situation which also held good for Johnson's response to Snowdon.[202]

Johnson's response to being on hospitable ground, then, was mediated by sociability. As such, it forms part of a use of literary reference to reassert the subordination of aesthetics to human concerns. Brownell (following Boswell) has written of Johnson's *nil admirari* response to natural wonders. The phrase derives from Horace, and Johnson's response was more literally Horatian than has been noted. Johnson's Horatian response at Slains Castle is only one instance; he also 'repeated the ode *Otium divos rogat*' (*Journal*, p. 128) on the voyage to Coll (*Odes*, II. xvi), the mariners' prayer for a peaceful sea. These were both references to aesthetic moments Johnson was responsive to, but he also deployed Horatian allusions in ways more redolent of the *nil admirari*. Thus, he wrote to Mrs Thrale that Coll was 'an obscure Island, on which *nulla campis*

Arbor æstivâ recreatur aurâ.' (*Letters*, ii. 100)[203]

Similarly, in writing of Boswell's desire to visit London he quoted, *In culpa est animus, qui se non effigit usquam*.[204] Other authors were used to develop the same response, as when Johnson called Mull 'a most dolorous country' (*Tour*, p. 318), echoing Milton.[205] More important was Johnson's response to the Firth of Forth. 'Water is the same every where.

Una est injusta caerula forma maris.' (*Tour*, p. 54)

The quotation, from Ovid's *Amores*, 'Unvaried still its azure surface flows', was used to deflate Boswell's enthusiasm for the prospect.[206]

Johnson was different from many of his contemporaries in his use of literary references with respect to landscape. Literature was commonly used to express enthusiasm about the landscape, as in Craddock's decontextualized evocation of Johnson's own famous literary landscape: 'the road to Harlech afforded great variety; there could scarce be more within the compass of ten miles. For the first three we surveyed "the Happy Valley" '.[207] Johnson's deflating use of Milton on Mull was in marked contrast to the more general usage: 'there is scarcely one eighteenth-century poem descriptive of exotic places...that is not fundamentally indebted to Milton'.[208] Similarly, Horace was most often cited in the landscape context to express support for the ideal of rural retirement found in his second *Epode*.[209] Thus, Boswell's *Account of Corsica* had cited line 2 of that epode, refracted through the lens of Rousseau:

> When we grew hungry, we threw stones among the thick branches of the chestnut trees which overshadowed us, and in that manner we brought down a shower of chestnuts...It was just being for a little while one of the 'prisca gens mortalium'['pristine race of mortals'].[210]

Johnson's use of Horace was inspired instead by the satirical conclusion of the second epode, which he had translated as a school exercise:

Thus did the us'rer Alphius praise,
With transports kindled, rural ease,
His money he collected stait,
Resolv'd to purchase a retreat.
But still desires of sordid gain
Fix'd in his canker'd breast remain:
Next month he sets it out again.[211]

The moral approach to physical landscapes

Johnson's use of classical quotation to deflate the pretensions of aesthetics is related to the same strategy in his didactic writings. Johnson, when viewing the actual landscape, redirected attention from aesthetics to morality and utility. In Wickins's garden, Johnson responded censoriously to a statue of the Venus De Medici: ' "Throw her", said he, "into the pond to hide her nakedness, and cool her lasciviousness." ' His final action in the garden connected this moral point to one relating to utility:

> I then observed him with Herculean strength tugging at a nail which he was endeavouring to extract from the bark of plum-tree; and having accomplished it, he exclaimed, 'There, Sir, I have done *some* good to-day; the tree might have festered.'[212]

The evidence for Johnson's turning encounters with the actual landscape into moral themes is not wholly anecdotal. To Queeny Thrale at Brighton he wrote,

> You...I suppose wander philosophically by the Seaside, and survey the vast expanse of the world of aters, comparing as your predecessors in contemplation have done its ebb and flow, it[s] turbulence and tranquillity to the vicissitudes of human life (*Letters*, iii. 288).

Johnson's use of this strategy of redirection was restricted, however, in the context of actual landscapes to his private writings. In empirical descriptions designed for publication, the redirection to issues of morality took a less analogical turn. Aesthetic appreciation was accepted only where, as at Slains Castle, it was not in 'violation of humanity' (*Journey*, p. 13). At Loch Ness aesthetic response was checked by Johnson's examination of a hut, reasserting 'our business was with life and manners' (p. 24). The moral turn in factual description, then, was away from the aesthetic towards the interlinked concerns of knowledge and utility. Hanway, for all her criticisms of Johnson as a traveller, was right that his aim was to be a *practical moralist*.[213]

Johnson's concern for utility was displayed in a persistent attention as a traveller to harvests. On the road to Aberdeen he noted that 'the harvest, which was almost ripe, appeared very plentiful' (*Journey*, p. 9) and likewise at Loch Ness (p. 22). For Siebert there is an 'irony in that Johnson allows himself to become excited by things that would be beneath the traveller's notice in any civilised country...even a corn-field!'[214] This judgement seems mistaken: Johnson's focus on utility in Scotland was part of a tradition from at least Munro (1594).[215] Also this was not only of interest in Scotland, suggesting that the country was not fully civilized, but applied to England, being a recurrent theme of the trips to Ashbourne. Utility was not a 'minor key' which 'serves as a contrast to the great tragic themes' of the *Journey*, for 'Johnson thought political economy a very serious subject indeed, highly relevant to human happiness.'[216]

A crude analysis could reduce Johnson's concern for harvests to part of the emerging money economy. Johnson wrote from Lichfield: 'the Harvest is abundant, and the weather a la merveille. No season ever was finer.

Barley, Malt, Beer, and Money. There is a series of Ideas' (*Letters*, iii. 56). But this was in a letter to Mrs Thrale, and 'the specificity of Johnson's letters often reflects a pragmatic interest in the fortunes of the Thrale brewery', in order 'to communicate a total identification with the pleasures and pains of his adopted family'.[217] Johnson cited rents and crop prices, figures adding factual precision to simple impressions of the traveller's eye. From Lichfield he wrote 'they have had in this country a very prosperous Hay harvest, but malt is five and sixpence a strike, or two pounds four shillings a quarter. Wheat is nine and sixpence a bushel'. That this was not simply bourgeois *utilitarian* calculation, but a Christian emphasis on *utility* becomes apparent by Johnson's closure of this paragraph: 'These are prices which are almost descriptive of a famine. Flesh is likewise very dear' (*Letters*, i. 400).

Johnson's concern for the poor as a traveller was consistent. He had advised Boswell in Holland to 'enquire how the poor are maintained in the Seven Provinces' (*Letters*, i. 240) and never forgot them in his own travels, as Piozzi recalled:

> I was wishing naturally but thoughtlessly for some rain to lay the dust as we drove along the Surry roads. 'I cannot bear (replied he, with much asperity and an altered look), when I know how many poor families will perish next winter for want of that bread which the present drought will deny them, to hear ladies sighing for rain, only that their complexions may not suffer for the heat, or their clothes be incommoded by the dust; – for shame! leave off such foppish lamentations, and study to relieve those whose distresses are real.'[218]

Johnson's hatred of barrenness appears closely related to this, such that fertility became yoked to beauty:

> an eye accustomed to flowery pastures and waving harvests is astonished and repelled by this wide extent of hopeless sterility. The appearance is that of matter incapable of form or usefulness (*Journey*, p. 31).

In this respect, Johnson's view was not dissimilar to that of the improvers,[219] but had roots in 'the ancient classical ideal which associated beauty with fertility'.[220] If the poor amounted to a dark side of the landscape,[221] Johnson consistently tried to illuminate it, and was not alone in this aim. Where he was more unusual was in his desire for a closer investigation of the dark side of the townscape, saying of London 'if you wish to have a just notion of the magnitude of this city, you must not be satisfied with seeing its great streets and squares, but must survey the innumerable little lanes and courts' (*Life*, i. 422).

Johnson's concern for the poor was also unusual in that it was based upon a sound knowledge of the rural trades they relied upon. In the *Tour*, Boswell recorded his 'attention on the useful arts of life' including 'the whole process of tanning, and of the nature of milk, and the various operations upon it, as making whey, &c' (*Tour*, p. 246).[222] Such knowledge also extended to industries, his review of Borlase being appreciative of the account of kelp production.[223] Both Hawkins and Piozzi attributed Johnson's concern for utility and consequent insensibility to natural aesthetics to his 'common' background; this may be a partial explanation, but, both being more interested in landscape aesthetics, they only judged Johnson's concern for the natural environment on the basis of their own. Where modern commentators have assumed the focus on harvests, rural trades and the poor must be satirical, or a counterpoint to more significant issues, Hawkins and Piozzi put it down to Johnson's origins. Both interpretations prioritize the aesthetic over the moral, whereas Johnson's intellectual hierarchy reversed that order. In this sense, Johnson's approach to the landscape was one seeking benevolent improvement, a 'tory view of landscape'.[224]

Religion: presence and absence in the landscape

The visible church, in the form of various religious sites, was a persistent interest of Johnson's when viewing the landscape. Johnson's descriptions of St Andrews and Arbroath are dominated by the sight of the ruined cathedral and monastery (*Journey*, pp. 3, 7–8). M'Nichol, noting Johnson had not visited Tarnaway castle where he would have found plentiful woodland, added 'it would not have taken him much out of his way, and he would have made a shift to visit a popish church, or even the ruins of one, at a greater distance'. M'Nichol, a Presbyterian minister, interpreted Johnson's interest in churches and religious ruins in the landscape as a sign of excessive attachment to the visual church, and thus as 'popish'.[225]

The denominational politics of Johnson's view of religion in the landscape are more complicated than this, or than the views of Hart and Curley. For Hart, Johnson's theme was 'the destruction of pre-Reformation Christian culture', and for Curley Johnson's travels to English bishoprics 'had the quality of a patriotic and religious act, strengthening his national pride and spiritual ties with a rich and passing Anglican heritage'.[226] Both neglect that Johnson's concern was not simply with church ruins but the continued vibrancy of the church in the landscape. On his way to Scotland, Johnson visited York Minster, 'an Edifice of loftiness, and elegance, equal to the highest hopes of architecture' (*Letters*, ii. 47). Also at York, he saw the 'ruins of an Abbey' but confessed 'I remember nothing of them distinct' (ibid., ii. 48), which suggests his was not simply a response of nostalgia to the decay of religion reflected in the landscape.[227]

Johnson's anger at 'the ruffians of reformation' (*Journey*, p. 3) who destroyed St Andrews cathedral does not reveal much about his denominational outlook on the relationship between religion and landscape. In 1770 Wesley had responded in a similar manner: 'I took a view of the small remains of the Abbey... The zealous Reformers they told us, burnt this down. God deliver us from reforming mobs!'[228] Much earlier, Sacheverell had said on Iona:

> I quitted the Abby with Indignation, to see so many noble Monuments of the Vertue and Piety, of those great and holy Men, Buried in their own Ruins; and so celebrated a Seminary of Learning and Religion sacrific'd to Zeal, Avarice, and Ignorance.[229]

Neither of these men shared Johnson's High Churchmanship,[230] yet Wesleyan Methodism did have close affinities with 'Orthodox' churchmanship,[231] both attacking the Latitudinarian emphasis on the design argument.[232] As such, for both Wesley and Johnson, a discussion of religious matters in a factual description was confined to discussing churches and ruins.[233]

That the face of the landscape did not have any theological resonance for Johnson can be shown by the absence of two types of argument connecting religion and landscape from his writings. First, he did not argue that the appearance of the landscape reflected the theological truth or administrative efficiency of a denomination. This argument was used most frequently on the Grand Tour in the Ecclesiastical territories surrounding Rome, but was also invoked in a British context. Walker, a Presbyterian minister and writer of reports on the Hebrides, spoke of the Catholic islands of Eigg and Canna as reflecting the weakness of their inhabitants' religion.

> For the Balefull influence of the Popish religion, wherever it is generally professed in the Highlands, is visible, even in the face of the Country. There, not only the morals and manners of the People, but the very Soil, is more rude and uncultivated.[234]

Johnson, by contrast, calling Canna a 'Popish Island,' argued the reverse, that it was more fruitful than the (protestant) island of Rum (*Journey*, p. 106). This was not due to sympathy with Roman Catholicism, but a refusal to connect religion with the fertility of the land. At the administrative level, the same argument could be used. At the close of Pennant's second Scottish tour, in Durham's county palatinate, he said: 'I have heard on my road many complain of the ecclesiastical government this county is subject to; but from the general face of the country, it seems to thrive wonderfully well under it.'[235] That Pennant posits such a connection, or sees the thriving countryside as a refutation of complaints against the system of ecclesiastical government, places him at a distance from Johnson.

The other connection which Johnson refused to make between religion and landscape is one which visualizes God in nature, either using the face of the land in a design argument or interpreting exceptional natural events as providential. The argument for seeing special providence in the operation of nature was one Johnson undermined at Slains Castle: 'storms, whether wished or not, will sometimes happen' (*Journey*, p. 13). At Loch Ness, Johnson theorized that aside in his scepticism about the claim that the Loch does not freeze, which he saw as a putative exception 'from the course of nature' (*Journey*, p. 23). The burden of proof lay with those who would assert such an exception. Reasoning from first principles, Johnson believes 'its profundity... can have little part in this exemption' from the course of nature, but if the Loch indeed does not freeze, Johnson attempts to invoke other natural processes to explain this.[236]

While Johnson firmly believed in explanation by the laws of nature, this did not lead him to hold up such laws as proof of God. Johnson's distance from the design argument was noted by Piozzi:

> Rousseau is not like Johnson when he thinks a mute & sublime Admiration of his works the best Worship of the Creator, altho' that Admiration should excite no Act of any sort, but end wholly in itself – Johnson thought that God Almighty sent us here to do something – not merely to *stare about*.[237]

This scepticism produced an absence in the *Journey*, noted by Gilpin (as a Latitudinarian) of any design argument:

> Dr Johnson says, the Scotch mountain has the appearance of matter *incapable of form, or usefulness*. As for it's *usefulness*, it may for any thing he can know, have as much use in the system of nature.[238]

Gilpin and Johnson deployed different notions of utility in landscape description, Johnson's being an immediate ability to aid human survival, Gilpin's a more cosmological contribution. In a lower key, but still deploying the design argument, Pennant could reflect while sitting on

> genuine basaltic columns...The ruins of the columns at the base made a grand appearance: these were the ruins of the creation: those of *Rome*, the work of human art, seem to them but as the ruins of yesterday.[239]

Johnson sat on a similar 'range of black rocks, which had the appearance of broken pilasters', but his response was less rhapsodic: 'we were easily accommodated with seats, for the stones were of all heights' (*Journey*, p. 123). It is possible that Pennant's use of the design argument was one of the things which caused Johnson to comment that 'Pennant has much in his notions that I do not like' (*Letters*, iii. 114).[240]

A similar pattern can be found with respect to Derbyshire. At the beginning of the period Fiennes had deployed the design argument to urge that tho' the surface of the earth looks barren yet these hills are impregnated with rich Marbles Stones Metals Iron and Copper and Coale mines in their bowells, from whence we may see the wisdom and benignitye of our greate Creator to make up the defficiency of a place by an equivolent as also the diversity of the Creation which encreaseth its beauty.[241]

The ugliness of the Peak District was not evidence of God's partial distribution of goodness as it contained mineral abundance. By the end of the period, changing tastes meant the Peak was regarded as a sublime glory. As such, a different version of the design argument was deployed:

it is the awe of the Allwise Creator of the universe, at whose words these mighty mountains with dreadful clash were rent assunder, that reaches us through the medium of his works.[242]

Throughout the shift in landscape aesthetics, Derbyshire remained the focus for some form of design argument. It is noticeable, then, that Johnson, whether like Fiennes saying of the 'horrid nakedness' of Loch Ness's rocks 'if I had not seen the Peak, [it] would have been wholly new' (*Journey*, p. 22), or, like Moore, arguing for the 'horrible profundity' (*Diaries*, p. 175) of Hawkestone Park, never connected either response to the design argument.

The argument so far has focused on an absence from Johnson's writings, but one presence in Boswell's *Tour* supports the notion of Johnson's scepticism of the design argument. At Inveraray, Boswell records Johnson's parody of Hervey's *Meditations among the Tombs*. As we saw in Chapter 4, Hervey's was one of a series of works in the meletetic genre started by Boyle. Johnson's playful response was contained in his 'Meditation on a Pudding' (*Tour*, p. 352):

let us seriously reflect of what a pudding is composed. It is composed of flour that once waved in the gold grain, and drank the dews of the morning....[of] milk, which is drawn from the cow, that useful animal, that eats the grass of the field, and supplies us with that which made the greatest part of the food of mankind in the age which the poets have agreed to call golden.

Johnson's 'Meditation' contained a reference to one of the greatest proponents of the argument from design, Thomas Burnet: 'it [a pudding] is made with an egg, that miracle of nature, which the theoretical Burnet has compared to creation'. The serious point behind the parody was that in

a passage concerning the moon...he could, in the same style, make reflections on that planet, the very reverse of Hervey's, representing her as treacherous to mankind (*Tour*, p. 351).

The design argument could only be a proof to those already convinced of God's existence.[243] Boyle believed that 'the World is the great Book, not so much of Nature, as of the God of Nature...crowded with instructive Lessons'.[244] Johnson was more sceptical: 'use the World likewise as a large book', to be read 'attentively, but do not believe it'.[245]

Theology was a system of propositions the merits of which could not be gleaned from the appearance of the landscape. In the joint concern for the fabric of the visible church and the diminution of the role of the proof of God from nature *vis-à-vis* other routes to faith, both of which were reflected in the characteristic presences and absences in Johnson's discussion of actual landscapes, Johnson's response to religion in his factual descriptions partook of two of the characteristics which distinguished eighteenth-century High-Churchmanship. As in Johnson's didactic works, so his empirical description betrays a High-Church view of landscape; the very different approaches being responses to the different modes of argumentation and topics of interest of moral and factual inquiry.

Comparison III: Mrs Thrale/Piozzi and landscape aesthetics

Mrs Thrale (later Piozzi), who travelled with Johnson to Wales and France, left extensive records, both published and private, of her response to landscape. In her aesthetic sensibilities she provides an interesting counterpoint to Hanway, Boswell and Johnson.

Piozzi was more attentive than Johnson to all sorts of prospects.[246] In her French journal, returning to Paris from Fontainbleau, she recorded

a Country beautiful, fruitful & highly cultivated. There are no Hedges in France, where I travelled at least – nor no Verdure: Corn fields, Towns finely scattered up & down, & built all of Stone as if they were meant to adorn as well as inhabit the Country – with Rivers perpetually winding in your Sight as if intended merely to amuse the Eye.[247]

Johnson had recorded the same details in a characteristically briefer way: 'the appearance of the country pleasant. No hills, few streams, only one hedge' (*Diaries*, pp. 238–39). The difference between the two modes of reporting reflected the different degrees of significance Piozzi and Johnson attached to the aesthetics of landscape. The same finding appears for the three other categories of prospect discussed earlier with respect to Johnson. Piozzi was mindful of the situation of towns in the landscape when approaching them:

we arrived at Rouen...the Situation of this Town is uncommonly delight-ful, somewhat resembling the Situation of Bath; surrounded by Hills of wonderful Beauty & adorned with Trees mingled among the Churches, as if it had originally been contrived merely to excite the Admiration of Travellers.[248]

As this suggests, Piozzi was also more inclined to comparative landscape aesthetics than Johnson.[249] Brownell has noted Piozzi's response to the situation of Welsh castles, seeing her descriptions as couched in 'the terms of a precocious student of picturesque landscape'.[250] Piozzi also used these descriptions as the basis for accomplished comparative descriptions in her tour to Europe a decade later:

> Had I told my companions of yesterday perhaps, that the view from *Madonna del Monte* reminded me of Chirk Castle Hill in North Wales, they would have laughed; yet from that extraordinary spot are to be distinctly seen several fertile counties, with many great, and many small towns, and a most extensive landscape...I think that the view has scarce its equal any where; and, if any where, it is here in the vicinity of Varese (*Observations*, ii. 228).

Finally, with respect to gardens, Piozzi is far more enthusiastic about prospects than Johnson. At Baron Hill her response was closer to Pennant's than Johnson's:

> a place of beautiful situation commanding the Castle, the streights, and the mountains, an assemblage scarcely to be mended even by the imagination.[251]

It is this sense of assemblage, of the connection between site and situation, which Johnson never showed. Especially important here are Piozzi and Johnson's responses to Hawkestone Park. Johnson's response was exceedingly long by the standards of his journals, probably to tease Piozzi about her sensitivity to landscape.[252] Such a humourous exercise in landscape aesthetics would not have been without precedent, for a year earlier Boswell recorded that

> Dr Johnson...bade me try to write a description of our discovery Inch Keith, in the usual style of travellers, describing fully every particular; stating the grounds on which we concluded that it must have once been inhabited, and introducing many sage reflections; and we should see how a thing might be covered in words so as to induce people to come and survey it (*Tour*, pp. 55–56).[253]

Even if Johnson's description (*Diaries*, pp. 174–75) is taken at face value, it does not show an aptitude for word painting, but a Burkean response: the focus is on the emotions of the spectator, rather than the prospect *per se*.[254]

Piozzi, by contrast, began with the park's ability to 'excite ideas of terror', but continued with a more descriptive approach:

> all the rough crags of Hawkestone, with whole promontorys of woodland stretching out into the beautiful meadows that compose the valley below; fill up the foreground. When the eye is tempted further a country of long extent and high cultivation detains it from the Welsh mountains, which, lying at a great distance, terminates the prospect.[255]

The sincerity of Johnson's description of Hawkestone is also questionable, given its enthusiastic connection of literature and landscape. Quoting *Paradise Lost* (xi. 642), Johnson said 'Hawkeston can have no fitter inhabitants than Giants of mighty bone, and bold emprise.'[256] Whether a parody or not, Piozzi's enthusiasm for landscape certainly was channelled through literary reference. She used Johnson's literary landscapes to capture her experience staying at the Borromean palace on Lake Maggiore.

> Our manner of living here is positively like nothing real, and the fanciful description of oriental magnificence, with Seged's retirement in the Rambler to his palace on the Lake Dambea, is all I ever read that could come in competition with it (*Observations*, ii. 218).

Attempting further description she, like Craddock, turned to the Happy Valley in *Rasselas*, though she may more specifically have had in mind the gardens in Chapters 6 and 20:

> The palace is constucted as if to realise Johnson's ideas in his Prince of Abyssinia: the garden consists of ten terraces; the walls of which are completely covered with orange, lemon, and cedrati trees, whose glowing colours and whose fragrant scent are easily discerned at a considerable distance (*Observations*, ii. 219).

Johnson's point in the Seged essays and the Happy Valley is lost in Piozzi's evocation of the two as earthly paradises.

Piozzi's fascination with landscape aesthetics was in conscious opposition to Johnson:

> my Eyes turn perpetually towards those glorious Productions of Nature, and I half scorn to think of anything but them...These four Days Journey from Port Bon Voison to Navalesa would be enough to make a Coxcomb of Dr Johnson.

In the same letter, she connected her love of nature to the glory of God's Creation, advising her correspondent to 'taste fresh Air at Sheene...in the more pleasing Contemplation of Gods Works unperverted by Man.'[257] Piozzi used the design argument on several occasions. At the outset of *Observations and Reflections*:

> when my eyes have watched the rising or setting sun through a thick crowd of intervening trees, or seen it sink gradually behind a hill which obstructed my closer observation, fancy has always painted the full view finer than I at last found it; and if the sun itself cannot satisfy the cravings of a thirsty imagination, let it at least convince us that nothing on this side of Heaven can satisfy them, and *set our affections* accordingly (*Observations*, i. 2).

This was undoubtedly one of the reflections of the title, but Johnson never engaged in such speculations in a factual genre. Such reflections, to the extent that they were not locative, were the digressions without connection he had criticized in Hanway's *Journal*. Piozzi also showed a belief in special providence in the natural world. Her carefully constructed description of arriving at Naples argued:

> sure the providence of God preserved us, for never was such weather seen by me since I came into the world; thunder, lightning, storm at sea, rain and wind, contending for mastery . . . (*Observations* ii. 1).[258]

Piozzi's use of the design argument could lead to a diminution in concern for the visible church by comparison as it did in Radcliffe's Latitudinarian Gothic (see Chapter 4). After her providential arrival, she said,

> the palaces and churches have no share in one's admiration at Naples, who scorns to depend on man, however skillful, for *her* ornaments; while Heaven has bestowed on her and her *contorni* all that can excite astonishment, all that can impress awe (*Observations*, ii. 6–7).

Interestingly, she criticized the situation of churches in the landscape as idolatrous:

> we have one [church] here in Italy in every district almost, as the rage for *worshipping on high places*, so expressly and repeatedly forbidden in scripture, has lasted surprisingly in the world (*Observations*, ii. 227).

Johnson's position was different: he did not analyse the relationship between churches and their position in the landscape and favoured the visible church in all its manifestations.

Piozzi's picturesque aesthetic did not preclude a concern for the poor (see *Observations*, i. 26), but the aesthetics of landscape clearly meant more to her than to Johnson. Piozzi's sensitivity to landscape aesthetics derived at least in part from the burden of theological proof in her doctrine which fell on landscape and the natural world, just as Johnson's denominational position led him to the opposite position vis-à-vis the natural world.

Concluding comment

'Travel was a leading instrument of that post-Reformation spirit of inquiry which valued empirical knowledge ... [T]he goal of inquiry was to understand mankind rather than inanimate objects.'[259] Johnson was one of the last representatives of this tradition: 'a kind of mainstream classical humanism seems to falter and stop about the year 1830'.[260] Yet this spirit remained vital through the bulk of the eighteenth century and was religious. Johnson's was a religiously informed theory of factual description, which then influenced the structure and content of his work. That he was not alone in this has been suggested by the comparative sketches of Hanway, Boswell and Piozzi, as well as asides about Forster, Wesley and Pennant. The differing emphases, in style, structure and content when describing the natural environment, were, at least in part, the result of different denominational positions. As such, the approach to landscape was still within the ambit of denominational discourse.[261]

Brownell has suggested Johnson's 'outlook on landscape' was part of the 'scientific aesthetic in European travel writing that Barbara Stafford has recently illustrated'.[262] That judgement needs to be modified. Stafford posits two positions: a picturesque aesthetic of reverie, which Johnson's approach had little in common with when compared to Hanway or Piozzi, and when his spoof 'Meditation' is considered; and a 'nominalistic impulse' (which Brownell ascribes to Johnson), to close description which can be 'precisely located in the historical context of a modern secularizing movement'.[263] Yet Johnson's approach traced a third position, which partook of the nominalistic impulse and worked in a theological context. Indeed, Johnson's denominational position positively encouraged factual description. In this, Johnson was part of the English Enlightenment which saw little conflict between science and religion,[264] and he has to be viewed within an English denominational context, rather than a European secularizing one.[265]

8
Life, Literature and Landscape: The Role of the Natural World in Johnson's Biographies and Biography, 1739–84

Having seen how Johnson used landscape imagery and ideas in didactic and empirical writings, I will now turn to the intermediate category of biography. Johnson's early biographies view landscape in terms accordant with the empirical approach discussed in Chapter 6, because he is writing historical and factual accounts. By contrast, the *Lives of the Poets*, being a strongly moral project, incorporated landscape themes in a recapitulation in the context of real lives of the doctrine of mind and place fleshed out in his essays (see Chapter 6). Johnson's edition of Shakespeare falls between these two positions, being an historical recovery of the beliefs about the natural world of a real person, which Johnson would have held to be morally pernicious if held in his own time.

Johnson's early biographies, 1739–61

Patriotic early biographies

Johnson's early biographies have received less attention than his other works. They have until recently being seen as 'in the last analysis, still a piece of hackwork'.[1] They are closely modelled on their sources, Blake's 'Life' from Birch's *General Dictionary* and the 'Life of Drake' from *Sir Francis Drake Revived* and *The English Hero*.[2] As such, the extent to which either biography reflected Johnson's own ideas is open to doubt.[3] In addition, both lives had a transparent political purpose, appearing in 1740 as part of the 'synthetic... hysteria'[4] against Spain. Thus the 'Life of Drake' served to recount the actions which 'laid the Foundation of that settled Animosity which yet continues between the two Nations'.[5] Johnson's biographies appeared surrounded by articles in the *Gentleman's Magazine* such as 'The Crown of England's Title to America prior to that of Spain'.[6] Indeed, Lieutenant Hudson's account of his attack on the Spaniards sounds startlingly similar to Johnson's buccaneering 'Life of Drake'.[7]

An initial analysis of the 'Life of Drake'[8] appears to confirm the scepticism about its worth as an indicator of Johnson's ideas on matters of landscape and the natural world. The figures who people the exotic lands of the 'Life' are the 'friendly savages' (*EBW*, p. 54) against whom Johnson later inveighed. He praised their

> natural Sagacity, and unwearied Industry [which] may supply the Want of such Manufactures, or natural Productions as appear to us absolutely necessary (*EBW*, p. 57).

What is more, this industry took place within a social framework favourably contrasted with that of Europe in general and the Robinocracy in particular:

> he that can temper Iron best, is among them most esteemed, and perhaps, it would be happy for every Nation, if Honours and Applauses were as justly distributed, and he were most distinguished whose Abilities were most useful to Society (*EBW*, p. 43).

This wisdom fed into a less irrational view of the natural world when Drake refused to eat an otter killed by the natives:

> there seems to be in *Drake*'s Scruple somewhat of Superstition, perhaps not easily to be justified; and the Negroe's Answer...will I believe be generally acknowledged to be rational (ibid.).

The noble savages are more natural and more rational, this being but one of a group of national/racial stereotypes the biography contains. Thus 'the Malice of the *Spaniards*' (*EBW*, p. 37) is asserted from the outset, as is their cruelty to native groups, unsurprisingly given the patriotic context of the 'Life' and the long history of voyage literature to the Americas asserting this.[9] Equally predictable given the 'real danger that France would intervene in the Spanish war',[10] was the stereotype of French cowardice which was validated by the instinctive sagacity of the natives:

> Nor did the *Symerons* treat them [the French] with that Submission and Regard which they paid to the *English*, whose Bravery and Conduct they had already tried (*EBW*, p. 47).

The noble savage, then, adds a third caricature (and one, like the others, rooted in the contemporary English political situation) to *London*'s abhorrence of 'a French metropolis' and the *Voyage to Abyssinia*'s portrait of Spanish rapacity.

Equally unusual in Johnson's *oeuvre* is the picture of the exotic lands the friendly savages populate. The Symerons find grass 'grows too high for them to reach; then the Inhabitants set it on fire, and in three Days it springs up again...so great is the Fertility of the Soil' (*EBW*, p. 44). Later in their journey, Drake's party find the soil fertile in another manner, being 'so Impregnated with Silver that five Ounces may be separated from an hundred Pound weight of common Earth' (*EBW*, pp. 59–60). This utility goes with an exotic landscape aesthetic, from the pastoral of 'cool Shades, and lofty Woods' (*EBW*, p. 43) to the tempestuous sublime of 'Skies blackened' in which 'winds whistled' (*EBW*, p. 52). The Magellan Straits combine pastoral, sublime and useful:

> The Land on both Sides rises into innumerable Mountains, the Tops of them are encircled with Clouds and Vapours, which being congealed fall down in Snow, and increase their Height by hardening into Ice, which is never dissolved; but the Valleys are, nevertheless, green, fruitful, and pleasant (*EBW*, p. 57).

The exaggeration of the figures peopling the 'Life', then, is matched by the imaginative colouring of their landscape.

Finally, the role of the natural world in the 'Life' is perceptibly different from Johnson's other writings in one respect. The natural world is occasionally connected with religion and providence, thereby taking on a role that was normally reserved by Johnson for emblematic landscapes in his essays. Drake's men's recovery from wounds suggested they were 'favoured by Providence' (*EBW*, p. 58), and at a climactic point:

> they arrived at the Top of a very high Hill...and from thence shew'd him [Drake] not only the North Sea, from whence they came, but the great *South Sea* [i.e. the Pacific], on which no *English* Vessel had ever sailed. This Prospect exciting his natural Curiosity and Ardour for Adventures and Discoveries, he lifted up his Hands to God, and implor'd his Blessing upon the Resolution, which he then formed, of sailing in an *English* Ship on that Sea (*EBW*, p. 43).

The abnormality of this deserves emphasis: viewed retrospectively, it is at least as exotic as the landscapes by which Drake is surrounded and their savage inhabitants. Each of these three elements disrupts the connections of landscape, humanity and religion which Johnson's other works construct.

'[T]he "Life of Drake" ...allows us to see the author confronting stories of natural wonders and alien peoples...Johnson seems to abandon much of his wonted scepticism.'[11] But this scepticism can only be 'wonted' in the light of his later career, not in 1740 when the 'Life' appeared. By comparing the 'Life' with its sources, I wish to suggest that Johnson's scepticism was already being deployed. Johnson's *nil admirari* pose was not yet fully consistent, but his editorial approach to landscape and the natural world was recognizably sceptical, such that his later connections of life and landscape would be developments not reversals of the pattern in the 'Life of Drake'.

Johnson's attitude to savages in the 'Life', even in the context of his sources, does appear different from his later position, yet forms part of the editorial logic. At several points Johnson adds reflections on the natives Drake encountered not derived from his source materials. The above-cited reflection on the rationality of eating otter, for example, was not in Bourne's *Sir Francis Drake Revived*[12] or Crouch's *English Hero*.[13] The purpose of these additions was *not* the romanticization of the noble savage, as was made clear by Johnson's longest addition. Those who believe in the superiority of the savage suppose 'he to whom Providence has been most bountiful [i.e. civilized Europe], destroys the Blessings by Negligence' (*EBW*, p. 63). In fact, savage and civilized nations are 'equally inclined to apply the Means of Happiness in their Power, to the End for which Providence conferred them'. Natural reason is distributed equally and desires the same outcome, namely the security of everyday existence in the face of natural vicissitudes, but natural resources are not equally available. Johnson's additions were designed to minimize the 'otherness' of the peoples Drake encountered, by showing that it was the same societal and personal urges which both groups in this cultural collision expressed, differently modified by natural environment and historical development. Johnson began by 'supposing Virtue and Reason the same in both [societies]' (*EBW*, p. 63) and punctuated his description of savage life with efforts to make their society intelligible to his readership on the basis of this assumption.[14] As such, Johnson's response to accounts of the Americas partakes of 'the principle of attachment', whereby the actions of alien groups were made recognizable to the West by seeing them as variants on Western practices: 'recognition of this kind, however much it distorted what the Indians may actually have intended by acting as they did, offered at least an initial identification of humanity'.[15] Johnson on savages, then, is perhaps less distant from his later views than has been suggested: the uniformity of mankind, surveyed from China to Peru, was asserted in the 'Life' as in *The Vanity of Human Wishes*.

Johnson dramatized the description of the natural world at two points in the 'Life'. First, when Drake was overcome in a small boat by 'so great an alteration in the weather, into a thick and misty fogge; together with an extream storm and tempest' (*WE*, p. 18). Johnson's version was far more evocative, suggesting action and immediacy not description:

on a sudden, the Weather changed, the Skies blackened, the Winds whistled ... Nothing was now desired but to return to the Ship, but the Thickness of the Fog intercepting it from their Sight, made the Attempt little other than desperate (*EBW*, p. 52).

In an earlier incident, Drake animated his party to build a raft because the way was not passable 'by land, because of the Hills, Thickets, and Rivers' (*FDR*, p. 79). In Johnson's version there was a more embellished description: 'to pass by Land was impossible, as the Way lay over high Mountains, thro' thick Woods, and deep Rivers' (*EBW*, p. 48). In both cases, Johnson's description increased the dangers posed by the natural world and thereby magnified Drake's bravery in facing them. These passages aside, Johnson's most common practice was to dampen down the exotic in his redaction. He removed passages on the arduousness of travel which he would later satirize in his *Idler* essay on Will Marvel. Thus, in the passage following Drake's exhortation, he manned a raft in which

> he sayled some three leagues sitting up to the waste continually in water, and at every surge of the wave to the armepits, for the space of six houres...with the parching of the Sunne, and...the beating of the Salt water (*FDR*, p. 80).

Johnson simply said Drake sailed 'with much Difficulty' (*EBW*, p. 48). However exotic the landscape as Johnson described it by his subsequent standards, a comparison with his sources suggests a sceptical principle of omitting travel lies and exaggerations. The island of Fogo, for instance, had by Johnson's account 'a Mountain...continually burning' (*EBW*, p. 51), but this is a considerable toning down of Bourne's account of

> a steepe upright hill, by conjecture at least six leagues, or eighteen English miles from the upper part of the water...The fire [of the volcano]...breaketh out with such violence & force, and in such main abundance, that besides that it giveth light like the Moone a great way off, it seemeth, that it would not stay till it touch the heavens themselves (*WE*, p. 10; cf. *EH*, p. 60).[16]

Even Johnson's description of the Strait of Magellan was more restrained than its original: he described the mountains in sublime terms, but did not go on to reflect that 'they may wel be accounted amongst the wonders of the world'; likewise the valleys are pastoral in Johnson's account, but not 'alwaies green...a place no doubt, that lacketh nothing, but a people to use the same to the Creator's glory' (*WE*, p. 36). As with the discussion of savages, Johnson's landscape descriptions appear as part of a sceptical outlook when viewed in their context rather than as part of his subsequent canon.

It was Johnson's addition which heightened the connection between landscape and providence in the passage where Drake first saw the Pacific. Bourne has no equivalent to Johnson's connective 'this Prospect exciting his natural Curiosity and Ardour for Adventures and Discoveries, he lifted up his Hands to God' (*EBW*, p. 43), beyond the disjunctive

[he] had seene that Sea of which he had hearde such golden reports: he besought Almighty God of his goodnesse . . . to Saile once in an *English* Ship in that Sea (*FDR*, p. 54).

Beyond this, however, Johnson consistently severed the linkage of God and nature asserted by the more providentialist accounts of Bourne and Crouch. The *World Encompassed* opened with a lofty exordium:

Ever since Almighty God commanded *Adam* to subdue the Earth, there hath not wanted in all Ages, some heroicall Spirits, which [have acted] in obedience to that high mandate (*WE*, p. 1).

Johnson omitted this, and refused to personify the conditions Drake's sailors faced, omitting in his account of the fifty-day storm they endured (*EBW*, p. 57) both the opening reflection that 'God by a contrary wind and intolerable tempest, seemed to set himself against us' (*WE*, p. 34) and the closing comment that

if the speciall providence of God himself had not supported us, we could never have endured that wofull state: as being invironed with most terrible and most fearful judgments round about (*WE*, p. 42).

At numerous other points in the maritime sections, Johnson omitted providential references.[17] Seen in this light, his reference to Drake's group being 'favoured by providence' (*EBW*, p. 58) appears to be the omission of an omission, not a conscious choice.

Johnson also removed the argument from design. This is particularly clear in the account of Drake's second Atlantic crossing, where Johnson retained a paragraph of speculation on the flying fish, but removed the framing comments of his source. In the *World Encompassed*, the easy passage left Drake's men to

behold the wonderfull works of God in his creatures . . . as if he had commanded and enjoyned the most profitable and the most glorious works of his hands to wait upon us . . . (*WE*, p. 12).

The passage on the flying fish in this account was closed with a reflection not in Johnson's version that 'if the Lord had not made them expert indeed, their generation could not have continued, being so desired a prey to so many' (*WE*, p. 13).[18] The biography of Drake intermediate between Bourne's and Johnson's, Crouch's *English Hero* (1716), had begun to weaken the providentialist frame of interpretion by dropping the exordium to the *World Encompassed*, but this had not been systematic, retaining the providential cessation of a storm at sea (*EH*, p. 27) and design arguments about the flying fish (*EH*, p. 62). Johnson's 'Life' was the first to consistently desacralize the interpretation of the environment in which Drake operated. The environment of the Americas was portrayed as a series of opportunities and obstacles, none of which were personified or interpreted as divine dispensation.

'[I]t is dangerous to assume that any particular sentiments expressed here or in any of his early biographies are Johnson's own',[19] but his editing is. However, the standard practice of translation in the period was far looser than the twentieth century would countenance,[20] so it is at least unclear the extent to which even Johnson's editing reflects his own intellect. Yet the editing does display a high degree of consistency in its reduction of exoticism and its weakening of the providential bonds of God, man and the natural world, both of which were consistent Johnsonian traits in later years. It could be maintained that the removal of passages on nature was simply a consequence of the need for drastic abridgement and the biographical rather than topographical aims of the piece. Such a view is consonant with the omission of scientific speculations on the cause of extreme cold in North America (*WE*, p. 66, omitted at *EBW*, p. 61), and on west to east sea currents which had been denied by geographers prior to Drake's passage through the Magellan Strait (*WE*, p. 35; omitted at *EBW*, p. 57). Such useful speculations were rarely passed over in Johnson's factual descriptions.[21] Yet this generic interpretation must itself be qualified. The 'Life' was *not* a straight biography, the passages of natural description and analyses of other societies being too long to sustain such a view. A particularly important example of this comes late in Johnson's 'Life' where he provided a long description of the country and inhabitants of an island Drake had discovered (*EBW*, pp. 62–63). Comparison with his source (*WE*, p. 80) shows Johnson rearranged the passage to describe the natural products of the island, then the housing and clothing of the inhabitants, before adding three paragraphs on the fallacy of the noble savage thesis. None of the passage, in its original or rearranged version, was of direct relevance to Johnson's biographical goals, and therefore its alteration must reflect some Johnsonian goal. In fact, the rearranged description, with its progression from the natural to the human world, followed by a philosophical reflection, is the earliest example of the method of structured description Johnson was to adopt in Scotland. Also, he removed descriptions of some exotic natural phenomena even when they aided the biographical narrative. In the attack on Nombre de Dios, Drake was delayed according to Bourne's account by 'a mighty shower of raine, with a terrible storme of thunder and lightning...which powred downe so vehemently (as it usually doth in those Countries)' (*FDR*, p. 14) that Drake's men had to take cover. Johnson implied that the delay was caused by rain wetting the assailants' bowstrings, where in his source, it was the violent storm which caused it. Johnson's account of the 'violent Shower' (*EBW*, p. 39), by underplaying its exotic nature, made a substantive change. A detail relevant to the biography was suppressed as part of Johnson's editorial policy on the exotic.

The 'Life', then, is far more sceptical than has been suggested previously. Johnson's approach was not consistent, not all providential asides being eliminated, and much that he included was perhaps material he would later have handled more cautiously. Yet by the removal of references to special providence and the assertion of equal providence for all societies, the legibility of God's writ in the natural world, so often asserted in the sources, was denied, the whole amounting to a movement of the presentation of nature in what would subsequently come to be seen as a Johnsonian direction.

Interestingly, Johnson's basic principles of interpretation of the natural world are emergent as early as 1740, during the period of his involvement with anti-Walpolian rhetoric, which lends tentative support to the view that Johnson's views on nature were drawn from something other than his politics, which would change dramatically whilst his view of nature remained stable. His High Churchmanship, important at least since his encounter with Law's *Serious Call* in 1729,[22] is a plausible candidate. This is not to deny the political context in which the 'Life of Drake' appeared, but to argue that the view of the natural world it contains was driven by other concerns.

Early scientific biographies

If the 'Life of Drake' establishes a certain pattern of intentions with respect to the treatment of the natural world, Johnson's biographies of Boerhaave and Morin do the same at the level of content rather than editing.[23]

Three important points emerge from an analysis of the scientific biographies. First, Johnson established an image of good scientific practice. Booklearning had to be matched by original observations. The result was the balanced approach of Boerhaave who

> neither neglected Observations of others, nor blindly submitted to cele-brated Names.... He examined the Observations of other Men, but trusted only to his own (*EBW*, p. 34).

This led to a paradigm of instructive travel observing closely the empirical realities of the natural and human worlds. Thus Johnson's 'Life of Dr Morin' says,

> he was sent to learn Philosophy at *Paris*. Whither he travelled on Foot like a Student in Botany, and was careful not to lose such an Opportunity of Improvement (*EBW*, p. 87).

The 'Life' of another distinguished botanist, Tournefort, in the *Medical Dictionary*, incessantly praises such instructive and observant travelling. Even if Johnson only wrote the opening paragraph,[24] this praised Tournefort for having 'carried Botany to a higher Degree of Perfection...by enriching it with numberless Discoveries' (*EBW*, p. 151) which, in the light of the subsequent narrative, must be viewed as the result of Tournefort's method of travelling.

More importantly, good observational science must be part of a moral life. This is particularly apparent in the 'Life of Boerhaave', where he opposed the 'chemical Enthusiasts' such as Paracelsus who, contrary to the tenets of empiricism, 'instead of enlightening their Readers with Explications of Nature have darken'd the plainest Appearances' (*EBW*, p. 30). Boerhaave was also said to have 'entirely confuted all the Sophistry of *Epicurus, Hobbes* and *Spinosa*' (*EBW*, p. 26). Each of these was associated with heterodoxy which spilled into the interpretation of the natural world, Spinoza being viewed as a pantheist who blurred the division of Creator and Creation, and Epicurus and Hobbes (the latter often seen as the modern Epicurus) were associated with the view that the earth was created by the random collision of atoms, eliminating the need for a Creator. The 'Life of Boerhaave' links these themes, associating non-observational science with those who would establish over-arching systems for the interpretation of nature. These systems were impious because they lacked the 'Sense of the Greatness of the Supreme Being, and the Incomprehensibility of his Works' (*EBW*, p. 30) which was coupled with the observational stance. It was, then, entirely consistent in Johnson's biography that Boerhaave's 'unwearied Observation of Nature' should lead him to his deathbed to 'a kind of experimental Certainty' of the 'spiritual and immaterial Nature of the Soul' (*EBW*, pp. 32, 33). Morin was likewise praised for his piety and, in the 'Life of Ruysch'[25] a parallel point was made; that science cannot usurp the place of religion (*EBW*, p. 146).

The third point is the vocational image of natural science in these biographies. Boerhaave was 'by Nature so well adapted' (*EBW*, p. 25) to scientific inquiry, this being the direction his 'natural Genius' (*EBW*, p. 27) took. Morin's vocation was more specific: 'Botany was the Study that appeared to have taken Possession of his Inclination, as soon as the Bent of his Genius could be discovered' (*EBW*, p. 87).[26] These biographies differ from Johnson's *Lives of the Poets* (1779–81) which rejected this deterministic view of genius.

None of the above can be taken to express Johnson's own attitudes to the scientific approach to nature.

There must be no mistaken belief that Johnson was here plying the trade of biographer as he did years later ... [M]any of the moral generalizations that a reader would confidently ascribe to Johnson are merely gleaned from his source.[27]

For Boerhaave, the source was Schultens's *Oratio Academica* and for the 'Life of Morin', Fontenelle's 'Eloge'. On the balance of observation and reading, Schultens reported Boerhaave's unwearied reading of books ('in Libris indefessus legebat') coupled with excursions in the fields deploying his 'eyes, hands and all senses'.[28] Similarly the 'Life of Tournefort', whether Johnson's or not, reflects Fontenelle's view of the scientist as heroic observer.[29] The stress on the moral life of the observer of nature was also derived directly from Schultens. Boerhaave was pictured battling all secular philosophers ('adversus omnes omnium secularus Philosophos') who would not engage in observational study.[30] Finally, the vocational notion of genius was an insistent theme, both in the Latin of Schultens, where Boerhaave's genius 'bursts forth' ('erupit, inquam, maximus ille genius'), and in Fontenelle's French, where Morin's genius for botany was repeatedly characterized in inspirational terms ('la goût de la Botanique', 'passion pour les Plantes').[31] It is worth noting an omission Johnson made in his translation of Schultens, who closed his funeral oration by speaking of Boerhaave's 'beautiful suburban estate whose extensive grounds he converted into a medical Paradise'. Schultens called this 'a foretaste of the beauty of the celestial Paradise', a conceit Johnson omitted from his version.[32] Johnson was never fond of such conceits, but its removal may simply reflect that 'Schultens's oration was spoken at an extremely formal ceremony'[33] and deployed an ornate rhetorical close inappropriate to the context in which Johnson's translation was to appear.[34]

Although these early scientific biographies cannot be seen as expressing Johnson's opinions, it would be unwise to dismiss their role in Johnson's development as a biographer. His later biographies of scientists and intellectuals from Barretier (1740) to Ascham (1761) show the consistent adoption of the same principles in his original comments. All these biographies agree with their predecessors as to the importance of close observation coupled with reading. Thus Barretier, applying himself to 'natural philosophy', laboured such that

> scarcely any Author, ancient or modern, that has treated on those Parts of Learning was neglected by him, nor was he satisfied with the Knowledge of what had been discovered by others, but made new Observations (*EBW*, p. 182).

Similarly, Johnson's 'Life of Browne' (1756) praised his enquiry into the quincunx for

> BROWNE has interspersed many curious observations on the form of plants, and the laws of vegetation; and appears to have been a very accurate observer of the modes of germination (*EBW*, p. 438).

Travel was portrayed as instructive in Johnson's original reflections which loom large for the first time in this biography: in both the *Posthumous Works* and the 'Life' prefixed to the 1736 edition of *Religio Medici*, Browne's travels to Europe had been noted,[35] but Johnson was the first to note that he left no records of this journey and 'what pleasure or instruction might have been received from the remarks of a man so curious and diligent' (*EBW*, p. 417). Equally characteristic was Johnson's refusal to accept Browne's characterization of his life as 'a miracle'. Adopting the *nil admirari* position, Johnson added that

> BROWNE traversed no unknown seas, or Arabian desarts: and, surely, a man may visit France and Italy, reside at Montpellier and Padua, and at last take his degree at Leyden, without anything miraculous (*EBW*, p. 424).

Johnson also continued the praise of scientific inquiry supportive of religion. Browne's 'Observations upon several plants mentioned in Scripture' were praised 'as they remove some difficulty from narratives, or some obscurity from precepts' (*EBW*, p. 440) it was important for the Christian to be certain of. Similarly, Johnson did not belittle Browne's working to answer 'two geographical questions' on the Dead Sea and the city of Troas (*EBW*, p. 445).[36] The one discontinuity was Johnson's omission of deterministic notions of genius after the early biographical translations for the *Gentleman's Magazine*.

Johnson's translations of Schultens and Fontenelle represent the hackwork necessary for the *Gentleman's Magazine*, but his lack of alterations may also reflect agreement with the image of natural science's position in a well-conducted life which his sources presented. Johnson's own biographies over the next twenty years marked no pronounced divergence from the pattern established in 1739–40. As such, the activities of the *Gentleman's Magazine* in the late 1730s and early 1740s begin to appear of importance in the forging of Johnson's attitude to the natural world.

Johnson's early biographies in context: the changing role of nature in the *Gentleman's Magazine*, 1736–47

Johnson developed his view of the natural world while closely involved with the *Gentleman's Magazine* (*GM*), in the period when the magazine 'broke the insular and narrowly humanistic bonds in which [the proprietor] Cave's limited vision had constrained it'.[37] One part of this transition was in the magazine's approach to the natural world.

Before Johnson's involvement, the *GM* incorporated issues of natural knowledge largely as a vehicle for religious and political reflections. In February 1737, for example, the 'Annals of a Modern Traveller' had a young man go on the Grand Tour, while throwing away money at home ordering 'all the old Planting about my Seat to be cut down, to make Room for new Improvements, adorned with *Canals, Cascades, Jets d'eaux*'. The traveller's profligacy was simply an emblem of his moral failings, which were confirmed in Rome where he met 'a *certain Person* [i.e. the Pretender], and kissed his Hand'.[38] Similarly in October of the same year, the *GM* reprinted an article on 'The King of Abyssimia's [*sic*] terrible Guard', a transparent attack on English standing armies, though claiming to be a report of 'a very great Traveller'.[39] The other reason for the appearance of nature in the *GM* was religious speculation on natural philosophy. In September 1736 the magazine reprinted from *Fog's Journal* 'Of the Deluge and the First Ages', which stressed our inability to reason from nature to an accurate account of the Creation.[40] In October 'R.Y.' replied that

> the Design of this *perplexed* Discourse is, by rendering the *plain History* of the *Creation, Flood, &c.* as incomprehensible as the most absurd *popish* Doctrines, to endeavour to introduce his beloved Popery into the *Island*.[41]

R.Y.[42] responded with his own account, which continued from October 1736 to April 1737 and triggered responses in May 1737 and February 1738. This debate had a politico-theological motivation, but more scriptural discussions of the natural world were also printed, including a series of questions as to whether heaven and hell were 'local' (i.e. real places), an attack on pantheism, and a rhapsodic 'Contemplation on Nature and her Works'.[43] Politics and religion, then, determined when the natural world appeared in the *GM*, and the form its appearance took.

Johnson's first piece appeared in the *GM* in May 1738, from which time he also had some editorial role which continued until mid-174?[44] 'In the eleven months of Johnson's editorial service' in 1738–9 'new ground was being broken in every second or third issue'.[45] One of the areas in which this was the case was the presentation of the natural world. There was an increase in items concerned with the natural knowledge to be gained from travel. Maupertuis's 'Observations on Lapland' was extracted,[46] being a factual account of climate, wildlife and inhabitants.[47] There was also an increase in geographical information. The pretext was often political, the native uprising in the Dutch East Indies leading to 'An Account of Batavia' in July 1741, but the accounts themselves, as of Batavia, gave geographical information.[48] Issues of geography could even emerge independently of other concerns, as in a series of three pieces on Dr Packe's

> Philosophico-Chorographical Chart of East Kent ... wherein are described the Progress of the Vallies, the Directions and Elevations of the Hills, and whatever is curious both in Art and Nature, that diversifies and adorns the Face of the Earth.[49]

The increased importance of the natural world in the *GM* did not go unnoticed, one author sending a contribution 'as I find, that Pieces of Natural Philosophy are generally acceptable'.[50] The place these issues had attained whilst Johnson had a hand in the editorial process was maintained subsequently. Thus, a follow-up piece to Maupertuis's, an 'Extract of M. Outhier's

Journal of a Voyage to the North, 1736' appeared in 1746. Other accounts, all highly factual and with little precedent prior to 1738 in the *GM*, included praise of Pococke's *Description of the East* (April/May 1743),[51] and Egede and Anderson's accounts of Greenland in 1743 and 1747 respectively.[52] Factual descriptions of towns continued to be included, and 1746 saw an article by Mr Bellin 'in relation to his Maps drawn for P. Charlevoix's History of New France, &c'. Of particular interest was 'A Journey up to Cross-fell Mountain' in 1747, designed to 'entertain such of your readers whose genius inclines them to the description of romantic scenes'. This was perhaps the first piece of unalloyed landscape aesthetics in the *GM* and was successful enough to warrant a second piece, 'A Journey to Caudebec Fells' in the November issue.

The changes mapped before, during and after Johnson's close connection with the *GM* are ones of emphasis. Yate could still write in 1743, and an article in Johnson's period on the language of the beasts was not far removed from one on the theology of insects in 1747. Moreover, Newtonian astronomical problems and articles in the 'Historical Chronicle' on extreme storms, floods and the like were staple fare throughout. Yet the changes in the treatment of the natural world, coming in a compressed period, should not be dismissed: coincident with Johnson's close connection with the *GM*, there were moves towards a desacralization of natural philosophy and a depoliticization of travel accounts. As such, the natural world emerged as an independent subject, rather than the carrier of other issues. The *GM* had started, largely due to its derivation of articles, by treating the natural world as the periodical essays of the early eighteenth century had done.[53] The late 1730s and early 1740s saw a change in this, prior to the *GM*'s self-pronounced withdrawal from partisan politics, signalled in Johnson's preface to the collected 1743 edition.[54] After Johnson's involvement, the *GM* showed its first signs of treating landscape as an aesthetic phenomenon in the realm of polite entertainment, closer to the approach of mid-century periodical essays, diverging from the factual and utilitarian approach it had forged in the preceding years.

Returning to Johnson, there are clear parallels between his presentation of the natural world in the early biographies and the general view of nature in the *GM*. No final assessment of Johnson's influence on the *GM* is possible, but with respect to the natural world the parallel between Johnson's and the *GM*'s position is too strong to ignore. Simultaneously, both acted to desacralize the discussion of nature, such that the 'Life of Drake' starts to take on a different appearance: as well as appearing in the midst of articles on the Spanish war, it was also surrounded by such pieces as a discussion of the 'Character of Mr Moore's Travels in Africa'. A second context for the 'Life' emerges, then, a political pretext allowing the presentation of speculations on the geographical diversity of the natural and human worlds. Political purpose, then, was matched by an independent drive to develop a factual approach to landscape and nature.

By the time Johnson wrote the early lives 'he had developed virtually all of the techniques for handling factual sources he was to use through the *Lives of the Poets*'.[55] Such a view also holds good at a thematic level for the interrelation of life and nature. The early biographies developed a position in which nature was not to be seen as demanding a providential interpretation, but, as the scientific biographies stressed, the study of nature was to be viewed within the context of a moral life. The 'Life of Browne' finalized the manner in which nature became relevant, biographical facts about natural knowledge being elaborated upon by moral reflections. However source dependent, these biographies show Johnson developing an attitude to the interrelation of life and landscape which received final expression in the *Lives of the Poets*.

Defenders of the text: Shakespeare's Natural Knowledge, 1723–1821

Between the 'Life of Ascham' (1761) and the *Lives of the Poets* (1779–81), Johnson had an intellectual encounter with one other biography, that of Shakespeare, an edition of whose works he published in 1765.

Johnson's 'Preface' to Shakespeare and the two poets of nature

The 'Preface' to Johnson's edition of Shakespeare has been intensively studied. For the present inquiry, two features should be noted. First, much of the argument is built using imagery of the natural world, and the effect is akin to Johnson's essays. Secondly, the prefaces to previous editions and the works of various Shakespearian critics had already explored this territory, but Johnson used these images with unusual frequency and clarity.[56]

The nature imagery in the 'Preface' was used for three purposes. First, it modelled the task and problems of the editor. Considering the exchanges of insults Shakespearian editors had engaged in, Johnson argued that

> sometimes truth and errour, and sometimes contrarities of errour, take each others place by reciprocal invasion. The tide of seeming knowledge which is poured over one generation, retires and leaves another naked and barren.[57]

Warburton in his 'Preface' had spoken similarly of having 'neither GRAMMAR nor DICTIONARY, neither Chart nor Compass, to guide us through this wide Sea of Words'.[58] Johnson, of course, provided the compass Warburton sought in 1755, yet could still fear the tides of critical fortune:

> to dread the shore which he sees spread with wrecks, is natural to the sailor. I had before my eye, so many critical adventures ended in miscarriage, that caution was forced upon me (*JOS*, p. 109).

The second use of natural imagery was to model the task of the critic. The critical faculty was portrayed as comparative:

as among the works of nature no man can properly call a river deep or a mountain high, without the knowledge of many mountains and many rivers; so in the productions of genius, nothing can be stiled excellent till it has been compared with other works of the same kind (*JOS*, p. 60).

Natural imagery also provided a way of representing Shakespeare's genius. Shakespeare was compared by Johnson to the local and the global:

Shakespeare's familiar dialogue is affirmed to be smooth and clear, yet not wholly without ruggedness or difficulty; as a country may be eminently fruitful, though it has spots unfit for cultivation: His characters are praised as natural, though their sentiments are sometimes forced, and their actions improbable; as the earth upon the whole is spherical, though its surface is varied with protuberances and cavities (*JOS*, pp. 70–71).

Upton shifted the ground from utility to aesthetics, but his imagery of Shakespeare's 'rough words' was similar: 'even in prospects (Nature's land-skips) how beautifully do rough and ragged hills set off the more cultivated scenes?'[59] Johnson's most elaborate comparison came closer still to Upton's:

The work of a correct and regular writer is a garden accurately formed and diligently planted, varied with shades, and scented with flowers; the composition of Shakespeare is a forest, in which oaks extend their branches, and pines tower in the air, interspersed sometimes with weeds and brambles, and sometimes giving shelter to myrtles and to roses; filling the eye with awful pomp, and gratifying the mind with endless diversity (*JOS*, p. 84)

This was one in a line of gardening similes in eighteenth-century Shakespearean studies. It was unsurprising, then, that Barclay chose to defend Johnson's edition from Kenrick's attack in terms of a gardening fad of the mid-century, for seeking 'to lead the reader through the inextricable mazes of a paradox, till you [Kenrick] bring him on an unexpected meaning, like a Chinese Hah! hah!'[60]

The final concern was Shakespeare's position as the 'poet of nature' (*JOS*, p. 75). The nature referred to was mainly human nature, but in establishing two models for Shakespeare as the poet of nature,[61] the 'Preface' also reflected two important approaches to Shakespeare's knowledge about landscape, nature and geography which guided his eighteenth-century editors.[62] First, Shakespeare could be the poet of an ideal form, the *natura naturans*, or nature as it should be.[63] The 'Preface' eulogized Shakespeare's 'just representations of general nature' (*JOS*, p. 61). Shakespeare was the poet of nature because 'his characters are not modified by the customs of particular places... they are the genuine progeny of common humanity' (*JOS*, p. 62). It follows from this that Shakespeare's characters 'are communicable to all times and to all places; they are natural, and therefore durable' (*JOS*, pp. 69–70).

Yet there was a quite distinct second sense of the 'poet of nature' in the 'Preface'. In this case the poet of nature was said to copy life as he saw it. The faithfulness here was to Elizabethan England, which is 'natural' in the sense of the *natura naturata*, or nature as it appears. Shakespeare's 'nature' in this case was not an underlying truth, but the projection of Elizabethan naturalness across time and space: Shakespeare 'gives to all nations the customs of England, and to all ages the manners of his own' (*JOS*, p. 433; cf. also p. 374). Given this, Shakespeare must be compared not to the underlying ideal but 'with the state of the age in which he lived' (*JOS*, p. 81).

In the 'Preface' Johnson's 'use of *nature* in different senses is confusing'.[64] Yet when referring to the natural world rather than human nature, Johnson's position was less perplexed, allying Shakespeare's natural knowledge with the second, time and place-bound sense of 'nature'. One paragraph of the 'Preface' dealt specifically with Shakespeare's natural knowledge, placing it in an empirical tradition: 'he was an exact surveyor of the inanimate world; his descriptions have always some peculiarities, gathered by contemplating things as they really exist' (*JOS*, p. 89). To place Shakespeare's natural knowledge so firmly in the Elizabethan context was, in fact, to make an important statement, challenging the approach of previous eighteenth-century editions, which attempted to establish Shakespeare as a natural philosopher of impeccable Christian orthodoxy.

Shakespeare's natural philosophy: the *natura naturans* editorial tradition, 1723–65

The evidence for eighteenth-century editors' approaches to Shakespeare's knowledge of nature is contained in the voluminous notes appended to their editions.[65] Theobald developed this approach inspired by Bentley, and through him by the later European humanism to which Johnson has been connected.[66] Humanism has traditionally been seen as unconcerned with the scientific apprehension of the natural world. In fact, however,

> the humanists did not confine themselves to strictly literary areas of study. After 1450 they often analysed scientific texts and produced results of interest to specialists in medicine and astronomy.[67]

Eighteenth-century editions of Shakespeare provide an example of the interlinkage of humanism and science in the proliferation of notes, conjectures and refutations relating to his approach to nature.

Pope started the tradition that Shakespeare was knowledgeable about the natural world: 'Nothing is more evident than that he had a taste of natural Philosophy, Mechanicks, ancient and modern History, Poetical learning and Mythology.'[68] Pope's reference to (natural) philosophy could have been taken in the first sense of Johnson's *Dictionary* entry as 'Knowledge natural or moral', and would be seen in these terms later in the century. But under the influence of Warburton, philosophy was understood in Johnson's second sense of 'Hypothesis or system upon which natural effects are explained.' As such, the first half of the century saw attempts to clarify Shakespeare's theorization of the natural world.

This theorization was a Christian one. Thus Warburton glossed a passage in *As You Like It* to make Shakespeare's approach similar to that of the Boyle lecturers:

> the Clown's reply, in a satire on *Physicks* or *Natural Philosophy*...is extremely just. For the Natural Philosopher is indeed as ignorant (notwithstanding all his parade of knowledge) of the efficient cause of things as the Rustic. It appears, from a thousand instances, that our poet was well acquainted with the Physicks of his time: and his great penetration enabled him to see the remediless defect of it.[69]

Warburton also ascribed two more specific systems an influence over Shakespeare's natural philosophy. First, Shakespeare was influenced by Aristotelian cosmology, with its division of vegetable, animal and rational souls and its belief in 'several Heavens one above another'.[70] Secondly, he anachronistically connected Shakespeare with the Cambridge Platonists. In a note on 'an invisible instinct' in *Cymbeline*,[71] he argued 'that the poet uses invisible for blind. And by blind instinct he means a kind of plastic nature, acting as an instrument under the Creator'.[72]

Recognition of man's ignorance could lead to attacks on vain systematizers of natural knowledge, rendering Shakespeare not unlike Boerhaave in Johnson's 'Life'. Shakespeare did make satirical reference to Paracelsus in *All's Well that Ends Well* (II, iii. 11–12), but other attacks Shakespeare was interpreted by Warburton as having made had less textual foundation. The most extraordinary example of Warburton's interpretation of Shakespeare's natural philosophy came in his notes to *King Lear*. When Edmund said, 'Thou, Nature, art my Goddess' (I, ii. 1), Warburton noted

> he makes his bastard an Atheist...this was the general title those Atheists in their works gave to Nature; thus *Vanini* calls one of his books *De admirandis* NATURAE *Reginae* DEAEQUE MORTALIUM *Arcanis*. So that the title here is emphatical.[73]

Warburton wished to deny that Shakespeare deified nature, as this would taint him with idolatry; his attack was probably also directed against Shaftesbury's rhapsodies on nature, which he criticized in a note to *Othello*.[74] Warburton then added a series of notes to make Shakespeare's orthodoxy even clearer. Edmund was also

> made to ridicule *judicial astrology* ... For this impious juggle had a religious reverence paid to it at that time. And therefore the best characters in this play acknowledge the force of the stars' influence.[75]

Shakespeare's aim was to ridicule the Elizabethan belief in astrology with 'the severest lash of satire': 'it was a tender point, and required managing ... [H]e, with great judgment, makes these pagans Fatalists',[76] which explained leaving the satire to an atheist. As such, Warburton's Shakespeare exposed atheistical praise of nature by Edmund, the credulity of pagan religion in Lear and Gloucester's fatalism, and the Elizabethan belief in astrology. *King Lear*, for Warburton, then, was an orthodox Christian statement of natural philosophy.

This picture allowed two further interpretations to be placed on his works. First, Shakespeare was seen as an acute theorizer. 'Warburton was liable to let his imagination run wild whenever the topic touched upon astronomy.'[77] Thus he saw a reference in *Love's Labour's Lost* to 'firey numbers' (IV, iii. 297–98) as

> alluding to the discoveries in modern astronomy; at that time greatly improving ... He calls them *numbers*, alluding to the Pythagorean principles of astronomy.[78]

Warburton also conjectured that *Troilus and Cressida*'s 'as plantage to the moon' (III, ii. 173) should be 'as planets to their moons' because

> other *Planets* besides the *Earth*, (before the Time of our Author,) were discovered to have their *Moons* ... *Jupiter* has four Moons, and *Saturn* five.[79]

Warburton retracted this note in his own edition, on the basis of its impossible chronology ('I did not reflect that it was wrote before *Galileo* had discovered the Satellites of *Jupiter*[80]), but his enthusiasm for Shakespeare's astronomical wisdom was unabated. This was part of a broader pattern of crediting Shakespeare with a complex theorization of the natural world. Thus *Macbeth*'s 'As whence the sun 'gins his reflection / shipwrecking storms' (I, ii. 25–26) was justified (rather than 'gives his reflection') as storms generally come from the east...This proves the true reading is *'gins*; the other reading not fixing it to that quarter. For the Sun may *give* its reflection in *any* part of its course above the horizon; but it can *begin* it only in *one*.[81]

Warburton cited Varenius and Halley in support of his contention, without pointing out that the *Geographia Generalis* was published in 1650 and Halley's work on trade winds in 1686; how Shakespeare, who died in 1616, was supposed to have acquired this knowledge, therefore, was unclear.[82]

Shakespeare was assumed to be in concordance with accurate theories of the natural world because of the second further interpretation of his work, that he was a close observer of nature. On this point, Theobald agreed with Warburton, inserting a remark that Shakepeare's 'short nooky Isle of Albion' (*Henry V*, III, v. 14) was

> as true and proper a Description of *Great Britain* as *Camden*, or the most exact Topographer, could have given...the very Situation of our Island![83]

Theobald, on this basis, refused to assign errors of geography in *The Taming of the Shrew* and *Julius Caesar* to Shakespeare.[84] He also altered lines where Shakespeare's imagery appeared to conflict with facts he could not but have observed in the natural world. For example *Henry V*'s 'Her vine...Unpruned dies' (V, ii. 41–2) 'must read, as Mr *Warburton* intimated to me, *lies*: For neglect of pruning does not kill the Vine, but causes it to ramify immoderately'.[85] The converse was also held: where Shakespeare's imagery appeared to conform with a point of natural knowledge, this could only be philosophical exactness on the playwright's part. Thus in *King Lear*, where things were said to sting 'venomously, that burning shame—' (IV, iii. 46), Warburton thought 'the metaphor here preserved with great knowledge of nature. The *venom* of poisonous animals being a high caustic salt, that has all the effect of *fire* upon the part'.[86] This observation was generalized by Theobald: '*Shakespeare*, 'tis well known, has a Peculiarity in thinking; and, wherever he is acquainted with Nature, is sure to allude to the most uncommon Effects and Operations.'[87]

Warburton was the prime mover behind this interpretation. 'Shakespeare's' view was suspiciously akin to that of many eighteenth-century Anglicans, notably Warburton himself. Warburton had stressed, like his Shakespeare, that

> of *our own* physical system, we know many particulars, (that is, we discover much of the *means*, but nothing of the *end*) and of the *universal* physical system we are entirely ignorant.[88]

Paralleling *King Lear*, Warburton attacked those who sought 'to overturn all ESTABLISHED RELIGION, founded in the belief of a Sovereign Master...

And on their ruins, [attempt] to erect NATURALISM.'[89] Warburton, again as in *Lear*, fought those who believed in 'the Dotages of Astrology': for him Christian revelation was antithetical to superstition, God's control of nature mostly being enacted by natural laws, a view stressed throughout his explanation of an earthquake in *Julian* (1751), and connected with his admiration of 'the Plastic-nature of Dr Cudworth' and the Cambridge Platonists.[90]

This parallel between Warburton's natural philosophy and that he imputed to Shakespeare is unsurprising given his view of historical studies. For him,

> while the *other Sciences* are daily Purging and Refining themselves from
> · the Pollutions of superstitious Error,...*History*, still the longer it runs,
> contracts the more Filth.[91]

Warburton's approach to history was to purify it in the light of experience:[92] history could be empirical, and thus 'the Historian has here the very same Advantages over the moral Philosopher, that the *Experimental Naturalist* has over the *Aristotelian*'.[93] It was to be expected, then, that Warburton's approach to Shakespeare's natural knowledge modernized him into an experimental naturalist rather than contextualizing him in terms of the 'uninformed senseless Heap of Rubbish'[94] of Elizabethan chroniclers who were, in fact, Shakespeare's source material.

The view of Shakespeare Warburton constructed was strongly criticized. Theobald had realized the weakness of Warburton's approach: in a note to the *Winter's Tale* (III, i. 2), Theobald refused to correct Shakespeare's reference to the Temple of Apollo being on the 'Isle of Delos' rather than at Delphi because 'the Groundwork and Incidents of his Play are taken from an old story' which had referred to the Isle of Delos.[95] Where Shakespeare's source's geography was wrong, Shakespeare would be. This was just the visible remnant of Theobald's efforts in correspondence to temper Warburton's conjectures.[96] The main criticisms, however, came from Edwards and Heath, one element being Warburton's approach to the natural world. Both pointed out that Warburton's own limited knowledge of the natural world had led him to make emendations where none was required. Thus the emendation in *Julius Caesar* of 'ravens, crows, and kites' (V, i. 84) to 'ravenous crows and kites' on the grounds that 'a raven and a crow is the same bird of prey' was mistaken: 'every crow-keeper in the country will tell him there is...a difference between a raven and a crow'.[97] At the level of critical principles also, Warburton's 'physiological criticism', was attacked. Thus his note on *Macbeth*'s 'the sun 'gins his reflection' was criticized on the grounds that 'in this island at least...storms and thunder do as frequently take their course from the North, and West, as from the East' and that natural philosophy 'was not the point Shakespeare had in view. He draws a similitude from a very common appearance'.[98] This criticism was underlain by a view of

Shakespeare's knowledge which recognized his observational skills but downplayed its philosophical organization: 'as much as I honour Shakespear, I cannot persuade myself that he was that adept Mr Warburton makes him'.[99] Heath also recognized that Warburton's editorial practice was driven by his Anglicanism, at one point admitting his notes 'passeth my comprehension...is it his religion which is alarmed by this expression?' Just as Shakespeare was not a natural philosopher, so the orthodoxy of his writing was not always of concern: 'Mr Warburton hath rejected the common reading...as he tells us it "is false divinity"...the poet in all probability did not intend to decide in this place.'[100]

Johnson on Shakespeare and nature

For all the virtuosity of Warburton's approach, there had been a different approach whose concern was simply to clarify the meaning of terms relating to the natural world and customary beliefs about it to which Shakespeare might have made reference. Its fragmentary beginnings are to be found in Pope's edition, the first to include explanatory footnotes. Pope, for example, included a note explaining that 'foyson' (*Tempest*, II, i. 169) was '*the natural juice of the grass or other herbs*'. Other notes included a geographical reference to Cain's Hill near Damascus (mentioned in 1 *Henry VI*), one elucidating the meaning of a term in falconry, and an explanation of an 'old idle notion that the hair of a horse dropt into corrupted water, will turn to an animal'.[101]

Theobald had a number of notes of this variety, clarifying what Shakespeare could have meant in his references to nature rather than what he should have meant (*natura naturata* rather than *natura naturans*). This was to be expected, given his historical principles of editing.[102] Thus, in explaining the phrase 'still-vext Bermuda' (*Tempest*, I, ii. 230), Theobald pointed out these islands had been discovered by the English in 1609, and that 'still-vext' probably referred to the fact that they

> are so surrounded with Rocks on all sides, that without a perfect Knowledge of the Passage, a small Vessel cannot be brought to Haven... [and] they are subject to violent Storms.[103]

This was part of a discussion of Elizabethan attitudes to travel as reflected in Shakespeare, with possible references to Raleigh and Frobisher being analyzed.[104] The explication of traditions relating to the natural and supernatural was also much expanded after Pope. Hanmer's edition, for example, explained a reference in *Timon* to the unicorn conquering itself from Gesner's *Historia Animalia*, and Warburton explained why rue was the herb of grace.[105]

In Theobald's edition, however, this approach was swamped by Warburton's speculations. Moreover, even if Theobald accepted the importance of historical elucidation, he did not do so consistently; his long note on the nature of the bat, with references to Gesner, Pliny and Albertus Magnus, attempting to resolve whether Ariel sung in the *Tempest* (V, i. 92) of flying on a bat 'after sunset' or 'after summer' suggested a vacillation between philosophical and historical modes of annotation and emendation with respect to Shakespeare's understanding of the natural world.

Johnson's notes in the main participate in the tradition of uncovering the allusions to the natural world Shakespeare could have been making, rather than claiming he held any systematic position.[106] Indeed, his first published Shakespearian note in *Miscellaneous Observations on...Macbeth* (1745) had recognized the need to recover 'the notions that prevailed at the time when this play was written' (*JOS*, p. 3) with regard to the supernatural. As this suggests, Johnson sought to elucidate the cosmologies current in the Elizabethan period as they affected Shakespeare's approach to the natural and supernatural world. His main original contribution was to discuss 'the pneumatology of that time' whereby 'every element was inhabited by its peculiar order of spirits' (*JOS*, p. 960; cf. p. 122–3). Johnson did mention other issues of natural philosophy, but rather to gloss the text than to assert a position of his own (or to call such a position Shakespeare's). A comparison between Johnson and Warburton on the theologically sensitive question of man's relationship with the animal kingdom is instructive. *Timon's* 'the strain of man's bred out / Into baboon and monkey' (I, i. 254–55), Johnson noted to mean 'man is exhausted and degenerated; his 'strain' or lineage is worn down into monkey' (*JOS*, p. 712). By contrast, when Warburton speculated on the same issue in response to the *Tempest's* 'any strange beast there makes a man' (II, ii. 30–31) he argued this was a

satire very just upon our countrymen, who have been always very ready to make Denisons of the whole tribe of the *Pitheii*, and complemented them with the *Donum Civitatis*, as appears by the names in use. Thus *monkey*, which the Etymologists tell us, comes from *monkin, monikin,* homunculus. *Baboon*, from *babe*.[107]

In 2 *Henry IV* Johnson did refer to 'an ancient opinion...that if the human race, for whom the world was made, were extirpated, the whole system of sublunary nature would cease', but believed of the passage which this note accompanied that there was 'no need to suppose it exactly philosophical' (*JOS*, p. 493). Finally, Johnson referred to the 'Pythagorean doctrine which teaches that souls transmigrate from one animal to another' (*JOS*, p. 253), but this was in response to a direct reference to Pythagoras in *As You Like It* (III, ii. 172–74).

The vitalist pneumatology of the natural world that Johnson discussed ramified into a number of other positions of relevance to Shakespeare. On omens Johnson could suspend his own scepticism to report the meaning of passages in Shakespeare. Thus to Gloucester's 'these late eclipses in the sun and moon / portend no good to us; tho' the wisdom of nature / can reason it thus and thus, yet nature finds itself / scourg'd by the sequent effects' (*Lear*, I, ii. 101 ff), Johnson simply added this meant 'though natural philosophy can give account of eclipses, yet we feel their consequences' (*JOS*, p. 667–8), a position the Newtonian Warburton would have found hard to leave uncorrected. Similarly Johnson found the captain's 'enumeration of prodigies' in *Richard II* (II, iv. 8 ff) 'in the highest degree poetical and striking' (*JOS*, p. 438). That Johnson was sceptical of natural omens was made clear, but only where Shakespeare gave him a legitimate opportunity. Hotspur (1 *Henry IV*, III, i. 23–25) had said: 'O, then the earth shook to see the heav'ns on fire, / And not in fear of your nativity. / Diseased Nature oftentimes breaks forth', a speech Johnson considered 'a very rational and philosophical confutation of a superstitious errour' (*JOS*, p. 475). Closely connected were 'astrological opinions' (*JOS*, p. 914; cf. also p. 312), which Johnson referred to only to clarify Shakespeare's meaning, rather than attributing to him any involved satire of astrology as Warburton had.

Given a cosmology which ascribed a 'general communication of one part of the universe with another, which is called sympathy and antipathy' (*JOS*, p. 1040), Johnson also illuminated a series of views in Shakespeare on the relations between human, natural and supernatural realms. In the vegetable world, Johnson was the first editor to note in response to 2 *Henry VI* (III, ii. 313–14) that

> the fabulous accounts of the plant called a *mandrake* give it an inferiour degree of animal life, and relate, that when it is torn from the ground, it groans (*JOS*, p. 589; see also p. 747).[108]

Transgressing from the vegetable to the human world, Johnson explained that the fern seed (referred to in 1 *Henry IV*, II, i. 86–87) was ascribed strange properties by 'those who perceived that fern was propagated by semination and yet could never see the seed' (*JOS*, pp. 464–65). Likewise Johnson explained a number of references to lore about the animal world: that the toad's head contained 'a stone, or pearl, to which great virtues were ascribed' (*JOS*, pp. 246–47); that a bear's offspring had to be 'lick[ed] into the form of bears' (*JOS*, pp. 604); and that the lion was supposed to display 'acts of clemency' (*JOS*, pp. 936–37). Ascending to the human world, Johnson gave a long account of witches and their connection with the natural and supernatural worlds to show

> with how much judgment Shakespeare has selected all the circumstances of his infernal ceremonies, and how exactly he has conformed to common opinions and traditions (*JOS*, p. 32).

Johnson, then, was reluctant to ascribe any systematic natural philosophy to Shakespeare, but did recognize that the pneumatological view of the world was part of an approach which led to a very different view of the natural world from his own, which had to be recovered to understand many passages in Shakespeare.

Johnson also dealt with Shakespeare's natural knowledge. His was the first edition to adopt the idea that Shakespeare's geographical and natural knowledge might be rather limited. Therefore Johnson, in contrast to Warburton and Theobald, could note that 'Shakespeare seldom escapes well when he is entangled with geography' (*JOS*, p. 220; see also p. 295). The point was not to ridicule Shakespeare by comparison with subsequent standards of natural knowledge,[109] but that Shakespeare did make errors by the standards of his own era, such as placing the Bohemia on the coast in *The Winter's Tale*.[110]

Given this limited knowledge, Johnson's strategy was to clarify the meaning of particular references to nature in the plays. These notes are 'the bric-a-brac of Johnson's commentary' on which 'the critic will perhaps not wish to linger long'.[111] For the historian of geographical knowledge they are of great interest, and much of that interest lies in their fragmentary nature, for this was in contrast to the approach previously adopted. Warburton had seen Shakespeare's close observation of nature as part of Christian rejection of dogmatizing systems and thence as part of a doctrine; the very unconnectedness of Johnson's comments suggests at least agnosticism as to what Shakespeare 'believed' about nature. Johnson's first bric-a-brac note was in *Miscellaneous Observations*, where he argued 'shough' (*Macbeth*, III, i. 95) must be falsely printed for 'slouths', 'a kind of slow hound' or 'shocks', a hairy hound (*JOS*, p. 26). His approach was historical: 'there is no such species of dogs as 'shoughs' mentioned by Caius *De Canibus Britannicus*'. Caius's book had been published in 1570, and thus provided a fair guide to canine terminology to which Shakespeare could have referred. It was because Shakespeare's reference could not have meant anything in his contemporaries' language that Johnson proposed the emendation. Such emendation, as previously stated, had begun in Pope's edition, and Johnson recognized his notes on the natural world as contributing to a tradition. For example, he quoted Pope's explanation in *Othello* that 'fitchew' meant polecat, adding a further gloss (*JOS*, p. 1041) and agreed with Hanmer's emendation in *Twelfth Night* (*JOS*, p. 318). Johnson added to the number of notes pointing to Shakespeare's close observation of nature: that Brutus's reference to the ferret's firey eyes (*Julius Caesar*, I, ii. 187), for example, was appropriate as 'a ferret has red eyes' (*JOS*, p. 825). He also suggested this close observation had its limits, particularly in the 'fiery glow-worm's eyes' in the *Midsummer Night's Dream* (III, i. 162):

I know not how Shakespeare, who commonly derived his knowledge of nature from his own observation, happened to place the glow-worm's light in his eyes, which is only in his tail (*JOS*, p. 151).

Johnson's notes also contributed to two related traditions of notes. First, that explaining Shakespeare's references to rural activities, the assumption being that Shakespeare was 'a skillful sportsman' (*JOS*, p. 733) who 'loves to draw his images from the sports of the field' (*JOS*, pp. 856–57). The other issue was 'how much voyages to the South-sea, on which the English had then first ventured, engaged the conversation of that time' (*JOS*, p. 254; cf. pp. 408, 1019). He was original in pointing out the strength of Elizabethan xenophobia (*JOS*, p. 332), and the debate over the immorality caused by travel (*JOS*, pp. 258–59).

It is in a sense unfortunate that Johnson's only note pertaining to the natural world frequently cited is his attack on vivisection which turns from Shakespeare's England to his own period (*JOS*, p. 881). That note is clearly of interest as an element of Johnson's attitude to the natural world, but is unrepresentative of his editorial approach. Johnson generally explained what Shakespeare could have meant to an audience whose view of the natural world was profoundly dissimilar, rather than using Shakespeare's authority as a vehicle for his own views. Johnson indicated the difference between his editing and Warburton's at a number of points, most explicitly in *Antony and Cleopatra* (I, ii. 117–19), where Warburton detected an allusion to the sun's course, and Johnson responded that Shakespeare was 'less learned than his commentator' (*JOS*, p. 840; cf. p. 507). Johnson also criticized Warburton's refusal to recognize the looseness of much of Shakespeare's imagery. Thus where Wolsey (*Henry VIII*, III, ii. 356) spoke of 'a killing frost' which 'nips his root', Warburton altered 'root' to 'shoot' on the grounds that a spring frost could not kill a tree's root. Johnson replied that

> vernal frosts indeed do not kill the 'root', but then to 'nip' the 'shoots' does not kill the tree or make it fall. The metaphor will not in either reading correspond exactly with nature (*JOS*, p. 652; cf. also pp. 878, 557).

Johnson, like Heath and Edwards, realized the limitations of Warburton's approach to the 'poet of nature', a point confirmed by Johnson and Heath's (apparently) independent criticism of Warburton's reading of 'fiery numbers' in *Love's Labour's Lost* as an allusion to astronomy rather than poetry.[112]

Yet notes such as that on vivisection cannot be ignored, as they mark the transitional nature of Johnson's *Shakespeare*. Against signs of adopting a historical approach to Shakespeare's natural knowledge must be balanced Johnson's continuation of many of Warburton's most extravagant notes with no editorial correction. As in Theobald's edition, Johnson's notes had to compete with those of Warburton, two principles of editing being textually juxtaposed. The transitional status of Johnson's edition is most apparent in his continuation of Warburton's notes on *King Lear* and their qualification by Steevens's cutting comment in the 'Appendix':

Dr *Warburton* (for the sake of introducing an ostentatious note) says, that Shakespeare has made his bastard an *Atheist*; when it is very plain that *Edmund* only speaks of *nature* in opposition to *custom*, and not (as he supposes) to the existence of a *God*.[113]

For all this ambivalence, however, Johnson was not trying to co-opt Shakespeare into expounding his view. If he had done, his version of the poet of nature would have had more similarities with Warburton's. Johnson was in no less doubt than Warburton that the natural philosophy of Shakespeare's time was wrong, but where Warburton felt a moral imperative to clear up history, Johnson was part of an emergent historical sensitivity in Shakespearian studies[114] and scholarship more generally.[115] Where Warburton wished to clean up the 'filth' of history, Johnson simply wished to clarify what the past's filth was. This attitude, which led to a profoundly different view of Shakespeare, and the different moral imperative of history on which it rested, was summarized in one of Johnson's most eloquent notes:

the dead it is true can make no resistance, they may be attacked with great security; but since they can neither feel nor mend, the safety of mauling them seems greater than the pleasure; nor perhaps would it misbeseem us to remember, amidst our triumphs over the 'nonsensical' and the 'senseless', that we likewise are men; that *debemur morti*, and as Swift observed to Burnet, shall soon be among the dead ourselves (*JOS*, p. 985).

Shakespeare and Elizabethan natural knowledge (*natura naturata*), 1765–1821

The historical approach Johnson had moved towards came to dominate later-eighteenth-century discussions of Shakespeare. This was in part due to Farmer, whose *Essay* was influential in rebutting the notion that Shakespeare was learned in 'the writings of the Ancients' such that erudite sources could 'discovered in every natural description and every moral sentiment'.[116] Johnson's assessment of the relationship between his movement towards a new editorial principle and Farmer's statement of that principle was shrewd: 'I knew in general that the fact was as he represented it; but I did not know it, as Mr Farmer has now taught it me, by *detail*...'[117] The next great editors, Steevens and Malone, agreed that Shakespeare's knowledge was limited and that the only way to recover that knowledge was by an analysis of Elizabethan documents.[118] The picture of Shakespeare's natural knowledge created by the notes to the several editions of these two men was akin to Johnson's in outline. Shakespeare was still 'a most acute observer of nature'.[119] He was also no great geographer, a point conceded by Johnson and repeated by Steevens, Malone and Tyrwhitt in Malone's 1790 edition.[120] The interpretative reversal was symbolized by the gap between Whalley's belief that 'were all the Arts to be lost' as much could be recovered from Shakespeare as from 'the *Georgics* of *Virgil* and Steevens's comment on his lack of knowledge 'de re Rusticâ' which he regarded 'in the gross, and little thought [of accuracy] when he meant to bestow some ornamental epithet'.[121] As this makes clear, Warburton's assumption that Shakespeare's imagery accorded with the precepts of natural science, undermined by Heath, Edwards and Johnson, was dropped by later editors. Thus Malone viewed the issue of whether Ariel spoke of flying on a bat after sunset or summer as insoluble, because it assumed the correct answer would be the one most in accord with the habits of the bat, whereas Shakespeare was 'seldom solicitous that every part of his imagery should correspond'.[122] There were dissenting voices: Ritson consciously reversed the historicist approach to Shakespeare's natural knowledge, arguing that *Romeo and Juliet* could not refer to a female nightingale singing as it is only the male that sings: 'the discovery is not, indeed, of the age of Shakespeare – but what of that?'[123] In a general comment to the same play, Ritson remarked that

whatever may be the temporary religion, Popish or Protestant, Paganism or Christianity, if its professors have the slightest regard for genius or virtue, Shakespeare, the poet of nature, addicted to no system of bigotry, will always be a favourite.[124]

Ritson's editorial principles, then, had much in common with Warburton's, even if the assessment of the religion of nature had been reversed in line with his own radicalism. Shakespeare editing was controlled by conservatives in the period from Johnson to Malone,[125] but whilst Ritson's discussion was transparently connected with his politics, the approach to Shakespeare's natural knowledge of Johnson *et al.* was not clearly defined by a High-Church Tory political theology. Respect for tradition may link Johnson, Burke and Shakespeare's later-eighteenth-century editors,[126] but this principle could only guide the procedure of historical enquiry, not the resultant picture of Shakespeare's learning.[127]

The assessment of Johnson in subsequent editions was positive, building as they did upon similar principles. Johnson was praised for 'his refutation of the false glosses of Theobald and Warburton, and his numerous explications of involved and difficult passages',[128] Malone being able to pay a greater compliment by retaining many of his notes, including those on the natural world. Malone's 1790 edition, for example, and its 1821 successor, maintained Johnson's note on the system of enchantment used in the *Tempest*, and a note to *Macbeth* on Shakespeare's knowledge of the supernatural.[129] Criticism of Johnson was of the limitations of his knowledge of Shakespearian natural history, rather than for his approach. Mason, for example, argued that Johnson's reference to the glow-worm was unnecessary: 'surely a poet is justified in calling the luminous part of a glow-worm the "eye" '.[130] Johnson himself had criticized Warburton on similar grounds for inadequately separating natural philosophy from natural imagery, and Mason's note simply extended Johnson's approach. The 1821 edition, with its increase in scope and research, added to the corrections of Johnson's notes, trying, for example, to clarify the meaning of the term 'harlock' in *King Lear* (IV, iii. 4); where Johnson had conceded 'I do not remember any such plant' (*JOS*, p. 693), it provided three possible explanations.[131]

The massive increase in information about Shakespeare's England which is the overwhelming impression of these later editions, shows itself in the realm of natural knowledge in a number of ways. First, there was increasing reference to specific books of travel and natural history that Shakespeare could have known about, notably Malone's extensive use of Hakluyt's *Voyages*. Secondly, more novel was the emergence of reference to specific natural incidents possibly alluded to by Shakespeare. Malone, for example, suggested a reference in *The Winter's Tale* to 'a fish that appeared upon the coast...it was thought she was a woman' (IV, iv. 273–77) could relate to 'a strange reporte of a monstrous *fish* that appeared in the form of a woman', a book recorded by the Stationers' Company in 1604.[132] Thirdly, Shakespeare's references to specific places were tracked down: Wincot, mentioned in *The Taming of the Shrew*, was identified as a village in Warwickshire, with the justification that 'the meanest hovel to which Shakespeare has an allusion, interests curiosity, and acquires an importance'.[133] Finally, there was an increased sensitivity to ascribing to Shakespeare anachronistic natural knowledge: on this basis Shakespeare's reference to 'Musk-Rose beds' (*Midsummer Night's Dream*, II, ii. 3) became less clear as 'what is at present called the *Musk Rose*, was a flower unknown to English botanists in the time of Shakespeare'.[134]

The expansion of specialist input into later editions increased annotations about the natural world. Among the contributors were: Joseph Banks, President of the Royal Society; Thomas Martyn, Professor of Botany at Cambridge; and Daines Barrington an FRS. and correspondent of Gilbert White. The 1821 edition required a sixteen page index of 'manners, customs, superstitions'.[135] Moreover, Steevens, known as the puck of editors for his malicious scholarly humour, as well as assigning obscene notes to pseudonymous contributors of his own creation, also created for them long notes of pedantry about the natural world, his *tour de force* being a two page note on the potato, filled with learned references and explaining little.[136] Footnotes on the natural world had to be a prominent feature to be thus satirized, and indeed one of 'Collins's' notes on the barnacle and its relation to the barnacle goose received a serious response.[137]

Editions of Shakespeare from Theobald to Boswell-Malone show the fruitful connection of humanism and science in eighteenth-century England. The decline in educated belief in witchcraft and astrology in the seventeenth century[138] left eighteenth-century editors with little understanding of the Elizabethan approach to the natural world. That incomprehension could lead to projects like Warburton's, but also led to a gradual accumulation of information reconstructing that worldview. At least with respect to the natural world, it was not Malone's edition which marked the decisive break in editorial principles,[139] but Johnson's in combination with the new view of Shakespeare's learning in Heath and Farmer. The notes to these editions show the interpenetration of 'scientific' and 'artistic' concerns in the period, and point to the importance of geographical and natural knowledge in the public sphere of eighteenth-century culture.[140] The eighteenth century's efforts to recover the Elizabethan worldview themselves, perhaps, need to be recovered from the obscurity in which subsequent divisions of knowledge have placed them.

Moral life, literature and nature in the *Lives of the Poets*

Biography

Johnson's last major project, the *Lives of the Poets* (1779–81), mapped out a rather different relationship between the biographer and his subject's understanding of nature from that in his *Shakespeare*. Where Johnson's *Shakespeare* moved towards a contextual understanding of Elizabethan cosmology, the *Lives* were more judgemental, attacking poets for their 'superstitions' about the natural world.

This change in attitude was to some extent related to the moral imperative to encourage rational behaviour in those who read the *Lives*, a task less relevant to the edition of *Shakespeare*. It was also, however, a shift justified by the decline in the Shakespearian worldview Johnson contributed to recovering. His *Lives* were of individuals living after the foundation of the Royal Society. Attitudes which Shakespeare could hold rationally had been conclusively exploded as far as Johnson was concerned by the era in which the poets he wrote of lived. Johnson, then, had some outline of the history of natural philosophy in his mind which guided his comments on the relationship between life, nature and landscape established by an author. This was indicated most clearly in Johnson's 'Life of Browne', where he suggested of Browne's *Pseudodoxia Epidemica* (1646), itself an anatomy of vulgar errors, including those on vegetables, animals and cosmology (books 2, 3 and 6), that

> It might now be proper... to reprint it with notes partly supplemental and partly emendatory, to subjoin those discoveries which the industry of the last age has made, and correct those mistakes which the author has committed, not by idleness or negligence, but for want of BOYLE's and NEWTON's philosophy (*EBW*, p. 430).

This history of natural philosophy surfaces at a number of points in the *Lives*, Johnson consistently praising new scientific knowledge,[141] and commending Dryden's praise of the Royal Society.[142] This position also resulted in a lack of sympathy for critics of science. Thus Johnson criticized Gay's play, *Three Hours after Marriage*, for its attempt 'to bring into contempt Dr Woodward, the Fossilist, a man not really or justly contemptible' (*Lives*, ii. 271–72). Retrospectively, Johnson found the 'Battle of the Books' hard to understand: Butler was

> among those who ridiculed the institution of the Royal Society, of which the enemies were for some time very numerous and very acrimonious; for what reason it is hard to conceive, since the philosophers professed not to advance doctrines but to produce facts; and the most zealous enemy of innovation must admit the gradual progress of experience, however he may oppose hypothetical temerity (*Lives*, i. 208–209).

As in the 'Life of Boerhaave', science was as much separated from vain systematizing as from superstition, and was not simply harmless but liberating.

With the growth of natural knowledge, Johnson found it harder to justify belief in astrology than in his *Shakespeare*. Poets such as Dryden were recognized to have believed in astrology but it 'will do him no honour in the present age' (*Lives*, i. 109), and Johnson certainly did not try to recover the mentality which rendered it credible. On the contrary, he praised poets who ridiculed astrology (*Lives*, ii. 218 and i. 216).

Given this, Johnson expected his subjects to establish a moral lifestyle which included a rational approach to the natural world and its influence on that lifestyle. Unsurprisingly, natural knowledge occupied a low position in the hierarchy of knowledge required to forge a moral lifestyle. Star knowledge was less important than self-knowledge, a point made in *Paradise Lost* and which Johnson praised extravagantly: 'Raphael's reproof of Adam's curiosity after the planetary motions ... may be confidently opposed to any rule of life which any poet has delivered' (*Lives*, i. 177). For the same reason, Johnson censured Milton's activities as a schoolmaster for encouraging the reading of 'authors that treat of physical subjects' (*Lives*, i. 99). Johnson's judgement was the fullest exposition of the hierarchy in which natural knowledge participated throughout the *Lives*:

> the truth is that the knowledge of external nature, and the sciences which that knowledge requires or includes, are not the great or frequent business of the human mind. Whether we provide for action or conversation, whether we wish to be useful or pleasing, the first requisite is the religious and moral knowledge of right and wrong; the rest is an acquaintance with the history of mankind ... we are perpetually moralists, but we are geometricians only by chance. Our intercourse with intellectual

nature is necessary; our speculations upon matter are voluntary and at leisure (*Lives*, i. 99–100).

Johnson's response to Milton's school was the reverse of Toland's. For Toland, much of what was 'commonly read in the Schools' was 'trivial',[143] where Johnson failed to find the utility of Milton's alternative. Given that Toland was a Deist and Johnson anything but, it is unsurprising that their views on the interrelationship between natural knowledge and religion should be diametrically opposed, Toland conjoining and Johnson separating them, nor that their views of Milton's academy should therefore be so divergent.

Returning to Johnson's hierarchy of knowledge, given the low status of natural knowledge, it was a corollary that to see nature as determining the lifeplan of a moral actor was an inversion of that hierarchy and an abnegation of moral responsibility. This led to one alteration from the pattern of the early biographies: the renunciation of the notion that the bent of an individual's genius determined their career. Johnson opened his 'Life of Cowley' with a famous definition of genius as 'a mind of large general powers, accidentally determined to some particular direction' (*Lives*, i. 2). That this thesis rejected environmental determinism was made clear in Johnson's attack on Pope's notion of man's ruling passion:

> to the particular species of excellence, men are directed not by an ascendant planet or predominating humour, but by the first book they read, some early conversation they heard, or some accident (*Lives*, iii. 174).

The refusal to countenance environmental determinism led to one of Johnson's most significant passages on the interrelation between man and nature. Johnson cited Milton's nephew, Phillips, on the poet's inability to write over the summer and Toland's reply, before commenting that

> this dependance of the soul upon the seasons, those temporary and periodical ebbs and flows of intellect, may, I suppose, justly be derided as the fumes of vain imagination (*Lives*, i. 136–37).

Such views were moral lethargy justifying itself:

> the author that thinks himself weather-bound will find, with a little help from hellebore, that he is only idle or exhausted; but while this notion has possession of the head, it produces the inability which it supposes (*Lives*, i. 137).

Johnson used this opportunity to attack other determinisms which reversed the lines of influence between man and nature. First, the notion that man's intellect was suffering as part of the 'decrepitude of Nature . . . that everything was daily sinking by gradual diminution' (*Lives*, i. 137). Secondly, Johnson attacked

an opinion that restrains the operations of the mind to particular regions, and supposes that a luckless mortal may be born in a degree of latitude too high or too low for wisdom or for wit (*Lives*, i. 137–38).

This whole approach marked a departure from previous biographies. The debate had started, as Johnson said, with Toland's refusal to accept Phillips's account that Milton had been unable to compose in the summer.[144] Newton followed Phillips's account, while Richardson preferred Toland's version, adding that Milton's fears that he lived at too cold a latitude to write well supported this.[145] Fenton was unable to decide in which season Milton could not write, but thought 'the great inequalities to be found in his composures are incontestable proofs' the story had some foundation.[146] Johnson changed the course of this argument; until his 'Life' the issue had been when Milton could not write and what could explain this. For Johnson the only relevant question was what species of delusion had led Milton to believe the mind's abilities to be determined by the exigencies of the natural world, be they season, climate, or natural senescence. The whole notion that Milton's 'muse was us'd to Revive as the Vegetable World does'[147] was one Johnson found worthless to debate at a factual level.[148] '[T]he opposition of the probable and the marvellous is in fact a controlling theme of the biography':[149] the nature of the mind, of the natural world, and of their interrelation made any causal account of Milton's pattern of composition which involved anything beyond the first category fantastic.

As environmental determinism was the abnegation of moral responsibility, so were efforts to use place as a refuge. Johnson's point was akin to the motto of *Rambler* 135, 'Place may be chang'd, but who can change his mind?' and the *Lives* display a similar range of delusions to the characters in Johnson's essays, crystallized in real lives. An escalation in delusions about place, rurality and nature as palliatives to mental distress can be seen in the *Lives*. At the lowest level was Savage's 'scheme of life for the country, of which he had no knowledge but from pastorals and songs' (*Lives*, ii. 410). Savage's friends raised funds for his retirement, where 'he imagined that he should be transported to scenes of flowery felicity' (*Lives*, ii. 410), but the delusion was shattered by a brief period in Swansea. Amidst the complex series of self-deceptions Savage perpetrated, the Arcadian one was brief and its effects minimal.

Pope, the chief sponsor of Savage's retirement, represented a more serious delusion about retirement in his garden at Twickenham. The syntax of Johnson's sentences reflect his complaint about Pope's retreat, namely its affectation:[150]

as some men try to be proud of their defects, he extracted an ornament from an inconvenience, and vanity produced a grotto where necessity enforced a passage (*Lives*, iii. 135).

Johnson's comments on Twickenham were novel. His view opposed praise of the grotto as 'one of the most elegant and romantic of retirements'.[151] But Johnson also opposed two lines of criticism in Popiana. Twickenham could be used to criticize Pope's work, in particular his translation of Homer which had given him the resources with which to improve it:

> Retire triumphant to thy *Twick'nam* Seat;
> That Seat! the Work of half-paid *Br—me*,
> And call'd by joking *Tritons, Homer's* Tomb.[152]

Twickenham also featured in straightforward personal invective:

> Horrid to view! retire from human Sight,
> Nor with thy Figure pregnant Dame affright.
> Crawl thro' thy childish Grot, growl round thy Grove,
> A Foe to Man, an antidote to Love.[153]

By contrast, Johnson's discussion of Pope, while pointed, suggested no personal dislike. The natural world was introduced to suggest a moral failing in Pope who, unlike Savage, persisted in his delusions as to the joys of rural life, due to his inability to see that these amusements were 'frivolous and childish' (*Lives*, iii. 135).

Johnson's discussion of Cowley's retirement achieved a greater reduction of the pretensions of place: 'if his activity was virtue, his retreat was cowardice' (*Lives*, i. 10), a more serious charge than that levelled at Pope. Johnson had satirized Cowley's retirement in *Rambler* 6 and he continued this in the 'Life'. The only other biography of Cowley to adopt such a tone was in Cibber's *Lives of the Poets*, a collection largely written by Shiels, an assistant in the preparation of Johnson's *Dictionary*, who has been shown to have taken a number of critical ideas from Johnson.[154] Whether Shiels's source was the *Rambler* or conversation, it seems likely the interpretation was Johnson's. More interesting is Johnson's relation to Sprat's 'Account' (1668), to which he said his own 'Life' was 'a slender supplement' (*Lives*, i. 18). Sprat, like Johnson, condemned Cowley's retirement, but on different grounds as 'a great disparagement to virtue'.[155] For Sprat no moral failing was shown by this retirement, because

he always professed, that he went out of the world, as it was man's, into the same world, as it was nature's, and as it was God's. The whole compass of the creation, and all the wonderful effects of the divine wisdom, were the constant prospect of his senses, and his thoughts.[156]

Johnson omitted this argument, thus reducing Sprat's picture of meditative Christian retreat to the recurrence of a delusion. Such an omission, as we have seen, was characteristic of Johnson's editorial and authorial response to the argument from design. The omission was noted at the time by Potter, a critic of Johnson's biographical writings, and attributed to his denominational position, eliding High Churchmanship and Catholicism as M'Nichol had in the *Journey*:

> he tells us that 'indeed, it must be some very powerful reason that can drive back to solitude him who has once enjoyed the pleasures of society.' That is, to induce a person to *retire* from the world as it is man's, into it as it is God's. Yet our author is an adorer of monasteries![157]

Given Johnson's aim to expose the folly of retirement, his omission of one part of Sprat's narrative is surprising. Sprat pointed out not only as Johnson did that Cowley caught a cold on his retirement, but that he died of 'a violent defluxion' caused 'by staying too long amongst his labourers in the meadows'.[158] Johnson's life made no connection between Cowley's retirement and his death. That final escalation of the dangers of human delusions about their relationship with the natural world was reserved for the 'Life of Shenstone'. Johnson began with a disclaimer:

> Whether to plant a walk in undulating curves, and to place a bench at every turn where there is an object to catch the view; to make water run where it will be heard, and to stagnate where it will be seen; to leave intervals where the eye will be pleased, and to thicken the plantation where there is something to be hidden, demands any great powers of mind, I will not enquire (*Lives*, iii. 350).

He added another commonplace disclaimer of the period that 'it must be at least confessed that to embellish the form of nature is an innocent amusement' (*Lives*, iii. 351). Yet Johnson did decide on Shenstone's powers, concluding 'his mind was not very comprehensive, nor his curiosity active' (*Lives*, iii. 354; and see iii. 359), suggesting a powerful mind was not required to be a landscape gardener, a point again noticed by Potter.[159] The innocence of the pleasure was also denied during the 'Life': his neighbour Lyttelton taking

> visitants perversely to inconvenient points of view, and introducing them at the wrong end of a walk to detect a deception; injuries of which Shenstone would heavily complain. Where there is emulation there will be vanity, and where there is vanity there will be folly (*Lives*, iii. 351–52).[160]

So far this appears little different from Twickenham, and Johnson 'at once shews what ideas he had of landscape improvement, and how happily he applied the most common incidents to moral instruction'.[161] Yet the 'Life of Shenstone' concludes:

> In time his expences brought clamours about him, that overpowered the lamb's bleat and the linnet's song; and his groves were haunted by beings very different from the fauns and fairies. He spent his estate in adorning it, and his death was probably hastened by his anxieties (*Lives*, iii. 352).

The progress from trivial follies to vanity concluded with a reminder of the vanity of human wishes, and the embellishment of nature appeared anything but innocent. Shenstone's decline was presented in a manner akin to Bob Cornice's route to prison in *Adventurer* 53, far divorced from his sources.[162] Graves responded to Johnson's 'Life': he remained within Johnson's scale of values, but reversed his judgements. For Graves, 'Shenstone had generally an eye to utility, as well as ornament, in his plans.' This was part of a refutation of Johnson's picture of the poet as an intellectual lightweight: 'he employed himself in the study of the mathematics, logic, natural and moral philosophy', and this was reflected in his garden, 'the planning and disposing of which certainly discovers no common degree of genius'.[163]

Graves's defence concluded with an explanation of Johnson's hostility to Shenstone's garden:

> Born in the city of Litchfield [*sic*]; confined in his youth, the season of fancy, in a bookseller's shop in Birmingham; after a short stay in the university, transplanted to the metropolis, there drudging for the press, and hackneyed in the ways of men, what leisure, or inclination, or opportunity could such a man have to attend to or study the beauties of nature, and the pleasures of a country life?[164]

This explanation was taken up by other critics of Johnson's *Lives*,[165] obscuring the role Johnson's biographies established for the natural world. Compared with the biographies on which he drew, Johnson rendered problematic the relationship between life and landscape. Johnson's own aesthetic response to landscapes was all but irrelevant to the manner in which they appeared in the *Lives*. Johnson 'finds value in many of those incidents in a man's life which do not display an exalted purpose or issue in great deeds',[166] and it was as the revealing detail that the natural world became significant in the biographies. Johnson's incorporation of the natural world into the biographies related not so much to a biographically comprehensible insensitivity to nature as to a Christian scepticism about the excessive role some allowed nature to play in their lives, as a false determinant or a consuming passion. Potter's quip about the 'adorer of monasteries' was probably a far more perceptive answer to the question as to 'why then are these infirmities recorded?' than his other response that 'the Doctor...sickens at the idea of any thing rural'.[167]

Literature

Delusions about nature were not only operative in the lives of poets, but also produced by the work of poets. This was clear in the case of Savage, whose only knowledge of life in the country was 'from pastorals and songs' (*Lives*, ii. 410). As it was an issue of moral importance to hold nature in subordination to human agency, so how to write about the natural world became an ethical concern. This was part of a general tendency in Johnson's critical writings: 'since his subjects are poets as well as men, moral reflection, in turn, continually extends into passages of literary criticism'.[168]

Johnson's criticism of the pastoral is well known. The poverty of Shenstone's mind was demonstrated by his *Pastoral Ballads*, where 'an intelligent reader acquainted with the scenes of real life sickens at the mention of the *crook*, the *pipe*, the *sheep*, and the *kid*' (*Lives*, iii. 356). It is the absence of a connection between pastoral representation and rural reality which annoyed Johnson, rather than (as Graves and Potter asserted) the representation of rurality *per se*. Most controversial was Johnson's response to *Lycidas*:

> Its form is that of a pastoral, easy, vulgar, and therefore disgusting: whatever images it can supply are long ago exhausted; and its inherent improbability always forces dissatisfaction on the mind (*Lives*, i. 163; see also ii. 217–18, 315).

The pastoral was supposed to be conventional, and Johnson marks the end of criticism responding to such types.[169] Yet his aims were not only critical; running alongside this were biographical examples of the follies pastorals could lead to. Innocent literary conventions became anything but innocent when they encouraged lifestyles which undermined the integrity of moral actors.

Johnson had one other complaint about the pastoral; its unneccesary attachment of current affairs to natural imagery. In his brief history of the pastoral, Johnson noted

> the speakers of Mantuan carried their disquisitions beyond the country, to censure the corruptions of the Church; and from him Spenser learned to employ his swains on topics of controversy (*Lives*, iii. 318).

Similarly, the criticism of *Lycidas*'s imagery led into censure of 'a grosser fault' by which 'with these trifling fictions are mingled the most awful and sacred truths, such as ought never to be polluted with such irreverent combinations' (*Lives*, i. 165). This was part of a broader attempt to define the limits the representation of nature presented to the discussion of more elevated topics. Poetry could defend a design argument about the natural world, but it could not go on to describe the dictates of Christianity via nature. This point was made in the 'Life of Waller', which created a chasm between nature which can be described and its Creator who cannot:

> a poet may describe the beauty and the grandeur of Nature, the flowers of the Spring, and the harvests of Autumn, the vicissitudes of the Tide, and the revolutions of the Sky, and praise the Maker for his works in lines which no reader shall lay aside. The subject of the disputation is not piety, but the motives to piety; that of the description is not God, but the works of God (*Lives*, i. 291).[170]

Johnson's criticism carefully monitored the relationship poets created between God and the natural world, censuring Yalden, for example, for suggesting that God marvelled at his own creation: 'infinite knowledge can never wonder. All wonder is the effect of novelty upon ignorance' (*Lives*, ii. 303). Moreover, even if poetry can describe the face of the earth, it cannot match the Mosaic account: 'the miracle of Creation, however it may teem with images, is best described with little diffusion of language: "He spake the word, and they were made" ' (*Lives*, i. 50). Johnson here followed a critical commonplace as to the sublimity of the *Bible*, derived from Longinus's *On the Sublime*. At the other end of the earth's chronology, Johnson spoke of Philips's hope to write 'a poem on *The Last Day*, a subject on which no mind can hope to equal expectation' (*Lives*, i. 314). Again, such a scene was not within the possibilities of poetic representation, a view different from Philips's first biographer, Sewell.[171]

For Johnson, then, poetic presentation of the natural world is limited to its appearance, its Creator being beyond poetry because he is beyond the scope of the human intellect. This was connected with the distancing of man from nature at the Fall: one of the sublimities of *Paradise Lost* was Milton's presentation of Adam and Eve 'on whose rectitude or deviation of will depended the state of terrestrial nature and the condition of all the future inhabitants of the globe' (*Lives*, i. 172; cf. also i. 51). After the Fall, man's view of the 'economy of nature' became a limited and low one, such that Johnson could criticize Pope (*Lives*, iii. 210) and the Metaphysical poets who sought to write 'as Epicurean deities making remarks on the actions of men and the vicissitudes of life, without interest and without emotion' (*Lives*, i. 20). It was Pope's attempt to adopt such an unhuman elevation which led to Johnson's criticism of the *Essay on Man*. For Johnson, Pope passed on commonplaces about man's relation to nature in a manner which simultaneously denied their importance:

> he tells us much that every man knows, and much that he does not know himself; that we see but little, and that the order of the universe is beyond

our comprehension, an opinion not very uncommon; and that there is a chain of subordinate beings 'from infinite to nothing', of which himself and his readers are equally ignorant (*Lives*, iii. 243).

Johnson had previously commented on the *Essay* in his translation of Crousaz's *Commentary* which repeatedly attacked Pope's conception of 'Nature'. Johnson believed Crousaz too 'watchful against Impiety', finding it where none was intended, but agreed Pope left 'a great Difficulty in annexing a reasonable Meaning to the term *Nature*'.[172]

Pope's *Essay* was vindicated by that great defender (and discoverer) of Anglican natural philosophy, Warburton. But even Warburton had to admit that Pope's expression was occassionally lax, as in the line 'If nature thunder'd in his op'ning ears' (*Essay on Man*, i. 201) where 'what is worse, he speaks of this [i.e. nature] as a *real object*'.[173] Johnson's strategy and Warburton's with respect to Pope's *Essay* differed in ways akin to their editions of Shakespeare: Warburton demonstrated the *Essay*'s orthodoxy in ways the poet had not thought of, while Johnson was prepared to see its heterodoxy as a sign of the poet's ignorance. Yet both agreed that nature, in poetry as elsewhere, should not be seen as a 'real object', a point also reflected in Johnson's criticism of personification: 'such an investment of the spirit's yearnings in the things of the world worked on dangerous moral grounds'.[174] For Johnson 'the parts of *Windsor Forest* which deserve least praise' (*Lives*, iii. 225) are the appearance of Father Thames and the nymph Lodona's metamorphosis into a river. Johnson also criticized Gray's *Prospect of Eton College* for 'the supplication to Father Thames' (*Lives*, iii. 434–35). The religious principles behind Johnson's criticisms were recognized by Tindal, who defended Gray's personification by pointing to *Psalm* 114: 'What ailed thee, O thou sea, that thou fleddest? thou Jordan, that thou wast driven back?' He added 'this is grounded on a fact which I am no more inclined to dispute or ridicule than Dr Johnson; but the demand is clearly poetical',[175] such that personification should not be conflated with the idolatry of nature worship.

An acceptable presentation of nature, for Johnson, must view it as a collection of potentialities akin to the picture created by Boyle, stripped of the errors of animism. Attention was turned to a nominalistic focus on the poet's ability to present individual images. Akin to empirical science, images of nature presented by the poet were criticized if stale (as in the pastoral) or too learned, rather than direct responses to sensory inputs: 'Cowley gives inferences instead of images, and shews not what may be supposed to have been seen, but what thoughts the sight might have suggested' (*Lives*, i. 51). Johnson's concern for authentic imagery of the natural world led him to reverse accepted judgements. Milton's

images and descriptions of the scenes or operations of Nature do not seem to be always copied from original form, nor to have the freshness, raciness, and energy of immediate observation (*Lives*, i. 178).

He particularly pointed out that 'the garden of Eden brings to mind the vale of Enna, where Proserpine was gathering flowers'. By contrast, opinion from Addison's *Spectator* 321, through to Newton's edition of *Paradise Lost* had accepted that 'he saw Nature Beautifully'.[176] Hume's praise was by eighteenth-century critical standards the greatest:

> if we compare our poet's topography of Paradise with Homer's description of Alcinous's gardens, or with Calypso's shady grotto, we may without affectation affirm, that in half the number of verses that they consist of, our author has outdone them.[177]

The exemplar of good imagery for Johnson was Dryden in whom 'every page discovers a mind very widely acquainted both with art and nature, and in full possession of great stores of intellectual wealth' (*Lives*, i. 417; see also ii. 229–30). With respect to landscape, Johnson praised Gay's *Shepherd's Week*, for its 'just representations of rural manners and occupations' (*Lives*, ii. 269) and Philips's *Cyder* for giving a useful knowledge of a rural trade, and therefore being 'at once a book of entertainment and of science' (*Lives*, i. 319).[178] Johnson sought in images of rurality the precise opposite to what he found in the pastoral: poetry which reflected what could actually be seen in the landscape rather than what previous poets had claimed to see.

A further question in the *Lives* was how to combine such images to produce literary landscapes. Poetry could not directly represent landscape as apprehended visually for 'verse can imitate only sound and motion' (*Lives*, i. 62). Hutcheson had argued that the poet was distinguished in that his 'Prospect of any of those Objects of natural Beauty, which ravish us even in his Description' was based upon a 'delightful Perception of the Whole'.[179] For Johnson this was exactly what poetic description of a prospect could not achieve due to the problems of converting the visual into the verbal. Thus Johnson defended Pope's *Windsor Forest* from the charge of lacking an ordered structure on the grounds that

> there is this want in most descriptive poems, because as the scenes, which they must exhibit successively, are all subsisting at the same time, the order in which they are shewn must by necessity be arbitrary (*Lives*, iii. 225; see also ii. 365).

Johnson's *Shakespeare* had been the first to highlight the Dover Cliff prospect in *Lear*, where he encountered the same problem: 'He that looks from a precipice finds himself assailed by one great image of irresistable destruction.' But the great image could not be represented verbally (which was the sublimity *Genesis* had acheived), and Shakespeare's answer, the enumeration of the choughs and crows, the samphire-man and the fishers, counteracts the great effect of the prospect, as it peoples the desert of intermediate vacuity, and stops the mind in the rapidity of its descent (*JOS*, p. 695).

In recognizing the roots of the problem of landscape description as lying in the conversion of visual images into poetic imagery and then into an ordered assemblage of that imagery, Johnson displayed considerable critical insight. When compared with his contemporaries, Johnson's position on literary landscapes appears decidedly original. It had become a commonplace to complain of the lack of particularity in Pope's *Windsor Forest*, a tradition starting in a letter of 1714 by Dennis and repeated by Joseph Warton.[180] It was in a review of Warton that Johnson had first signalled his opposition to this view, arguing that one 'must inquire whether Windsor Forest has in reality any thing peculiar',[181] but it was only in the *Lives* that the principle behind the query, whether the ordering of elements in a poetic description could convey the visual specificity of the prospect, was fully formulated.[182]

In a variation on the theme, Johnson commented of Thomson's *Seasons* that within each season no narrative justification for the ordering of incidents could be found, but that there was no remedy. The main source for Johnson's 'Life', by Murdock, had said 'the *Seasons* are placed in their natural order',[183] a consideration which did not impinge on Johnson's problem. In response to Johnson, Stockdale said 'the poet surveys . . . Summer's morning, noon, evening, and night as they succeed one another, in the course of nature.'[184] While this addressed Johnson's point as to the ordering of description within a season, it did not tackle the ordering of a poetical prospect in space.[185]

A possible solution to the problem of verbal landscape representation was suggested a year after the 'Life of Thomson' by Gilpin's *Observations*. Gilpin criticized Dyer's *Grongar Hill* for failing to give a landscape 'which melts a variety of objects into one rich whole' by gradually giving further distant items 'still fainter colours' akin to painterly aerial perspective.[186] For Gilpin, Dyer had attempted to be a landscape painter in writing but failed to copy their technique adequately, but this did not show that word 'painting' was a contradiction in terms. Scott responded that Gilpin's criticism was inappropriate: 'His [Dyer's] hill's extensive view would probably have afforded *several* complete landscapes, but it is not clear that he aimed at producing *any*'.[187] Whether Dyer was influenced by painting or not,[188] it is clear that Scott's remarks were appropriate to Johnson's criticism of Dyer and of prospect poetry as a whole. For Johnson, *Grongar Hill* was 'not indeed very accurately written, but the scenes which it displays are so pleasing, the images which they raise so welcome to the mind' (*Lives*, iii. 345). His criticism did not consider that eighteenth-century prospect poetry had attempted to produce coherent painterly landscapes.[189] Gilpin's *Observations*, and the formal picturesque which they announced, had been circulating in manuscript since 1771, but clearly his ideas made no impact on Johnson's criticism.

The terms in which Johnson praised descriptive poetry were rather different from Gilpin's, as revealed in his assessment of *Grongar Hill* where he praised the insertion of 'the reflections of the writer so consonant to the general sense and experience of mankind' (*Lives*, iii. 345). In praising the moralizing of prospects, Johnson was part of a widely held critical opinion. The same criterion was applied to Denham's *Cooper's Hill* which for Johnson defined the genre of 'local poetry' as including 'such embellishments as may be supplied by historical retrospection or incidental meditation' (*Lives*, i. 77). The most sustained achievement in this field was Blackmore's *Creation* where 'in his descriptions both of life and nature the poet and the philosopher happily co-operate' (*Lives*, ii. 254; also ii. 243). Yet the ideal was not simply didactic, for the reflections had to arise naturally from the prospect. As *Lycidas's* greatest fault had been to engraft religious politics onto the lament of shepherds, so *Cooper's Hill* was only partially successful in the genre it established, because 'the digressions are too long, the morality too frequent' (*Lives*, i. 78).

. Johnson's attitude to literary landscapes could be argued to be more 'sceptical' than that of many contemporaries. Yet this reflected less his own responsiveness to landscape than a recognition that the conversion of enthusiasm for landscape into words was by no means simple. The ability of words to give a direct representation of a visual prospect was minimal for Johnson. His own solution was rooted in a nominalistic view of nature, of which the poet could present true images, and in an empiricist association of ideas, where reflections naturally arising from a prospect could be attached to images, which dignified the image, given the intellectual hierarchy in which natural knowledge participated.

Conclusion

Folkenflik has observed that Johnson's biographies are 'profoundly influenced by Christian and classical conceptions of man that had not been central to biographical writings before his appeared'.[190] The presentation of the natural world, both biographically and critically, partook of these influences. The most insistent theme was a Christian voluntarism which environmental determinism threatened to undermine. Even where not seen as determinative, the belief that change of place could reduce the problems created by the interplay of human wills was a similar delusion, upsetting the balance of mind and place towards the latter. Accordant with this, Johnson's critical comments refused to view nature as a real entity capable of acting. Man as a fallen moral actor was enmired in the prospect, not an Epicurean deity surveying it, and this was paralleled in man's limited ability to paint prospects from an elevated perspective in words.

Johnson's life and landscape

The *Lives of the Poets* presented, *inter alia*, Johnson's public doctrine of a moral life lived with respect to nature. Johnson's private doctrine of life in the natural world, and the connection forged in Johnson's own biography between life and landscape, mind and place, is another matter. This redirects attention from intellectual history towards the affective and personal construction of meaning in the land.[191]

Johnson's early life: transience and affirmation

There is a dichotomy in Johnson's early relationship with and thoughts about nature, land and landscape. On the one hand, Johnson's school exercises, translations of Horace and Virgil, point to two themes. The Horatian exercises make conventional parallels between the natural world and the brevity of human life:

> Your shady groves, your pleasing wife,
> And fruitfull fields, my dearest friend,
> You'll leave together with your life,
> Alone the cypress shall attend.[192]

Johnson's translations of Virgil's eclogues (*Poems*, pp. 5–7) involved him in pastoral conventions which he was to attack throughout his adult life, but certainly influenced his own first poem 'On a Daffodill'. In this Johnson has the natural world respond to the poem's subject, Cleora, in precisely the way he was to criticize in Waller in the *Lives* (i. 285):

> Cleora's smiles a genial warmth dispense;
> New verdure ev'ry fading leaf shall fill,
> And thou shalt flourish by her influence.[193]

The poem also closed conventionally, with nature now pointing to man's future rather than simply responding:

> And ah! behold the shriveling blossoms die,
> So late admir'd and prais'd, alas! in vain!
> With grief this emblem of mankind I see,
> Like one awaken'd from a pleasing dream,
> Cleora's self, fair flower, shall fade like thee...[194]

Johnson's youthful literary labours, then, affirmed the importance of nature only inasmuch as it provided an emblem for mankind. Yet in Johnson's early life, a different picture emerges. We find him taking occasional pleasure in his connection with the environment: 'his only amusement was in winter, when he took a pleasure in being drawn upon the ice by a boy barefooted' (*Life*, i. 48), a pleasure which continued in his undergraduate days by 'sliding in Christ-Church meadow' (*Life*, i. 59). But the main function of the natural world for Johnson was as a foil to his depression: 'Johnson, upon the first violent attack of this disorder, strove to overcome it by forcible exertions. He frequently walked to Birmingham and back again' (*Life*, i. 64).[195] In Johnson's early life, then, a relationship between man and nature was established quite contrary to that suggested by his writings: rather than nature gaining its significance through mankind, the natural world was one way of affirming himself. Johnson was later to refute Berkeley by 'striking his foot with mighty force against a large stone, till he rebounded from it' (*Life*, i. 471),[196] and in his 'forcible exertions' Johnson was trying to rebound from nature to show his mental torment could be alleviated by the stability of the physical. This was not the aesthetic pull of landscape, but a recognition that in its sheer physicality, the land could push the mind out of itself and into the world of the sensory.

Johnson's private doctrine: place and mind

Johnson's need to affirm his existence through nature continued into his London life. In his early days, he wandered the streets with the poet Savage, something not brought about by poverty.[197] Other instances include Johnson's 'frisk' with Beauclerk and Langton in 1752, where they 'walked down the Thames, took a boat, and rowed to Billingsgate' (*Life*, i. 251), and Boswell and Johnson's trip to Greenwich in 1763 where they 'were entertained with the immense number and variety of ships that were lying at anchor, and with the beautiful country on each side of the river' (*Life*, i. 458). Towards the close of this Greenwich ramble, the *Life* records the following exchange:

> [Johnson] "Is not this very fine?" Having no exquisite relish of the beauties of Nature, and being more delighted with 'the busy hum of men', I answered, "Yes, Sir; but not equal to Fleet-street." JOHNSON. "You are right, Sir" (*Life*, i. 461).

This conversation was not in Boswell's *London Journal*, and was probably inserted during the composition of the *Life*.[198] Even if Johnson did say this, a recurrent pattern can be observed in these rambles combining sociability and the pleasure of temporary respite from urban routine. It is hard to view otherwise the picture of Johnson at Windsor:

> one Sunday, when the weather was very fine, Beauclerk enticed him, insensibly, to saunter about all the morning. They went into a church-yard, in the time of divine service, and Johnson laid himself down at his ease upon one of the tomb-stones. "Now, Sir, (said Beauclerk) you are like Hogarth's Idle Apprentice" (*Life*, i. 250).

Clearly, the interconnection of lives, landscapes and literature continued to help Johnson in mid-life to take solace from his 'vile melancholy' (*Life*, i. 35). Recognition of the importance of occasional movement and of place to mental stability led Johnson to a rather less austere view of the connection of mind and place in his private advice to friends than in his published writings. The gap between Johnson's public and private pronouncements on this issue amounts not to deceitfulness, but the recognition that a person cannot always live up to what they believe to be worthy, coupled with a fear that such a failure may set a bad example. No matter how much his essays and biographies attempted to encourage mental constancy regardless of place, Johnson loved the experience of travelling; 'the very act of going forward was delightful to him'.[199] As such, Johnson privately could reverse his position on mind and place: 'some benefit may be perhaps received...from mere change of place'.[200] This was said in Johnson's last years as he battled for his health, but Tyers recorded for an earlier period that 'change of air and of place were grateful to him, for he loved vicissitude'.[201] In advice to friends Johnson also recommended change of place as a relief from mental pressures. This was manifested most strongly in the series of letters Johnson sent to John Taylor on his estrangement from his second wife:

> I cannot but think it would be prudent to remove from the clamours, questions, hints, and looks of the people about you...

> I once more advise removal from Ashbourne as the proper remedy...

> I cannot but think that by short journeys, and variety of scenes you may dissipate your vexation (*Letters*, i. 230, 230, 237).

Johnson even advised Taylor to indulge in 'sports of the field abroad, improvements of your estate or little schemes of building' (*Letters*, i. 242).[202] In a sense, the position adopted was the same as in the essays, that all schemes of gardening and the like were simply attempts to redirect the mind from more painful prospects or from indolence, but in private Johnson accepted this could be necessary.

Johnson's later life: the affirmation of transience

Johnson's last years saw the sociable sallies into the natural world change their character. As death robbed his landscapes of the figures who had made them meaningful, so Johnson saw in familiar physical surroundings the landscape of loss. He wrote of this in 1783, saying of London, whose 'busy hum' had died down, 'the neighbourhood is impoverished. I had once Richardson and Laurence in my reach. Mrs Allen is dead. My house has lost Levet...' (*Letters*, iv. 160). The same letter points to a source of comfort Johnson found in the London of his old age: 'I have however watered the garden both yesterday and to day' (*Letters*, iv. 161).[203] This was a very different sort of pleasure in gardening from that of Pope or Shenstone, for Johnson's was a quite private activity, presumably to allow calm thought and gentle exercise, not a public display of affectation.

The fusion of loss, memory and landscape was clearest in Lichfield. At the height of his career, Johnson had written with some contempt of his native town: 'I found the streets much narrower and shorter than I thought I had left them' (*Letters*, i. 206). Fifteen years later, Johnson's attitude, as he himself admitted had changed: 'in age we feel again that love of our native place and our early friends, which in the bustle or amusements of middle life were overborn and suspended' (*Letters*, v. 6–7). Even in 1769 Johnson had expressed his dislike of the cutting down of trees in Lichfield (*Letters*, i. 327), a traditional theme in laments about transience and landscape.[204] Johnson still found pleasure in physical exertion, although this was more gentle than in his youth. A few months before his death, Johnson could recover something of the pleasure occasioned by his mid-life rambles:

> last evening, I felt what I had not known in a long time, an inclination to walk for amusement. I took a short walk, and came back neither breathless nor fatigued (*Letters*, iv. 400–401).

This may reflect the calmness which Johnson exhibited in his last year after what he considered a miraculous release from illness in February 1784,[205] but even before this Johnson's personal relationship with the land had become more affective and mnemonic:

> A gentleman of Lichfield meeting the Doctor returning from a walk, inquired how far he had been? The Doctor replied, he had gone round Mr Levet's field (the place where the scholars play) in search of a rail that he used to jump over when a boy, "and", says the Doctor in a transport of joy, "I have been so fortunate as to find it": I stood, said he, "gazing upon it some time with a degree of rapture, for it brought to my mind all my juvenile sports and pastimes, and at length I determined to try my skill and dexterity; I laid aside my hat and wig, pulled off my coat, and leapt over it twice"[206]

Johnson's more quiescent relationship with the natural world in later life was in part forced upon him by declining health: a lifetime of rebounding from nature had its inevitable effect in terms of human attrition. Johnson's rambles were very different from those of his middle years:

> I have this day taken a passage to Oxford for Monday. Not to frisk as you express it with very unfeeling irony, but to catch at the hope of better health. The change of place may do something (*Letters*, iv. 49–50).

This hope that movement, change of place and of climate might aid his health was a recurrent feature of Johnson's last letters, being invoked on his journeys to Rochester and Heale (*Letters*, iv. 173, 191). Johnson himself recognized the ironic disparity between his geographical search for health and his earlier writings, which had tried to separate human well-being from the concerns of place:

> my Journey has at least done me no harm, nor can I yet boast of any great good. The weather, you know, has not been very balmy. I am now reduced to think, and am at last content to talk of the weather. Pride must have a fall (*Letters*, iv. 358).

Boswell made much of this (*Life*, iv. 271), but Johnson's public attacks on environmental determinism had never suggested that the individual, especially in the ill-health of old age, could not succumb to inclement weather or need to avoid it. The denial of fatalism was not the denial of human dependency on the environment, 'but how low is he sunk whose strength depends upon the weather?' (*Letters*, iv. 353)

Contemporary with these biographical interconnections of life and landscape, of meaning and mortality, was a series of private Latin poems and prayers which returned Johnson to the themes of his youthful compositions, now reflecting rather than reversing his biographical experience of the natural world. At this time of his life, Johnson could translate Horace's *Ode* IV,vii with full feeling:

> The snow dissolv'd no more is seen,
> The fields, and woods, behold, are green,
> The changing year renews the plain,
> The rivers know their banks again,
> The spritely nymph and naked grace
> The mazy dance together trace.
> The changing year's succesive plan
> Proclaims mortality to man. [207]

This was written during Johnson's last visit to Lichfield which also produced an original Latin poem, '*In rivum a mola Stoana Lichfeldiae Diffluentem*' (On the stream flowing away from the Stowe Mill at Lichfield). This was where Michael Johnson had taught his son to swim, and Johnson provided a complex reflection on the relationship between nature and human life:

> Now the old shadows have been destroyed by harsh axes, and the naked bathing places are open to distant eyes. But the unwearied water goes on its perpetual course, and where it used to flow hidden, it now flows in the

open. Whatever the haste of a stranger carries off, or old age wears away, may your life also, Nisus, move serenely on.[208]

Unlike the Horace translation, nature's lesson to man here was not just transience but transcendence in 'its perpetual course'. This probably reflects Johnson's hope for continuity in change as a Christian facing his own mortality. One of his last Latin prayers, written in January 1784, also spoke of God having planted

> Pure virtue's seeds, with large hand on me shed,
> Mature, till crops a beauteous prospect spread.[209]

At the close of his life, then, Johnson's biography and his landscape imagery came together (unlike in his youth), expressing a concern for a prospect more important than the earthly. This had always been the guiding principle directing his response to landscapes, literary and real, across a variety of genres. Johnson's ambivalence about attitudes to landscape in eighteenth-century England had been a principled and intelligent one; we can only hope his belief in a more important beauteous prospect was as well founded.

9
Conclusion: The Unfamiliar Prospect of Eighteenth-Century Landscape Studies

The adoption of a linguistic contextual approach to eighteenth-century discussions of ideas relating to landscape and nature has led to some results which, in the light of previous contextualizations, are unexpected. In terms of the discursive connections made between landscape ideas and other conceptual realms,[1] three points become clear. First, the sheer number of ideas to which discussions about landscape were connected in various textual genres in the period. In order to gain an *historical* overview of landscape ideas it is essential to recognize this diversity, and any theorization should avoid prioritizing one discursive context, unless it can be shown that the actors engaged in making such 'moves' prioritized it themselves. Second, the overwhelming importance of religious and moral deployments of landscape ideas throughout the eighteenth century is apparent; such ideas were both more resistant and more popular in the literature of the period than conventional secularizing historiographies of the Enlightenment would allow.[2] In an attempt to recognize both points, and third, it would appear that landscape ideas participated in a hierarchy of knowledge where they were of little intrinsic significance, but could gain importance by analogically naturalizing moral and religious truths. This led to the moralizing of prospects, but also to empirical viewing of the land concerned with its fertility (as with the improvers) and with the well-being of its inhabitants as a reflection of the quality of governance. By taking part in this structure of thought, landscape ideas were not, in fact, an independent element in the division of knowledge of the period.

Moving from the discursive level to the linguistic, a parallel result was found in an analysis of Johnson's *Dictionary*.[3] This is a liminal document, reflecting both Johnson's personality and the social history of words, and, by extension, both Johnson's ideas on the natural world and the genealogy of the language of natural description. The landscape vocabulary was given moral and religious resonance by Johnson's construction of the juxtaposition of definitions and illustrative quotations, which was in turn bolstered by the hierarchy of knowledge adopted in the *Dictionary* as a whole. The language of natural description was elevated both conceptually, by the extensive use of the Boyle lecturers, and verbally, by the use of religious sources to illustrate terms such as 'prospect'.

Chapter 4 adds specificity to the contention that landscape description was 'religious' in the eighteenth century, showing that specific theological positions could led to characteristic patterns in the deployment of landscape imagery. It shows the genealogy of the single most important theologically inspired canon of landscape writers in the era, namely the Latitudinarian tradition stretching from Robert Boyle at the Restoration to Ann Radcliffe in the era of the French Revolution. It is against this context that Johnson's different theological sensibilities led him to such a distinctively different use of landscape imagery in his *oeuvre*.

What emerges from these three linguistic contextual engagements is a very different view of landscape ideas from that currently held. Most modern studies have implicitly assumed that landscape formed its own field of knowledge, as it would after the 1790s, with the increasing independence of aesthetics signalled in Romanticism and the autonomy of scientific endeavour. Viewed on this assumption, it appears that many incongruous intellectual debates and, of particular interest, political issues were *added on* to landscape discussions. Yet this is to view eighteenth-century attitudes through a subsequent division of knowledge. Landscape did not form an independent element in the field of knowledge prior to the picturesque, at which point it was being *detached from* intellectual realms, discovering not regaining an autonomous position. Of course, some in the eighteenth century attempted unencumbered landscape descriptions, but few felt landscape

> — had not need of a remoter charm,
> By thought supplied, nor any interest
> Unborrow'd from the eye.[4]

When this view became important in the 1790s, it was coupled with a shift in the nature of political debate, and the intellectual structure in which landscape ideas had participated collapsed. It is noticeable that most recent studies of the politics of landscape in eighteenth-century England have focused on the period after the French Revolution stretching to the Napoleonic Wars and Reform Bill, from Gilpin onwards in literature, and from late Gainsborough to Turner in painting.[5] In this period, the politics of landscape was indeed the 'ideological' addition of one issue to another, but for the bulk of the eighteenth century the division of knowledge gave such a process a very different significance. Landscape ideas were only discussed because of their ability to illuminate truths from more important spheres. This was not simply a false consciousness, as the nature of this illumination was debated, both by those who feared that the prospect could deceive (as in attractive Italy with its Catholic 'despotism' and rugged Switzerland with its 'ideal' government), and by those who were sceptical of nature's analogical utility, preferring the direct discussion of moral and religious issues. In what appears to be a paradox, to treat landscape ideas contextually for eighteenth-century England they must be submerged into other discourses. In fact, this is not surprising as the initial presupposition, that landscape was its own discourse, turns out to be false for this period's division of knowledge. As such, a contextual reading of landscape ideas must run them together with natural knowledge,[6] and recognize that both in turn gained much of their meaning by being able to participate in other discursive spheres, a process which began with the Boyle lectures of the late seventeenth and early eighteenth centuries,[7] and continued throughout the period. Most discussions of eighteenth-century landscape ideas have projected back the presuppositions which emerged in the last few years of the century, and thus contextualized landscape in one division of knowledge as it could only be understood by a subsequent division. The present linguistic contextual approach has attempted to recover the position of natural knowledge in the previous division of knowledge, by recovering the hierarchical presuppositions behind the discursive connections made for the bulk of the eighteenth century.

Fitting Johnson's view of landscape ideas into this pre-1790 patterning of knowledge, his patriot rhetoric of the late 1730s allowed landscapes to act ambiguously as both physical representations and allegories of moral and political health. He established a series of binaries built on town and country, on present and past (both in *London*), on Catholic and Protestant, on Spanish and English (both in the *Voyage to Abyssinia*), and on civilized and savage (in the 'Life of Drake'), all suggesting that place, geography and landscape played an important role in moral action and moral vision. As such, place determined mind, although the degree to which this was a physical and literal notion as opposed to an emblematic one was confused even in Johnson's own presentation, as was the mechanism by which it was supposed to operate.

After the 1730s Johnson deployed two main approaches to landscape, dividing the emblematic from the empirical. First, there was his didactic use of the language of natural description, a phenomenon to be observed in the juxtapositions of the *Dictionary*, but more clearly and more personally in his essays, sermons and biographies.[8] After the collapse of his patriot rhetoric and geography, the language of landscape became detached from physical landscapes and was left without a referent. This allowed Johnson to use this language as a flexible tool in moral instruction ranging from essay-length allegories such as the *Vision of Theodore* to the repeated language of 'paths', 'prospects' and 'tempests' in moral contexts. Such an approach derived from the elaborate use of topographical imagery in moral exhortation which Johnson found in Restoration sermons, both in turn being grounded in the advice of Roman rhetoric on the use of tropes and figures to persuade an audience. The moral life these figures exhorted demanded, *inter alia*, a certain doctrine concerning the correct way to live in relation to the natural world. This is noticeable in Johnson's essays, but more so in the biographies. Just as natural knowledge occupied a low position in the intellectual hierarchy of the eighteenth-century division of knowledge, so to believe that nature determined human action was an error, reversing that hierarchy, and to believe that place could remove mental unease was both an imbalance of the hierarchical relationship between mind and place, and a temporal delusion, mistaking a temporary palliative for a permanent consolation that man's earthly condition could not offer. As with the didactic language, so this doctrine of landscape was rooted in Christianity and classicism. Johnson's views had parallels in the Restoration homilists (notably those who were sceptical of natural evidences as a route to faith) who had been important sources for the *Dictionary*. This position was also culled from Horace's *nil admirari*, to which Johnson frequently returned as a way of expressing his Christian deflation of the pleasures attached to place when they threatened to invert the relationship between nature and man as a moral actor. Both sources aided Johnson to develop an ascetic approach to landscape: ascetics, by draining the world of its temptation, allows it to return as a figurative language of moral persuasion, a technique adopted by Johnson and his Christian sources with respect to landscape imagery.[9]

As the post-patriot divorce of the actual from the emblematic in landscape facilitated a didactic approach, so it allowed actual landscapes to be viewed empirically.[10] Johnson refused to consider the physical environment allegorically in factual genres, and thus his travel accounts avoided design arguments, preferring to see the land as a collection of potentialities for human survival and sociability. This nominalistic view encouraged close inquiry, coupled with the clear and structured relation of findings. When viewed empirically, then, the natural world only participated in the higher concerns of human existence through its causal influence on modes of living and could only link to religion through discussion of the visible church. This approach surfaced not only in Johnson's own travel accounts, but in his critical comments on others, and in his editorial work in the 'Life of Drake' and the *Voyage to Abyssinia*. Moreover, Johnson demanded that the natural knowledge of others be viewed empirically, not mythologized. Thus Johnson's edition of Shakespeare opposed Warburton's approach to history, which converted Shakespeare's knowledge of the natural world into a prefigurement of his own.[11]

Did anything link these two very different approaches to landscape together in Johnson's intellect? The suggestion I have made is that Johnson's High Churchmanship is the moving force behind both discursive deployments of landscape ideas. The stress on scriptural and traditionary routes to faith left landscape and nature with little theological function to perform for Johnson. As such, landscape imagery could become an exhortatory tool for Johnson when freed from patriot rhetoric. This also explains Johnson's fear of the enjoyment of landscapes slipping from innocent pleasure to vicious error: as the face of nature was unimportant in the persuasion to Christian faith, so an excessive focus on it could lead to the transgression of nature into a religion via its personification. Johnson avoided the possibility of such a transgression being derived from his writings by using landscape ideas and imagery in a transparently emblematic manner, a move which paralleled his Anglican regulation of the meaning of the word 'nature' in the 1773 edition of the *Dictionary*. As such, nature could be an inducement to faith, evidentially and rhetorically, but to Johnson it was more important still that it was never confused with the object of faith itself, as his 'Life of Waller' made clear. As the emblematic use of landscape ideas could not be confused with actual nature, so the reverse also held. It was a religious demand that physical landscapes be viewed empirically and described clearly, thus avoiding the 'medley' of the didactic and the empirical which had so annoyed Johnson in Jonas Hanway and Soame Jenyns. Much could be learnt from the natural world, Johnson supporting the new science, but this knowledge was not contained in glib moral analogies derived from nature's instructive book, as the attack on the philosopher of nature in *Rasselas* showed. This was also apparent in Johnson's criticism of literary landscapes. The poet could nominalistically describe images and be led by association to reflections, but these reflections were the product of the human intellect, not implanted in nature itself as active powers: Pope weed not the could ask moral questions, for anything else would upset the hierarchy in which natural knowledge participated. A nominalistic *nil admirari* approach to actual landscapes and their artistic representation, then, was not a 'secular' one.

The suggestion that Johnson's denominational position structured his use of landscape ideas across a range of genres can be strengthened by looking at some of his High-Church contemporaries. I have already shown Johnson's connections with a number of the followers of the natural philosopher and theologian John Hutchinson,[12] and these links were conceptual as well as personal.

Hutchinson's doctrine started from an assumption of the absolute supremacy of scripture in providing all varieties of knowledge:

> nothing can be known of the *Essence* [i.e. God], of the Persons of that Covenant but by Revelation... The Powers in this System [of the universe] were also made known by Revelation.[13]

For Hutchinson, with the truths of revelation unshakeable, inquiry into nature could follow two routes, both of which acted to confirm scripture. First, because he held with Locke that all our ideas derived from sensations, it followed that 'All the Ideas of Divinity are formed from the Ideas in Nature.' As such, humans in a fallen state could not understand revelation were it not for the fact that all the Ideas of the Essence existing, of the Personality, of their Operations, &c are revealed to us in the Scriptures by Words which raise Ideas taken from Things, or are emblematically represented by Things which God has created, formed, or fitted, and by Scripture constituted.[14]

For this reason, Hutchinson repeatedly engaged in allegorical readings of natural phenomena to explain their scriptural meanings.[15] Yet as he himself admitted:

there are two sorts of human Learning ... That which others have already learned, which comes by Instructions from Writings, Words, or Examples; and that which has not yet been learned, which is acquired by Observations, and Comparisons of Opinions, Actions or Things.[16]

This explains Hutchinson's second line of inquiry respecting nature, his natural philosophy. His approach was rigorously atomistic and elemental: 'it should be the Work of a Naturalist to shew how one Sort of Matter is moved or acted upon by another here, and so backward or upward'.[17] As this suggests, and as his 'Observations ... in the Year 1706' confirms,[18] Hutchinson's approach was based upon close observation of the structure of the earth.

Hutchinson kept these two ways of viewing nature, the allegorical and the empirical, separate, believing that blurring them would lead towards an impious ascription of active powers to a personified Nature. It was on these grounds that he opposed Newtonian natural philosophy, viewing the explanation of gravity as giving active powers to what was purely passive matter, so that it was an occult philosophy, not a mechanical one:

the moderns who have aspired to be accounted Wise ... suppose each part able to move itself, or go by itself, or at least by innate Powers, each to move the other; and if that be once allowed, the next set of Philosophers will teach it to talk, and the next to Reason'.[19]

In Johnson's generation, the two great followers of Hutchinson were Johnson's friend, George Horne, and William Jones of Nayland. On the basis that 'nature must be compared with itself; and the scripture must be compared with itself',[20] they continued to view nature as useful, first, as an analogy to illuminate and visualize the truths of scripture, and secondly, as a site of scientific inquiry confirming the truth of *Genesis*'s account of the Creation. In the analogical realm, Jones wrote his *Book of Nature* to catechize children as 'the whole world is a picture, and all the things we see with our eyes speak something to the mind'.[21] Jones's educative writing has a tone akin to Johnson's in the *Vision of Theodore* allegory, which was also aimed at a young audience. Jones, like Johnson the essayist, could also create characters in his sermons whose folly was displayed in the realm of landscape taste. Jones claimed to 'remember an example of a gentleman', whose life story was very similar to that of Bob Cornice in *Adventurer* 53:

> he had laid out large sums of money in beautifying a seat which did not belong to him; and he was shewing a friend what waters and plantations he had added, and how much farther he intended to carry his improvements; while the officers of justice were then actually in the house, to apprehend him as a debtor.[22]

In the empirical realm Jones, who was a Fellow of the Royal Society, continued to push Hutchinson's criticism of the invocation of occult qualities, preferring to explain action at a distance by the intermediate action of electricity:

> we should never treat of this globe, as if it were exposed to certain artificial forces, independent of the natural forces of the elements in which it is involved, and by which it is governed.[23]

As with Johnson, it was Jones's conviction about the truths of revelation which led him to a causal and empirical approach to nature. Many have pointed to the link between materialism in science and radical religious and political beliefs, but it would appear that High-Church Tories could also support an entirely causal approach to the natural world, on the grounds that anything else tended to give nature active powers independent of God. The difference between the two groups lay in their causal priorities. As we have seen, Johnson criticized Brydone for arguing that scientific analysis of the lavas of Vesuvius proved the earth to be older than the scriptural account suggested. William Jones also took up this same question in his capacity as a scientist, arguing that there were only six lava layers at Herculaneum despite records of twenty-eight eruptions, so that:

> we may soon run wild into very strange speculations, if we oppose our own views of natural appearances, which are very contracted and imperfect, to the truth of historical records.[24]

Given that Jones believed, with Johnson, that the Bible was 'unquestionably the most ancient writing,' his appeal to historical records and Johnson's to 'the accumulated evidence of the history of the world'[25] both demanded that causal analysis in general, and of Vesuvius in particular, bear in mind that truth derived from revelation, not *vice versa* as the deist materialists held due to their failure to make the categorical division of the realms of nature and of grace that the High Churchmen did.

William Jones's High-Church empiricism was also visible in his travel account, *Observations in a Journey to Paris*. Although he did not focus on the natural world, his prefatory comment that 'we see even the same things with different eyes, according to our several interests and dispositions'[26] was confirmed, his own disposition to High Churchmanship being evident in the narrative. As we have seen, Donald M'Nichol considered Johnson to be tainted by a 'Popish' fascination with the visible fabric of the church in his *Journey to the Western Islands*. He could have made this point *a fortiori* of Jones, whose *Observations* is studded with descriptions of church interiors, as at St Omer:

> their internal magnificence, variety of ornament, the perfect cleanliness of the place from the roof to the floor, and the brightness of the furniture, is such as a protestant in England can have no idea of.[27]

Jones, like Johnson, was clearly fascinated by churches and church ornament, and in both cases it is denominational interests which make the balance of their interests in the landscape different from that of other travellers.

This is not to claim that Johnson was a Hutchinsonian,[28] nor to deny the numerous differences between the opinions expressed by William Jones and John Hutchinson and those of Johnson. But it would appear that the distinctive emphasis all three placed upon scriptural revelation led them to adopt a closely comparable position on the significance of the natural world. Nature had an emblematic role to play as a vehicle for moral and religious instruction, and it was also, as God's handiwork, worthy of scientific study, provided this did not lose its moorings in a broader Christian way of life, which could occur if nature's role as a moral vehicle was forgotten and nature, personified, became the arbiter as it was for the philosopher of nature in *Rasselas* and for materialist scientists. Landscape was also a concept whose use was divided into these two realms of the didactic and the empirical, its role in sermons and essays being categorically divided from its presentation in travel accounts. It is this patterning of argument into two realms which defines the High-Church approach to landscape, which links Johnson, Jones and Hutchinson, and which was still being deployed by Newman in his *Idea of a University* (1852).

To emphasize denominational dynamics as the key to Johnson's approach to natural knowledge appears to be at odds with the contexts evoked by the bulk of studies of landscape and nature in the eighteenth century in recent years. Yet in the context established by the analyses of the discourses of landscape and the language of natural description it is less surprising. If natural knowledge remained in an intellectual structure which prioritized religious and ethical debates, it is understandable that Johnson's approach was driven by his faith. Johnson's well-known scepticism about the fashion for landscape appreciation simply expresses the degree to which the routes to faith he stressed made landscape a distraction from more important issues and his refusal, also to be found in William Law and Jeremy Taylor, to view such distractions as innocent. As he used the language of landscape as a rhetoric of moral persuasion, so his satire was of a piece with classical rhetorical advice that opponents should be mocked to persuade others to adopt your position.

Johnson was in no way 'representative' of the broader eighteenth-century argument with respect to landscape and nature (the Latitudinarians of Chapter 4 have far more claim to that mantle), although his scepticism was echoed by other members of his circle, and can tentatively be connected to a Tory and High-Church nexus developed in the period after 1760.[29] But the motivations behind his position do have a wider applicability and suggest that future studies of landscape ideas in the pre-1790 period need to be sensitive to two things. First, the denominational differences which led to different approaches to landscape and the natural world: some discursive contours of this map have been sketched,[30] and individuals have been plotted onto it by comments *en passant* on Knight, Hanway, Boswell, Piozzi and Warburton,[31] but the quantitatively more important Low-Church views of nature formulated by the Boyle lecturers need careful analysis, particularly how they were transmitted into the mid- and late-eighteenth century. Second, the classical elements of debates about nature need to be recovered. The Horatian *nil admirari* approach to mind and place was most important to Johnson, and to the wider Augustan tradition.[32] Little work, however, has yet been done on the continued importance of classical geography and natural knowledge, as conveyed via the humanist fusion of science and textual scholarship, to eighteenth-century attitudes. Both of these foci are essential, precisely because of the position landscape and natural knowledge occupied in the eighteenth-century division of knowledge. As these ideas lacked an autonomous rationale, they were part of public intellectual life, not isolated preserves amenable to modern disciplinary history. Landscape ideas reflected the presuppositions which drove intellectual life more generally, which in eighteenth-century England were based on the twin pillars of classicism and Christianity.

Notes

Chapter 1

1. Edward Relph, *Rational Landscapes and Humanistic Geography* (London, 1981), p. 23.
2. Denis Cosgrove, Landscape studies in geography and cognate fields of the humanities and social sciences, *Landscape Research*, 15.3, 1990, pp. 1–6, has made a similar division.
3. Raymond Williams, *The Country and the City* (London, 1973), p. 121.
4. Denis Cosgrove, *Social Formation and Symbolic Landscape* (Beckenham, 1984), p. 1; he here invokes the ideas of John Berger, *Ways of Seeing* (London, 1972). Cosgrove and Jackson have called landscape 'a particular way of composing, structuring and giving meaning to an external world whose history *has to be* understood in relation to the material appropriation of land.' D. Cosgrove and P. Jackson, New Directions in cultural geography, *Area*, 19, 1987, pp. 95–101, at p. 96 (emphasis added).
5. Denis Cosgrove, Place, Landscapes and the Dialectics of Cultural Geography, in *Canadian Geographer*, 22, 1978, pp. 66–72.
6. Cosgrove, *Social Formation*, p. 2 (emphasis added).
7. See also Ann Bermingham's *Landscape and Ideology: the English rustic tradition, 1740–1860* (London, 1986) where she says 'this study examines how rustic painting *intersected* with the agrarian revolution, how the genre *echoed* prevailing cultural values that *linked it directly* to enclosure' and that she 'uses historical information to speculate about the cultural meaning of the development', p. 2 (emphasis added).
8. Cosgrove, *Social Formation*, p. 63.
9. Ibid., p. 64.
10. Ibid., Chapter 9.
11. Chris Fitter, *Poetry, Space, Landscape: toward a new theory* (Cambridge, 1995).
12. Bermingham, *Landscape and Ideology*, p. 1.
13. W.J.T. Mitchell, Imperial Landscape, in *idem* (ed.) *Landscape and Power* (Chicago, 1994), pp. 5–34. Yet he admits imperialism was not a necessary condition for the rise of landscape (pp. 9–10) in the case of the Dutch, whose landscape painting emerged in part as a nationalistic expression of anti-Spanish imperialist feeling. Nor does it appear a sufficient condition: the colonial exploits of fifteenth- and sixteenth-century Spain led to no important school of landscape painting. Mitchell has subsequently suggested a less literal link of landscape and imperialism, where it is the strategy of universalizing Western categories of landscape that amounts to an imperial strategy; it is hard to see this as intended to be a contextual-historical suggestion. See W.J.T. Mitchell, Gombrich and the Rise of Landscape, in A. Bermingham and J. Brewer (eds) *The Consumption of Culture, 1600–1800: image, object, text* (London, 1995), pp. 103–18.
14. See Richard Wollheim, *Art and Its Objects* (2nd edn, Cambridge, 1980 [Canto ed. 1992])§62.

318

15. 'Though I have often used the language of intention in my essays... this must be taken as a metaphor only, used to call attention to what the polite wished to believe about the society of the countryside... whether that wish was conscious or not'. John Barrell, *The Dark Side of the Landscape: the rural poor in English painting, 1730–1840* (Cambridge, 1980), p. 18.

16. Wollheim, *Art and Its Objects*, §63.

17. Cosgrove, *Social Formation*, p. 10.

18. Barrell, *Dark Side*, p. 1.

19. Barrell, *Dark Side*, p. 16. See also Michael Rosenthal, *British Landscape Painting* (London, 1982) and Anne Janowitz, *England's Ruins: poetic purpose and the national landscape* (Oxford, 1990), p. 64.

20. Stephen Daniels, Marxism, culture and duplicity of landscape, in R. Peet and N. Thrift (eds) *New Models in Geography* (2 vols, London, 1989), ii. 196–219, commenting on Cosgrove's position in *Social Formation*, at p. 206. Similarly, and drawing on Althusser, Bermingham, *Landscape and Ideology* (p. 3 and p. 31) states 'I propose that there is an ideology of landscape and that in the eighteenth and nineteenth centuries a class view of landscape embodied a set of socially and, finally, economically determined values to which the painted image gave cultural expression.' She later talks of 'the cultural directive to screen out labour from a representation of gentry'. Fabricant has posited a similar role for landscape gardens: by allowing the lower classes into parks the ruling classes 'reinforced their hegemony, by enlisting the complicity of the ruled in the fiction of their *inclusion* in an increasingly *exclusionary* society'. Her view appears to be similarly conspiratorial in its discussion of threat to a 'system'. Carole Fabricant, The literature of domestic tourism and the public consumption of private Property, in Felicity Nussbaum and Laura Brown (eds) *The New Eighteenth Century: theory – politics – English Literature* (London, 1987), pp. 254–75, at p. 257 and p. 270.

21. Barrell *Dark Side*, p. 18.

22. David Solkin, *Richard Wilson: the landscape of reaction* (London, 1982), p. 34.

23. On the importance of genre, see E.D. Hirsch, *Validity in Interpretation* (New Haven, 1967); T. Todorov, *Genres in Discourse* (Cambridge, 1990); and Richard Wollheim, *Art and Its Objects* §32. Wollheim has also pointed out that whilst painting has to represent the particular, it need not paint particular things as opposed to particular kinds of things (see *Painting as an Art*, London, 1987, pp. 69–71).

24. See Francis Haskell, *History and its Images: art and the interpretation of the past* (New Haven, 1993).

25. Solkin, *Richard Wilson*, p. 49. The creation of collections was a process hedged in by the conventions of taste, and as such was insulated from a concern for social actualities. Painting helped the 'cultural unification of the upper ranks of English society' with 'the other ranks of society scarcely figuring in their understanding of the "nation"'. Iain Pears, *The Discovery of Painting: the growth of interest in the arts in England, 1680–1768* (New Haven, 1988), p. 3.

26. Thus the painter Jonathan Richardson in *An Essay on the Theory of Painting* (London, 1725 edn, pp. 118–19) argued that 'Every Picture should be so contriv'd, as that at a Distance, when we cannot discern what Figures there are, or what they are doing, it should appear to be composed of Masses, Light, and Dark; the Latter of which serve as Reposes to the Eye. The Forms of These Masses must be agreeable, of whatsoever they consist, Ground, Trees, Draperies, Figures, etc.' This lack of concern with the actual in portrayal was converted into a way of seeing: 'Sir Joshua Reynolds told me, that when he and [Richard] Wilson the landscape

painter were looking at the view from Richmond terrace, Wilson was pointing out some particular part, and in order to direct his eye to it, 'There,' said he, 'near those houses – there! where the figures are! – Though a painter, said Sir Joshua, I was puzzled. . . . for I did not at first conceive that the men and women we plainly saw walking about, were by him only thought of as figures in the landscape.' Uvedale Price, *An Essay on the Picturesque* (London, 1796 edn) note to i. 379.

27. John Barrell, *The Idea of Landscape and the Sense of Place, 1730–1840: an approach to the poetry of John Clare* (Cambridge, 1972), pp. 22, 59.

28. At the beginning of the century, Pope, Gay and Philips discussed the extent to which the pastoral should incorporate rural realities: see Richard Feingold, *Nature and Society: later-eighteenth century uses of the Pastoral and Georgic* (Hassocks, Sussex, 1978). At the close of the period the debate continued, Repton arguing that labour should be excluded from view in gardens (in Theory and Practice of Landscape Gardening [1804] in John Nolen, ed., *The Art of Landscape Gardening* [London, 1907] p. 138), and Cowper (The Task, in *Poetical Works*, ed., H.S. Milford, Oxford, 1967, i. 219–51) refusing to romanticize the lot of the rural poor. As such, Barrell's dark side of the landscape was a theorized omission rather than a meaningful absence. Where the poor were removed from the prospect, this was often noticed and condemned, for which see, Louis Simond *Journal of a Tour and Residence in Great Britain, During the Years 1810 and 1811* (2 vols, Edinburgh, 1815), i. 222 and ii. 86.

29. Thus Cosgrove, *Social Formation*, draws on Perry Anderson and Robert Brenner in general, and on Christopher Hill for the notion of the English Civil War as a revolution. Fitter, *Poetry, Space, Landscape*, also draws on Anderson, which explains some of the similarities between his view and Cosgrove's. Barrell, *Dark Side*, is reliant on E.P. Thompson and D. Hay.

30. Barrell, *Dark Side*, p. 3. Similarly, Rosenthal, *British Landscape Painting*, p. 22 can speak of a 'proletariat' in the seventeenth century, and Solkin, *Richard Wilson*, pp. 56–57, of conflicts of interest between middle and gentry classes in the eighteenth century.

31. Barrell, *Dark Side*, p. 5. See also Solkin, *Richard Wilson*, p. 25; and Cosgrove, *Social Formation*, p. 58.

32. Cosgrove, *Social Formation*, p. 199. Such a view seems to derive from C.B. Macpherson, *The Political Theory of Possessive Individualism: Hobbes to Locke* (Oxford, 1962). His work is specifically cited by Fabricant, Literature of Domestic Tourism, p. 258.

33. Cosgrove, *Social Formation*, p. 198. See also Williams, *Country and City*, p. 39 for similar language and argument.

34. See A.O. Hirschman, *The Passions and the Interests: political arguments for capitalism Before its Triumph* (Princeton, 1977); K. Tribe *Land, Labour and Economic Discourse* (London, 1978); J.G.A. Pocock, *The Machiavellian Moment: Florentine political thought and the Atlantic republican tradition* (Princeton, 1975); and idem, *Virtue, Commerce and History: essays on political thought and History, chiefly in the eighteenth century* (Cambridge, 1985).

35. This view is supported by D. Winch, *Adam Smith's Politics: an essay in historiographical revision* (Cambridge, 1978) and *Riches and Poverty: an intellectual history of political economy in Britain, 1750–1834* (Cambridge, 1996).

36. See Boyd Hilton, *The Age of Atonement: the influence of evangelicalism on social and economic thought, 1795–1865* (Oxford, 1988); A.M.C. Waterman, *Revolution, Economics and Religion: christian political economy, 1798–1833* (Cambridge, 1991); and Richard Brent, God's providence: liberal political economy as natural theology

at Oxford, 1825–62, in Michael Bentley (ed.) *Public and Private Doctrine: essays in British History presented to Maurice Cowling* (Cambridge, 1993), pp. 85–107.

37. See Pierre Force, *Self-Interest before Adam Smith: a genealogy of economic science* (Cambridge, 2003).

38. See James J. Sack, *From Jacobite to Conservative: reaction and orthodoxy in Britain, c. 1760–1832* (Cambridge, 1993), esp. pp. 178–87; and Nigel Everett, *The Tory View of Landscape* (New Haven, 1994).

39. See R.S. Neale, *Class in English history, 1650–1800* (Oxford, 1981).

40. See P. Corfield, Class by name and number in eighteenth-century Britain, in *History*, 72, 1987, pp. 38–61. See also: S. Welleck, 'Class versus rank': the transformation of eighteenth-century English social terms and theories of production, in *Journal of the History of Ideas*, 47, 1986, pp. 409–31; Susie I. Tucker, *Protean Shape. A Study in Eighteenth-Century Vocabulary and Usage* (London, 1967) pp. 158–62; and J.C.D. Clark, *English Society: Ideology, Social structure and Politcal practice during the ancien régime* (Cambridge, 1985), p. 90. For Thomas Paine's lack of class consciousness, see Jack Fruchtman, *Thomas Paine and the Religion of Nature* (Baltimore, 1993), p. 123.

41. D. Hay, Property, authority and the criminal law, in D. Hay *et al.* (eds) *Albion's Fatal Tree: Crime and Society in Eighteenth-Century England* (London, 1975), pp. 17–63, p. 26.

42. P.J.M. King, Decision-Makers and Decision-Making in the English Criminal Law, 1750–1800, in *Historical Journal*, 27, 1984, pp. 25–58; J. Beattie, *Crime and the Courts in England, 1660–1800* (Oxford, 1986); and J.H. Langbein, Albion's Fatal Flaws, in *Past and Present*, 98, 1983, pp. 96–120. Hay's later work, War, Dearth and Theft in the Eighteenth Century: the record of the English courts, in *Past and Present*, 95, 1982, pp. 117–60, has also pointed to a more subtle interpretation.

43. A.O. Hirschman, *The Rhetoric of Reaction: Perversity, Futility, Jeopardy* (Cambridge, Mass., 1991), p. 21. This attitude has also been recognized by Paul Langford, *Public Life and the Propertied Englishman, 1689–1798* (Oxford, 1991), p. 8, and by J.C.D. Clark, *The Language of Liberty, 1660–1832: political discourse and social dynamics in the Anglo-American World* (Cambridge, 1994), p. 190. What this might mean for explanation in landscape studies is starting to be addressed: see Tom Williamson, *Polite Landscapes: Gardens and Society in Eighteenth-Century England* (Baltimore, 1995).

44. Barrell, *Dark Side*, p. 82.

45. See Clark, *English Society;* and A.O. Lovejoy, *The Great Chain of Being: a study in the history of an idea* (Cambridge, Ma., 1936).

46. Cosgrove, *Social Formation*, p. 62.

47. See A.W.B. Simpson, *An Introduction to the History of the Land Law* (Oxford, 1961), pp. 46–5, 84–85.

48. J.M. Neeson, *Commoners: common right, enclosure and social change in England, 1700–1820* (Cambridge, 1993) Ch. 1.

49. James Tully, *A Discourse on Property: John Locke and his adversaries* (Cambridge, 1980), p. 170; and John Dunn, *The Political Thought of John Locke: an historical account of the argument of The Two Treatises of Government* (Cambridge, 1969).

50. See Richard Tuck, *Natural Rights Theories: their origin and development* (Cambridge, 1979) and T.A. Horne, *Property Rights and Poverty: political argument in Britain, 1605–1834* (Chapel Hill, N.Carolina, 1990).

51. See Everett, *Tory View.*

52. Denis Cosgrove, Prospect, perspective and the evolution of the landscape Idea, in *Transactions of the Institute of British Geographers*, N.S. 10, 1985, pp. 45–62, p. 46.

53. Ibid., pp. 48–49. The same argument is made by Ronald Paulson, *Breaking and Remaking: aesthetic practice in England, 1700–1820* (New Brunswick, 1989), p. 266 and for gardens by Chandra Mukerji, Reading and Writing with Nature: a Materialist Approach to French Formal Gardens, in J. Brewer and R. Porter (eds) *Consumption and the World of Goods* (London, 1993), pp. 439–61, esp. p. 450 and p. 456.

54. See Martin Warnke, *Political Landscape: an art history of Nature* (trans., London, 1994), Ch. 3.

55. Cosgrove, *Social Formation*, p. 26.

56. Ibid., p. 27. The same argument is used by Fitter, *Poetry, Space, Landscape*, pp. 194–97.

57. See Barrell, *Idea of landscape*, Ch. 1; and James Turner, *The Politics of Landscape: rural scenery and society in English poetry, 1630–1660* (Oxford, 1979).

58. In seventh-century Italy, the Edict of Rothari had to deal with the problem of violence between shepherds over the moving of land markers: Vito Fumagalli, *Landscapes of Fear: perceptions of nature and the city in the middle ages* (trans., Cambridge, 1994), pp. 18–19. Aristophanes's, *Clouds* (c. 423 BC) has a student explain that geometry is useful for 'sharing out allotments of land' *Lysistrata/The Acharnians/The Clouds* trans., Alan Sommerstein (Harmondsworth, 1973), p. 120. The realities of ancient land surveys are discussed in O.A.W. Dilke, *Greek & Roman Maps* (Baltimore, 1985), pp. 87–101 where it is clear that it was the need to allocate property which led to such cartographic labours.

59. '[P]erspective theory . . . is within contestation; it is open to appropriation, can be made to speak in the service of competing and contradictory interests.' Peter de Bolla, *The Discourse of the Sublime: readings in history, aesthetics and the subject* (Oxford, 1989), p. 218.

60. Eighteenth-century writers were clearly aware of the antiquity of perspective. See George Turnbull, *A Treatise on Ancient Painting* (London, 1740), pp. 69–70; Daniel Webb, *An Inquiry into the Beauties of Painting* (2nd edn, London, 1761), pp. 115–16; and on Pope's views on this topic, see Morris Brownell, *Alexander Pope and the Arts of Georgian England* (Oxford, 1978), p. 80. Perhaps the most commonly cited source was Vitruvius, *De Architectura* (trans., F. Granger, London, 1934), i. 27. See also Hubert Damisch, *The Origin of Perspective* (Cambridge, Mass., 1994), pp. xvi, 444–47.

61. Cosgrove, Evolution of the landscape idea, p. 48.

62. The eighteenth century was clearly sensitive to this division of the analogical from the literal sense of 'property' and emphasized that 'appropriation' by the eye of taste was open to all, even if ownership was restricted to the few. See Alexander Gerrard, *An essay on taste* (London, 1759), p. 193; and the *Guardian*, no. 49 (by George Berkeley); and Addison, *Spectator*, no. 414. The same point was put in verse by William Shenstone:

> Bonds, contracts, feoffments, names unmeet for prose,
> The towering Muse endures not to disclose;
> Alas! her unrevers'd decree
> More comprehensive and more free,
> Her lavish charter, taste, appropriates all we see.

Rural Elegance, in Alexander Chalmers, *Works of the English Poets*, xiii, p. 282.

63. For the limitation of perpective to one moment, see James Harris, *Three Treatises* (London, 1744), pp. 76–78; James Ussher, *Clio: or, a discourse on taste* (3rd edn, London, 1772), p. 157; and Joshua Reynolds, *Discourses on Art*, ed. R. Wark

(New Haven, 1975), p. 60. For the recognition of different viewpoints on a single object in the landscape, see Barbara Stafford, *Voyage into Substance: art, science and the illustrated travel account, 1760–1840* (Cambridge, Mass., 1984), p. 400.
64. Cosgrove, *Social Formation*, p. 26.
65. See Edward Relph, *Place and Placelessness* (London, 1976) and *idem, Rational Landscapes*.
66. Cosgrove, Evolution of the Landscape Idea, p. 48.
67. See Barrell, *Idea of Landscape*, p. 173; and *idem*, Geographies of Hardy's Wessex, in *Journal of Historical Geography*, 8, 1982, pp. 347–61.
68. The centrality of the eighteenth century to the emergence of a 'visual' culture of landscape appreciation is at the heart of Peter de Bolla, *The Education of the Eye: painting, landscape, and architecture in eighteenth-century Britain* (Stanford, 2003).
69. Barrell, *Idea of Landscape*, p. 80 See also Carole Fabricant, *Swift's Landscapes* (Baltimore, 1982) for a visual but not pictorial landscape sensibility.
70. Bertrand H. Bronson, When was Neoclassicism? in his *Facets of the Enlightenment: studies in English Literature and its Contexts* (Berkeley, 1968), pp. 1–25, p. 9. He refers to Handel's, *L'Allegro, Il Penseroso, ed il Moderato* (1740).
71. For more on this topic, see Chapter 3.
72. Ernst Curtius, *European Literature in the Latin Middle Ages* (trans., London, 1953), p. 198.
73. The romanticism of the notion of the insider in landscape studies is highlighted by Eric Hirsch, Introduction: Landscape: between place and space, in Eric Hirsch and Michael O'Hanlon (eds) *The Anthropology of Landscape: perspectives on place and space* (Oxford, 1995), pp. 1–30, at pp. 13–16. Perhaps it would be better to follow Alison's non-judgemental approach, in which habit simply led to different associations being connected to natural phenomena by different actors: see Archibald Alison, *Essays on the Nature and Principles of Taste* (Dublin ed., 1792), p. 125.
74. This is what Conal Condren, *The Language of Politics in seventeenth-century England* (London, 1994), p. 13 calls the Piltdown Effect: 'we first dig our vocabulary well into the earth, and then uncover it again as a set of causes or foundations for what was on the surface'. In landscape studies this point has been recognized as 'an implicit circularity' by Nicholas Green, Looking at the landscape: class formation and the visual, in Eric Hirsch and Michael O'Hanlon, *The Anthropology of Landscape: perspectives on place and space* (Oxford, 1995), pp. 31–42, p. 33.
75. See Cosgrove, *Social Formation*, opening chapters. Barrell, *Dark Side*, speaks of 'a leftist like me' (p. 29) and Solkin, Rosenthal and Bermingham all ally themselves with a similar politics.
76. Cosgrove, *Social Formation*, p. 66.
77. Rosenthal, *British landscape painting*, p. 10.
78. 'It is almost as if there is something built into the grammar and logic of the landscape concept that requires the elaboration of a pseudohistory, complete with a prehistory, an originating moment that issues in progressive historical development.' Mitchell, Imperial Landscape, p. 12. Mitchell's recognition of this makes his own efforts to link landscape history to imperialism all the more hard to explain (see n. 13).
79. Solkin, *Richard Wilson*, p. 103; Rosenthal *British Landscape Painting*, pp. 48, 54.
80. Andrew Hemmingway, *Landscape Imagery and Urban Culture in Early Nineteenth-Century-Britain* (Cambridge, 1992), p. 11.
81. Barrell, *Dark Side*, p. 2.

82. See Chapter 2.
83. Cosgrove, *Social Formation*, p. 214.
84. Williams, *Country and City*, p. 124; see also Rosenthal, *British Landscape Painting*, p. 45.
85. See John Barrell, *The Birth of Pandora and the Division of Knowledge* (London, 1992), p. 1.
86. Solkin, *Richard Wilson*, pp. 126–27; see also Hugh Prince, Art and Agrarian Change, 1710–1815, in D. Cosgrove and S. Daniels (eds) *The Iconography of Landscape: essays on the symbolic representation, design and use of past environments* (Cambridge, 1988), pp. 98–118.
87. Barrell, *Dark Side*, pp. 12–16, 87.
88. Rosenthal, *British Landscape Painting*, p. 58.
89. Solkin, *Richard Wilson*, pp. 69, 102–103, 106.
90. Roger Scruton, *The Philosopher on Dover Beach: Essays* (Manchester, 1990), p. 292.
91. Cosgrove, *Social Formation*, p. 4.
92. See for example, Cosgrove, *Social Formation*, pp. 65–68.
93. Daniels, Duplicity of landscape, p. 217.
94. Stephen Daniels, *Fields of Vision: landscape imagery and national identity in England and the United States* (Cambridge, 1993), p. 245; see also D. Cosgrove and S. Daniels, Spectacle and text: landscape metaphors in cultural geography, in J. Duncan and D. Ley (eds) *Place/Culture/Representation* (London, 1993), pp. 57–77, p. 59.
95. See in particular his The myth and the stones of Venice: an historical geography of a symbolic landscape, in *Journal of Historical Geography*, 8, 1982, pp. 145–69.
96. Denis Cosgrove, *The Palladian Landscape: geographical change and its cultural representations in sixteenth-century Venice* (Leicester, 1993), p. 8.
97. Ibid., p. 7.
98. John Barrell, *The Political Theory of Painting from Reynolds to Hazlitt: 'the body of the public'* (New Haven, 1986) and idem, *Birth of Pandora*. For Pocock, see n. 34.
99. Roy Porter has noted this danger in his review of a collection of essays edited by Barrell: 'the historical framework . . . invoked by the contributors seems a bit too rarified and too directly derived from J.G.A. Pocock *et al.*'. Review of *Painting and the Politics of Culture*, in *Journal of Historical Geography*, 19, 1993, pp. 244–45, p. 245.
100. Daniels, *Fields of Vision*, p. 243.
101. Barrell's is perhaps the more Foucauldian use of the term 'discourse' as an overarching way of structuring knowledge, where Daniels looks at lower level 'discourses'.
102. For Wright, see *Fields of Vision*; for De Loutherbourg, see Loutherbourg's chemical theatre: *Coalbrookdale by Night*, in J. Barrell (ed.) *Painting and the Politics of Culture: new essays on British art, 1700–1850* (Oxford, 1992), pp. 195–230.
103. U. Eco *et al.*, *Interpretation and Overinterpretation* (Cambridge, 1992).
104. Ibid., p. 128.
105. Daniels, *Fields of Vision*, p. 8.
106. Barrell, Introduction to *Painting and the Politics of Culture*, p. 2.
107. Barrell, *Birth of Pandora*, p. xvi.
108. Daniels, *Fields of Vision*, p. 5.
109. Cosgrove, *Palladian landscape*, p. 5.
110. Barrell, *Birth of Pandora*, p. xiii; see also Daniels, *Fields of Vision*, introduction; Hemmingway *Landscape Imagery*, Preface; and Fabricant, Literature of domestic tourism, p. 275.

111. Barrell, *Birth of Pandora*, p. xvi.
112. The phrase is from Frank Kermode, *Forms of Attention* (Chicago, 1985). See Chapter 2.
113. Daniels, Loutherbourg's chemical theatre, p. 195.
114. Barrell, *Painting and the Politcs of Culture*, p. 4.
115. See, for example, Cosgrove, Myth and Stones of Venice.
116. Daniels, *Fields of Vision*, pp. 244–45.
117. Barrell, *Birth of Pandora*, p. xiv; see also Barrell, *Political Theory of Painting*, p. 45.
118. A. Howkins, J.M.W. Turner at Petworth: agricultural improvement and the politics of landscape, in Barrell (ed.) *Painting and the Politics of Culture*, pp. 231–51.
119. Herbert Butterfield, *The Whig Interpretation of History* (Cambridge, 1931), p. 95.

Chapter 2

1. The importance of this is also suggested by Richard Wollheim's discussion of art as a 'form of life' with its own practices. See *Art and Its Objects* (2nd edn, Cambridge, 1980) §45 ff. esp. §53. See also Michael Baxandall, *Patterns of Intention: on the Historical Explanation of Pictures* (New Haven, 1985), pp. 44–47, on previous pictures establishing the problem situation for subsequent work.
2. See p. xx.
3. James Tully (ed.) *Meaning and Context: quentin skinner and his critics* (Cambridge, 1988), p. 37.
4. Ibid., p. 99.
5. Thus, in political thought this was central to Tully's response to C.B. Macpherson's thesis. Literary studies have seen an increased concern with history in analyses of publishing history and also, at least at a rhetorical level, amongst the new historicists: see, for example, H. Aram Veeser (ed.) *The New Historicism* (London, 1989); more generally, see J.J. McGann (ed.) *Historical Studies and Literary Criticism* (Wisconsin, 1985); and J.R. de J. Jackson, *Historical Criticism and the Meaning of Texts* (London, 1989). The history of philosophy has also seen the permeation of such ideas; see Alasdair Macintyre, *After Virtue: a study in moral theory* (2nd edn, London, 1985). In geography, a number of calls for contextualization have been made: see Felix Driver, The historicity of human geography, in *Progress in Human Geography*, 12, 1988, pp. 497–506; David N. Livingstone, *The Geographical Tradition: episodes in the history of a contested enterprise* (Oxford, 1992); and V. Berdoulay, The contextual approach, in D.R. Stoddart (ed.) *Geography, Ideology and Social Concern* (Oxford, 1981), pp. 8–16.
6. See Edward Relph, *Rational Landscapes and Humanistic Geography* (London, 1981).
7. See Chapter 1.
8. This does not, of course, preclude a contextual approach to the history of appropriations of an idea: for a classic study of this variety, see J.G.A. Pocock, *The Machiavellian Moment: Florentine political thought and the Atlantic republican tradition* (Princeton, 1975).
9. Tully, *Meaning and Context*, p. 44.
10. See for example Bertrand H. Bronson, When was neoclassicism? in his *Facets of Enlightenment: studies in English literature and its contexts* (Berkeley, 1968), pp. 1–25, at pp. 23–24; Ronald Paulson, *Breaking and Remaking: aesthetic practice in England, 1700–1820* (New Brunswick, 1989), p. 4.
11. The meaning-significance division was one which E.D. Hirsch, *Validity in Interpretation* (New Haven, 1967) developed. Similar binaries are proposed

in: E.H. Gombrich, *Symbolic Images: studies in the art of the Renaissance II* (London, 1972), introduction; T. Todorov, *Genres in Discourse* (Cambridge, 1990); and Umberto Eco, *Interpretation and Overinterpretation* (Cambridge, 1992).

12. Denis Cosgrove, *Social Formation and Symbolic Landscape* (Beckenham, 1984).

13. John Barrell, The public prospect and the private view, in Simon Pugh (ed.) *Reading Landscape: country – city – capital* (Manchester, 1990), pp. 19–40; and *idem, The birth of Pandora and the Division of Knowledge* (London, 1992).

14. Similar reservations, built from a Skinnerian perspective, are expressed by John Dixon-Hunt, *Gardens and the Picturesque: studies in the history of landscape architecture* (Cambridge, Mass., 1992), pp. 171–74.

15. Tully (ed.) *Meaning and Context*, p. 48.

16. Ibid., pp. 77–78.

17. Ibid., p. 91.

18. John Barrell, *The Dark Side of the Landscape: the rural poor in English painting, 1730–1840* (Cambridge, 1980), pp. 6, 16–18. See also David Solkin, *Richard Wilson: the landscape of reaction* (London, 1982) who argues that Wilson would not have been able to articulate the interests his pictures served (p. 34), and that his patrons may not have been able to say why they found his pictures so appealing (p. 70), but that in both cases class interests explain matters.

19. Michael Oakeshott, *Experience and its Modes* (Cambridge, 1933), p. 145. See also F.H. Bradley: 'history is the testimony of the past to the past' in *The Presuppostions of Critical History* (Oxford, 1874), p. 26.

20. Michael Oakeshott, History and the social sciences, in Institute of Sociology, *The Social Sciences: their relation in theory and in teaching* (London, 1936), pp. 71–81, at pp. 72–75.

21. See Michael Oakeshott, The voice of poetry and the conversation of mankind, *in Rationalism in Politics and other Essays* (rev. and expanded edn, Indianapolis, 1991), pp. 488–541.

22. See Isaiah Berlin, The concept of scientific history, in *idem, Concepts and Categories: philosophical essays* (Oxford, 1978), pp. 103–42. See also Arthur Schopenhauer, On History, in *The World and Will and Representation*, trans., E.F.T. Payne (2 vols, New York, 1966 ed.), ii. pp. 439–46.

23. Oakeshott, *Experience*, p. 103.

24. For Oakeshott and idealism, see David Boucher, The creation of the past: British idealism and Michael Oakeshott's philosophy of history, in *History and Theory*, 23, 1984, pp. 193–214. My approach to Oakeshott's view of history prefers to dissociate it from Hegelianism in favour of the view of history as a constructed form of life: see Ludwig Wittgenstein *Philosophical Investigations*, trans., G.E.M. Anscombe (3rd edn, Oxford, 1967) and Nelson Goodman *Ways of Worldmaking* (Indianapolis, 1978). The similarities between Oakeshott and Wittgenstein have been pointed out by Robert Grant, *Oakeshott* (London, 1990), pp. 14, 49, 111–12, and by Harwell Wells, The philosophical Michael Oakeshott, in *Journal of the History of Ideas*, 55, 1994, pp. 129–45 at p. 136 and p. 141.

25. Oakeshott, *Experience*, p. 159. Wittgenstein also argues that 'misunderstandings concerning the use of words, [are] caused, among other things, by certain analogies between the forms of expression in different regions of language'. (*Philosophical Investigations*, §90, cf. also p. 232).

26. J.L. Auspitz, Michael Joseph Oakeshott (1901–90), in A. Jesse Norman (ed.) *The achievement of Michael Oakeshott* (London, 1993), pp. 1–25, at p. 13. This view has affinities with Macintyre's notion of 'practices' in *After Virtue*, p. 187.

27. Oakeshott *Experience*, pp. 131–32; see also Michael Oakeshott *On History and other Essays* (Oxford, 1983) p. 67: 'there is no *explanans* of a different character from the *explanadum*'.
28. Ibid., p. 142.
29. J.G.A. Pocock *Politics, Language and Time: essays in political thought and history* (New York, 1971) saw Oakeshott's view as akin to Taoist or Heraclitan views of history as flux. Oakeshott's source for this view was Aristotle, *Physics*, V 227a (see *On History*, p. 113).
30. Oakeshott *On History*, p. 94. Oakeshott provided an example of this view of historical identities being passages of differences in his analysis of Christianity: see, The importance of the historical element in Christianity, in *idem* (ed. Timothy Fuller) *Religion, politics and the moral life* (New Haven, 1993), pp. 63–73. This vision of historical identity has much in common with Wittgenstein's 'family resemblances' which he pictured as a rope in which no one fibre runs through the whole course (*see Philosophical Investigations*, §66–67).
31. Oakeshott, *Experience*, p. 90.
32. Ibid., pp. 107–108.
33. R.G. Collingwood, *The Idea of History* (rev. edn, Jan van der Dussen, Oxford, 1993), pp. 269–70; see also Arthur C. Danto *Analytical Philosophy of History* (Cambridge, 1965), p. 105.
34. Michael Oakeshott, *The Voice of Liberal Learning: Michael Oakeshott on Education* (ed. Timothy Fuller, New Haven, 1991), p. 44.
35. Oakeshott, *Experience*, p. 106.
36. See P.L. Janssen, Political thought as traditionary action: the critical response to Skinner and Pocock, in *History and Theory*, 24, 1985, pp. 115–46, at p. 119. This view distinguishes Oakeshott and Skinner from the approach to 'otherness' of Edward Said, *Orientalism: Western conceptions of the orient* (London, 1978), and Stephen Greenblatt, *Marvelous Possessions: the wonder of the new world* (Oxford, 1991), for whom the otherness of cultures in time and space appears to be insuperable.
37. D. Harlan, Reply to David Hollinger, in *American History Review*, 94, 1989, pp. 622–27, p. 624.
38. Boucher, The creation of the past, p. 206, argues that Oakeshott's project is different in seeing history as a constructed activity from Skinner's view of history as recovery. But Skinner's methodological work shows him to be aware of the constructedness and conditionality of historical inquiry. Equally, it is hard to see how a concern like Oakeshott's for the past for its own sake can avoid a process of recovery. The very fact that this is a recovery shifts the effort from the practical to the historical mode, but Skinner has not denied this. As such, Skinnerian historical recovery and Oakeshottian historical construction can be seen as partners rather than the opposites.
39. Tully (ed.) *Meaning and Context*, p. 252.
40. J.G.A. Pocock, *Virtue, Commerce and History: essays on political thought and history, chiefly in the eighteenth century* (Cambridge, 1985), p. 8.
41. Ibid., p. 28.
42. Oakeshott moved towards such a discursive/linguistic view in his later writings on history, arguing in *On History*, p. 30, that past utterances must be understood in terms of 'universes of discourse'.
43. Pocock, *Virtue, Commerce and History*, p. 9.
44. See Skinner's discussion of Defoe's *The Shortest Way with Dissenters* in Tully (ed.) *Meaning and Context*, pp. 270–71.

45. Skinner is explicit about this (ibid., p. 105, 276): Skinner is distinguished from Michel Foucault, *Archaeology of Knowledge* (trans., London, 1972), pp. 95–96, in his refusal to deny the importance of authorial intention in history. His position is closer to Wittgenstein's view of intention as a term our public language designates as 'privately' sanctioned (see *Philosophical Investigations*, §247). As such, the third person perspective of communication which Kant called philosophical anthropology is the relevant one to an historical inquiry into meaning.

46. John Dunn, *Political Obligation in Historical Context: essays in political thought* (Cambridge, 1980), p. 18.

47. J.G.A. Pocock, A discourse of sovereignty: observations on the work in progress, in N. Phillipson and Q. Skinner (eds) *Political Discourse in Early Modern Britain* (Cambridge, 1993), pp. 377–428, at p. 394.

48. See J. Arthos, *The Language of Natural Description in Eighteenth-Century Poetry* (London, 1949).

49. See Charles Batten, *Pleasurable Instruction: form and convention in eighteenth-century travel literature* (Berkeley, 1978).

50. See Barrell, *The Dark Side of the Landscape*.

51. M. Bevir, The Errors of Linguistic Contextualism, in *History and Theory*, 31, 1992, pp. 276–98, at p. 293.

52. J.E. Toews, Intellectual history after the linguistic turn: the autonomy of meaning and the irreducibility of Experience, in *American History Review*, 92, 1987, pp. 879–907, at p. 906.

53. Tully (ed.) *Meaning and Context*, p. 68. See also Pocock, *Politics, Language and Time*, p. 36.

54. See Janssen, Political Thought, p. 141; and Butterfield, *Whig Interpretation*, pp. 21–22.

55. See Chapter 1.

56. Carole Fabricant, *Swift's Landscapes* (Baltimore, 1982); and Barrell, *Idea of Landscape*.

57. See Keith Thomas, *Man and the Natural World: changing attitudes in England, 1500–1800* (Harmondsworth, 1983); and Charles W.J. Withers, Geography, natural history and the eighteenth-century enlightenment: putting the world in place, in *History Workshop Journal*, 39, 1995, pp. 137–63.

58. D. Harlan, Intellectual history and the return of literature, in *American History Review*, 94, 1989, pp. 581–609, at pp. 594–5.

59. R.E. Atkinson, *Knowledge and Explanation in History: an introduction to the philosophy of History* (London, 1978), p. 75: 'the notion of "the whole truth"...is incoherent. It is not a huge totality that is very difficult to encompass; it is not a finite, homogeneous totality at all.' See also Danto, *Analytical Philosophy*, p. 18.

60. Tully, (ed.) *Meaning and Context*, p. 68.

61. Dominick LaCapra, *Rethinking Intellectual History: texts, contexts, language* (Ithaca, 1983) and Hayden White, *The Content of the Form: narrative discourse and historical representation* (Baltimore, 1987).

62. Oakeshott, *Experience*, pp. 91, 93. Wittgenstein also made this point; language is formed by usage and it makes no sense to speak of a division of objects from ways of designating them (*Philosophical Investigations*, §293). But as Kripke points out, 'Wittgenstein proposes a picture of language based, not on truth conditions, but on assertability conditions or justification conditions: under what circumstances are we allowed to make a given assertion?' (Saul A Kripke, *Wittgenstein on Rules and Private Language*, Oxford, 1982, p. 74.) The assertability conditions for language can be discovered as they lie in public usage (see *Philosophical*

Investigations, §77). Foucault, *Archaeology*, also saw history as the recovery of conditions in which statements could be made (pp. 57–60).

63. J.V. Femia, An historicist critique of 'revisionist' methods for studying the history of ideas, in Tully (ed.) *Meaning and Context*, pp. 156–75, at p. 158. See also Harlan, *Intellectual History*, p. 603.

64. This point was made by Karl Popper in *The Poverty of Historicism* (London, 1957 [1986 ed.]) p. 150.

65. Femia, *Historicist Critique*, p. 174.

66. Kenneth Minogue, Method in intellectual history: Quentin Skinner's *Foundations*, in Tully (ed.) *Meaning and Context*, pp. 176–93.

67. Tully (ed.) *Meaning and Context*, p. 76.

Chapter 3

1. Keith Thomas, *Religion and the Decline of Magic: studies in popular beliefs in sixteenth- and seventeenth-century England* (Harmondsworth, 1971).

2. Christopher Hill, *The English Bible and the Seventeenth-Century Revolution* (Harmondsworth, 1993). The language of landscape in the eighteenth century is addressed in Chapter 5.

3. John Gascoigne, *Cambridge in the Age of the Enlightenment: Science, Religion and Politics from the Restoration to the French Revolution* (Cambridge, 1989), p. 2; see also Larry Stewart, *The Rise of Public Science: rhetoric, technology, and natural philosophy in Newtonian Britain, 1660–1750* (Cambridge, 1992).

4. See John Walsh and S. Taylor, Introduction: the Church and Anglicanism in the 'long' eighteenth century, in J. Walsh *et al.* (eds) *The Church of England, c. 1689 to c. 1833: from Toleration to Tractarianism* (Cambridge, 1993), pp. 1–64, pp. 42–43; and Sheridan Gilley, Christianity and enlightenment: an historical survey in *History of European Ideas*, 1, 1981, pp. 103–21 at pp. 103–105.

5. Simon Patrick, *A Brief Account of the New Set of 'Latitude Men'* (London, 1662), p. 24.

6. Ian Harris (ed.) *Edmund Burke: pre-revolutionary writings* (Cambridge, 1993); and *idem*, Paine and burke: god, nature and politics, in Michael Bentley (ed.) *Public and Private Doctrine: Essays in British History Presented to Maurice Cowling* (Cambridge, 1993), pp. 34–62. For the linkage of the Deity and the Sublime, see Margaret Nicholson, *Mountain Gloom and Mountain Glory: the development of the aesthetics of the Infinite* (Ithaca, 1959); and Ernst Tuveson, Space, deity and the 'natural sublime' in *Modern Language Quarterly*, 12, 1951, pp. 20–38.

7. See Brian Allen, Jonathan Tyers's Other Garden, in *Journal of Garden History*, 1, 1981, pp. 215–38; see also David Jacques, *Georgian Gardens: in the realm of nature* (London, 1983), p. 17; and Edward Bevan *Box-Hill; a descriptive poem* (London, 1777), pp. 26–27.

8. See John Dixon Hunt, *The Figure in the Landscape: poetry, painting and gardening during the eighteenth century* (Baltimore, 1976), p. 93; and Ronald Paulson, *Breaking and Remaking: aesthetic practice in England, 1700–1820* (New Brunswick, 1989), p. 191. Young's words come from *Genesis*, iii. 8. For the importance of Pope to early eighteenth century gardening ideas, see Peter Martin, *'Pursuing innocent pleasures': the gardening world of Alexander Pope* (Hamden, Connecticut, 1984) and Morris Brownell, *Alexander Pope and the Arts in Georgian England* (Oxford, 1978).

9. Timothy Nourse, *Campania Foelix: or, a discourse of the benefits and improvements of husbandry* (London, 1700), pp. 339–40. See also the prefaces to John Lawrence's *The Clergyman's Recreation: shewing the pleasure and profit of the art of gardening*

(4th edn, London, 1716) and *idem The Gentleman's Recreation: or the second part of gardening improved* (2nd edn, London, 1717).

10. Martin Battestin, *The Providence of Wit: aspects of form in Augustan literature and the arts* (Oxford, 1974), pp. 34–48, quote on pp. 40–41.

11. Keith Thomas, *Man and the Natural World: changing attitudes in England, 1500–1800* (Harmondsworth, 1983), p. 180.

12. E.G. Rupp, *Religion in England, 1688–1791* (Oxford, 1986), p. 259.

13. See Robin Butlin, Ideological contexts and the reconstruction of Biblical landscapes in the seventeenth and early eighteenth centuries: Dr Edward Wells and the historical geography of the Holy Land, in Alan R.H. Baker and Gideon Biger (eds) *Ideology and Landscape in Historical Perspective: essays on the meanings of some places in the past* (Cambridge, 1992), pp. 31–62.

14. Stan A.E. Mendyk, *'Speculum Britanniae': regional study, antiquarianism, and science in Britain to 1700* (Toronto, 1989), p. 8.

15. Lawrence, *The Gentleman's Recreation*, Preface.

16. Gascoigne, *Cambridge in the Age of the Enlightenment*, p. 308.

17. Peter de Bolla, *The Discourse of the Sublime: readings in history, aesthetics and the subject* (Oxford, 1989).

18. For the purposes of this discussion, I have focused mainly upon aesthetic treatises, garden manuals and treatises, travel writing, and topographical and nature poetry, although some plays, novels, sermons and periodical essays are also referenced.

19. Edward Young, *Night Thoughts*, ed. S. Cornford, (Cambridge, 1989), iv. 704.

20. Martin Battestin, *The Providence of Wit*, p. 50.

21. James Hutton, Theory of the earth, in *Trans. of the royal society of Edinburgh* 1 1788, pp. 209–304. See Clarence Glacken, *Traces on the Rhodian Shore: nature and culture in Western thought from ancient times to the end of the eighteenth century* (Berkeley, 1967), Part IV; and David N. Livingstone, *The Geographical Tradition: episodes in the history of a contested enterprise* (Oxford, 1992), p. 119.

22. Samuel Bowden, *Poetical Essays on Several Occasions* (2 vols, London 1733–35), ii. 93–94.

23. William Cowper, *The Task*, vi. 240–42, in H.S. Milford (ed.) *Poetical Works* (London, 1967).

24. Sir Richard Blackmore, *Creation; a philosophical poem* (London, 1712), Argument to Book I. See also John Gilbert Cooper, *The Power of Harmony* (London, 1749), Richard Savage's *The Wanderer; a vision in five cantos* (London, 1729), David Mallett's *The excursion; a poem* (London, 1728) and Isaac Watts's *Looking upward*. Shaftesbury in *Characteristics of men, manners, opinions, times* (6th edn, London, 1737) and Joseph Spence in *Crito: or, a dialogue on beauty by 'Sir Harry Beaumont'* (London, 1752) have prose rhapsodies on the same theme. The degree of Christian sincerity behind these expressions is decidedly variable, Watts and Shaftesbury being at the two extremes.

25. Anon, *The rise and progress of the present taste in planting parks, pleasure grounds, gardens, and so on.* (1767) in John Dixon-Hunt and Peter Willis (eds) *The Genius of the Place: the English landscape garden, 1620–1820* (2nd edn, Cambridge, Mass., 1988), p. 300. The same image was used for the same gardener by William Whitehead in On the late improvements at Nuneham, The seat of the Earl of Harcourt, in William Mason (ed.) *Memoirs of the Life and Writings of William Whitehead* (York, 1788), pp. 75–79.

26. William Derham, *Physico-theology: or, a demonstration of the being and attributes of God, from His works of creation* (4th edn, London, 1716), p. 36.

27. For example John Wansleben, *The Present state of Egypt; or, a new relation of a late voyage into that kingdom* (London, 1678), p. 3.

28. Jonas Hanway, *A journal of eight days journey from Portsmouth to Kingston upon Thames. With miscellaneous thoughts, moral and religious* (2nd edn, 2 vols, London, 1757), i. 162. Childry, in the *Britannia Baconica* (1661) thought Stonehenge 'planted *ab initio*' and Henry Browne *An illustration of Stonehenge and Abury, in the County of Wilts, pointing out their origin and character, through considerations hitherto unnoticed* (Salisbury, 1823) similarly argued that Stonehenge was a survival from the period before the Flood.

29. Cowper, *The Task*, v. 559–69. Further references to the disintegration of all things can be found in poetry: Mallett *Excursion*, Bowden *Poetical essays on several occasions*, and James Kirkpatrick *The Sea-Piece. A Narrative, Philosophical and Descriptive* (London, 1750). In travel writing, see Thomas Story, *A journal of the life of Thomas Story* (London, 1747) and in gardening George Ritso, *Kew Gardens. A poem* (London, 1763).

30. See Thomas Burnet *The sacred theory of the Earth* (4th edn, 2 vols, London, 1719 [1st Latin edn, 1681; 1st English ed., 1690]), ii. 120–22. For the continued importance of the idea, see *The Freethinker: or, essays of wit and humour* (3rd edn, 3 vols, London, 1739), no. 183 (Dec., 21, 1719). As late as 1794 Gray could still use this argument: 'the volcanic nature of the country about Rome, like Sodom, shall 'be utterly burnt with fire.' See J. Black, *The British Abroad: the grand tour in the eighteenth century* (Stroud, 1992), p. 246.

31. Thomas Gibbons, *Juvenalia: poems on various subjects of devotion and virtue* (London, 1750), p. 239. For the same on the London earthquake, see [Z. Grey] *A chronological and historical account of the most memorable earthquakes that have happened in the world, from the beginning of the Christian period to the present year 1750* (Cambridge, 1750), Preface. The Lisbon earthquake five years later led to the same response in John Wesley's *Serious thoughts occasioned by the late earthquakes in Lisbon* (London, 1755) and at a more mundane level from Thomas Turner, see David Vaisey (ed.) *The diary of Thomas Turner, 1754–65* (Oxford, 1984 [reprint, East Hoathly, 1994]), pp. 22–26. The use of analogical reasoning between the moral and natural worlds facilitated this line of argument (see p. xx).

32. Story, *Travels*, p. 444 and Cowper, *Task*, Book II.

33. John Milton, *Paradise Lost*, John T. Shawcross (ed.) *The Complete Poetry* (New York, rev. edn 1971), viii, 66–9.

34. R.D. Havens, *The influence of Milton on English poetry* (Cambridge, Mass., 1922)

35. James Thomson *The Seasons*, ed. James Sambrook (Oxford, 1981), Winter, 579–83.

36. Anon ('Eliezer') *On Mr Harward's observatory*, in *Gentleman's Magazine*, XX, June 1750, p. 277.

37. For other examples of the play of prospect and heavenly prospect, see George Berkeley, *Guardian*, no. 70, June, 1, 1713, James Usher, *Clio: or, a discourse on taste* (3rd edn, London, 1772); Cowper, *Task*, v. 761; and John Scott, *Critical essays on some of the poems, of several English poets* (London, 1785) p. 109. Joseph Butler, *The analogy of religion, natural and revealed* (1736 [London, 1906]), pp. 12–13, 18–19 discusses this.

38. Natural religion was seen by High Churchmen as weakening 'the centrality of the Christian Gospel and tended to end in its advocates either materializing God or deifying matter'. Nigel Ashton, Horne and Heterodoxy: the defence of Anglican beliefs in the late Enlightenment, in *English Historical Review*, 108, 1993, pp. 899–907, at p. 901. See also Simon Schaffer, The consuming flame: electrical showmen and

Tory mystics in the world of goods, in J. Brewer and R. Porter (eds) *Consumption and the World of Goods* (London, 1993), pp. 489–526 for similar Tory/High-Church fears about the dangers of Whig visualizations of God in nature earlier in the century, again centring on Newtonianism.

39. For the High-Church sympathies of eighteenth-century Evangelicalism, see Peter B. Nockles, *The Oxford Movement in Context: Anglican high churchmanship, 1760–1857* (Cambridge, 1994), pp. 32, 48–49. This group also praised the Hutchinsonians; see *idem*, Church parties in the pretractarian Church of England, 1750–1833: the 'orthodox' – some problems, in J. Walsh *et al. Church of England*, pp. 334–59, at p. 345. For the Methodists, see John Wesley *A survey of the wisdom of God in the Creation* (3 vols, Bristol, 1763–70): 'men run into absurdity, concerning Spiritual things, when not content with this analogical knowledge, they argue from things Natural to the intrinsic Nature of the Supernatural' (ii. 212). He also praised Hutchinson (ii. 136–39 and 244–45).

40. See, for example, Carole Fabricant, The aesthetics and politics of landscape in the eighteenth century, in Ralph Cohen (ed.) *Studies in Eighteenth-Century British Art and Aesthetics* (Berkeley, 1985), pp. 49–81; and Ronald Paulson, *Breaking and Remaking*.

41. Jean Hagstrum, *The Sister Arts: the tradition of literary pictorialism and English poetry from Dryden to Gray* (Chicago, 1958), p. 147.

42. John Dyer, *The Ruins of Rome*, ll. 16–20, in *The Works of the English poets. With Prefaces, Biographical and Critical by Samuel Johnson* (London, 1790) vol. 58. See also Dyer's *Grongar Hill*, ll. 114–30.

43. Joseph Warton, *The Enthusiast; or, the lover of nature* (London, 1744), p. 19; Robert Dodsley *Public virtue: a poem* (Dublin, 1754) pt. iii, 1–7. Butler, *Analogy*, p. 19 criticized such comparisons, as ignoring that man's bodily transience was not evidence that his reasoning faculty would then cease.

44. See Maren-Sofie Røstvig, *The Happy Man: studies in the metamorphosis of a classical ideal*, vol. 2, *1700–60* (Oslo, 1958).

45. See Richard Graves, *Columella: or, the distressed anchoret* (2 vols, London, 1779). Vicesimus Knox, On the peculiar danger of falling into indolence in a literary and retired life, in his *Essays, moral and literary* (2 vols, London, 1782), ii. 189–93 summarized the argument of its title.

46. See p. xx.

47. Anon, *Dawley Farm (By an admirer of Ld. Bolingbroke)*, in the *Gentleman's Magazine*, vol. i, June 1731, p. 262. The examples of this genre are legion – for example, Samuel Humphreys, *Cannons. A poem* (London, 1728), and the anonymous *Chatsworth. A poem* [no date, no place – Aubin dates it 1788; S.M. Bod. Gough Derby 2(8)] As this late date suggests, the genre was remarkably persistent. In novels Grandison House in Samuel Richardson's *Sir Charles Grandison* (1754) and Squire Allworthy's house in Henry Fielding's *Tom Jones* (1749) perform a similar emblematic function, as does Smollett's *Expedition of Humphrey Clinker* (1771) in the contrasting visualizations of good and corrupt households in the estates of Dennison and Baynard.

48. Anon, *An answer to the writer of Dawley Farm. A poem* in the *Gentleman's Magazine*, vol. i, July, 1731, p. 306.

49. Lawrence, *Clergyman's Recreation*, Preface.

50. Uvedale Price, *An essay on the picturesque* (London, 1796), p. viii. Other defences of landscape gardening as harmless can be found in Nourse *Campania Foelix*; Joseph Heely, *Letters on the beauties of Hagley, Envil, and the Leasowes* (2 vols, London,

1777), i. 5–8; William Gilpin, *Three essays: on picturesque beauty; on picturesque travel; and on sketching landscape* (London, 1792), p. 41; and George Mason, *An Essay on Design in Gardening* (2nd edn, London, 1795), p. 194.

51. John Wesley, *Journal*, ed. Nehemiah Curnock (8 vols, London, 1909–16), vi. 127–8. A similar response to the statues at Stourhead came from Hanway, *Journal*, i. 140, and is implied more generally in the *Connoisseur* no. 113 (4 vols, 5th edn, Oxford, 1767), iv. 66–69. For a satirical repetition of this argument as part of a critique of methodist enthusiasm, see Richard Graves's *The Spiritual Quixote: or, the summer's ramble of Mr Geoffry Wildgoose. A comic romance* (Dublin, 1774, 2 vols). Methodist iconoclasm was invoked genuinely as the reason that so many antiquities were lost in Wales in Henry Windham, *A Gentleman's Tour Through Monmouthshire and Wales* (new edn, London, 1781), p. 108.

52. Cited in F.C. Mather, *High Church prophet. Bishop Samuel Horsley (1733–1806) and the Caroline tradition in the later Georgian church* (Oxford, 1992), p. 150.

53. [Henry Coventry] *Philemon to Hydaspes: relating a conversation with Hortensius upon the subject of false religions* (London, 1736) pp. 13–14. See also J. Gwynn, *An essay on Design* (London, 1749), p. iii; and Cooper, *Letters*, pp. 52–54.

54. Thomas Percival, On the advantages of a taste for the general beauties of nature and of art, in his *Moral and Literary Dissertations* (Warrington, 1784). Samuel Hall, An attempt to shew, that a taste for the beauties of nature and the fine arts, has no influence favourable to morals, in *Memoirs of the Literary and Philosophical Society of Manchester*, 1, 1785, pp. 223–40.

55. I have used the 'B' text, draft IV of Brendan O Hehir's *Expans'd hieroglyphicks: a critical edition of Sir John Denham's Cooper's Hill* (Berkeley, 1969), lines inserted by Denham between ll. 188, 189.

56. Oliver Goldsmith, *The Traveller* (1764), ll. 123–26 in Tom Davis (ed.) *Oliver Goldsmith. Poems and Plays* (2nd edn, London, 1990). See also Joseph Addison, *a letter from Italy* (London, 1701); and Patrick Brydone, *A Tour through Sicily and Malta, in a series of letters to William Beckford, Esq.* (Dublin edn, 1773), pp. 206–207.

57. Goldsmith, *The Traveller*, l. 176. See also George Keate, The Alps (1763) in *Poetical Works* (2 vols, London, 1780).

58. Dyer, *Ruins of Rome*, ll. 511–12.

59. Tobias Smollett, *Travels through France and Italy* (2 vols, London 1766) ii. 196, 197–98. See also Josiah Tucker, *Instructions for Travellers* (Dublin, 1758).

60. Daniel Defoe, *Caledonia, A Poem in Honour of Scotland, and the Scots Nation* (London, 1707), p. 6.

61. Gilbert White, *The Natural History and Antiquities of Selborne* (1788) ed. Richard Mabey (Harmondsworth, 1977) p. 11. See also James Boswell *Life of Johnson*, ed. G.B. Hill and rev., L.F. Powell (6 vols, Oxford, 1934–50), iii. 160.

62. The *locus classicus* for this is Shakespeare's *Richard II*, Act III scene iv.

63. For other examples of the precise linkage of high politics and landscape, see the *Gentleman's Magazine* (1731) for an exchange of opinions about Bolingbroke's Dawley Farm (discussed above) and William Mason's *The English Garden* (London, 1782), Book IV. For further discussion, see pp. 60–69.

64. Keith Tribe, *Land, Labour and Economic Discourse* (London, 1978)

65. The phrase comes from Marcus Terentius Varro *Rerum Rusticorum*, trans. William D. Hooper, revised Harrison B. Ash (London, 1935), p. 185. This theme is also to be found in Pliny, *Natural History* Book XVIII; Vitruvius *On Architecture*, Book I, Ch. II and IV; Cato *On Agriculture*, sections CXXXVI–VII; and Columella *On Agriculture*, Book I.

66. Robert Lloyd, *The Cit's Country Box* (1757), ll. 1–2, in his *Poems* (London, 1762).
67. Richard Graves, *The Cascade* in his *Euphrosyne: or, amusements on the road of life* (London, 1776), p. 263. See also George Colman and David Garrick *The clandestine marriage* (1766), Act II, and Joseph Craddock, *Village Memoirs: in a series of letters between a clergyman and his family in the country and his son in town* (3rd edn, London, 1775). For more on this theme, see James Raven, *Judging New Wealth: Popular Publishing and Responses to Commerce in England, 1750–1800* (Oxford, 1992).
68. William Chambers, *A dissertation on Oriental Gardens* (2nd edn, London, 1773). Heely *Letters*, i. 39 called Brown a 'shilling-a-day fellow'. G. Mason *An essay*, pp. 124–25 said an improver needed to live in a place, not just visit it to design tastefully.
69. Nourse, *Campania Foelix*, pp. 15–16. Even in this case, Nourse's discussion is highly derivative from the remarks of Roman writers on how to control and oversee servants and workmen.
70. Charles Crawford, *Richmond Hill. A Poem* (London, 1777) l. 125.
71. See M. Andrews, *The Search for the Picturesque: landscape aesthetics and tourism in Britain, 1760–1800* (Aldershot, 1989), Ch.1. On the origins of the linkage of patriotism and landscape in Virgil's *Georgics*, see R.A. Aubin, *Topographical poetry in XVIIIth century England* (London, 1936), p. 36. I use the term patriotism to mean simply the attempt to distinguish one country from others, rather than to suggest 'proto-nationalism'.
72. Fortescue, *Devonia* in his *Dissertations*, pp. 38–39. See also Dyer's *The Fleece*, on which see Stephen Daniels, The implications of industry: Turner and Leeds, in Simon Pugh (ed.) *Reading landscape: country-city-capital* (Manchester, 1990), pp. 66–80.
73. Bowden, *Antiquities and curiosities in Wiltshire and Somerset*, in his *Poetical essays on several occasions*, i. 108.
74. Smollett *Travels*, ii. 131. See also his comparison of British and Italian gardens at ii. 111–14. For the same tactic in the four nations context, see John Macky, *A Journey through Scotland* (London, 1723), which consistently makes comparisons in favour of Scotland.
75. William Cowper, *The Task* (1785), ii. 206, 212–14.
76. 'John English' (Pseud? no date [New Cambridge Bibliography, 1660–1800, col. 1402 suggests 1762]) *Travels through Scotland*, p. 19. English also argues 'Shetland' derives from 'Shit-Land'. Similar attacks were launched on Wales by Swift in his *The Briton Described, or, a Journey thro' Wales* and in the anonymous *A trip in North Wales [by a barrister of the Temple]* (both of which can be found in John Torbuck (ed.) *A collection of Welsh travels and memoirs of Wales*, London, 1738).
77. Giuseppe Baretti, *An Account of the Manners and Customs of Italy: with observations on the mistakes of some travellers with respect to that country* (2 vols, London, 1768), i. 2. For a similar point in relation to travel accounts of Scotland, see Edward Topham, *Letters from Edinburgh; Written in the years 1774 and 1775* (London, 1776); and Samuel Patterson's play, *Another Traveller* (1767).
78. Goldsmith, *The Traveller*, ll. 77–80.
79. Laurence Sterne, *A Sentimental Journey through France and Italy by Mr Yorick* (2 vols, London, 1768), i. 86.
80. Thomas, *Man and the natural world*, pp. 300–3 also points to the 'messiness' of the relation between ideas of landscape and nature and society.
81. Joseph Addison, *Remarks on Several Parts of Italy &c in the Years 1701, 1702, 1703* (5th edn, London, 1736), Preface, not paginated.
82. See, for example, Charles Burney, *The Present State of Music in France and Italy; or, the journal of a tour taken through these countries* (2nd edn, London, 1773), p. 266.

Burney's *Present state* is also worthy of note in passing as an example of linking landscape to music. This reinforces the point that landscape was tied to many other modes of experiencing and understanding our surroundings.

83. Addison, *Remarks*, p. 159.

84. Richard Pococke, *A description of the East, and some other countries* (2 vols, London, 1743), ii. 43–44. For a similar example, see Thomas Shaw, *Travels, or observations relating to the Barbary and the Levant* (2nd edn, London, 1757), p. 314.

85. Richard Jago *Edge-Hill, or the rural prospect delineated and moralized* (London, 1767), bk. iv. 387–558. See also Richard Rolt, *Cambria. A poem illustrated with historical, critical and other explanatory notes* (London, 1749) which lives up to its name. Other particularly good examples are two poems by Thomas Maude: *Wensley-Dale; or, rural contemplations, a poem* (3rd edn, London, 1780) and *Verbeia; or Wharfdale, a poem descriptive and didactic* (York, 1782).

86. Robert Castell, *The villas of the ancients illustrated* (London, 1728). On gardening and the ancients, see also Stephen Switzer, *Ichnographia Rustica: or, the nobleman, gentleman, and gardener's recreation* (London, 1718), Chapter 1 ; Horace Walpole, On modern gardening in his *Anecdotes of painting in England* (Strawberry Hill, 1771), vol. 4, Ch. vii; and G. Mason, *An Essay*, pp. 15–52.

87. For example, at Stourhead in Wiltshire where Kenneth Woodbridge, Henry Hoare's paradise, in *Art Bulletin*, 47, 1965, pp. 83–116, has argued that the entire garden symbolises the *Aeneid*. For some excellent sceptical comments on this thesis, see James Turner, The structure of Henry Hoare's Stourhead, in *Art Bulletin*, 61, 1979, pp. 68–77; and Malcolm Kelsall, The iconography of Stourhead, in *Journal of the Warburg and Courtauld Institute*, 46, 1983, pp. 135–43.

88. John Campbell, *The travels of Edward Brown, Esq.* (2 vols, London, 1753), i. xii. The tension between 'pure' description and other ways of seeing landscape is dealt with more fully later in this chapter. The attack on antiquarianism was part of the debate between the ancients and moderns: see Joseph Levine, *The Battle of the Books: history and literature in the Augustan age* (Ithaca, 1991).

89. Laurence Sterne, *The life and opinions of Tristram Shandy, Gentleman. The Florida edition of the works of Laurence Sterne* (Florida, 1978), 3 vols, eds., Melvyn and Joan New [based on the 1768 edition], ii. 580.

90. Arthur Young, *A six weeks tour, through the southern counties of England and Wales* (2nd edn, London, 1769) and *A six months tour through the north of England* (2nd edn, 4 vols, London, 1771), and Joseph Marshall *Travels* (2nd edn, 4 vols, London, 1773). See Richard Feingold, *Nature and Society: later eighteenth century uses of the pastoral and georgic* (Hassocks, Sussex, 1978), Ch. 2 and 3.

91. See for example Stephen Duck, *A description of a journey*, in his *Poems on several occasions* (London, 1736) p. 217. See also his *Caesar's Camp: or, St George's Hill. A poem* (London, 1755), p. 4; Goldsmith *The deserted village* in *Poems and Plays*, ll. 275–76 and ll. 295–97; George Pearch (ed.) *A collection of poems in four volumes by several hands* (London, 1770) i. 67; Keate *Alps*, ii. 59; Switzer, *Ichnographia Rustica*; and Richard Graves, *Recollection of Some Particulars of the Life of the Late William Shenstone, esq.* (London, 1788), p. 65, for whom utility converted the 'picturesque' into the 'moral landscape'.

92. Ernst Curtius, *European Literature in the Latin Middle Ages* (trans., London, 1953), p. 185, shows how this connection made by Cato, Columella, and so on was transmitted down the ages.

93. William Worthington, *The scripture theory of the Earth* (London, 1773), p. 395, quoting *Ezekiel* xxxvi, 35. See also pp. 388, 393. This connection could also take

on a denominational aspect, a fertile prospect being seen as proof of a healthy denomination, particularly in the Irish context: see Toby Barnard, Art, architecture, artifacts and Ascendancy, in *Bullán*, 1.2., 1994, pp. 17–34, at p. 25–26; and Gerard McCoy, 'Patriots, Protestants and Papists': religion and Ascendancy, 1714–60, in *Bullán*, 1.1, 1994, pp. 105–18, at pp. 111–12.

94. Francis Hutcheson, *An inquiry into the origin of our ideas of beauty and virtue* (3rd edn, London, 1729), p. 12.

95. See Adam Smith, *The theory of moral sentiments* (1759), eds., D.D. Raphael and A.L. Macfie, Vol. 1 of The Glasgow Edition of the works of Adam Smith (Oxford, 1976), p. 188. See also David Hume, *A treatise of human nature* (1739/40) eds., L. Selby-Bigge and P.H. Nidditch (2nd ed., Oxford, 1978) Bk. II, pt. I, viii–ix: *Enquiry concerning the principles of morals* (1751) ed. L. Selby-Bigge and P.H. Nidditch (3rd edn, Oxford, 1975), para. 200; and *Of the Standard of Taste* (1757) in *Essays: moral, political and literary*, ed. E.F. Miller (2nd edn, Indianapolis, 1987): and Alexander Gerard *An essay on taste* (London, 1759), p. 43. Edmund Burke, *A Philosophical Enquiry into the Origin of Our Ideas of the Sublime and Beautiful* (2nd edn, 1759) ed. J.T. Boulton (Oxford, 1987), agreed that fitness was not the cause of beauty in his more physically based aesthetic, see Part III, Sect. VI and VII.

96. On the strength of interart comparisons in eighteenth-century England see Hagstrum, *The sister arts* and Murray Roston, *Changing Perspectives in Literature and the Visual Arts, 1650–1820* (Princeton, 1990).

97. Daniel Defoe, *A Tour through the whole island of Great Britain* ed. Pat Rogers (Harmondsworth 1971), p. 157; Giuseppe Baretti, *A Journey from London to Genoa through England, Portugal, Spain, and France* (4 vols, London, 1770) iv. 54; and J. Georg Forster, *A Voyage round the World in his Britannic Majesty's Sloop Resolution, commanded by Capt. James Cook, during the years 1772, 3, 4 and 5* (2 vols, London, 1777), i. 540.

98. See Andrews, *The search for the picturesque*, Christopher Hussey, *The picturesque: studies in a point of view* (London, 1927) and Elizabeth Manwaring *Italian landscape in eighteenth-century England* (New York, 1925).

99. Thomas Herring, *Letters from the late most reverend Thomas Herring* (London, 1777), p. 42.

100. See Humphrey Repton, *Sketches and hints on landscape gardening* (London, 1795) Ch. viii, and *Theory and Practice of Landscape Gardening* (London, 1803), Ch. ix; and William Marshall, *Planting and Rural Ornament* (2nd edn, London, 1796), pp. 267–77.

101. Edward Wright, *Some observations made in travelling through France, Italy &c. In the years 1722, 3, 4 and 5.* (London, 1730), pp. 169–70.

102. Barbara Maria Stafford, *Voyage into substance: art, science, nature, and the illustrated travel account, 1760–1840.* (Cambridge, Mass., 1984), p. 156; and John Brown, *A Description of the Lake at Keswick* (London, 1770). Michel Foucault, *The order of things* (trans., London, 1970), p. 132 spoke of this as a project of 'the nomination of the visible'. For painterly attempts to achieve the same ends, see Charlotte Klonk, *Science and the Perception of Nature: British landscape art in the late eighteenth and early nineteenth centuries* (New Haven, 1996).

103. Thomas, *Man and the natural world*, p. 228; see also D.C. Streatfield, Art and nature in the English landscape garden: design theory and practice, 1700–1818, in D.C. Streatfield and A.M. Duckworth, *Landscape in the Gardens and Literature of Eighteenth-Century England* (Los Angeles, 1981), pp. 1–87, p. 53.

104. Support for the priority of the visual came from Addison (though he never ignored the other senses) in his *Spectator* Papers on The pleasures of the imagination (for a detailed analysis, see Chapter 4); James Harris, *Three Treatises* (London, 1744), pp. 67–68; Burke, *Enquiry*, p. 123; Lord Kames, *Elements of Criticism* (2nd edn, 3 vols, Edinburgh, 1763), i. 1 and 242, and Usher *Clio*, p. 165. Others, such as Hutcheson, *An inquiry*; Coventry *Philemon*, p. 9; and Gerard, *An Essay*, p. 80 gave equal emphasis to the other senses. On the debate, see Peter de Bolla, *The Education of the Eye: painting, landscape, and architecture in eighteenth-century Britain* (Stanford, 2003).

105. Anthony Champion, To a friend. The scene Dovedale, 1756. In, *Miscellanies in verse and prose*, ed., William Henry (London, 1801), p. 48; James Woodhouse, *Poems On Several Occasions* (2nd edn, London, 1766), p. 112; Wright, *Some observations*, p. 21. See also White, *History of Selbourne*, esp. letter 3 to Daines Barrington. On the many senses beyond the visual in which the sea was appreciated, see Alain Corbin, *The Lure of the Sea: the discovery of the seaside in the Western world, 1750–1850* (trans., Harmondsworth, 1995), pp. 94–95.

106. Cowper, *Task*, i. 181–82.

107. Goldsmith, *Deserted Village* ll. 113–24. See also Thomson *Seasons*, spring, 197–202; Cowper, *Task* i, 197–206; Anne Finch, *A Nocturnal Reverie*, ll. 32–36, in Myra Reynolds (ed.) *The poems of Anne Finch, Countess of Winchelsea* (Chicago, 1903).

108. William Hurn, *Heath Hill: a descriptive poem in four cantos* (Colchester,1777), i. 15–19 and iv. 17–18.

109. Forster, *A Voyage Round the World*, preface; see also Brown, *Description*, p. 6.

110. Maude, *Wensley-Dale* p. vi; Fortescue *Dissertations*, p. i.

111. Henry Fielding, *The Journal of a Voyage to Lisbon* (London, 1755), p. ii. See also ——Loyd, *A Month's Tour in North Wales, Dublin, and Its Environs* (London, 1781) [Bod. SM., Gough Wales, 20(3)], Preface.

112. Swift, *Briton Described*, preface; Sterne, *Tristram Shandy*, ii. 579–80 – the irony of Sterne's words is redoubled when compared to his *Sentimental Journey*, where Mr Yorick spends innumerable chapters describing his first hour and a half in Calais.

113. Defoe, *Tour*, p. 156.

114. For more on these themes in Defoe's *Tour*, see Pat Rogers, *The Text of Great Britain: theme and design in Defoe's tour* (Newark, NJ., 1998), Chapters 4, 5 and 9.

115. Aubin, *Eighteenth Century Topographical Poetry*, p. 46. For gardening Maynard Mack, *The Garden and the City: retirement and politics in the later poetry of Pope, 1731–43* (Toronto, 1969), p. 21 wrote that 'gardening and the pursuits associated with them at this period could "mean" a great deal'.

116. Bertrand H. Bronson, When was Neoclassicism? in his *Facets of the Enlightenment: studies in English literature and its contexts* (Berkeley, 1968), pp. 1–25, quote on p. 4. Lawrence Lipking in *The ordering of the arts in eighteenth-century England* (Princeton, 1970), pp. 3–5 makes much the same point.

117. J.C.D. Clark, *English Society, 1688–1832: ideology, social structure and political practice during the ancien régime.* (Cambridge, 1985), p. 43. This is not to argue that all engaged in landscape debates were religious, but that the religious origins (pp. 41–43) and language (see Chapter 5) of landscape discourse meant men were 'constrained in their tactical options by the nature of their chosen rhetoric, whatever the sincerity of its articulation'. Clark, *The Language of Liberty*, p. 35.

118. Clark, *English Society*, p. 79; cf. Thomas, *Man and the Natural World*, p. 60.

119. E.R. Wasserman, Nature moralized: the divine analogy in the eighteenth century, in *J. of English Literary History*, 20, 1953, pp. 39–76. Quote p. 40. 'All the *Integral*

Parts of *Nature*, have a beautiful *Resemblance, Similitude,* and *Analogy* to one another, and to their Almighty *Original,* whose Images [are] more or less expressive according to their several Orders and Gradations, in the *Scale* of Beings...' George Cheyne, *Philosophical Principles of Religion: natural and reveal'd* (2nd edn, London, 1716), p. 5; see also Burnet *Sacred Theory,* i. 146 and 342; and Butler *Analogy,* pp. 101–102.

120. See A.O. Lovejoy, *The Great Chain of Being: a study in the history of an idea* (Cambridge, Mass., 1936). For good examples of this scale, see Matthew Hale, *The Primitive Origination of Mankind, Considered and Examined According to the Light of Nature* (London, 1677), p. 266 and pp. 310–11; Thomas Burnet, *Sacred theory,* i. 416–17 and ii. 101 and 300; and George Berkeley *Guardian* no. 62 (May 22, 1713).

121. William Cockin, *Ode to the Genius of the Lakes in the North of England* (London, 1780), p. 17, stanza XVII. Frances Reynolds, *An enquiry concerning the principles of taste, and the origins of our ideas of beauty, &c* (London, 1789), pp. 36–40 also spoke of taste as seeming 'to comprise three orders or degrees in its universal comprehension': first 'those objects which immediately relate to divinity'; second, taste and moral virtue where 'sentiment must be intimately related to moral excellence'; and finally items of 'general ornament and honour'.

122. Spence, *Crito,* pp. 57–58; and Goldsmith, *The Traveller,* ll. 423–24. Samuel Johnson elaborated on this theme at the close of this poem, adding the lines 'Still to ourselves in every place consign'd, / Our own felicity we make or find' (ll. 431–32). For Johnson *in propria persona* on the relationship between mind and place, see Chapter 6.

123. Thomson *Seasons,* Autumn, ll. 670–72. 'Religious in subject matter, didactic in intent. That description fits most published writings in the late seventeenth and early eighteenth centuries in England – so fully as almost to constitute a definition of taste, desire and habit.' J. Paul Hunter, *Before Novels: the cultural contexts of eighteenth-century English fiction* (New York, 1990), p. 225. As the dates of many of the works discussed here suggest, this remained an important part of landscape debates until at least the 1790s.

124. Jago, *Edge-Hill,* p. vii; see also Maude, *Wensley-Dale,* p. vi.

125. Joseph Warton, *An Essay on the Genius and Writings of Pope* (London, 1762, edn), pp. 30–1. See also Thomas Paget's (1736) comments on Denham, cited in Aubin, *Topographical Poetry,* p. 38. James Woodhouse, *Poems,* p. 98, praises Thomson's *Seasons* in the same manner. See also Scott, *Critical Essays,* pp. 301–302: responding to Thomson's *Seasons* (Spring, ll. 169–70), on forests 'seeming impatient for rain' he argued that this 'if not too poetically bold, is at least misplaced; it should have immediately followed that of "the rivers seeming forgetful of their course" [Spring, 159–61]; the process would then have been, from inanimate to animate matter, from water and earth, to birds, beasts, and man; this would have been a climax.' As such literary criticism praised not simply the elevation of theme, but the ordering of this elevation by a preexistent hierarchy.

126. As E. Audra and A. Williams, Introduction to Windsor Forest, in *The Poems of Alexander Pope, the Twickenham edition,* Vol. 1 (London, 1961), p. 133, point out 'it is not simply that the poem offers one a scene from nature and then injects into it a moral or ethical prescription; the two elements are rather fused in the act of perception, for the poet in this instance is discovering meanings inherent in nature, not adding one thing to another.'

127. Fortescue, *A poem written on Castle-Hill* in *Dissertations,* p. 122.

128. '[A] thing inanimate acquires a certain elevation by being compared to a sensible being'. Kames, *Elements of criticism*, iii 77–78.
129. John Barrell, *The Political Theory of Painting from Reynolds to Hazlitt: 'the body of the public'* (New Haven, 1986), p. viii; for the Ciceronian origins of the eighteenth-century view of politics as for the 'common good', see Peter Miller, *Defining the Common Good: empire, religion and philosophy in eighteenth-century Britain* (Cambridge, 1994).
130. J.G.A. Pocock, A discourse of sovereignty: observations on the work in progress, in Nicholas Phillipson and Quentin Skinner (eds) *Political Discourse in Early Modern Britain* (Cambridge, 1993), pp. 377–428; quote on p. 381. James J. Sack, *From Jacobite to Conservative: reaction and orthodoxy in Britain, c. 1760–1832.*(Cambridge, 1993), Chs. 5, 8 and 9, has also discussed the subordination of monarchical to religious allegiance. This was by no means a discursive position exclusive to the nascent right: James Bradley, *Religion, Revolution and English Radicalism: non-conformity in eighteenth century politics and society* (Cambridge, 1990) argues that dissenters 'understood politics in moral terms' (p. 140).
131. For Tory adherence to a moral ideal of 'benevolent improvement,' see Nigel Everett, *The Tory view of landscape* (New Haven, 1994). The Evangelicals also looked at the land in these terms: when Wilberforce visited the Cheddar Gorge, he returned saying ' "Miss Hannah More, something must be done for Cheddar." Wilberforce had an even sharper eye for people in need than for grand scenery and had been appalled by what he had seen of poverty and squalor.' Rupp, *Religion in England*, p. 534.
132. See Clark, *English society* and Frank O'Gorman, *Voters, Patrons and Parties: the unreformed electorate of Hanoverian England,1734–1832* (Oxford, 1989).
133. 'The development of English writing about the arts was determined largely by theories designed to counter three indictments: poetry is immoral; the skills involved in painting are manual; music is an idle (or sacrilegious) amusement. As a result, critics throughout the seventeenth and eighteenth centuries are likely to argue that poetry is supremely and primarily moral, that painting is the result of intellectual contemplation, and that music is important (or holy) work. Whatever the rights and wrongs of these arguments, critical debate was founded upon them.' Lipking, *Ordering of the Arts*, p. 31.
134. Joshua Reynolds, *Discourses on Art* (1797) ed. Robert R. Wark (New Haven, 1975) p. 70. For Reynolds's vacillations on this subject, see John Barrell, The public prospect and the private view: the politics of taste in eighteenth-century Britain, in Simon Pugh (ed.) *Reading Landscape* (Manchester, 1990), pp. 19–40.
135. On periodical essays and landscape, see Chapter 6, p. xx. Further mockery of poor taste can be found in George Colman and David Garrick, *The Clandestine Marriage* (1766), Act II; and Craddock, *Village Memoirs*. For agreement that this form of satire became more important after mid-century, see James Raven, *Judging New Wealth: popular publishing and responses to commerce in England, 1750–1800* (Oxford, 1992).
136. Graves, *Columella*, i. 69.
137. Hester Piozzi, *Anecdotes of the late Samuel Johnson, LLD, during the last twenty years of his life* ed. Arthur Sherbo (London, 1974), p. 93.
138. I do not wish to argue that this was the only reason for scepticism: Sterne's satire of fashion, for example, was clearly unrelated to this nexus as he was a Latitudinarian (see Chapter 4); the importance of Methodist and Evangelical scepticism

has already been pointed out; and many Whig improvers such as Defoe refused to subordinate utility to aesthetics.

139. Though he was also a 'whig in politics' (D.N.B.).

140. R.M. Greaves, Religion in the university of Oxford, 1715–1800, in L.S. Sutherland and L.G. Mitchell (eds) *The history of the university of Oxford, vol v, The Eighteenth Century* (Oxford, 1986), pp. 401–24. Quote on p. 401.

141. Clark, *English Society*, p. 51, n. 27 notes that 'Westminster had a distinctly Tory, High-Church tone.' It is noticeable that Westminster and Christ Church, Oxford, were closely linked in the battle of the books on the side of the Ancients, who tended to criticize the worth of natural science due to its distance from humanistic concerns, see Levine, *Battle of the Books*, p. 54.

142. Gascoigne, *Cambridge in the Age of the Enlightenment*, p. 101

143. Ibid.

144. Thomas Patten, 'my excellent friend' (Boswell *Life*, iv. 162 and 508). George Horne (*Life*, ii. 445) who wrote *A fair, candid, and impartial state of the case between Sir Isaac Newton and Mr Hutchinson* (Oxford, 1753) and essay 13 in *Olla Podrida* in 1787 in praise of Johnson. *Olla Podrida* is itself of interest, being an Oxonian periodical essay which also satirized cits (number 28) and criticized the wealthy for spending money on their gardens rather than refurbishing churches (number 33). Nathaniel Wetherell, Master of University College, mentioned *Life*, ii. 440–41 and *Letters of Samuel Johnson*, ed. Bruce Redford (5 vols, Oxford, 1992–94) ii. 304–308. Nockles, *Oxford Movement*, p. 15, notes that 'it was such ties of patronage, family, and kinship as well as of friendship that forged a sense of union among High Churchmen.' Similarly Sack, *Jacobite to Conservative*, p. 2 points out that the 'right' mentality in the late eighteenth century was not a coherent party, but 'a peculiar form of rhetoric . . . collective likes and, especially, dislikes'.

145. Nockles, *Oxford Movement*, p. 10. On the collapse of the Tory party, see Linda Colley, *In defiance of oligarchy: the Tory party 1714–60* (Cambridge, 1982); J.C.D. Clark, *The dynamics of change: the crisis of the 1750s and English party systems* (Cambridge, 1982); and Eveline Cruickshanks, *Political Untouchables: the Tories and the '45* (London 1979).

146. These categories are laid out in Hume's *Treatise*. Allan Ramsay *The investigator* (London, 1762) is another strong statement of the case.

147. R.P. Knight, *The landscape. A Didactic Poem* (2nd edn, London, 1795), pp. 101, 104. On Knight's radicalism and its role in his landscape *oeuvre*, see Andrew Ballantyne, *Architecture, Landscape and Liberty: Richard Payne Knight and the Picturesque* (Cambridge, 1997); for Repton's 'county Whiggism' at this time, see Stephen Daniels, *Humphry Repton: landscape gardening and the geography of Georgian England* (New Haven, 1999), p. 77 ff.

148. Anne Janowitz, The Chartist Picturesque, in S. Copley and P. Garside (eds) *The Politics of the Picturesque: literature, landscape and aesthetics since 1770* (Cambridge, 1994), pp. 261–80, at p. 262.

149. J.G.A. Pocock, Political thought in the English-speaking Atlantic, 1760–90: (ii) Empire, revolution and the end of early modernity, in *idem* (ed.) *The Varieties of British Political Thought, 1500–1800* (Cambridge, 1993), pp. 283–317, at p. 283.

150. John Gascoigne, *Joseph Banks and the English Enlightenment: useful knowledge and polite culture* (Cambridge, 1994), pp. 249–51.

151. Richard Yeo, *Defining Science: William Whewell, Natural Knowledge and Public Debate in early Victorian Britain* (Cambridge, 1993), p. 33. See also Gascoigne, *Cambridge in the Age of the Enlightenment*, Epilogue.

152. Of course, interaction between the sciences and the arts continued, but at an increasingly general level as the technicalities of science increased: see John Wyatt *Wordsworth and the Geologists* (Cambridge, 1995), pp. 214–15.
153. William Wordsworth, Lines composed a few miles above Tintern Abbey (1798), in *Poetical Works*, ed. T. Hutchinson and E. de Selincourt (Oxford, 1936) ll. 81–83. See also *The Prelude*, ii. 201–203 for Wordsworth learning to seek nature 'for her own sake'.
154. Aubin, *Topographical Poetry*, p. 108.
155. This point is also made by Martin Warnke, *Political Landscape: An Art History of Nature* (trans., London, 1994), pp. 120–22.

Chapter 4

1. See John Walsh and Stephen Taylor, Introduction: the church and Anglicanism in the 'long' eighteenth century in John Walsh *et al.*, eds, *The Church of England, c. 1689–1833: from Toleration to Tractarianism* (Cambridge, 1993), pp. 1–64, at pp. 35–43; and Richard Kroll 'Introduction' in Richard Kroll *et al.*, eds, *Philosophy, Science and Religion in England, 1640–1700* (Cambridge, 1992), pp. 1–28, at pp. 2–4.
2. See John Gascoigne, *Cambridge in the Age of the Enlightenment: science, religion and politics from the restoration to the French revolution* (Cambridge, 1989), Chapter 2.
3. John Tillotson, *Works* (3 vols, London, 1735), i. 40.
4. The theological origins of Latitudinarianism are discussed in W.M. Spellman, *The Latitudinarians and the Church of England, 1660–1700* (Athens, Ga., 1993), Ch. 1.
5. Tillotson, *Works*, i. 301.
6. As well as the references in Notes 1, 2 and 4, see Norman Sykes, *From Sheldon to Secker: Aspects of English Church History 1660–1768* (Cambridge, 1959), esp. Ch. 5; John Spurr, *The Restoration Church of England, 1646–89* (New Haven, 1991); and Isabel Rivers, *Reason, Grace and Sentiment: a study of the language of religion and ethics in England 1660–1780, Volume 1: From Whichcote to Wesley* (Cambridge, 1991), Ch. 2.
7. Robert Hurlbutt, *Hume, Newton and the Design Argument* (Lincoln, Ne., 1965), p. 47.
8. Tillotson, *Works*, ii. 551.
9. See Rogers B. Miles, *Science, Religion and Belief: the clerical virtuosi of the royal society of London, 1663–1687* (New York, 1992). Michael Hunter rightly emphasizes that not all clergymen who were active in the Royal Society were of a Low-Church orientation: see 'Latitudinarianism and the "ideology" of the early Royal Society' in Kroll *et al.*, *Philosophy, science and religion*, pp. 199–229, at pp. 210–12.
10. Tillotson, *Works*, ii. 304; Samuel Clarke, *Sermons* (10 vols, 3rd edn, London, 1732) i. 9.
11. Isaac Barrow, *Works* (3 vols, 5th edn, London, 1741), iii. 289.
12. Clarke, *Sermons*, vii. 224.
13. Ibid., vii. 354; see also vii. 306.
14. For Boyle's self-constructed image as a Christian scientist, see Steven Shapin, *A Social History of Truth: civility and science in seventeenth-century England* (Chicago, 1994), pp. 156–70. Jan Wojcik argues in *Robert Boyle and the Limits of Reason* (Cambridge, 1997), pp. 6–7, 38–40, 114–15, 215–17 that Boyle was differentiated from the Latitudinarians by his emphasis on the limitations of reason. This is not convincing, as the Latitudinarians all emphasised the limits of reason. Certainly,

most of the Latitudinarian theologians and writers discussed in this chapter mentioned Boyle in complimentary terms and saw themselves as working within the same idiom.

15. See Hurlbutt, *Hume and Design Argument*, part 2.
16. Robert Boyle *Works*, ed. Thomas Birch (5 vols, London, 1744), 1. 423.
17. Ibid., i. 424.
18. Ibid., iv. 535 and 537.
19. Richard Bentley, *Eight Sermons Preach'd at the Honourable Robert Boyle's lecture* (6th edn, Cambridge, 1735), p. 180.
20. Hurlbutt, *Hume and Design Argument*, calls Boyle's and Newton's approaches the arguments *to* and *from* design respectively.
21. Barrow, *Works*, ii. 134; see also Tillotson, *Works*, ii. 551.
22. John Ray, *The Wisdom of God manifested in the Works of the Creation* (London, 1691), p. 63.
23. For the importance of Burnet in the development of landscape aesthetics, see: Marjorie H. Nicholson, *Mountain Gloom and Mountain Glory: the development of the aesthetics of the infinite* (Ithaca, 1959).
24. Bentley, *Eight Sermons*, pp. 297–98; see also William Derham, *Physico-theology: or, a demonstration of the being and attributes of God, from his works of creation* (4th edn, London, 1716), pp. 71–72.
25. Barrow, *Works*, ii. 71; see also Tillotson, *Works*, i. 12; and Clarke, *Sermons*, i. 16 and vii. 316–17.
26. Boyle, *Works*, ii. 147, 149, 156, 145 and 161. J. Paul Hunter discusses meletetics at greater length in *Before Novels: the cultural contexts of eighteenth-century English fiction* (New York, 1990), pp. 201–208.
27. Derham, *Physico-theology*, p. 112.
28. Barrow, *Works*, iii. 192.
29. William Paley, *Works* (London, 1827), p. 552. On Paley, see D.L. Le Mahieu, *The Mind of William Paley: a philosopher and his age* (Lincoln, Ne., 1976).
30. Paley, *Works*, p. 442; and Le Mahieu, *Mind of Paley*, p. 79.
31. On this group from a largely political perspective, see Richard B. Sher, *Church and University in the Scottish Enlightenment: the moderate literati of Edinburgh* (Princeton, 1985).
32. James Sambrook, *James Thomson, 1700–1748: a life* (Oxford, 1991), p. 15.
33. James Thomson, *The Seasons*, in *The Seasons and the Castle of Indolence*, ed. James Sambrook (Oxford, 1972) 'Hymn on the Seasons' ll. 1–3 and ll. 37–40. For the parallel of the Spring and the Resurrection see 'Winter' ll. 1042–44.
34. Peter Smithers, *The Life of Joseph Addison* (Oxford, 1954), p. 427. Edward A. and Lillian D. Bloom are closer to the mark in *Joseph Addison's Sociable Animal: in the Market Place, on the Hustings, in the Pulpit* (Providence, 1971) seeing that Addison wrote as an Anglican (pp. 173–74), but they still distance him from Latitudinarianism on the mistaken assumption that it was purely about a rational proof of God (p. 176).
35. Joseph Addison and Richard Steele, *The Spectator*, ed. Donald F. Bond (5 vols, Clarendon Press: Oxford, 1965), iv. 117 (*Spectator* 458). All further references to this edition are incorporated in the text.
36. Smithers, *Life of Addison*, p. 103.
37. See Richard Ashcraft, 'Latitudinarianism and toleration: historical myth versus political history' in Kroll *et al.*, *Philosophy, Science and Religion*, pp. 151–77.
38. Joseph Addison, *The Freeholder*, ed. James Leheny (Oxford, 1979), pp. 69–70.

39. See *The Drummer; or, the Haunted House* (1716) in *The Miscellaneous Works of Joseph Addison: Volume 1: Poems and Plays*, ed. A.C. Guthkelch (London, 1914), pp. 431–90.
40. *Freeholder* 24, p. 140; see also nos. 1, 22, 27 and 32.
41. Drummer, in *Miscellaneous Works*, p. 441 (Act I, scene.i).
42. On this rhetoric of the *via media*, see J.G.A. Pocock, 'Within the Margins: the definitions of Orthodoxy' in Roger Lund, ed., *The Margins of Orthodoxy: Heterodox Writing and Cultural Response, 1650–1750* (Cambridge, 1995), pp. 33–53.
43. *Freeholder* 5, p. 59.
44. In *Miscellaneous Works*, pp. 284–89. For a translation, see *The Works of the Right Honourable Joseph Addison* (6 vols, London, 1854–56), vi. 583–85.
45. *Spectator* 531. See also Addison's youthful encaenia speech at Oxford, *Nova Philosophia Veteri Preferenda Est*, in *Miscellaneous Works of Joseph Addison: Volume 2*, ed. A.C. Guthkelch (London, 1914), pp. 467–69. Again, the 1854–6 edition of Addison's *Works* provides a translation, vi. 607–12.
46. For Boyle as a Christian layman, see *Works*, iii. 405. Addison's similar position was emphasised by his literary executor, the poet Thomas Tickell, see *Works* (1854–56 edition), i. v.
47. Boyle, *Works*, iv. 40 and 45.
48. The practice of writing a more devotional 'Saturday paper' was explicitly recognised in *Spectator* 513.
49. Richard Steele and Joseph Addison, *The Tatler*, ed. Donald F. Bond (3 vols, Oxford, 1987), ii. 206–207. Addison believed in the possibility of spirits (which he distinguished from an irrational belief in ghosts): see *Spectator* 12, 110 and 519. In this, Addison was in line with most Latitudinarians, including Boyle and Tillotson.
50. *The Guardian* ed. John Stephens (Lexington, Ky., 1982), p. 362.
51. Martin Battestin with Ruthe Battestin *Henry Fielding: a life* (London, 1989), pp. 273–74, 332–35.
52. Martin Battestin, 'Tom Jones: the argument of design,' in Henry Knight Miller *et al.*, eds., *The Augustan Milieu: essays presented to Louis A. Landa* (Oxford, 1970), pp. 289–319.
53. Ibid., p. 302.
54. Henry Fielding, *The History of Tom Jones, A foundling*, ed. John Bender and Simon Stern (Oxford, 1996), p. 37 (I. iv).
55. Ibid., pp. 94–95 (II. viii).
56. See Janet Butler, 'The garden: early symbol of Clarissa's complicity', in *Studies in English Literature*, 24, 1984, pp. 527–44.
57. Samuel Richardson, *Clarissa, or, the History of a Young Lady*, ed. Angus Ross (Harmondsworth, 1985), pp. 350–53; p. 849; and p. 449.
58. See Lansing Hammond, *Laurence Sterne's Sermons of Mr Yorick* (New Haven, 1948).
59. Laurence Sterne, *Sermons*, ed. Melvyn New (Gainesville, Florida, 1996), pp. 12–13, 347–50.
60. Ibid., pp. 247–50, 423.
61. For which, see Melyvn New, *Laurence Sterne as Satirist: A reading of Tristram Shandy* (Gainesville, Florida, 1969), pp. 1–2, 16. See also Mark Loveridge, *Laurence Sterne and the Argument about Design* (London, 1982).
62. Laurence Sterne, *A Sentimental Journey*, ed. Ian Jack (Oxford, 1984), p. 29.
63. Howard Forster, *Edward Young: the poet of the Night Thoughts, 1683–1765* (Alburgh, Norfolk, 1986), pp. 5–7.
64. Edward Young, *Night Thoughts*, ed. Stephen Cornford (Cambridge, 1989) viii. 1386–90. Further references are incorporated in the text.

65. Henry Pettit, ed., *The Correspondence of Edward Young* (Oxford, 1971), pp. 246–47.
66. Edward Young, *Conjectures on Original Composition*, in *Works* (6 vols, London, 1774–78) v. 118.
67. See also Young, *Works*, iv. 272.
On Young and design, see: Isabel, St John Bliss 'Young's *Night Thoughts* in relation to contemporary Christian apologetics' in *Proceedings of the Modern Language Association of America*, 49, 1949, pp. 37–70
68. See Young, *Correspondence*, pp. 70–71, 600.
69. Ibid., p. 312.
70. James Hervey, *Works* (6 vols, London, 1807), v. 341. See also Henry Clark's assessment of Hervey in *The History of English Nonconformity, Volume 2* (London, 1913), p. 215.
71. James Hervey, *Meditations and Contemplations* (2 vols, 3rd edn, London, 1748) note to i. 114. This work is hereafter cited in the text. For further praise of Addison by Hervey, see *Works*, v. 312 and v. 362.
72. 'The Life of the Rev. Mr James Hervey' prefaced to his *Works* correctly recognised the origins of his approach in the *Meditations*: 'Of this kind of writing we had before an example from no less a man than the great philosopher Mr Boyle, in his *Occasional reflections on several subjects.*' (*Works*, i. p. iv.)
73. For Hervey's hostility to Shaftesbury, see *Works*, vi. 94.
74. Ibid., iv. 387–89.
75. Ibid., vi. 137–38.
76. See Bertrand Bronson, 'The Pre-Romantic or post-Augustan mode' in his *Facets of the Enlightenment* (Berkeley, 1968), pp. 159–72; and Northrop Frye, 'Towards defining an age of sensibility' in *English Literary History*, 23, 1956, pp. 144–52.
77. See John Sitter, *Literary Loneliness in Mid-Eighteenth-Century England* (Ithaca, 1982); and John Mullan, *Sentiment and Sociability: the language of feeling in the eighteenth century* (Oxford, 1988).
78. For which see: Christopher Hussey, *The Picturesque: studies in a point of view* (London, 1927); Malcolm Andrews, *In Search of the Picturesque: landscape aesthetics and tourism in Britain, 1760–1800* (Aldershot, 1989).
79. See Carl Paul Barbier, *William Gilpin: his drawings, teaching, and theory of the picturesque* (Oxford, 1963). A more balanced assessment of Gilpin's life can be found in William D. Templeman's *The Life and Work of William Gilpin (1724–1804). Master of the Picturesque and Vicar of Boldre*, in *University of Illinois Studies in Language and Literature*, 24, 1939.
80. William Gilpin, 'An account of the Revd. Mr Gilpin, vicar of Boldre in New Forest, written by himself' in *Gilpin Memoirs*, ed. William Jackson for the *Cumberland and Westmoreland Antiquarian and Archaeological Society* (London, 1879), pp. 147–48.
81. Ibid., pp. 123, 131.
82. Stephen Copley and Peter Garside, eds., *The Politics of the Picturesque: literature, landscape and aesthetics since 1770* (Cambridge, 1994); and Kim Ian Michasiw, 'Nine revisionist theses on the Picturesque' in *Representations*, 38, 1992, pp. 76–100.
83. Barbier, *William Gilpin*, p. 104.
84. W.H. Grove, *A Memoir of Gilpin*, p. 206, cited in Templeman, *Life and Work of Gilpin*, p. 164.
85. William Gilpin, *Sermons Preached to a Country Congregation* (3 vols, Lymington, 1799–1804), i. 29; see also William Gilpin, *Observations on the Western Parts of England* (London, 1798), p. 47.
86. William Gilpin, *Dialogues on Various Subjects* (London, 1807), p. 330. This work is hereafter cited in the text.

87. Gilpin, *Memoir*, p. 110. For this movement more generally, see Sykes, *Seldon to Secker*.
88. Gilpin, *Sermons*, iii. 398.
89. Ibid., i. 181–82; see also i. 22–23.
90. See [William Gilpin] *Three Dialogues on the Amusements of Clergymen* (London, 1796), p. 165.
91. Gilpin, *Sermons*, ii. 318–20.
92. Ibid., i. 249–50.
93. Ibid., i. 240–50. Gilpin recapitulated this homiletic argument in a didactic context in, 'On the moral uses that may be drawn from husbandry,' in the *Dialogues*.
94. William Gilpin, *Moral Contrasts: or, the power of religion exemplified under different characters* (Lymington, 1798), pp. 31–32, 38, 80.
95. William Gilpin, *Three Essays: on Picturesque Beauty; on Picturesque Travel; and on Sketching Landscape* (London, 1792), pp. 46–47.
96. Ibid., p. 47.
97. William Gilpin, *Observations ... of Cumberland and Westmoreland* (2 vols, London, 1786), i. xx.
98. William Gilpin, *Observations ... [in] the High-lands of Scotland* (2 vols, London, 1789), ii. 33–34.
99. Gilpin, *Observations ... Cumberland*, ii. 120.
100. William Gilpin, *Remarks on Forest Scenery* (2 vols, London, 1791) i. 103 and 103–105n.
101. William Gilpin, *Observations on the River Wye* (London, 1782), p. 18.
102. Gilpin, *Western Tour*, p. 177.
103. William Gilpin, *Observations on Several Parts of the Counties of Cambridge, Norfolk, Suffolk, and Essex. Also on Several Parts of North Wales* (London, 1809), pp. 174–75.
104. *Guardian* 70, ed. Stephens, pp. 261–62.
105. Gilpin uses the phrase in his poem, 'On landscape painting: a poem' in *Three Essays*, l.538 and in the North Wales Tour, *Observations ... Cambridge*, p. 176.
106. Gilpin, *Observations ... Cambridge*, p. 175.
107. Gilpin, *Observations ... High-lands*, ii. 148.
108. See William Gilpin, *Observations on the coasts of Hampshire, Sussex and Kent* (London, 1804), pp. 4–5.
109. Hervey, *Works*, v. 362.
110. Gilpin, *Observations ... High-lands*, i. 27 and ii. 17–18.
111. Gilpin, *Three Essays*, pp. 52–53.
112. Gilpin, *Western Tour*, p. 33n.
113. This fragment is reprinted in Barbier, *William Gilpin*, pp. 177–80.
114. See Walter Scott, *Prose Works* (Edinburgh, 1834–36), iii. 337–89; and [T.N. Talfourd] 'Memoir of the Life and Writings of Mrs Radcliffe' in *Gaston de Blondeville, or The Court of Henry III. Keeping Festival in Ardenne, A Romance* (4 vols, London, 1826).
115. Many of the claims in the previous paragraph are summarized in Robert Miles, *Ann Radcliffe: The Great Enchantress* (Manchester, 1995); see also Mary Poovey, Ideology and *The Mysteries of Udolpho*, *Criticism*, 21, 1979, pp. 307–30.
116. Ann Elwood, *Memoirs of the Literary Ladies of England* (2 vols, London, 1843) ii. 169; see also [Talfourd] 'Memoir', pp. 6, 13.
117. [Talfourd] 'Memoir', p. 105. For modern concurrence, see Marilyn Butler, *Jane Austen and the War of Ideas* (2nd edn, Oxford, 1987), p. 30.
118. For the foundations of Latitude in the same parts of London, see Margaret Jacob, *The Newtonians and the English Revolution 1689–1720* (Hassocks, Sussex, 1976) pp. 49–50; and Gascoigne, *Cambridge in the Age of the Enlightenment*, pp. 42–48.

119. Review of *The Mysteries of Udolpho, The Gentleman's Magazine*, 64, 1794, p. 834.
120. Ann Radcliffe, *A Sicilian Romance*, ed. Alison Milbank (Oxford, 1993), p. 58. Hereafter referred to in the text as *SR*. See also Ann Radcliffe, *The Mysteries of Udolpho*, ed. Bonamy Dobrée (Oxford, 1966), pp. 113–14. Hereafter cited in the text as *U*.
121. Ann Radcliffe, *The Italian, or, The Confessional of the Black Penitents. A Romance*, ed. Frederick Garber (Oxford, 1981), pp. 62–63. Further references in the text as *I*. See also *Udolpho*, 4–5 and 475; and Ann Radcliffe, *The Romance of the Forest*, ed. Chloe Chard (Oxford, 1986), p. 265. Further references in the text as *RF*.
122. Ann Radcliffe, *A Journey made in the Summer of 1794, Through Holland and the Western Frontier of Germany, with a return down the Rhine: To which are added Observations during a Tour to the Lakes of Lancashire, Westmoreland, and Cumberland* (London, 1795), pp. 408, 477.
123. William Cowper, *Poetical Works*, ed. H.S. Milford (Oxford, 1967), *The Task*, Book 2.
124. [Talfourd] 'Memoir,' p. 53, extracted from Radcliffe's private journals.
125. As Colin Haydon puts it, by the 1780s 'influential sections of the Anglican Church wanted more toleration for the papists'. Colin Haydon, *Anti-Catholicism in eighteenth-century England c. 1714–80: A Social and Political Study* (Manchester, 1993), pp. 243–44.
126. Diego Saglia, Looking at the other: cultural difference and the traveller's gaze in *The Italian, Studies in the Novel*, 28, 1996, pp. 12–37 at p. 19.
127. See Radcliffe, *Gaston de Blondeville*, ii. 282–83, ii. 287 and iii. 40.
128. [Talfourd] 'Memoir', p. 115. See also Scott, 'Mrs Radcliffe', pp. 369–71.
129. Julia Kavanagh, *English Women of Letters* (2 vols, London, 1863), i. 308.
130. E.J. Clery, *The Rise of Supernatural Fiction, 1762–1800* (Cambridge, 1999), pp. 111, 113.
131. See also *Gaston de Blondeville*, ii. 313–14.
132. [Talfourd] 'Memoir', p. 39.
133. For which, see J.C.D. Clark, *English Society 1688–1832: ideology, social structure and political practice during the ancien régime* (Cambridge, 1985), Ch. 6
134. Martin Fitzpatrick, 'Latitudinarianism at the Parting of the Ways: a suggestion' in Walsh *et al.*, *Church of England, 1689–1833*, pp. 209–27, at p. 227.
135. Ibid., p. 226.
136. J.C.D. Clark, *The Language of Liberty, 1660–1832: political discourse and social dynamics in the Anglo-American World* (Cambridge, 1994).

Chapter 5

1. See Paul J. Korshin, Johnson and the Renaissance dictionary *Journal of the History of Ideas*, 35, 1975, pp. 300–12; De Witt T. Starnes and Gertrude E. Noyes, *The English Dictionary from Cawdrey to Johnson, 1604–1755* (Chapel Hill, 1946); and Richard Yeo, *Encyclopaedic Visions: scientific dictionaries and enlightenment culture* (Cambridge, 2001).
2. On the decline of the theory of the divine origin of words, see Murray Cohen, *Sensible Words: linguistic practice in England, 1640–1785* (Baltimore, 1977).
3. See Cohen, *Sensible Words* for Locke's theory, which is derived from Book III of his *Essay Concerning Human Understanding* (1689). On Johnson and Locke's theory of language, see Elizabeth Hedrick, Locke's theory of language and Johnson's *Dictionary, Eighteenth-Century Studies*, 20, 1986–87, pp. 422–44; James McLaverty, From definition to explanation: Locke's influence on Johnson's

Dictionary *Journal of the History of Ideas*, 47, 1986, pp. 377–94; and Robert DeMaria, Jr., The theory of language in Johnson's Dictionary, in Paul J. Korshin (ed.) *Johnson after two hundred years* (Philadelphia, 1986), pp. 159–74.

4. Preface, para. 67.

5. On the Dictionary's predecessors, see Starnes and Noyes, *English Dictionary*; on its lack or originality, see James H. Sledd and Gwin J. Kolb, *Dr Johnson's Dictionary: essays in the biography of a book* (London, 1955), p. 4.

6. The first English dictionary to systematically divide definitions was Benjamin Martin's *Lingua Britannica reformata: or, a new universal English dictionary* (2nd edn, London, 1754).

7. Although Thomas Wilson's *A Christian Dictionarie* (1612) did use quotations; see A.D. Horgan, *Johnson on Language: an introduction* (London, 1994), pp. 123–24.

8. Preface, paras. 57 and 58. Robert DeMaria, Jr *Johnson's Dictionary and the Language of Learning* (Oxford, 1986), p. 16. See also Johnson's prayer on beginning the second volume of the Dictionary, in *Diaries, Prayers, Annals*, ed. E.L. McAdam, with Donald and Mary Hyde (New Haven, 1958), p. 50 for the connection between writing and Christian dictates in Johnson; also excellent on this theme is Paul Fussell, *Samuel Johnson and the Life of Writing* (London, 1972). On Johnson's choice of quotations for moral purposes, see also E. San Juan Jr, The actual and the ideal in the making of Johnson's *Dictionary*, *University of Toronto Quarterly*, 34, 1964–65, pp. 146–58; and H.J. Jackson, Johnson and Burton: The *Anatomy of Melancholy* and the *Dictionary of the English Language*, *English Studies in Canada*, 5, 1979, pp. 36–48.

9. J.P. Hardy, *Samuel Johnson: a critical study* (London, 1979), p. 122; James L. Clifford *Dictionary Johnson: Samuel Johnson's middle years* (London, 1979), p. 148, makes a similar claim.

10. John Wain, *Samuel Johnson* (London, 1974), p. 188; Robert DeMaria *The Life of Samuel Johnson: a critical biography* (Oxford, 1993) suggests the *Dictionary* is Johnson's most personal work.

11. These figures come from DeMaria, *Life of Samuel Johnson*, p. 112.

12. Allen Reddick, *The Making of Johnson's Dictionary, 1746–1773* (Cambridge, 1990), pp. 9–10.

13. Starnes and Noyes, *The English Dictionary*, p. 183.

14. All subsequent quotations from the *Dictionary* will italicize the word the illustration is cited under.

15. For Johnson's annotation of Matthew Hale's *The Primitive Origination of Mankind, Considered and Examined according to the Light of Nature* (London, 1677), see Gwin J. and Ruth A. Kolb, The selection and use of the illustrative quotations in Dr Johnson's *Dictionary*, in Howard D. Weinbrot (ed.) *New Aspects of Lexicography: literary criticism, intellectual history, and social change* (Carbondale, 1972), pp. 61–72, at p. 63. For Johnson's reading habits in general see Robert DeMaria, *Samuel Johnson and the Life of Reading* (Baltimore, 1997).

16. For Johnson's use of physico-theology, see W.K. Wimsatt, *Philosophic Words: A Study of Style and Meaning in the Rambler and Dictionary of Samuel Johnson* (New Haven, 1948), esp. p. 34; A.D. Atkinson, Dr Johnson and some physico-theological themes, *Notes and Queries*, 197, 1952, pp. 16–18, 162–65, 249–53. For his use of Thomson, see Reddick, *Making of Johnson's Dictionary*, pp. 136–40, and Thomas B. Gilmore, Implicit criticism of Thomson's *Seasons* in Johnson's *Dictionary*, *Modern Philology*, 86, 1988–89, pp. 265–73.

17. James Boswell, *Life of Samuel Johnson*, ed. G.B. Hill and L.F. Powell (6 vols, Oxford, 1934–50), ii. 468, discussed also in another context in Chapter 6. For the Bible as the oldest book, see also Samuel Johnson, *Sermons*, ed. Jean Hagstrum and James Gray (New Haven, 1978), p. 159; for the Mosaic chronology and its connection with the appearance of the natural world, see *Sermons*, p. 55 and the late Latin poem, 'Septem aetates', in *Diaries*, p. 376.

18. Nehemiah Grew, *Cosmologia Sacra: or, a Demonstration of the Universe as it is the creature and kingdom of God* (London, 1701), p. 95.

19. DeMaria, *Language of Learning*, p. 223. DeMaria argues more generally that the *Dictionary* focuses on the fundamentals of religion rather than theological disputes.

20. For Johnson and theodicy, see *Sermons*, p. 168; Richard B. Schwartz, *Samuel Johnson and the Problem of Evil* (Madison, 1975), which also reprints the Jenyns review; and Paul Alkon, *Samuel Johnson and Moral Discipline* (Evanston, 1967), Ch. 2.

21. See also *Johnson on Shakespeare*, ed. Arthur Sherbo (2 vols, New Haven, 1968), p. 700, where Johnson glossed Lear's 'And take upon's the mystery of things, / As if we were God's spies.' He argues this means 'as if we were angels commissioned to survey and report the lives of men, and were consequently endowed with the power of prying into the original motives of action and the mysteries of conduct.'

22. A.O. Lovejoy, *The Great Chain of Being: a study in the history of an idea* (Cambridge, Mass., 1936). For Johnson's ambiguous views on the great chain, see Alkon, *Johnson and Moral Discipline*, pp. 48–49; and Isobel Grundy, *Samuel Johnson and the Scale of Greatness* (Leicester, 1986) p. 22.

23. On hearing of a monkey supposed to be able to undo knots, 'the Doctor he treated the possibility of it with derision, and insisted that I must be mistaken in the matter, "For, Sir," (said he) "you might as well tell me that the monkey can extract metal from ore as to perform an exploit of a nature that required the intervention of reason to effect it."' Daniel Astle in *Yale edition of the Private papers of James Boswell (Research edition), Volume 2: The correspondence and other papers of James Boswell relating to the making of the Life of Johnson*, ed. Marshall Waingrow (London, 1969), p. 184. In this particular, Johnson was a representative figure, for 'in early modern England, we find anxiety, latent or explicit, about any form of behaviour which threatened to transgress the fragile boundaries between man and the animal creation.' Keith Thomas, *Man and the Natural World: changing attitudes in England, 1500–1800* (Harmondsworth, 1983), p. 38.

24. DeMaria, *Language of Learning*, p. 40.

25. On the significance of Johnson's use of juxtaposition, see Reddick, *Making of Johnson's Dictionary*, p. 9, and DeMaria, *Language of Learning*, p. 150.

26. Wimsatt, *Philosophic Words*, p. 113. See also Richard B. Schwartz, *Samuel Johnson and the New Science* (Madison, 1971), p. 56.

27. 'One of the most persistent strains of higher meaning which runs through the *Dictionary* is the union of the scientific and the religious.' W.K. Wimsatt, Johnson's *Dictionary*, in F.W. Hilles (ed.) *New Light on Dr Johnson. Essays on the Occasion of his 250th birthday* (New Haven, 1959), pp. 65–90, p. 83.

28. The same point was made using one of the period's favourite examples of a heathen being forced to admit the existence of God from the study of the natural world: 'admirable artifice! wherewith Galen, tho' a mere *naturalist*, was so taken, that he could not but adjudge the honour of a hymn to the wise creator.' (More) The notion that the use of the eyes in astronomy would refute the atheist was taken from Plato's *Timaeus* 47a and b, and from *Genesis* 1.14.

29. This is, as I suggest above, part of the standard refutation of Epicurus in this period.

30. John Ray, *The Wisdom of God as Manifested in the Works of the Creation* (London, 1691), Preface.
31. See, for example, Samuel Clarke, *Sermons* (10 vols, London, 1731–32), i. 16, 282–83, vii. 316–19; John Tillotson, *Works* (3 vols, 5th edn, London, 1735), ii. 551; and Isaac Barrow, *Works* (3 vols, 5th edn, London, 1741), ii. 66–74.
32. As such, the design argument became part of the *Dictionary*'s attack on the Manichean heresy (see DeMaria, *Language of Learning*, p. 234). This was a standard part of Latitudinarian homiletics, see Rolf P. Lessenich, *Elements of Pulpit Oratory in Eighteenth-Century England (1660–1800)* (Cologne, 1972), pp. 165–67.
33. Derham, *Physico-theology*, p. 87. The same point is made in Ray, *Wisdom of God*, p. 22; Grew, *Cosmologia Sacra*, pp. 47–48; Hale, *Primitive Origination*, pp. 365–66; and Clarke, *Sermons*, ii. 369–70.
34. DeMaria, *Life of Samuel Johnson*, p. 119.
35. Hooker was also cited under 'supernatural': 'there resteth either no way unto salvation, or if any, then surely a way which is *supernatural*, a way which could never have entered into the heart of a man, as much as once to conceive or imagine, if God himself had not revealed it extraordinarily; for which cause we term it the mystery or secret way of salvation'.
36. Richard Baxter, *The reasons of Christian religion* (London, 1667), p. 193. The same argument is made in Henry Hammond, *A Practical Catechism* (London, 1677) and in South's claim that Adam could perceive God's nature by natural reason in nature (*Sermons*, i. 56).
37. Richard Hooker, *Of the Laws of Ecclesiastical Polity* (3 vols, London, 1830 [1st publ., 1593–1661]). This point has also been observed in a different context by Nicholas Hudson, Three steps to perfection: *Rasselas* and the philosophy of Richard Hooker, in *Eighteenth-Century Life*, 14, 1990, pp. 29–39, at pp. 30, 36.
38. Hooker, *Ecclesiastical Polity*, i. 117. Johnson cites this under 'attainment'.
39. DeMaria, *Language of Learning*, Ch. 2.
40. Joseph Glanville, *Scepsis Scientifica: or, confest ignorance, the way to Science* (London, 1665), pp. 52, 67.
41. Clarke, *Sermons*, i. 267.
42. *Life*, iv. 299.
43. On this subject, see J.G.A. Pocock, Within the margins: the definition of orthodoxy, in R.D. Lund (ed.) *The Margins of Orthodoxy: heterodox writing and cultural response, 1660–1750* (Cambridge, 1995), pp. 33–53.
44. Thomas Baker, *Reflections Upon Learning, wherein is shewn the insufficiency thereof, in its several particulars in order to evince the usefulness and necessity of revelation* (3rd edn, London, 1700), Preface.
45. Emphasis added.
46. Nathan Bailey, *Dictionarium Britannicum: or, a more complete universal etymological English dictionary than any extant* (London, 1730).
47. DeMaria, *Language of Learning*, pp. 73–74.
48. Clarke, *Sermons*, ii. 255–56.
49. Johnson defined 'enthusiast' as 'one who vainly imagines a private revelation; one who has a vain confidence of his intercourse with God'. While compiling the *Dictionary*, Johnson also wrote a 'Life of Cheynel' (1751), attacking Protestant zealotry in the Civil War.
50. Baker, *Reflections*, Preface. Johnson himself utilized this argument in his 'Life of Garth' to explain Garth's possible deathbed conversion to Catholicism: see *Lives of the Poets*, ed. G.B. Hill (3 vols, Oxford, 1905), ii. 63.

51. 'The works of nature are no less exact, than if she did both behold and study how to express some absolute shape or mirror always present before her.' Johnson cites this passage under 'mirror'.
52. 'Who the guide of nature, but only the God of nature? "In him we live, move, and are" [Acts xvii, 28]. Those things which nature is said to do, are by divine art performed, using nature as an instrument; nor is there any such art or knowledge divine in nature herself working, but in the guide of nature's work'. Cited under 'guide' in the *Dictionary*.
53. Cohen, *Sensible Words*.
54. Susie I. Tucker, *Protean Shape. A Study in Eighteenth-Century Vocabulary and Usage* (London, 1967), p. 178.
55. See Chapter 6.
56. Grundy, *Scale of Greatness*, p. 22, also notes that 'Johnson defines 'to excel' in terms which collect together his various physical images for intellectual or spiritual superiority'.
57. John Wilkins, *Of the Principles and Duties of Natural Religion* (7th edn, London, 1715 [1st edn, 1675]), p. 111.
58. DeMaria, *Language of Learning*, p. 80.
59. On this theme, see also Barbara Stafford, *Body Criticism: imaging the unseen in enlightenment art and medicine* (Cambridge, Mass., 1991), pp. 417–36.
60. South, *Sermons*, i. 51 and 56. Johnson cites part of the first passage under 'vegete'.
61. South, *Sermons*, i. 12. For a discussion of Johnson's *Vision of Theodore*, see Chapter 6.
62. Thomas Burnet, *The Sacred Theory of the Earth* (2 vols, 4th edn, London, 1719), Preface to Book IV. Johnson cited part of this passage under 'hazy'.
63. DeMaria, *Language of Learning*, p. 241; DeMaria, *Life of Samuel Johnson*, p. 124. See also Robert DeMaria, Jr, The Politics of Johnson's Dictionary, *Proceedings of the Modern Language Association of America*, 104(1), 1989, pp. 64–74, p. 69.
64. Wimsatt, *Philosophic Words*, p. 37.
65. Clarke, *Sermons*, i. 57–58.
66. See Chapter 3, p. xx.
67. For example, John Whale, Romantics, explorers and picturesque travellers, in Stephen Copley and Peter Garside (eds) *The Politics of the Picturesque: Literature, Landscape and Aesthetics since 1770* (Cambridge, 1994), pp. 175–95, at p. 190; Ian Ousby, *The Englishman's England: taste, travel and the rise of tourism* (Cambridge, 1990), p. 153, points out that 'picturesque' was only defined in the 1801 supplement to the *Dictionary*.
68. *Life*, ii. 90.
69. On Johnson's scepticism over landscape aesthetics, see Morris R. Brownell, *Samuel Johnson's Attitude to the Arts* (Oxford, 1989), pp. 153–79, and Chapter 7.
70. W.K. Wimsatt, Jr, Samuel Johnson and Dryden's DuFresnoy, *Studies in Philology*, 48, 1951, pp. 26–39, at p. 30.
71. For the same argument more generally, see *Life*, ii. 166.
72. This quotation came from *Spectator* no. 489, not the pleasures of the imagination series.
73. Benjamin Martin, *Lingua Britannica Reformata: or, a new universal English dictionary* (2nd edn, London, 1754). The *OED* in turn reverses this on historical grounds.
74. Unless Martin's terse sense 7 of 'scene' is excepted: 'face, or appearance'.
75. Bailey, *Dictionarium Britannicum, sub* 'prospect'.

76. Brownell, *Johnson's Attitude to the Arts*, pp. 178–79. See also DeMaria, *Language of Learning*, p. 102.
77. Wimsatt, *Philosophic Words*, p. 37.
78. Peter DeBolla, *The Discourse of the Sublime: readings in History, aesthetics and the subject* (Oxford, 1989).
79. Elizabeth Hedrick, Fixing the language: Johnson, Chesterfield, and the Plan of a *Dictionary*, *English Literary History*, 55, 1988, pp. 421–42, at p. 438 argues this for the *Dictionary tout court*.
80. DeMaria, *Life of Samuel Johnson*, p. 122.
81. Clifford, *Dictionary Johnson*, p. 145; see also Arthur Sherbo, Dr Johnson's revision of his *Dictionary*, in *Philological quarterly*, 31, 1952, pp. 372–82.
82. Joshua Reynolds, *Discourses on Art*, ed. Robert R. Wark (New Haven, 1975), pp. 69–70. See above, Chapter 3.
83. On Reynolds's view of landscape painting and its place in the hierarchy of genres, see John Barrell, The public prospect and the private view: the politics of taste in eighteenth-century Britain, in Simon Pugh (ed.) *Reading Landscape: country – city – capital* (Manchester, 1990), pp. 19–40.
84. Reddick, *Making of Johnson's Dictionary*, p. 9.
85. J.E. McGuire, Boyle's conception of nature, *Journal of the History of Ideas*, 33, 1972, pp. 523–42, at p. 525.
86. Steven Shapin, Social uses of science, in George Rousseau and Roy Porter (eds) *The Ferment of Knowledge: Studies in the Historiography of Eighteenth-Century Science* (Cambridge, 1980), pp. 93–139, p. 102. See also Barbara Stafford, *Voyage into Substance: art, science, nature and the illustrated travel account, 1760–1840* (Cambridge, Mass., 1984).
87. Alasdair Macintyre, *After Virtue: a study in moral theory* (2nd edn, London, 1985), p. 234, is incorrect in ascribing to Johnson a Christian stoicism in which 'nature for many writers becomes what God had been for Christianity.' Johnson's position on nature was the reverse of this, and constantly attacked the elision of meaning between God and nature. See also Chapter 6.
88. Nathan Bailey, rev. Joseph Nicol Scott *A New Universal Etymological English Dictionary* (new edn, London, 1764).
89. Reddick, *Making of Johnson's Dictionary*, p. 139. See also Gilmore, Implicit criticism of Thomson's *Seasons*.
90. In his 'Life of Thomson' Johnson said that Thomson's 'florid and luxuriant' diction gave his images 'splendour, through which perhaps they were not always easily discerned.' *Lives*, iii. 300. Later in life, Johnson grew more fearful of the ambiguity about nature this might foster.
91. Reddick, *Making of Johnson's Dictionary*, pp. 144, 164.
92. J.C.D. Clark, *English Society, 1688–1832: ideology, social structure and political practice during the ancien régime* (Cambridge, 1985), pp. 314–15.
93. Marjorie Morgan, *Manners, morals and class in England, 1774–1858* (London, 1994), p. 81 points out that 'late eighteenth-century grammarians and lexicographers waged a campaign to rid language of ambiguity,' this being because 'at a time when traditional political, social, and economic authorities were being questioned and undermined, ambiguity in any form was particularly alarming'. Johnson's last lexicographical efforts were an early part of this process.
94. James J. Sack, *From Jacobite to Conservative: reaction and orthodoxy in Britain, c. 1760–1832* (Cambridge, 1993).
95. See especially Denis Cosgrove, *Social Formation and Symbolic Landscape* (Beckenham, 1984) pp. 16–17; John Barrell, *The Idea of Landscape and the Sense of Place, 1730–1840*:

an approach to the poetry of John Clare (Cambridge, 1972), pp. 1–2, 21–25; *idem*, *English Literature in History, 1730–80*. An Equal, Wide Survey (London, 1983); Carole Fabricant, Binding and Dressing Nature's loose tresses: The Ideology of Augustan Landscape Design, in Roseann Runte (ed.) *Studies in Eighteenth-Century Culture*, Vol 8 (Wisconsin, 1979), pp. 109–35; and James Turner, *The Politics of Landscape: Rural Scenery and Society in English Poetry, 1630–1660* (London, 1979), pp. 42–46.

96. Barrell, *English Literature in History*, p. 31. As I show in Chapter 3, pp. 61–63, the language of landscape emerged from religious origins, rather than the other way around as Barrell implies. See also Chapter 6, pp. 195–99.

97. See also Carole Fabricant, The Aesthetics and Politics of Landscape in the Eighteenth Century, in Ralph Cohen (ed.) *Studies in Eighteenth-Century British Art and Aesthetics* (Berkeley, 1985), pp. 49–81, at p. 49.

98. Barrell, *Idea of Landscape*, pp. 24–25.

99. Fabricant, Binding and dressing nature's loose tresses, pp. 109–10.

Chapter 6

1. For Johnson and patriot/country rhetoric, see Donald J. Greene, *The Politics of Samuel Johnson* (2nd edn, Athens, Ga., 1990); Howard Weinbrot, *The Formal Strain: studies in Augustan imitation and satire* (Chicago, 1969), p. 185; John Butt, Johnson's practice in poetical imitation, in Frederick W. Hilles ed., *New Light on Dr. Johnson: Essays on the Occasion of his 250th birthday* (New Haven, 1959), pp. 19–34, at p. 22; and Mary Lascelles, Johnson and Juvenal, in *idem*, pp. 35–55, at p. 39. For patriot rhetoric more generally, see Isaac Kramnick, *Bolingbroke and his Circle: the politics of nostalgia in the age of walpole* (2nd edn, Ithaca, 1992), and especially Christine Gerrard, *The Patriot Opposition to Walpole: politics, poetry, and national myth, 1725–42* (Oxford, 1994).

2. John P. Hardy, Johnson's London: the country vs. the city, in R.F. Brissenden (ed.) *Studies in the Eighteenth Century. Papers presented at the David Nichol Smith memorial seminar* (Canberra, 1968), pp. 251–68, at p. 257.

3. Samuel Johnson, London, in *Poems*, ed. E.L. McAdam with George Milne (New Haven, 1964), l. 2. All subsequent references are to this edition, and will be referenced by line in the text.

4. The notion of a 'true Briton' was important to the patriot vocabulary used to oppose Walpole, and Johnson reverted to it in his political pamphlet *Marmor Norfolciense*. [Samuel Johnson] Marmor Norfolciense (1739), in *Political Writings*, ed. Donald J. Greene (New Haven, 1977), p. 28.

5. This theme is ably discussed also in Rajani Sudan, Foreign bodies: contracting identity in Johnson's *London* and the *Life of Savage*, in *Criticism*, 34, 1992, pp. 173–92.

6. John Wain, Dr. Johnson's Poetry, in his *A House for Truth: critical essays* (London, 1972), pp. 105–29, at p. 116.

7. For the eighteenth-century reading of Juvenal's third satire, see Howard Weinbrot, Johnson's London and Juvenal's third satire: the country as ironic norm, *Modern Philology*, 73, 1976 (supplement), S. 56–65, at S. 60.

8. See Thomas Kaminski, *The Early Career of Samuel Johnson* (New York, 1987), p. 99.

9. John A. Vance, *Samuel Johnson and the Sense of History* (Athens, Ga., 1984), p. 25.

10. Johnson, *Political Writings*, p. 49.

11. On the importance of Elizabeth to patriot rhetoric, see Gerrard, *Patriot Opposition*, Ch. 6 and Peter Miller, *Defining the Common Good: empire, religion and philosophy in eighteenth-century Britain* (Cambridge, 1994), p. 159. A genealogy of images of Elizabeth is provided by: Michael Dobson and Nicola Watson, *England's Elizabeth: an afterlife in fame and fantasy* (Oxford, 2002).

12. Simlarly, the degeneration of London into a French metropolis is something we should 'see' (l. 91).

13. Johnson, *Political Writings*, p. 15. This essay first appeared in the *Gentleman's Magazine* for July, 1738, two months after the publication of *London*. Whether this essay was authored by Johnson is unclear: see J.D. Fleeman, *A Bibliography of the Works of Samuel Johnson* (2 vols, continuously paginated, Oxford, 2000), p. 66 for scepticism.

14. On Johnson and techniques of comparison, see Isobel Grundy, *Samuel Johnson and the Scale of Greatness* (Leicester, 1986). The strategy adopted in 'Eubulus' was not uncommon in the eighteenth century, as Jeremy Black, *The English Press in the Eighteenth Century* (Philadelphia, 1987), p. 129 points out.

15. J.G.A. Pocock, *Virtue, Commerce and History: essays on political thought and History, chiefly in the eighteenth century* (Cambridge, 1985).

16. John Hardy in *Samuel Johnson: a critical study* (London, 1979), p. 56, suggests that the palace referred to may be Houghton Hall, seat of Robert Walpole.

17. Johnson, *Political Writngs*, p. 25.

18. Weinbrot, *Formal Strain*, p. 188; see also Edward Tomarken *Johnson, Rasselas and the Choice of Criticism* (Lexington, 1989), p. 125.

19. Johnson's relationship with Jacobitism is a subject of considerable controversy at present: see J.C.D. Clark, *English Society, 1688–1832: ideology, social structure and political practice during the ancien régime* (Cambridge, 1985) pp. 186–89; and Howard Erskine-Hill, The political character of Samuel Johnson, in Isobel Grundy (ed.) *Samuel Johnson: New Critical Essays* (London, 1984), pp. 107–36 for his sympathy for the Jacobites, and Greene *Politics of Samuel Johnson*, Preface to the second edition, for a rejoinder. The debate has continued in J.C.D. Clark, *Samuel Johnson: literature, religion and English cultural politics from the restoration to romanticism* (Cambridge, 1994); John Cannon, *Samuel Johnson and the Politics of Hanoverian England* (Oxford, 1994), Ch. 2; *The Age of Johnson*, Volumes 7 and 8 (1996–97); and Jonathan Clark and Howard Erskine-Hill (eds) *Samuel Johnson in Historical Context* (London, 2002). For a contemporary view, see John Hawkins *Life of Samuel Johnson* (2nd edn, London, 1787), who slips between seeing Johnson's work of the 1730s as Jacobite (p. 72), as country rhetoric (p. 60), or both (p. 78). This may reflect the fact that Johnson's writings show an 'unusual oscillation between Whig Patriot Idealism and stubborn Jacobite resentment' (Gerrard *Patriot Opposition*, p. 232).

20. Weinbrot, *Formal Strain*, pp. 177–78; see also D.V. Boyd, Vanity and vacuity: a reading of Johnson's verse satires, in *English Literary History*, 39, 1972, pp. 387–403, at p. 393.

21. See Greene, *Politics of Samuel Johnson*, p. 141. Looking back from 1773 Johnson said 'Pulteney was as paltry a fellow as could be. He was a Whig who pretended to be honest...'. James Boswell, *Journal of a tour to the Hebrides*, ed. Frederick A. Pottle and Charles Bennett (London, 1936), p. 340. Johnson's rapid disillusionment with Pulteney was also recorded by John Hawkins, *Life*, p. 506. See Vance, *Johnson and the Sense of History*, p. 20; Walter J. Bate, *Samuel Johnson* (London, 1975; paperback

edn, 1984), p. 296, and S.C. Roberts, *Dr Johnson and Others* (Cambridge, 1958), pp. 46–47, for the shift from the political to the moral at this time.

22. Samuel Johnson, The Vanity of Human Wishes, in *Poems*. All subsequent references are to this edition, with line numbers given in the text.

23. Marshall Waingrow, The mighty moral of Irene, in Frederick Hilles and Harold Bloom (eds) *From Sensibility to Romanticism: essays presented to Frederick A Pottle* (Oxford, 1965), pp. 79–92, at p. 83.

24. Samuel Johnson, Irene, in *Poems*. All subsequent references are to this edition, with act, scene and line numbers given in the text.

25. The Vision of Theodore, hermit of Teneriffe, in *Rasselas and Other Tales*, ed. Gwin J. Kolb (New Haven, 1990). All subsequent references are to this edition, page numbers being given in the text.

26. See Nalini Jain, *The Mind's Extensive View: Samuel Johnson as a critic of poetic language* (Strathtay, 1991), p. 24.

27. See Bernard Einbond, *Samuel Johnson's Allegory* (The Hague, 1971), p. 57.

28. See Boyd, Vanity and Vacuity, p. 398.

29. Tomarken, *Choice of Criticism*, p. 139.

30. On eighteenth-century belief in the dangers of the visual, see Barbara M. Stafford, *Artful Science: enlightenment Entertainment and the eeclipse of visual education* (Cambridge, Ma., 1994); and *idem*, *Body Criticism: imaging the unseen in enlightenment art and medicine* (Cambridge, Mass., 1991), Ch. 5.

31. This description has similarities with the Happy Valley in *Rasselas*. Both gratify every sense, yet neither leads to human happiness. See pp. 202–203.

32. Susie I. Tucker and Henry Gifford, Johnson's poetic imagination, in Review of English Studies, n.s. 8, 1957, pp. 241–48, at p. 248. See also Keith Stewart, Samuel Johnson and the ocean of life: variations on a commonplace, in *Papers in Language and Literature*, 23, 1987, pp. 305–17. For this image in Johnson's work, see, for example, *Irene*, III. i. 46, and IV. i. 59.

33. W.B. Carnochan, *Confinement and Flight: an essay on english literature of the eighteenth century* (Berkeley, 1977) p. 165. See also Boyd, Vanity and Vacuity, p. 403, and Lawrence Lipking, Learning to read Johnson: *The Vision of Theodore* and *The Vanity of Human Wishes*, in *English Literary History*, 43, 1749, pp. 517–37, p. 532.

34. Frederick Hilles, Johnson's Poetic Fire, in Hilles and Bloom, *From Sensibility to Romanticism*, pp. 67–77, at p. 70.

35. Temples or temple of happiness – Johnson is not consistent, cf. pp. 204, 208.

36. Weinbrot, *Formal Strain*, p. 216.

37. Lashing the wind is also an image of man's limitations in the Preface to the *Dictionary*, para. 85. The Vinerean law lectures of Sir Robert Chambers (on which Johnson collaborated) also argue that 'it is impossible for a reasonable being ... to fight the air because the wind is cold, or to beat a tree because it stands in his way; and it is equally impossible for reason to wish any evil but for the sake of good.' Sir Robert Chambers (with Samuel Johnson), *A Course of Lectures on the English Law, 1767–73*, ed. Thomas M. Curley (2 vols, Oxford, 1986), i. 310–11. This imagery comes from Herodotus *History*, trans., David Greene (Chicago, 1987) 7.35 (p. 482).

38. Mahomet, however, does not show any sign of turning to faith, preferring revenge. As he admits, he has not 'quit the sceptre of dominion' (V, xii. 49).

39. On this subject, see Robert DeMaria, Jr, *Johnson's Dictionary and the Language of Learning* (Oxford, 1986), pp. 70–74. For the broader context in English history,

see Keith Thomas, *Religion and the Decline of Magic: studies in popular beliefs in sixteenth- and seventeenth-century England* (Harmondsworth, 1971).

40. Bertrand H. Bronson, *Johnsonian Agonistes and Other Essays* (Berkeley, 1965), p. 133.

41. Robert DeMaria, *The Life of Samuel Johnson: a critical biography* (Oxford, 1993), pp. 49–50, points out that 'pious philosophers, combining fundamental Christian beliefs with ancient wisdom and natural science ... are embodiments of the Johnsonian hero'.

42. Reprinted in James T. Boulton (ed.) *Johnson: the critical heritage* (London, 1971), esp. pp. 52–53.

43. See Boulton, *Critical Heritage* for contemporary comments to this effect. The idea has been repeated recently by, amongst others, Bate, *Samuel Johnson* and Patrick O'Flaherty, Towards an understanding of Johnson's Rambler, in *Studies in English Literature, 1500–1900*, 18, 1978, pp. 523–36.

44. This was the norm in eighteenth-century tales. See Martha Conant, *The Oriental Tale in England in the Eighteenth Century* (New York, 1908), p. 233.

45. Thomas Curley, *Samuel Johnson and the Age of Travel* (Athens, Ga., 1976); and Robert Mayhew, *Geography and Literature in Historical Context: Samuel Johnson and eighteenth-century English conceptions of Geography* (Oxford, 1997).

46. Samuel Johnson, *Rambler*, ed. W.J. Bate and Albrecht B. Strauss (New Haven, 1969), v. 297. All subsequent references are to this edition, by volume and page number in the text.

47. Carey McIntosh, *The Choice of Life: Samuel Johnson and the world of fiction* (New Haven, 1973), p. 94.

48. For the background to these essays, see Arthur Sherbo, The making of Ramblers 186 and 187, in *Proceedings of the Modern Language Association of America*, 67, 1952, pp. 575–80.

49. The crocodile makes its reappearance in an Ethiopian setting at Seged's palace of Dambea on the fourth day. Here, however, the crocodile is designed as an unexpected intervention, leading Seged 'to contemplate the innumerable casualties which lie in ambush on every side to intercept the happiness of man' (v. 301–302).

50. John Bunyan, *The Pilgrim's Progress*, ed. Roger Sharrock (Harmondsworth, 1965), p. 161.

51. On the status of *Rambler* 65 as an allegory, see Einbond, *Johnson's Allegory*, p. 70. A similar structure of Oriental tale blending into allegory is achieved in *Idler* 99, where Ortogrul of Basra is shown a vision of the torrent of rapidly-made money, and the gentle stream of slowly-accumulated wealth.

52. A similar interplay of elevated and low-level prospects played its part in *Rambler* 102, 'The Voyage of Life'.

53. Einbond, *Johnson's Allegory*, p. 9.

54. W.K. Wimsatt, *Philosophic Words: a study of style and meaning in the Rambler and Dictionary of Samuel Johnson* (New Haven, 1948), p. 113; and Grundy, *Johnson and the Scale of Greatness*, p. 4.

55. John C. Riely, The pattern of imagery in Johnson's periodical essays, in *Eighteenth-Century Studies*, 3, 1970, pp. 384–97, at p. 393. See also pp. 384–85.

56. McIntosh, *Choice of Life*, p. 108. For the seventeenth century metaphorical language of moral topography, see Christopher Hill, *The English Bible and the Seventeenth-Century Revolution* (Harmondsworth, 1993), pp. 115–22. For its eighteenth-century continuation, see Paul Fussell, *The Rhetorical World of Augustan Humanism: ethics and imagery from Swift to Burke* (Oxford, 1965), pp. 268–75.

57. On Johnson and competition, see Grundy, *Johnson and the Scale of Greatness*, Ch. 10, esp p. 107.
58. See *Rambler* 110 for the 'precipice of destruction'.
59. See also Cali in *Irene*, III, i. 22.
60. O'Flaherty, Johnson's Rambler, p. 529; see also Stewart, Johnson and the Ocean of Life, p. 308.
61. Samuel Johnson Adventurer, in *Idler and Adventurer*, ed. W.J. Bate, J.M. Bullitt and L.F. Powell (New Haven, 1963), p. 392. All subsequent references to the *Idler* and *Adventurer* are to this edition, given by volume number and page number in the text.
62. Riely, Pattern of Imagery, p. 394.
63. For an analysis of fluvial imagery in *Rasselas*, see Charles L. Campbell, Image and symbol in *Rasselas*: narrative form and the flux of life, in *English Studies in Canada*, 16, 1990, pp. 263–78. For Johnson invoking this image in his life, see *Life*, v. 279.
64. See also *Rambler* 89.
65. Riely, Pattern of Imagery, p. 395.
66. The origins of the images used here are discussed in Ernst Curtius, *European Literature in the Latin Middle Ages* (trans., London, 1953).
67. Grundy, *Johnson and the Scale of Greatness*, p. 156.
68. For Johnson and *deceptio visus*, see Robert Folkenflik, Samuel Johnson and art, in Paul Alkon and Robert Folkenflik (eds) *Samuel Johnson: pictures and words* (Los Angeles, 1984), n. 83, p. 116. In Wickins's garden, Johnson said 'don't tell me of deception; a lie, Sir, is a lie, whether it be a lie to the eye or a lie to the ear.' *Johnsonian Miscellanies*, ed. G.B. Hill (2 vols, Oxford, 1897), ii. 427.
69. See Ariel Sachs, *Passionate Intelligence: imagination and reason in the work of Samuel Johnson* (Baltimore, 1967), p. 79.
70. Grundy, *Johnson and the Scale of Greatness*, p. 160.
71. For vision imagery in the *Lives of the Poets*, see Martin Maner, *The Philosophical Biographer: Doubt and Dialectic in Johnson's Lives of the Poets* (Athens, Ga., 1988), pp. 27–30.
72. Richard B. Schwartz, *Samuel Johnson and the New Science* (Madison, 1971), p. 56.
73. *Rambler* 111 is precisely the same, starting with the observation 'that late springs produce the greatest plenty' and hopes to apply this to 'the younger and sprightly part of my readers' (iv. 226). Under two weeks previously, in *Rambler* 108, Johnson had used a similar structure: opening with the Lucretian point as to the many uninhabitable parts of the earth, Johnson transferred this to the many hours unavoidably lost to our life as moral actors (iv. 210).
74. For more on this theme, see Alan McKillop, Local attachment and cosmopolitanism – the eighteenth-century pattern, in F. Hilles and H. Bloom (eds) *From Sensibility to Romanticism: essays presented to Frederick A Pottle* (Oxford, 1965), pp. 191–218, esp. pp. 195–98.
75. *Idler* 37, 'Iron and Gold', achieves a similar interplay of the natural and the moral.
76. In *Rambler* 99, for example, Johnson develops his views on human sociability by looking at animals.
77. See A.T. Elder, Irony and humour in the Rambler, in *University of Toronto Quarterly*, 30, 1960, pp. 57–71.
78. See also *Rambler* 61 (Mr Frolick) and *Idler* 49 (Will Marvel).
79. For Johnson on cits, see also *Life*, ii. 120 and ii. 337.
80. This resembles Johnson's parody of Thomas Warton's poetry (*Life*, iii. 159).
81. See also *Rambler* 138 for the 'economist' Mrs Busy who 'has turned a large manor into a farm'.

82. Given the recent focus on Johnson's connection with humanism (see Clark *Samuel Johnson*; and DeMaria, *Life of Samuel Johnson*), it is worth pointing out that this theme is also to be found in Erasmus, *Praise of Folly*, trans., Betty Radice (Harmondsworth, 1971; rev. edn, 1993), p. 61, and Thomas More *Utopia*, trans., Robert M Adams (2nd edn, New York, 1992) p. 39. The ultimate source for this theme is Horace. For Horace's significance to Johnson on landscape, see Chapter 8.

83. For the opposition to political economy in eighteenth-century landscape debates, see Nigel Everett, *The Tory View of Landscape* (New Haven, 1994).

84. Daniel Astle recognized that Johnson's deflation of his romantic response to Needwood Forest in Staffordshire was to 'display that satyrical turn for which he was so famous – a turn which I never thought originated from a splenetic disposition, but purely from an honest indignation at the follies and vices of mankind'. Marshall Waingrow (ed.) *Yale edition of the private papers of James Boswell (Research edition): Volume 2: Correspondence relating to the making of the Life of Johnson* (London, 1969), p. 180.

85. Grundy, *Johnson and the scale of greatness*, p. 91 and p. 98; Patrick O'Flaherty, *Johnson's Idler: the equipment of a satirist*, in *English Literary History*, 37, 1970, pp. 211–25; and McIntosh *Choice of life*, Ch. 3.

86. See W.K. Wimsatt, *Philosophic words*; idem, *The prose style of Samuel Johnson* (New Haven 1941); Donald J. Greene, Pictures to the mind: Johnson and imagery, in Mary Lascelles *et al.* (eds) *Johnson, Boswell and their Circle: essays presented to L.F. Powell* (Oxford, 1965), pp. 137–58; and Peter T. Koper, Samuel Johnson's rhetorical stance in the *Rambler*, in *Style*, 12, 1978, pp. 23–34.

87. See Leopold Damrosch, Jr, *The Uses of Johnson's Criticism* (Charlottesville, 1976), pp. 78–92.

88. As the motto of *Adventurer* 69 puts it *fere libenter homines id quod volunt credunt*: men willingly believe what they wish to be true.

89. Similarly, Dick Shifter in *Idler* 71, who had 'received from his favourite authors very strong impressions of a country life', despite setbacks on his first day, 'rose early in the morning, surveyed the landscape, and was pleased'.

90. On the seasonality of Johnson's essays, see James Woodruff, Johnson's *Rambler* in contemporary context, in *Bulletin of Research in the Humanities*, 85, 1982, pp. 27–64, esp. p. 35.

91. He elsewhere called such retirement 'the State of a Mill without Grist'. See *Thraliana: the diary of Mrs Hester Lynch Thrale (later Mrs Piozzi), 1776–1809*, ed. Katharine Balderston (2 vols, Oxford, 1959), p. 171.

92. In a similar vein, though not related to gardens, is the landlady's experience in *Rambler* 161: 'the poor woman was teased for seven weeks by innumerable passengers, who obliged her to climb with them every hour up five stories, and then disliked the prospect' (v. 93).

93. The message is repeated in *Idler* 30, speaking of 'money and time' as 'the heaviest burthens of life': 'one hurries to New-market; another travels over Europe; one pulls down his house and calls architects about him; another buys a seat in the country' (ii. 92–95).

94. The dream of the lottery-obsessed shopkeeper in *Rambler* 181 is instructive in this regard: 'the great delight of my solitary hours was to purchase an estate, and form plantations' (v. 188).

95. The pattern is the same in a domestic context for Mercator (*Adventurer* 102) and Tim Ranger (*Idler* 62 and 64).

96. This bears an obvious parallel to Johnson's 'Life of Shenstone', where he is haunted by duns in his garden, the Leasowes. See Chapter 8, p. xx.
97. See also Frolick (*Rambler* 61) who 'has reeled with giddiness on the top of the monument'.
98. On Johnson's character names, see E.A. Bloom, Symbolic names in Johnson's periodical essays, in *Modern Language Quarterly*, 13, 1952, pp. 333–52.
99. For Johnson, nabobs could only come to dominate the countryside in areas where money was already the ruling principle, and as such supposed rural innocence had already disappeared. See *Life*, v. 106.
100. McIntosh, *Choice of Life*, p. 151.
101. See *Rambler* 142 and 51.
102. See *Rambler* 53 and *Adventurer* 53. Everett, *Tory View*, pp. 33–34, recognizes Johnson's ambivalence with respect to the opposition to political economy and its impact on the landscape. In his refusal to believe in a rural nostalgia, Johnson is also clearly distinguished from the Tory view Everett sketches.
103. For Johnson on animal cruelty, see Grundy, *Johnson and the Scale of Greatness*, p. 151; see also John Wiltshire, *Samuel Johnson and the Medical World: the doctor and the patient* (Cambridge, 1991) and Keith Thomas, *Man and the Natural World: changing attitudes in England, 1500–1800* (Harmondsworth, 1983).
104. As Johnson later argued in his *Journey to the Western Islands of Scotland*, ed. J.D. Fleeman (Oxford, 1985), p. 33 'honesty is not greater where elegance is less'.
105. Thomas Macaulay, Review of Croker's edition of Boswell's *Life*. Reprinted in Boulton (ed.) *Critical Heritage*, pp. 423–31.
106. As Morris Brownell, *Samuel Johnson's Attitude to the Arts* (Oxford, 1989) tends to.
107. See also *Life*, iii. 253.
108. See also the 'vernal speculation', Rambler 5, where the view of nature 'has always a certain prospect of discovering the sovereign author of the universe'. See Chester Chapin, *The Religious thought of Samuel Johnson* (Ann Arbor, 1968), p. 76.
109. It is generally agreed that Johnson believed in a chain of being, but not in any full theorisation of this. See Paul Alkon, *Johnson and Moral Discipline* (Evanston, 1967); and Richard B. Schwartz, *Samuel Johnson and the Problem of Evil* (Madison, 1976). Johnson's views are most clearly expressed in his review of Soame Jenyns's *Free Enquiry*, which is discussed briefly in section 'Jonson's life and landscape'.
110. Robert Walker, Johnson in the 'Age of evidences', in *Huntington Library Quarterly*, 44, 1980–81, pp. 27–42; Chapin *Religious Life*, p. 80. Alkon, *Johnson and Moral Discipline*, p. 48; Nicholas Hudson, *Samuel Johnson and Eighteenth-Century Thought* (Oxford, 1988), p. 22.
111. The same design argument from human limitations is made in *Rambler* 128 and *Adventurers* 50 and 107.
112. See Schwartz, *Johnson and Science*, p. 18; and *idem, Johnson and Evil*, p. 69.
113. On this, see Hudson, *Johnson and Eighteenth Century thought*, p. 104. The London earthquakes of February and March 1750 were not represented in the *Rambler* papers which commenced at the beginning of March 1750.
114. Ibid., p. 111.
115. See also the allegory of Rest and Labour, *Rambler* 33, discussed above.
116. For example, *Life*, i. 437, iii. 246, and iv. 210.
117. The classic analysis of this aspect of Johnson's thought is Walter Bate, *The Achievement of Samuel Johnson* (New York, 1955).
118. See also *Rambler* 9 on the discovery of the method of making glass.

119. Bate, *Achievement of Johnson*, pp. 78–79; and Curley, *Johnson and the Age of Travel*, pp. 53, 68.
120. See Samuel Johnson, *Letters*, ed. Bruce Redford (5 vols, Oxford, 1992–4), i. 206 for Johnson on his return to Lichfield; and *Diaries, Prayers, Annals*, ed. E.L. McAdam with Donald and Mary Hyde (New Haven, 1959) p. 206 for Mrs Thrale's return to her birthplace.
121. Schwartz, *Johnson and Science*, pp. 113–15.
122. It is in these terms that the virtuoso naturalist Quisquilius is defended in *Rambler* 83.
123. Schwartz, *Johnson and Science*, Ch. 5.
124. For other examples, see *Idler* 17 and 55.
125. Katharine Balderston, Dr Johnson and William Law, in *Proceedings of the Modern Language Association of America*, 75(1), 1960, pp. 382–94, at p. 391. In fact, High Churchmen more generally were less condemnatory of retirement in the eighteenth century; see Peter B. Nockles, *The Oxford Movement in Context: Anglican high churchmanship, 1760–1857* (Cambridge, 1994), pp. 188–90.
126. Maren-Sophie Røstvig, *The Happy Man: Studies in the Metamorphosis of a Classical Ideal, vol 2: 1700–60* (Oslo, 1958), sees this social ideal of retirement as the final collapse of the *beatus ille* tradition. The role of Johnson in this is recognized, p. 237.
127. See Maurice Quinlan, *Samuel Johnson: a layman's religion* (Madison, 1964), p. 164.
128. Johnson cited Horace's *Epistles* for a similar idea in *Letters*, iii. 215: *in culpa est animus, qui se non effugit usquam* (What is at fault is the mind, which never escapes from itself) [*Epistles*, I, xiv. 15]. For the importance of the classics to Johnson's position in the 'Anglo-Latin' tradition, see Clark, *Samuel Johnson*, Ch. 1. See also *Life*, ii. 195 for Johnson arguing 'What is *climate* to happiness?' See also Johnson's gloss on the lines in *Cymbeline* 'Return he cannot, nor / Continue where he is: to shift his being / Is to exchange one misery with another' where he saw 'shift his being' as meaning 'to change his abode': *Johnson on Shakespeare*, ed. Arthur Sherbo (2 vols, New Haven, 1968), p. 881.
129. See *Life*, i. 426, and ii. 358. Voitle, *Johnson the Moralist*, p. 25; and Hudson, *Johnson and Eighteenth-Century Thought*, pp. 85–98. For the eighteenth century's scepticism about environmental determinism in general, see John Gascoigne, *Joseph Banks and the English Enlightenment: useful knowledge and polite culture* (Cambridge, 1994), pp. 141–44; and Fussell, *Rhetorical World*, pp. 95–97.
130. McIntosh, *Choice of Life*, p. 1. To understand Johnson's view of landscape, it is important to see the positions of the moral and the political in his thought. For him, 'polity being only the conduct of immoral [or indeed moral] men in public affairs' (Review of Soame Jenyns, *Literary Magazine* 1757, p. 306), politics is a branch of moral debate.
131. *Life*, i. 201.
132. Ibid., i. 420.
133. Walter Graham, *English Literary Periodicals* (New York, 1930), p. 119, pp. 140–43.
134. Donald Greene, Samuel Johnson, journalist, in Donavan H. Bond and W. Reynolds McLeod (eds) *Newsletters to Newspapers: eighteenth-century journalism* (W. Virginia, 1977), pp. 87–101, at p. 92.
135. Edward A. and Lillian D. Bloom, *Joseph Addison's Sociable Animal: in the market place, on the hustings, in the pulpit* (Providence, 1971), p. 191.
136. [Addison] *The Spectator*, ed. Donald F. Bond (5 vols, Oxford, 1965). All subsequent references are to this edition, and are included in the text.

137. The argument from design was also used in Tatlers 100 and 109, and in Guardian 103. In this last paper, Addison was led from a firework display to God's stars, considered 'as a sky-rocket discharged by a hand that is Almighty'. Richard Steele *et al.*, *The Guardian* (Philadelphia, 1831), p. 146. All future references are to this edition by page number, and are in the text. Johnson's 'O.N. on the fireworks' was led to rather less devotional reflections by the celebrations of the Peace of Aix-la-Chappelle (1748–98): 'it will resemble the war of which it celebrates the period. The powers of this part of the world... have set Europe in a flame'. Samuel Johnson, O.N. on the fireworks, in *Political Writings*, pp. 111–15, quote on p. 114. The letter initially appeared in the *Gentleman's Magazine*.

138. The argument was made more fully a week previously in *Spectator* 387.

139. See also *Spectators* 37, 110 and 425, and *Guardian* 125.

140. He made the same claim in *Spectator* 262.

141. See also *The Champion* (2 vols, London, 1741), no. 31

142. This argument was repeated by Berkeley in his philosophical writings: see The Principles of Human Knowledge, Part 1, paras. 109 and 152–53 in *Philosophical Works*, ed. M.R. Ayers (rev. edn, London, 1989). The argument was copied in William Webster's *The Weekly Miscellany*, no. 20 (2 vols, 2nd edn, London 1738).

143. *Grumbler*, no. 10, April 19, 1715.

144. See Nicholas Joost, The Authorship of the Freethinker, in Richmond P. Bond (ed.) *Studies in the Early English Periodical* (Chapel Hill, 1957), pp. 103–34.

145. For anti-Catholicism in the English Press, see Black, *English Press in the Eighteenth Century*, p. 253.

146. *The Freethinker* (3rd edn, 3 vols, London 1739, nos. 1–159) [then individual numbers].

147. The *Freethinker's* tone here is akin to that of Probus Britanicus in *Marmor Norfolciense* but with a very different political purpose. See also *Grumblers* 10 and 11 on the 1715 uprising and natural omens, and *Museum* 16 (3 vols, London, 1746) on the 45 and omens.

148. Clark, *English Society*, p. 178; and J. Paul Hunter, *Before Novels: the cultural contexts of eighteenth-century English fiction* (New York, 1990), pp. 217–24 on the Providence book tradition.

149. No. 56. See also nos. 181 and 183, where Burnet's theory that the Conflagration will start in the 'Whorish Babylon' of Rome is supported. For the origins of this interpretation of the messenger's delay, see Quentin Skinner, *The Foundations of Modern Political Thought* (2 vols, Cambridge, 1978), ii. 106–107.

150. Tatler 240, in *The Tatler* ed. Donald F. Bond (Oxford, 1987). All references are to this edition.

151. Hudson, *Johnson and Eighteenth-Century Thought*, p. 114.

152. A similar position was adopted by the *Champion* for 3 June 1740.

153. The same argument is made in *Museum* 25.

154. Steele does not acknowledge his source, but it is noted by the editor.

155. It can be compared to the tale of Omar in *Idler* 101.

156. See also *Spectator* 4 and *Tatler* 89.

157. For example *Spectators* 155 and 210; *Plain Dealer* (2 vols, London, 1730) 32; *World* (6 vols, London 1755), no. 44; and *The Citizen of the World* (2 vols, London 1762), no. 92.

158. For example, *Spectators* 205, 224, 316 and 408; *Freethinker* 97; and *Plain Dealer* 20, 25, 32, and 106.

159. For example, *Spectators* 237, 626 and 635; *Weekly Miscellany*, 63.

160. Other good examples are the Mount of the Muses (*Spectator* 514), the Temple of Fame (*Tatler* 81) and the Mount of Fortune (*Plain Dealer* 18).
161. See also *World* 121, and *Museum* 18.
162. See Earl Wasserman, Johnson's *Rasselas*: implicit contexts, in *Journal of English and Germanic Philology*, 74, 1975, pp. 1–25; and Lipking, Learning to read Johnson.
163. Other early essays on cits include Steele's *Tatler* 176 and *Grumbler* 22: 'when a Citizen goes to his Seat for a little fresh Air, he does not consider the Country either like a Philosopher, or a Poet... if you mention the *Golden Age* to him, he understands it in a literal Sense, and regrets his not living in the glorious Days of universal Traffick.'
164. See Stafford, *Artful Science*, Ch. 1 and Jeremy Black, *The British Abroad: the grand tour in the eighteenth century* (Stroud, 1992), Ch. 11.
165. Joost, Authorship of Freethinker, p. 119. See for example, nos. 5, 6, 28; and *Tatler* 93.
166. See also *Guardian* 34. On the demands made as to the nature of travel, see Charles Batten, *Pleasurable Instruction: form and convention in eighteenth-century travel literature* (Berkeley, 1978).
167. *Spectator* 1 and 117.
168. See also *Tatlers* 19 and 37 for attacks on foxhunting.
169. See also *Tatler* 221, *Freethinkers* 113 and 115, and *The Censor* (2nd edn, 3 vols, London, 1717) 91.
170. John Dixon-Hunt and Peter Willis (eds) *The Genius of the Place: the English landscape garden, 1620–1820* (paperback edn, Cambridge Mass., 1988), p. 274, on *The World*.
171. See *World* 1 and *Connoisseur* 2. The pattern of the well-travelled eidolon was established by Mr Spectator in *Spectator* 1.
172. See also number 207 which advocated travel to China as 'the majority of our young travellers return home entirely divested of the religion of their country... [and] the doctrine of Confucius might have a good effect upon them'.
173. Cicero *De Officiis* 1 39 139. See also *Citizen of the world*, 92.
174. Ibid., 68. See Philip C. Almond, *Heaven and Hell in Enlightenment England* (Cambridge, 1994), pp. 105–106, and Edward Said *Orientalism: western conceptions of the orient* (London, 1978), pp. 68–70, 96, 102. The equation of Vauxhall and Mahomet's paradise started in *Spectator* 383.
175. See David H. Solkin, *Painting for Money: the visual arts and the public sphere in eighteenth-century England* (New Haven, 1993), Ch. 4 on Vauxhall.
176. There are numerous references to this in Eliza Haywood's *The Female Spectator* (3 vols, London, 1745). This association was already a venerable one in literature concerning the metropolis: see Cynthia Wall, *The Literary and Cultural Spaces of Restoration London* (Cambridge, 1998), pp. 150–67.
177. *The Connoisseur* (4 vols, 5th edn, Oxford, 1767), no. 113.
178. *World* 172.
179. *World* 132 and 134; *Connoisseur* 9 and 35.
180. *World* 162.
181. James Gray *Johnson's Sermons: a study* (Oxford, 1972), p. 48, and Stephen Lynn, Johnson's *Rambler* and Eighteenth-Century Rhetoric, in *Eighteenth-Century Studies*, 19, 1986, pp. 461–79.
182. The sermons certainly range in their writing from 1745 to 1777, but the composition of most is imposible to date with precision, see Samuel Johnson, *Sermons*, vol. xiv, Yale edition of the works of Samuel Johnson, ed. Jean H. Hagstrum and

James Gray (New Haven, 1978), pp. 315–16. All further references are to this edition, cited by page number in the text.

183. Gray, *Johnson's Sermons*, p. 216.
184. For Johnson's position on habits, see Paul Alkon, Robert South, William Law, and Samuel Johnson, in *Studies in English Literature*, 6, 1966, pp. 499–528.
185. Alkon *Johnson and Moral Discipline*, p. 195.
186. See *Spectator* 455; and *Freethinkers* 29, 74, 102, 139.
187. The language of uprooting has traditionally been associated with 'radicalism' (to the roots) and was certainly so used by Paine: see S. Daniels, The political iconography of Woodland in later Georgian England, in D. Cosgrove and S. Daniels (eds) *The Iconography of Landscape: essays on the symbolic representation, design and use of past environments* (Cambridge, 1988), pp. 43–82; and Jack Fruchtman, *Thomas Paine and the Religion of Nature* (Baltimore, 1993), Ch. 1 and Ch. 3. Johnson's conservative use of this imagery does follow a tradition: 'the imagery of radicality, however, was also deployed to focus on what was fundamentally important and thus worthy of conservation'. Conal Condren, *The Language of Politics in Seventeenth-Century England* (London, 1994), p. 156. For a seventeenth-century example, see Hill, *English Bible*, p. 151.
188. Chambers (and Johnson) *Law Lectures*, i. 83–84; see also E.L. McAdam, *Dr Johnson and the English Law* (Syracuse, 1951) p. 82 on this passage. For this natural language of conservatism, see Ian Harris, Paine and Burke: God, Nature and Politics, in Michael Bentley (ed.) *Public and Private Doctrine: essays in British history presented to Maurice Cowling* (Cambridge, 1993), pp. 34–62. This 'discourse of custom' as John Barrell, *The Political Theory of Painting from Reynolds to Hazlitt: 'The Body of the Public'* (New Haven, 1986), esp. pp. 136–58, styles it, is also found in sermons 12 and 24.
189. Alkon, *Johnson and Moral Discipline*, p. 183.
190. Wimsatt, *Philosophic Words*, p. 113.
191. For Johnson on natural law and natural evidence, see Hudson, *Johnson and Eighteenth-Century Thought*, p. 46, p. 21; Gray, *Johnson's sermons*, p. 68; Chapin *Religious Life*, p. 76; and Donald Greene, Samuel Johnson and Natural Law, in *Journal of British Studies*, 2, 1962, pp. 59–75.
192. Johnson makes the same point in the *Journey*, p. 118. A very similar argument is put forward by John Ray in the Preface to his *The Wisdom of God Manifested in the Works of the Creation* (London, 1691).
193. See Chapin, *Religious Life*, p. 80; see also *Sermons*, p. 29 and p. 107.
194. See Hudson, *Johnson and Eighteenth-Century Thought*, p. 139, p. 147.
195. For Johnson on church ceremonial, see Chester F. Chapin, Religion and the Nature of Johnson's Toryism, in *Cithara Essays in the Judaeo-Christian tradition*, 29, 1990, pp. 38–54, at p. 41; and G.M. Ditchfield, Dr Johnson and the dissenters, in *Bulletin of the John Rylands Library*, 68, 1986, pp. 373–409, at p. 380.
196. Johnson, *Political Writings*, p. 335.
197. As such, natural imagery partakes in the far more general trend for late eighteenth-century Toryism to use the frameworks established by early eighteenth century Whiggism, see James J. Sack *From Jacobite to Conservative: reaction and orthodoxy in Britain c. 1760–1832* (Cambridge, 1993), and Nockles, *Oxford Movement in Context*, Ch. 1.
198. Johnson, *Political Writings*, p. 436, see also pp. 445–47.
199. For the general response of the English establishment to American claims to politcal representation, see J.C.D. Clark, *The Language of Liberty, 1660–1832: political*

discourse and social dynamics in the Anglo-American World, (Cambridge, 1994). For the long history of the geographical nature of political debate in the Anglo-American context, see Jack P. Greene, *Peripheries and Center: constitutional development in the extended polities of the British Empire and the United States, 1607–1788* (Athens, Ga., 1986). Miller, *Defining the Common Good,* is particularly useful on this point: Johnson's 'metrocentric' argument about the irrelevance of place was that taken by the defenders of colonization of America (pp. 240–41, 382–83), whilst those supporting American independence argued that liberty required government responsive to locality (p. 260). For the latter view, see Thomas Paine, *Rights of Man, Common Sense and other Political Writings,* ed. Mark Philp (Oxford, 1995), p. 24 and Richard Price, *Political Writings,* ed. D.O. Thomas (Cambridge, 1991), pp. 30–35. Miller sees this as a debate in political rhetoric over Cicero's common good; the point made by Horace's *Epistles* about the relation of mind and place may also have been relevant to defenders of the Empire such as Johnson.

200. Rolf P. Lessenich, *Elements of Pulpit Oratory in Eighteenth Century England (1660–1800)* (Cologne 1972), pp. 11–12. See also W. Fraser Mitchell, *English Pulpit Oratory: from Andrewes to Tillotson* (London, 1932), p. 136, p. 349, and pp. 376–77; and John Spurr, *The Restoration Church of England, 1646–89* (New Haven, 1991), pp. 391–93.

201. James Downey, *The Eighteenth-Century Pulpit: a study of the Sermons of Butler, Berkeley, Secker, Sterne, Whitefield, and Wesley* (Oxford, 1969), p. 19.

202. Bloom and Bloom, *Addison's Sociable Animal,* pp. 176–78.

203. For other examples see Robert South, *Twelve Sermons Preached Upon Several Occasions* (6 vols, 4th edn, London 1718), iii. 223 and iv. 155; Samuel Clarke, *Sermons,* ed. John Clarke (10 vols, London, i–ii, 3rd edn, 1732, iii–x, 2nd edn, 1731), i. 267 and vii. 196–97.

204. John Rogers, *Nineteen Sermons on Several Occasions* (London, 1735), p. 96; see also Clarke, *Sermons,* vi. 209. See also South, *Twelve Sermons,* ii. 258; and Richard Baxter, *The Divine Life* (London, 1664), p. 38.

205. Richard Bentley, *Eight Sermons Preach'd at the Honourable Robert Boyle's Lecture, in the First Year MDCXCII* (6th edn, Cambridge, 1735), p. 11. On this metaphor, see Stephen Daniels and Denis Cosgrove, Spectacle and Text: Landscape Metaphors in Cultural Geography, in James Duncan and David Ley (eds) *Place/ Culture/Representation* (London, 1993), pp. 57–77.

206. Barrow, *Works,* i. 2.

207. John Tillotson, *Works* (3 vols, 5th edn, London 1735), i. 77–78.

208. Jeremy Taylor, *A Course of Sermons for all the Sundays of the Year* (London, 1826 [1st edn, 1653]), ii. 71. Both Taylor and Johnson may have derived this image from Horace, *Satires,* I. i. 54–60.

209. See Quentin Skinner, *Reason and Rhetoric in the Philosophy of Hobbes* (Cambridge, 1996).

210. Quintilian, *Institutio Oratoria,* trans., H.E. Butler (London, 1921), Bk. IX. II. 44. See Skinner *Reason and Rhetoric,* pp. 182–88.

211. Lessenich, *Elements of Pulpit Oratory,* p. 165; and Spurr *Restoration Church,* p. 251.

212. Clarke, *Sermons,* i. 14; see also South, *Twelve Sermons,* i. 19, 51; and Tillotson, *Works,* ii. 551–54.

213. Clarke, *Sermons,* vi. 223.

214. Barrow, *Works,* ii. 138–39.

215. Richard Baxter, *The Reasons of the Christian Religion* (London, 1667), p. 192. See also, *idem, Divine Life,* pp. 9–10. See Mitchell, *English Pulpit Oratory,* p. 270, on Baxter and natural philosophy.

216. Barrow, *Works*, ii. 67.
217. South, *Twelve Sermons*, i. 44.
218. South, *Twelve Sermons*, ii. 203.
219. Clarke, *Sermons*, vi. 306. Boyle's *Free Enquiry* said the same; see Chapter 4 and Chapter 7.
220. Taylor, *Course of Sermons*, ii. 405.
221. Baxter, *Divine Life*, p. 35.
222. Taylor, *Twelve Sermons*, i. 159–60.
223. Baxter, *Divine Life*, p. 84. Baxter alludes here to the parable of the talents, Matthew xxv. 14–30, which was also vital to Johnson's religious life (see Paul Fussell, *Samuel Johnson and the Life of Writing*, London, 1972, pp. 95–97; and Charles Pierce, *The Religious Life of Samuel Johnson*, London, 1983, p. 105) and undoubtedly influenced his response to retirement in *Rambler* 135 and *Adventurer* 126.
224. Rogers, *Nineteen Sermons*, p. 86.
225. Taylor, *Course of Sermons*, i. 520.
226. Mitchell, *English Pulpit Oratory*, p. 313.
227. Hudson, *Johnson and Eighteenth-Century Thought*, p. 8.
228. See Gwin J. Kolb, The 'Paradise' in Abyssinia and the 'Happy Valley' in *Rasselas*, in *Modern Philology*, 56, 1958, pp. 10–16; Donald Lockhart, 'The Fourth Son of the Mighty Emperor': The Ethiopian Background of Johnson's *Rasselas*, in *Proceedings of the Modern Language Association of America*, 78, 1963, pp. 516–28; and Arthur Weitzman, More light on *Rasselas*: the background to the Egyptian episodes, in *Philological Quarterly*, 48, 1969, pp. 42–58.
229. For example, see W.K. Wimsatt, In Praise of *Rasselas*: Four Notes (converging), in Maynard Mack and Ian Gregor (eds) *Imagined Worlds: essays on English novels and novelists in honour of John Butt* (London, 1968); Gwin J. Kolb, The structure of *Rasselas*, in *Proceedings of the Modern Language Association of America*, 66, 1951, pp. 698–717; Harold Pagliaro, Structural patterns of control in *Rasselas*, in John Middendorf (ed.) *English Writers of the Eighteenth Century* (New York, 1971), pp. 208–29; and Emrys Jones, The artistic form of *Rasselas*, in *Review of English Literature*, n.s. 18, 1967, pp. 387–401.
230. Campbell, Image and Symbol in *Rasselas*, p. 267.
231. Frederick Keener, *The Chain of Becoming: the philosophic tale, the novel, and a neglected realism of the Enlightenment: Swift, Montesquieu, Voltaire, Johnson and Austen* (New York, 1983).
232. Samuel Johnson, Rasselas, in *Rasselas and Other Tales*, ed. Gwin J. Kolb (New Haven, 1990), p. 76. All further references are to this edition, by page number in the text.
233. See Bate, *Achievement of Johnson*, Ch. 2, and Sachs, *Passionate Intelligence*, Ch. 1.
234. Ellis Cornelia Knight, *Dinarbas; A Tale: being a continuation of Rasselas, prince of Abyssinia*, ed. Lynne Meloccaro (London, 1994), p. 126. Future references are to this edition and are inserted in the text.
235. See also Pekuah after her capture by the Arab, p. 135, and the group on their return to Cairo, looking at the Nile by moonlight, p. 154.
236. This is noted by Mahmoud Manzalaoui, Rasselas and some mediaeval ancillaries, in Magdi Wahba (ed.) *Bicentenary Essays on Rasselas* (Cairo, 1959), pp. 59–73, at p. 62.
237. *Rasselas* was published in the same year as Burke's *Enquiry into the Origin of the Sublime and the Beautiful*, which Johnson called 'an example of true criticism' (*Life*, ii. 90). Burke probably wrote the review of *Rasselas* in the *Annual Register*.

See Alkon, Illustrations of Rasselas, p. 19, and W.B. Conarchan *Confinement and Flight*, pp. 154–56.

238. See Kolb, Paradise in Abyssinia, Lockhart, Fourth son, and Marlene R. Hansen, The Happy Valley: a version of hell and a version of pastoral, in *New Rambler*, 14, 1973, pp. 24–30.

239. Tomarken, *Choice of Criticism*, pp. 43–45.

240. Tomarken, *Choice of Criticism*, p. 45 points out that 'Johnson liked to retire to a summer house at Streatham and ... visited Shenstone's garden, the Leasowes', but this is of no moment: Johnson's visits to Streatham only began with his acquaintance with the Thrales in 1765, and his only recorded visit to the Leasowes came in 1774 (see *Life*, i. 390 for Thrales; and *Diaries*, pp. 218–19 for the Leasowes). None of this could have influenced his descriptions in 1759.

241. For a parallel to this in Boethius's *Consolation of Philosophy*, see James Woodruff, Rasselas and the tradition of Menippean satire, in Isobel Grundy (ed.) *Samuel Johnson: new critical essays* (London, 1984), pp. 158–85, at pp. 177–8.

242. Robert Walker, *Eighteenth-Century Arguments for Immortality and Johnson's Rasselas* (Victoria, 1977), pp. 36–41.

243. Nekayah responds similarly to life in the lower orders: 'with these girls she played as with inoffensive animals' (p. 92).

244. See *Dinarbas*, Chs. 16 and 17, and note 1 to Ch. 16.

245. Walker, *Arguments for Immortality*, pp. 46–47.

246. Agostino Lombardo, The Importance of Imlac, in Wahba (ed.) *Bicentenary Essays*, pp. 31–49 at p. 48.

247. Said, *Orientalism*, p. 65.

248. Ibid., p. 1, 12.

249. Ibid., p. 42.

250. Ibid., p. 45. See also Gascoigne, *Banks and English Enlightement*, p. 171, and Thomas *Man and the Natural World*, p. 135.

251. See Chapter 6, p. xx.

252. Geoffrey Tillotson, Time in *Rasselas*, in Wahba (ed.) *Bicentenary Essays*, pp. 97–103, at p. 102; Kolb Structure of *Rasselas*, pp. 705–706.

253. Tomarken, *Choice of Criticism*, p. 102; and Walker, *Arguments for Immortality*, p. 43.

254. Walker, *Arguments for Immortality*, p. 55; and Richard B. Schwartz, Johnson's philosopher of nature: *Rasselas*, Chapter 22, in *Modern Philology*, 74, 1976, pp. 196–200, at p. 198.

255. Gwin Kolb, Rousseau and the background to the 'life led according to nature' in Chapter 22 of *Rasselas*, in *Modern Philology*, 73 , 1976, (supplement), S. 66–73; and R.P. Kaul, The philosopher of nature in *Rasselas* XXII, in *Indian Journal of English Studies*, 3, 1962, pp. 116–20. Miller, *Defining the Common Good*, pp. 136–37, 143–45, 284, points out that opposition to the notion of a life led according to nature came from High Churchmen in particular.

256. The same contradiction, ascribing human motivations of happiness to animals and prioritizing animal instincts in man, was also attacked by Johnson in his review of Jenyns's *Free Enquiry*: 'Perfection or imperfection of unconscious beings has no meaning as referred to themselves ... *Pope* might ask the *weed* why it was less than the *Oak*, but the *weed* would never ask the question for itself.' *Literary Magazine*, 1757, p. 172. Johnson returned to the theme in his 'Life of Pope' (*Lives*, iii. 243–44) when attacking the 'penury of knowledge' in the *Essay on Man*, see Chapter 8. Johnson's attitude to this issue once more shows his reluctance to see the boundaries in the chain of being blurred; see Chapter 5.

257. Edward Bloom, *Samuel Johnson in Grub Street* (Providence, 1957), p. 187.
258. Skinner, *Reason and Rhetoric*, pp. 198–211.
259. See Gwin Kolb, Johnson's dissertation on flying, in Hilles (ed.) *New Light on Dr Johnson*, pp. 91–106; and Louis A. Landa, Johnson's feathered man: 'A Dissertation on the Art of Flying' considered, in W.H. Bond (ed.) *Eighteenth-Century Studies in Honour of Donald Hyde* (London, 1970), pp. 161–78.
260. 'Life cannot be surveyed with the same safety as Nature' (*Letters*, iv. 198). See also Johnson's 'Life of Pope' (*Lives*, iii. 210) for his attack on Pope's pretence to look 'with gay indifference' on human actions, and his 'Life of Cowley' (*Lives*, i. 20) for an attack on the idea that the poet can adopt the position of an Epicurean deity, overlooking the prospect of human misery. However common the theatre metaphor in landscape viewing – see Daniels and Cosgrove, Spectacle and text – it is rare in Johnson. Where he uses it in the review of Jenyns, the point is to deflate the hubris of the human overview, by discussing superior beings looking on man's suffering as man does at a cockpit.
261. On this discussion see Walker, *Arguments for Immortality*, pp. 49–50.
262. Keener, *Chain of Becoming*, p. 239.
263. See in particular Schwartz, *Johnson and Evil*, and Sachs, *Passionate Intelligence*, Ch. 2. Quotes from the review are by page number in the text.
264. Hudson, Three steps to perfection, p. 29. For *Rasselas* and religious sources, see also Thomas Preston, The Biblical Context of Johnson's Rasselas, *Proceedings of the Modern Language Association of America*, 84, 1969, pp. 274–81.
265. Everett, *Tory view*, pp. 12–17.
266. See Chapter 6.
267. Steven Scherwatzky, Johnson, *Rasselas* and the politics of empire, in *Eighteenth-Century Life*, 16, 1992, pp. 103–13, at p. 112.
268. *Life*, i. 68. John Hawkins, *Life Of Johnson* (2nd edn, London, 1787), p. 564, pointed to the influence of Law on Johnson. He also said of Johnson's evaluation of Jeremy Taylor that 'he placed the author at the head of all divines that have succeeded the fathers' (p. 542).
269. On this see, Nockles, *Oxford Movement in context; idem* Church parties in the pre-Tractarian Church of England, 1750–1833: the 'Orthodox' – some problems of definition and identity, in John Walsh, Colin Haydon and Stephen Taylor (eds) *The Church of England, c. 1689–1833: from Toleration to Tractarianism*, pp. 334–59; and Kenneth Hylson-Smith, *High Churchmanship in the Church of England: from the sixteenth century to the late twentieth century* (Edinburgh, 1993).

Chapter 7

1. A.D. Horgan, *Johnson on Language: an introduction* (London, 1994).
2. See Chapter 6.
3. W.K. Wimsatt, *Philosophic Words: a study of style and meaning in the Rambler and dictionary of Samuel Johnson* (New Haven, 1948), p. 113.
4. Barbara Stafford, *Voyage into Substance: art, science, nature and the illustrated travel account, 1760–1840* (Cambridge, Mass., 1984), p. 156.
5. Thomas Curley, *Samuel Johnson and the Age of Travel* (Athens, Ga., 1976).
6. Morris Brownell, *Samuel Johnson's Attitude to the Arts* (Oxford, 1989).
7. Samuel Johnson, *Voyage to Abyssinia*, ed. Joel J. Gold (New Haven, 1985), p. 3. All further references (included in the text) are to this edition, by page number.

8. Robert Walker, Johnson in the 'Age of Evidences', in *Huntington Library Quarterly*, 44, 1980–81, pp. 27–42.
9. Paul Fussell, *The Rhetorical World of Augustan Humanism: ethics and imagery from Swift to Burke* (Oxford, 1965).
10. Joel J. Gold, Introduction to *Voyage to Abyssinia*, p. liii–iv.
11. Joel J. Gold, Johnson's Translation of Lobo, in *Proceedings of the Modern Language Association of America*, 80, 1965, pp. 51–61, at p. 59.
12. See Chapter 6, p. xx.
13. For which more generally, see Colin Haydon, *Anticatholicism in Eighteenth-Century England, c. 1714–80: A Political and Social Study* (Manchester, 1993).
14. Percy Adams, *Travellers and Travel Liars, 1660–1800* (Berkeley, 1962), p. 196.
15. Richard Helgerson, *Forms of Nationhood: the Elizabethan writing of England* (Chicago, 1992), Ch. 4.
16. I owe this distinction to Horgan, *Johnson on Language*, pp. 192–93, n. 43.
17. Gold, Introduction, p. xiv-vi.
18. See pp. 230–40.
19. See pp. 243–57.
20. Whilst the reviews pose few problems, the prefaces were not necessarily written to express Johnson's opinions, but rather those of the author Johnson wrote the preface for. As such, the prefatory material is seen as expressing Johnson's opinions only to the extent that it coincides with his opinions expressed elsewhere, particularly in his contemporaneous book reviews.
21. Samuel Johnson, Review of 'Essay on the Description of China in Two Volumes Folio'. From the French of Pere Du Halde, in *Gentleman's Magazine*, XII, 1742, pp. 320–23, 353–57, 484–86, at p. 320. Further references by page number are incorporated in the text.
22. Samuel Johnson, Review of 'The Civil and Natural History of Jamaica', by Patrick Browne, in *The Literary Magazine, or Universal Review*, IV. 1756, pp. 176–85, at p. 176. Similarly, Johnson called Borlase's *Observations* 'one of the most pleasing and elegant pieces of local enquiry' on the basis of its accuracy. See Samuel Johnson, Review of 'Observations on the ancient and present state of the islands of Scilly', in *Literary Magazine*, II. 1756, pp. 91–97, at p. 91.
23. *Samuel Johnson's Prefaces and Dedications*, ed. Allen T. Hazen (New Haven, 1937), pp. 170–71. All future references are to this edition, in the text. These qualifications were particularly important to travel in America, because in Johnson's opinion 'there is little to be observed [there] except natural curiosities'. Samuel Johnson, *Letters*, ed. Bruce Redford (5 vols, Oxford, 1992–94) i. 203. All further references are to this edition and are incorporated in the text.
24. Joseph Levine, *The Battle of the Books: history and literature in the Augustan age* (Ithaca, 1991), p. 2. The extent to which geography was a science in Johnson's lifetime is questioned in Robert Mayhew, *Enlightenment Geography: the political languages of British geography, 1650–1850* (London, 2000).
25. John Gascoigne, *Joseph Banks and the English Enlightenment: useful knowledge and polite culture* (Cambridge, 1994). The assertion that geography provided useful knowledge had been a commonplace since Strabo's *Geography* (c. 7 B.C.), Book 1.1. See *The Geography of Strabo* (Loeb Library, 8 vols, London, 1917–32), i. 3–5, 29–33, 39.
26. James Boswell, *Life of Johnson*, ed. G.B. Hill and rev. L.F. Powell (6 vols, Oxford, 1934–50), ii. 225. All further references are to this edition, cited in the text.
27. See also Johnson's comment on slavery in *Johnson on Shakespeare*, ed. Arthur Sherbo (2 vols, New Haven, 1968), p. 227.

28. In the *World Displayed*, Johnson's assessment was that whilst 'much knowledge has been acquired, and much cruelty been committed . . . there is reason to hope that out of so much evil good may sometime be produced, and that the light of the gospel will at last illuminate the sands of Africa, and the desarts of America . . . ' (*Prefaces*, p. 228).

29. Samuel Johnson, *Journey to the Western Islands of Scotland*, ed. J.D. Fleeman (Oxford, 1775), p. 16. All further references are to this edition and are incorporated in the text.

30. Mary Hanway, *A Journey to the Highlands of Scotland, with Occasional Remarks on Dr Johnson's Tour* (London, 1775), p. 49.

31. Stafford, *Voyage into substance*, p. 402.

32. M. Hanway, *Journey*, pp. vii–viii.

33. Fussell discusses this Augustan pattern of argument in travel in *Rhetorical World of Augustan Humanism*; and *idem*, Patrick Brydone: the eighteenth-century traveller as representative man, in *Bulletin of the New York Public Library*, 66, 1962, pp. 349–63.

34. Johnson, *Review of Patrick Browne*, IV. p. 176.

35. Much the same is said at Aberdeen: 'such an unnecessary description would have the appearance of a very frivolous ostentation' (p. 10).

36. At the distance of a decade, Johnson could still say he 'got an acquisition of more ideas by it than by any thing' (*Life*, iv. 199), and Boswell describes Johnson's recollection of his first important tour, with Reynolds to Devon, in the same terms as an 'accession of new ideas' (ibid., i. 377).

37. When shown 'natural curiosities' (as opposed to useful nature), his response was dismissive: 'There are so many more important things, of which human knowledge can give no account, that it may be forgiven us, if we speculate no longer on two stones in *Col*' (*Journey*, p. 104).

38. James Boswell, *The Journal of a Tour to the Hebrides with Samuel Johnson, LLD.*, p. 13 (see also p. 313) Vol 5 of Hill/Powell *Life of Johnson*, hereafter cited as *Tour* in the text.

39. Curley, *Johnson and Age of Travel*, p. 69.

40. Quentin Skinner, Meaning and understanding in the history of ideas, reprinted in James Tully (ed.) *Meaning and Context: quentin skinner and his critics* (Cambridge, 1988), pp. 29–67, at p. 46.

41. Although further discussion in this chapter will suggest there were differences between the doctrines of the two.

42. Curley, *Johnson and the Age of Travel*, at p. 68, p. 69, and p. 70.

43. George C. Brauer, *The Education of a Gentleman: Theories of Gentlemanly Education in England, 1660–1775* (New York, 1959), Ch. 6.

44. James Howell, *Instructions and Directions for Forreine Travell* (London, 1650 edn), p. 17.

45. See Fussell, *Rhetorical World of Augustan Humanism*.

46. Howell, *Instructions*, pp. 9–11; Josiah Tucker, *Instructions for Travellers* (Dublin edn, 1758), p. 5.

47. The phrase is in Tucker, *Instructions*, p. 3; cf. Howell, *Instructions*, p. 32.

48. Howell, *Instructions*, pp. 94–95; Tucker, *Instructions*, p. 3.

49. Curley, *Johnson and the Age of Travel* (pp. 76–78) has briefly explored the same approach, his concern being largely with the style rather than the content of the works Johnson referred to. Pat Rogers, *Johnson and Boswell: The Transit of Caledonia* (Oxford, 1995), pp. 74–82, does the same for Johnson's comments on Baretti and Burney.

50. A procedure also followed by Blainville, Twiss and Brydone, all of whom Johnson commended (*Life*, ii. 345–46).
51. John Bell, *Travels from St Petersburgh in Russia, to various parts of Asia* (2 vols, Edinburgh, 1788 [1st edn, 1763]) i, p. vii. P. de Blainville, *Travels through Holland, Germany, Switzerland, but especially Italy* (Eng. trans., 3 vols, London, 1757) was also particularly insistent in this respect, continually correcting the errors of previous travellers such as François Misson at the cascade of Terni (ii. 303–304) and Gilbert Burnet at the gardens of Frescati (ii. 346), 'even to the point of distorting his own descriptions of these places by the bulk of his censures. His insistent corrections are also noted by Adams, *Travellers and Travel Liars*, p. 179.
52. Patrick Browne, *The Civil and Natural History of Jamaica* (2nd edn, London, 1789), p. viii.
53. Blainville, *Travels*, iii. 461.
54. Richard Twiss, *Travels through Portugal and Spain, in 1772 and 1773* (London, 1775), p. i–ii. Johnson himself had said to Boswell 'there is a good deal of Spain that has not been perambulated' (*Life*, i, 409–10) and was said by Giuseppe Baretti, *A Journey from London to Genoa through England, Portugal and Spain* (4 vols, London, 1770), i, p. vi, to have 'exhorted me to write daily, and with all possible minuteness...as I was setting out on my first journey to Spain...', a view confirmed in Johnson's letter to Baretti (*Letters*, i. 200), saying 'I wish you had staid longer in Spain, for no country is less known to the rest of Europe.'
55. Blainville, *Travels*; quotes at i, p. iii, p. v, p. vi, and pp. vi–viii.
56. John George Keysler, *Travels through Germany, Bohemia, Hungary, Switzerland, Italy, and Lorrain* (trans., 4 vols, London, 1756), i. 7.
57. See, for example, Patrick Brydone, *A Tour through Sicily and Malta, in a series of letters to William Beckford, Esq.* (2 vols [cont. paginated] Dublin, 1773 edn), p. 49.
58. Bell, *Travels*, i. 366.
59. Keysler, *Travels*, i. 305. Johnson's praise of Charles Burney's musical travels relied on his second trip of 1773 to Germany which had discussed the landscape and its connection to human happiness: see Roger Lonsdale, *Dr Charles Burney: a literary biography* (Oxford, 1965), pp. 129–30 and p. 126.
60. Bell, *Travels*, i. 402 ff.
61. Keysler, *Travels*, ii. 447. See also Brydone, *Tour*, pp. 88–89.
62. Blainville, *Travels*, iii. 484, records how 'in vain the Head and Blood of the mighty Saint *Januarius* were exposed several times; Mount *Vesuvius* being so firmly determined to burst forth, it laughed at the Relicks of the Holy Bishop'.
63. Twiss, *Travels*, p. 174
64. Ibid., p. ii and p. 338.
65. Ibid., p. 193. Similarly, Keysler, *Travels*, iv. 163, commented on the fossils and spars in the mines at Ilmenauer, that 'some have gone so far as to imagine they can distinguish in such stones...*Noah's* ark, a crucifix, *Moses's* face, a death's head, *Luther's* portrait and the like; which may be urged on occasion, as *argumenta ad hominem* [i.e. versus Roman Catholics], but have no other use or foundation in nature.'
66. Gascoigne, *Banks and the English Enlightenment*, esp. pp. 36–37.
67. Keysler, *Travels*, i. 113.
68. Bell, *Travels*, ii. 57. See also above, note 10.
69. Blainville, *Travels*, i. 261–62.
70. Keysler, *Travels*, i. 114.
71. See Chapter 5.

72. *Life*, iii. 356. See also *Life*, ii. 346 for Johnson's positive assessment of Brydone as a travel writer.
73. Brydone, *Tour*, pp. 23, 60, 68, 91, and 184.
74. Peter B. Nockles, *The Oxford Movement in Context: Anglican high churchmanship 1760–1857* (Cambridge, 1994), p. 104.
75. J.H. Newman's, *Idea of a University, Defined and Illustrated* (eds Martin Svaglic, Notre Dame, 1982) rehearsed similar arguments as late as 1852.
76. As at Rosetto [Rashid]: 'The fine country of Delta, on the other side of the Nile, and two beautiful islands a little below the town, make the prospect very delightful; the country to the north is improved with most pleasant gardens of oranges, lemons, and citrons . . . and when the fields are green with rice, which is cultivated here, it adds a great beauty to the country.' Richard Pococke, *A Description of the East, and some other Countries* (2 vols, London, 1743) i. 14.
77. Ibid., For example the temple of Jupiter at Thebes, i. 91–96; and correcting Strabo, i. 88.
78. Richard Chandler, *Travels in Asia Minor: or, An Account of a Tour made at the expense of the Society of Dilettanti* (Dublin, 1775), p. viii.
79. Ibid., p. 5
80. Nathaniel Wraxall, *A Tour through some of the Northern Parts of Europe* (Dublin, 1776), p. 212. See also pp. 54–56 for a prospect in similar circumstances.
81. Georg Forster, *A Voyage Round the World, in his Britannic majesty's sloop Resolution, commanded by Capt. James Cook, during the years 1772, 3, 4, and 5* (2 vols, London, 1777) i. pp. xi–xii.
82. Ibid., i. 30.
83. John Wesley, *Journal*, ed. Nehemiah Curnock (8 vols, London, 1909–16) vi. 521, later commented 'I employed some part of my leisure time in reading Mr Forster's *Voyage Round the World*. In many parts of this one would think he was almost persuaded to be a Christian.'
84. Forster, *Voyage*, p. xii.
85. As with the flying fish incident, Johnson's complaint is only that this is inappropriate to a factual account. Forster's 'justifying the ways of God' came from Milton's *Paradise Lost*, in which context Johnson approved of the aim (*Lives of the Poets*, ed. G.B. Hill, Oxford, 1905, i. 171).
86. Forster, *Voyage*, i. 3; see also i. 100.
87. Jonas Hanway, *An Historical Account of the British Trade over the Caspian Sea: with a journal of travels* (4 vols, London, 1753) i. p. vii and xi.
88. James Stephen Taylor, *Jonas Hanway: founder of the Marine Society: charity and policy in eighteenth-century Britain* (London, 1985), p. 24.
89. Hanway, *Historical Account*, i. 261.
90. Ibid., i. 238.
91. Ibid., i. 174.
92. Jonas Hanway, *A Journal of Eight Days Journey from Portsmouth to Kingston upon Thames* (2nd edn, 2 vols, London, 1757), i. 95.
93. Ibid., i. 5–6, and i. 3.
94. Ibid., i. 22.
95. Samuel Johnson, Review of A Journal of Eight Days . . . To which is added an essay on tea, in *Literary Magazine*, VII, 1756, pp. 335–42, and XIII, 1757, pp. 161–67, at p. 335.
96. Hanway, *Journal*, i. 9 and i. 143.
97. Ibid., i. 317–18.
98. See Chapter 6.

99. Hanway, *Journal*, i. 65–66. For Johnson's dislike of Shaftesbury, see his citation of Gray in *Lives*, iii. 432.

100. Taylor, *Jonas Hanway*, p. 139.

101. Charles Batten, *Pleasurable Instruction: form and convention in eighteenth-century travel literature* (Berkeley, 1978), p. 110. Johnson's contemporary biographer Sir John Hawkins, by no means sympathetic to Johnson's High Churchmanship, could attack Hanway's writing 'for [the] triteness and inanity of the sentiments'. John Hawkins, *The Life of Samuel Johnson* (2nd edn, London, 1787), p. 360.

102. See also Johnson, *Journey*, p. 48.

103. For discussions of Johnson on money and emigration in the *Journey*, see Kevin Hart, Economic Acts: Johnson in Scotland, in *Eighteenth-Century Life*, 16, 1992, pp. 94–110; Karen O'Brien, Johnson's view of the Scottish Enlightenment in Journey to the Western Islands of Scotland, in *The Age of Johnson: a scholarly annual*, 4, 1991, pp. 59–82; and John B. Radner, The significance of Johnson's changing views of the Hebrides, in John Burke and Donald Kay (eds) *The Unknown Samuel Johnson* (Madison, 1983), pp. 131–49. Of more importance to the present discussion, however, is the position of this debate within the broader patterning of Johnson's consecutive argument.

104. See Johnson, *Journey*, p. 85 for initial impressions on settlement on Skye.

105. Arthur Sherbo, Some animadversions to Patrick O'Flaherty's Journey to the Western Islands of Scotland, in *Studies in Burke and his Time*, 13, 1971, pp. 2119–27 at p. 2124.

106. Again, the Ossian controversy is beyond the scope of this discussion except for its structural situation in Johnson's argument.

107. Thomas Jemielity, 'More in notions than in facts': Samuel Johnson's Journey to the Western Islands, in *The Dalhousie Review*, 49, 1969, pp. 319–30. See also Patrick O'Flaherty, Johnson in the Hebrides: philosopher becalmed, in *Studies in Burke and his time*, 13, 1971, pp. 1986–2001.

108. Curley, *Johnson and the Age of Travel*, pp. 197–200; at p. 197.

109. *Literary Magazine*, IV, pp. 193–97. Reviewing *Philosophical Transactions*, vol. xlix.

110. Robert Boyle, *General Heads for the Natural History of a Country, Great or Small; Drawn out for the use of Travellers and Navigators* (London, 1692 ed.), p. 9.

111. O'Flaherty, Johnson in the Hebrides, p. 2001; see also Ann Schalit, Literature as product and process: two differing accounts of the same trip, in *Serif: The Kent State University Library quarterly*, 4, 1967, pp. 10–17; and Greg Clingham *Boswell's The Life of Johnson* (Cambridge, 1992), p. 34.

112. Richard Schwartz, Johnson's Journey, in *Journal of English and Germanic Philology*, 69, 1970, pp. 292–303, at p. 297.

113. See also Francis Hart, Johnson as philosophic traveller: the perfection of an idea, in *Journal of English Literary History*, 36, 1969, pp. 679–95, at pp. 693–94.

114. Radner, Changing views, in Burke and Kay – full reference above, n. 105; see also Curley, *Johnson and the Age of Travel*, pp. 184, 188, 199.

115. Radner, Changing Views, p. 138.

116. As such, Edward Tomarken, Travels into the unknown: Rasselas and A Journey to the Western Islands of Scotland, in Burke and Kay (eds) *Unknown Samuel Johnson*, pp. 150–67, is right to say 'locality . . . is particularly important in the Hebrides because the landscape is a formidable obstacle to development.' (p. 163.). But this does not prove that Johnson's views underwent great change during the journey, only that the principles he went with – unsurprisingly given their demand for close observation – were responsive to situation.

117. Thomas Pennant, *A Tour in Scotland and Voyage to the Hebrides; MDCCLXXII* (2 vols, Chester, 1774), i. 305. We know Johnson had read this volume as he responds to an error in it (at i. 270) in the *Journey* (p. 108). On their relation see Ralph E. Jenkins, 'And I travelled after him': Johnson and Pennant in Scotland, in *Texas Studies in Literature and Language*, 14, 1972, pp. 445–62 and a good corrective by Thomas Jemielity, Thomas Pennant's Scotland Tours and *A Journey to the Western Islands of Scotland*, in Prem Nath (ed.) *Fresh Reflections on Samuel Johnson: Essays in Criticism* (New York, 1987), pp. 312–27.

118. William Borlase, *Observations on the Ancient and Present state of the Islands of Scilly* (Oxford, 1756), p. 70.

119. Robert G. Walker, Johnson and the Trees of Scotland, in *Philological Quarterly*, 61, 1982, pp. 98–101.

120. For this controversy, see Helen L. McGuffie, *Samuel Johnson in the British press, 1749–84. A Chronological Checklist* (New York, 1976). For the importance of trees to eighteenth-century debates, see Stephen Daniels, The Political Iconography of Woodland in late Georgian England, in Denis Cosgrove and Stephen Daniels (eds) *The Iconography of Landscape: essays on the symbolic representation, design and use of past environments* (Cambridge, 1988), pp. 43–82.

121. Rogers, *Transit of Caledonia*, Ch. 5, has used the same technique, but his focus of attention was more general than that of the present discussion.

122. Philip Edwards, *The Story of the Voyage: sea narratives in eighteenth-century England* (Cambridge, 1994), p. 6.

123. Johnson, *Journey*, p. xxxviii.

124. For like reasons, the content of Johnson's two Latin odes to Mrs Thrale from Scotland made them inappropriate. Johnson also made such personal references in his French and Welsh journals, which undoubtedly would have been excised, such as the following at the Palais Bourbon: 'as I entred my Wife was in my mind. She would have been pleased; having now nobody to please I am little pleased'. Samuel Johnson *Diaries, Prayers, and Annals*, ed. E.L. McAdam with Donald and Mary Hyde (New Haven, 1958), p. 238. Both Johnson's French and Welsh journals are in this work, which is hereafter referenced in the text.

125. Hart, Johnson as Philosophic Traveller, p. 693.

126. Batten, *Pleasurable Instruction*, p. 39.

127. Also noted by Rogers, *Transit of Caledonia*, pp. 123–29.

128. At Dunvegan Johnson discussed the barrenness of the face of the land (*Letters*, ii. 91), whilst at Talisker he discusses drink (ii. 93), food (ii. 95–96) and fuel (ii. 96–97).

129. Johnson had displayed this skill previously in his translation of Lobo's *Voyage to Abyssinia*, where his 'sense of structure led him to rearrange details, ideas, or sentences for more logical or more effective organization'. Gold, Johnson's Translation of Lobo, p. 55.

130. Two days later the same process occurs in the journey from Aberdeen to Slains Castle: where in the letters 'we went thither on the next day (Aug. 24) and found a house not old' (*Letters*, ii. 61), in the *Journey* 'the road beyond Aberdeen grew more stony, and continued equally naked of all vegetable decoration' (*Journey*, p. 13).

131. Hart, Johnson as Philosophic Traveller, p. 694.

132. See also *Diaries*, p. 241: 'trees on the road some tall, none old, many very young and small'.

133. For other instances of Johnson reading rather than viewing whilst travelling, see *Life*, iv. 109, 111–19; *Johnsonian Miscellanies*, ed. G.B. Hill (2 vols, Oxford, 1897), ii. 429 n. 2; *Diaries*, pp. 163–64, 351; and *Letters*, iv. 51, 352.

134. Brownell, *Johnson's Attitudes to the Arts*, p. 169, rightly points to the precision of this description.
135. Stafford, *Voyage into Substance*, p. 1.
136. Curley, *Johnson and the Age of Travel*, p. 105.
137. Howell, *Instructions*, p. 33. Howell, therefore was certainly not 'the specific Renaissance source' for Johnson's views on the deterministic influence of the environment on human society.
138. See O'Flaherty, Johnson in the Hebrides, p. 1992; Hart, Johnson as Philosophic Traveller, esp. p. 690; and Arthur Sherbo, Johnson's intent in the *Journey to the Western Islands of Scotland*, in *Essays in Criticism*, 16, 1966, pp. 382–97.
139. Jemielity, Pennant's tours, in Nath (ed.) *Fresh Reflections*, p. 315.
140. Thomas Pennant, *A Tour in Scotland MDCCLXIX* (2nd edn, London, 1772), pp. 179–81.
141. Ibid., pp. 182–84.
142. See Johnson, *Journey*, p. 296, *sub* Monday 30 August.
143. See for example, Pennant, *Tour and Voyage*, i. 200–12 for a history of the Hebrides, and i. 243–56 for the antiquities of Iona.
144. Martin Martin, *A Description of the Western Islands of Scotland* (London, 1703), p. 135, then pp. 135–38. Similarly Walker's *Report on the Hebrides* divided its reflections on the face of the country in Skye by topographical features. Margaret McKay (ed.) *The Rev. Dr Walker's Report on the Hebrides of 1764 and 1771* (Edinburgh, 1980), p. 204.
145. Stan A.E. Mendyk, *Speculum Britanniae: regional study, antiquarianism, and science in Britain to 1700* (Toronto, 1989), p. 46. See pp. 218–20 for Mendyk's view of Martin Martin.
146. Jemielity, More in notions than facts, p. 324.
147. Jeremy Black, *The British Abroad: the grand tour in the eighteenth century* (Stroud, 1992), p. 289.
148. Donald Munro, *Description of the Western Isles of Scotland, called Hybrides*, in *Miscellania Scotica: a collection of tracts relating to the History, antiquities, topography, and literature of Scotland, Vol II* (Glasgow, 1818), p. 126.
149. Batten, *Pleasurable Instruction*, p. 95.
150. John Knox, *A Tour Through the Highlands of Scotland, and the Hebrides Islands in MDCCLXXXVI* (London, 1787), p. lxxxii.
151. Louis Simond, *Journal of a Tour and Residence in Great Britain, During the Years 1810 and 1811* (2 vols, Edinburgh, 1815), i. 296.
152. On Coll (*Tour*, pp. 300–301) a short description started with the physical characteristics of the island before discussing its population, rank structure and the education available to islanders.
153. James Boswell, *An Account of Corsica* (Glasgow, 1768). Harbours, pp. 16–20; towns pp. 21–26.
154. Ibid., Rivers pp. 35–37; animals pp. 38–44; vegetation pp. 44–48; and mines pp. 50–53.
155. For more information on the two works, see L.F. Powell, Boswell's original journal of his tour to the Hebrides and the printed version, in *Essays and Studies*, 23, 1938, pp. 58–69.
156. Frederick Pottle and Charles Bennett (eds) *Boswell's Journal of a Tour to the Hebrides* (London, 1936). All references in the text as *Journal*.
157. Though it should be noted that Boswell was more sentimental in this respect than Johnson, retaining at least one anecdote about his daughter Veronica (*Tour*, pp. 25–26) which his publisher Malone repeatedly asked to be removed. See

Peter Baker *et al.* (eds) *The Yale Edition of the Private Papers of James Boswell. Research Edition – Correspondence Volume 4. The Correspondence of James Boswell with David Garrick, Edmund Burke, and Edmond Malone* (London, 1968).

158. Frank Brady and Frederick Pottle, *Boswell on the Grand Tour: Italy, Corsica and France, 1765–6* (London, 1955), p. 23, see also pp. 52,58; and Frederick Pottle (ed.) *Boswell on the Grand Tour: Germany and Switzerland, 1764* (London, 1953), p. 7, see also pp. 11,176. Other commentators agree: see Frederick Pottle, *James Boswell: the Earlier Years, 1740–69* (London, 1966), p. 198; Patrick Anderson, *Over the Alps: reflections on travel and travel writing, with special reference to the grand tours of Boswell, Beckford and Byron* (London, 1969), p. 24.

159. Boswell, *Grand Tour, Italy*, p. 127.

160. Clingham, *Boswell's Life*, p. 84.

161. Richard B. Sher, Scottish divines and legal lairds: Boswell's Scots presbyterian identity, in Greg Clingham (ed.) *New Light on Boswell: critical and historical essays on the occasion of the bicentenary of the life of Johnson* (Cambridge, 1991), pp. 28–55, at p. 31.

162. Boswell, *Grand Tour, Germany*, p. 211 ff.

163. Ibid., p. 215. Pottle, *Boswell the Earlier Years*, p. 165 , 182 also noted this.

164. See also his comment at Loch Ness, *Journal*, p. 99.

165. Boswell, *Grand Tour Italy*, p. 115.

166. Marlies K. Danziger, Self-restraint and self-display in the authorial comments in the Life of Johnson, in Clingham (ed.) *New Light on Boswell*, pp. 162–73, p. 169.

167. Boswell did remove this from the published *Tour*.

168. Boswell, *Corsica*, p. 1. See Thomas Curley, Boswell's Liberty-loving Account of Corsica, in Clingham (ed.) *New Light on Boswell*, pp. 89–103.

169. Ibid., p. 27.

170. See R.W. Ketton-Cremer, Johnson and Countryside, in M. Lascelles *et al.* (eds) *Johnson, Boswell and their Circle. Essays Presented to Lawrence Fitzroy Powell* (Oxford, 1965), pp. 65–75; William Hutchings and William Ruddick, Samuel Johnson and landscape, in Nalini Jain (ed.) *Re-Viewing Samuel Johnson* (London, 1991), pp. 67–81; and Alison Hickey 'Extensive Views' in Johnson's Journey to the Western Islands of Scotland, in *Studies in English literature, 1500–1900*, 32, 1992, pp. 537–53.

171. Brownell, *Johnson's Attitude to the Arts*.

172. Donald M'Nichol, *Remarks on Dr Samuel Johnson's Journey to the Hebrides* (London, 1779), p. 93.

173. Johnson also expressed his pleasure on this occasion to Boswell: 'If this be not *Roving Among the Hebrides*, nothing is' (*Tour*, p. 333).

174. Brownell, *Johnson's Attitude to the Arts*, p. 178.

175. Adams, *Travellers and Travel Liars*, p. 15.

176. Henry Skrine, *Two Successive Tours Throughout the Whole of Wales* (London, 1795), p. 205. See also Joseph Craddock, *Letters from Snowdon: descriptive of a tour through the northern counties of Wales* (London, 1770), p. 205.

177. Thomas Pennant, *A Tour in Wales* (2 vols, London, 1784), ii. 253. See also George Lyttelton, *An Account of a Journey into Wales* [1756], in Henry Windham, *A Gentleman's Tour Through Monmouthshire and Wales* (new edn, London, 1781), p. 239.

178. Richard Pococke, *The Travels Through England of Dr Richard Pococke* [1750–7], ed. James J. Cartwright (2 vols, London, 1888–89), i. 224; and Arthur Young, *A Six Months Tour Through the North of England* (2nd edn, 4 vols, London, 1771) iii. 286 ff.

179. It must be borne in mind that Johnson's journals in Wales and France were never intended for publication, but neither were some of the sources here quoted, such as Pococke and Lyttelton.

180. C. Bruyn Andrews (ed.) *The Torrington Diaries. Containing the Tours Through England and Wales of the Hon. John Byng* (4 vols, London, 1934–38), i. 46.

181. See Esther Moir, *The Discovery of Britain: the English tourists, 1540 to 1840* (London, 1964), p. 18.

182. John Chandler (ed. and modernized English) *John Leland's Itinerary: Travels in Tudor England* (Stroud, 1993), p. 437. T.K. Meier, Pattern in Johnson's A Journey to the Western Islands, in *Studies in Scottish Literature*, 5, 1967, pp. 185–93 is right to point out that Johnson 'discusses the military significance of various caves, rocky coasts, mountains and other terrain features' (p. 190) but in his approach to towns he noticeably avoids an originally military way of seeing.

183. '[S]ome have used to get on the top of the highest Steeple, where one may view with advantage, all the Countrey circumjacent, and the site of the City, with advenues and approaches about it, and so take a Landskip of it.' (Howell, *Instructions*, p. 17.) This is another divergence between Howell's recommendations and Johnson's practice of travel and description.

184. Wesley, *Journal*, iv. 452 disagreed.

185. John Macky, *A Journey Through Scotland* [vol. 3 of his *Journey Through Great Britain*, 1714–23] (London, 1723), p. 83; see also Pennant, *Voyage and Tour*, ii. 188.

186. M'Nichol, *Remarks*, p. 15. For a corrective to the general stereotype of Johnson as a Scotophobe, see Murray Pittock, Johnson and Scotland, in Jonathan Clark and Howard Erskine-Hill (eds) *Samuel Johnson in Historical Context* (London, 2002), pp. 184–96.

187. See Constantia Maxwell, *The English Traveller in France, 1698–1815* (London, 1932), pp. 7–8; and Antoni Maczak, *Travel in Early Modern Europe* (trans., Ursula Phillips, Cambridge, 1995), pp. 257–63.

188. Johnson's only description of the prospect of a city on approaching it was in the fictional context of *Rambler* 14. He deployed it there to show that the appearance of a thing at a distance could be deceptive.

189. Brownell, *Johnson's Attitude to the Arts*, pp. 165–68.

190. Craddock, *Letters from Snowdon*, p. 13; Pennant, *Tour in Wales*, ii. 36.

191. Simond, *Journal*, i. 221.

192. Thomas Jemielity, Dr Johnson and the uses of travel, in *Philological Quarterly*, 51, 1972, pp. 448–59, at p. 448.

193. Richard Warner, *A Tour Through the Northern Counties of England, and the Borders of Scotland* (2 vols, Bath, 1802), i. 140.

194. Paget Toynbee (ed.) Horace Walpole's journals of visits to country seats, &c, in *Walpole Society*, 66, 1927–28 (Oxford, 1928), pp. 9–80, at p. 65; see also Joseph Warton's letter to his brother Thomas in *The Correspondence of Thomas Warton*, ed. David Fairer (Athens, Ga., 1995), p. 188.

195. Edmund Burke, *A Philosophical Enquiry into the Origin of our Ideas of the Sublime and Beautiful*, ed. James T. Boulton (rev edn, Oxford, 1987), p. 82.

196. See E.P. Thompson, Rough Music, reprinted in *Customs in Common* (London, 1991), pp. 467–538.

197. G.B. Hill, *Footsteps of Dr Johnson* (1890, reprint Menston, 1973), p. 245.

198. Hawkins, *Life*, p. 318.

199. Stafford, *Voyage into Substance*, pp. 367–78.

200. This passage is also notable for Johnson's persistent concern for travel instructing by the acquisition of 'images' (see pp. 219–21), for the connection of beauty with utility (see pp. 250–53), and for Johnson speculating on whether the experience could be incorporated into the sublime – 'it was horrible, if barrenness and danger could be so.' The quote is the last line of Act II, scene i of Addison's *Cato*.

201. For the more general connections of the *Journey* and Homer, see Thomas Preston, Homeric allusion in A journey to the Western islands of Scotland, in *Eighteenth-Century Studies*, 5, 1971–72, pp. 545–58.

202. On this occasion Mrs Thrale recorded 'Mr Johnson says he would not have the images he has gained since he left the vale erased for £100.' A.M. Broadley (ed.), Mrs Thrale's unpublished journal of her tour in Wales with Dr Johnson, in his *Dr Johnson and Mrs Thrale* (London, 1910), pp. 155–219, p. 205. This comment does not justify Curley's suggestion, *Johnson and Age of Travel*, p. 100, that Johnson was 'sensitive to the sublimity. . . . of awesome mountains.' The images Johnson would not have erased are, like those of Raasay and Dunvegan, of civility in hostile conditions: 'Mrs Wynn sung Welsh songs to a harper's accompaniment, and Roberts, the vicar, provided dinner at the far end of the lake' (*Diaries*, note to p. 208).

203. 'Lifeless plains where no tree revives under the summer breeze'. Horace *Odes and Epodes*, trans., C.E. Bennett (London, 1938) *Ode*, I. xxii. 17–18.

204. 'What is at fault is the mind, which never escapes from itself'. Horace Satires, *Epistles and Ars Poetica*, trans., H. Rushton Fairclough (London, 1926), *Epistle*, I. xiv. 13. The motto to *Rambler* 135, discussed above in Chapter 6, is identical in its message about the relation of mind and place. For the importance of Horace and the classical traditon to Johnson, see J.C.D. Clark, *Samuel Johnson: literature, religion and English cultural politics from the restoration to romanticism* (Cambridge, 1994), Ch. 1; Barry Baldwin, *The Latin and Greek Poems of Samuel Johnson: Text, Translation and Commentary* (London, 1995), where Horace is the most cited author in the index (p. 285); and Robert C. Olson, *Motto, Context, Essay. The Classical Background of Samuel Johnson's Rambler and Adventurer Essays* (New York, 1984). Rogers, *Transit of Caledonia*, p. 25, points out that the mere fact of travelling to Scotland was 'a reversal of the values of their classically based education,' but how Johnson and Boswell then responded to this was strongly influenced by the classics.

205. See Hill, *Footsteps*, p. 218.

206. The importance of classical literature to the eighteenth-century readership's understanding of landscape is perhaps exemplified by the exchanges between Edmond Malone and Boswell over the need to clarify the Ovidian passage in the *Tour*, which is the only editorial exchange of any relevance to the presentation of landscape in that work. See Baker *et al.* (eds) *Correspondence*, pp. 204, 210, 270.

207. Joseph Craddock, *An Account of Some of the Most Romantic Parts of North Wales* (London, 1777), p. 36. A note references Johnson's *Rasselas*.

208. Tony Bareham, 'Paths to guide imaginations flight': some eighteenth-century poets abroad, in John McVeagh (ed.) *All Before Them: English literature and the wider world, Vol 1 1660–1780*, (London, 1990), pp. 247–62, p. 249. See also Malcolm Andrews, *The Search for the Picturesque: landscape aesthetics and tourism in Britain, 1760–1800* (Aldershot, 1989), p. 13; and R.D. Havens *The Influence of Milton on English Poetry* (Cambridge, Mass., 1922), p. 236 ff.

209. See Maren Sofie Røstvig, *The Happy Man: studies in the metamorphoses of a classical ideal* (2nd edn, 2 vols, Oslo, 1962–71).

210. Boswell, *Account*, p. 286.
211. Samuel Johnson *Poems*, ed. E.L. McAdam with George Milne (New Haven, 1964), p. 10, ll. 73–79. The relationship between Horace on the landscape and Johnson continued right up to the month before Johnson's death, when he made a verse translation of Horace's *Ode*, IV. vii. (see *Poems*, pp. 343–44). See Chapter 8, pp. 303–308.
212. Hill (ed.) *Johnsonian Miscellanies*, ii. 427–29. For other similar and not wholly reliable anecdotes, Hester Piozzi, *Anecdotes of the Late Samuel Johnson, LLD, During the Last Twenty Years of His Life*, ed. Arthur Sherbo (Oxford, 1974), pp. 147–48; and Waingrow (ed.) *Making of the Life*, p. 184.
213. Hanway, *Journey* p. 65. See also Curley, *Johnson and the Age of Travel*, pp. 53, 75, 200.
214. Donald Siebert, Johnson as satirical traveller: A journey to the Western Islands of Scotland, in *Tennessee Studies in Literature*, 19, 1974, pp. 137–48, at p. 141.
215. See p. xx.
216. Jeffrey Hart, Johnson's A Journey to the Western Islands: History as Art, in *Essays in Criticism*, 10, 1960, pp. 44–59, at p. 55; and Donald Greene, Response, in ibid., pp. 476–80, at p. 477.
217. Bruce Redford, *The Converse of the Pen: acts of intimacy in the eighteenth-century familiar letter*, (Chicago, 1986), pp. 218–19.
218. Piozzi, *Anecdotes*, p. 95.
219. See Richard Feingold, *Nature and Society: later eighteenth-century uses of the pastoral and georgic* (Hassocks, 1978), Ch. 2.
220. Keith Thomas, *Man and the Natural World: changing attitudes in England, 1500–1800* (Harmondsworth, 1983), p. 255.
221. John Barrell, *The Dark Side of the Landscape: The Rural Poor in English Painting, 1730–1840* (Cambridge, 1980).
222. Johnson also displayed a knowledge of butchery (*Tour*, 247), of threshing and of thatching (*Tour*, 263). See also Hawkins, *Life*, p. 469
223. *Literary Magazine*, II. pp. 95–97.
224. Nigel Everett, *The Tory View of Landscape* (New Haven, 1994).
225. M'Nichol, *Remarks*, p. 54. While Clark, *Samuel Johnson* has suggested the Journey was the expiation of Johnson's guilt over his failure to support the Jacobites in the '45, M'Nichol prefers to slur Johnson as a Catholic rather than a Jacobite, suggesting indeed (p. 144) that Johnson feared the Jacobites rather than being one. For Johnson's concern for church fabrics being Popish, see also William Seward's letters in John Nichols (ed.) *Illustrations of the Literary History of the Eighteenth Century* (8 vols, London, 1817–58), vii. 327–28. For the suggestion that the Jacobite elements of the tour to Scotland came from Boswell, not Johnson, see Rogers, *Transit of Caledonia*, Ch. 6.
226. J. Hart, Johnson's Journey, p. 47; Curley, *Johnson and the Age of Travel*, p. 96.
227. For other churches in the landscape, see *Diaries*, pp. 254, 255–56 in France; and pp. 172, 180, 181, 182, 184, 186, 193, 204, 206, 212, 214, 218 for Wales.
228. Wesley, *Journal*, v. 365–66 at Arbroath; see also at Elgin, vi. 506.
229. William Sacheverell, *An Account of the Isle of Man* (London, 1702), pp. 141–42.
230. I have been unable to find any biographical details about William Sacheverell. He was governor of the Isle of Man 1693–4 and his (probable) father was a Tory in parliament, which could suggest some affinity to Johnson. No relation of these Sacheverells to Henry Sacheverell the High Churchman is known to exist. See D.N.B. for Sacheverell's father and see J.G. Cumming's edition of the

Account of the Isle of Man (Douglas, 1859) for what few details are available on the author.

231. J.C.D. Clark, *English Society, 1688–1832: ideology, social structure and political practice during the ancien régime* (Cambridge, 1985), p. 235; and Nockles, *Oxford Movement in Context*, pp. 48–49.

232. Henry D. Rack, *Reasonable Enthusiast: John Wesley and the rise of methodism* (2nd edn, London, 1992) notes that for Wesley '[natural] facts lie within the compass of our senses, but causes do not... like the High-Church Hutchinsonians he distrusted Newton's hypotheses, though he was equally sceptical about their own curious theories. But his empiricism, too, had a religious purpose.' (p. 348)

233. For more on this topic, see Robert Mayhew, The Denominational Politics of Travel Writings: The Case of Tory Anglicans in the 1770s, in *Studies in Travel Writing*, 3, 1999, pp. 47–81.

234. McKay (ed.) *Walkers Reports*, p. 4.

235. Pennant, *Tour and Voyage*, ii. 335.

236. Johnson was recorded by Mrs Piozzi (in *Thraliana: The Diary of Mrs Hester Lynch Thrale (later Mrs Piozzi), 1776–1809*, ed. Katharine C. Balderston 2 vols, Oxford, 1951) as sceptical of all exceptional natural events, Johnson himself admitting 'I did not give Credit a long time to the Earthquake at Lisbon' (p. 468).

237. Ibid., ii. 716.

238. William Gilpin, *Observations, Relative Chiefly to Picturesque Beauty, Made in the Year 1776, on Several Parts of Great Britain; particularly the high-lands of Scotland* (2 vols, 2nd edn, London, 1792), ii. 120–21.

239. Pennant, *Tour and Voyage*, i. 291.

240. If this were the case, he would not have been alone. John Wesley, in common with Johnson, found Pennant 'judicious... in most respects' but associated his sceptical views on witchcraft – therefore possibly with those on the operation of natural laws – with ones 'I cannot give up to all the Deists in Great Britain'. Wesley, *Journal*, vi. 109.

241. Celia Fiennes, *Journeys* [1685–1703], Christopher Morris (ed.) (rev edn, London, 1949), pp. 96–97. The same argument was used by Charles Leigh, *The Natural History of Lancashire, Cheshire, and the Peak in Derbyshire* (Oxford, 1700) Bk III. p. 40. The list of subscribers includes Mr Michael Johnson – this could be Johnson's father, who had been a bookseller in Lichfield since 1681. See James L. Clifford, *Young Samuel Johnson* (London 1955 [1962 edn]), p. 13.

242. Henry Moore, *Picturesque Excursions from Derby to Matlock Bath, and its vicinity* (Derby, 1818), p. 41. See also John Hutchinson, *Tour Through the High Peak of Derbyshire* (Macclesfield, 1809), p. 3. A slightly different connection of religion and the Derbyshire landscape, but still one predicated on the beauty of the area was given by Edwards: 'it will be a delightful exercise to compare the sublime and beautiful scenery... with the descriptions of the *Holy Land* given by the pen of inspired writers.' John Edwards, *The Tour of the Dove; or a Visit to Dovedale &c. A Poem* (2nd edn, London, [1825]), p. 107.

243. As Newman said 'Physical Theology is a most jejune study, considered as a science, and really is no science at all, for it is ordinarily nothing more than a series of pious or polemical remarks upon the physical world viewed religiously.' In *Idea of a University*, p. 46. From a very different perspective on religion, this was also one of Hume's ('Philo's') points in the *Dialogues Concerning Natural Religion* (1779).

244. Cited in J. Paul Hunter, *Before Novels: the cultural contexts of eighteenth-century English fiction* (New York, 1990), p. 202.

245. Piozzi, *Thraliana*, i. 188.

246. See James L. Clifford, *Hester Lynch Piozzi (Mrs Thrale)* (2nd ed., corrected, Oxford, 1968), p. 114.

247. Hester Piozzi, *Mrs Thrale's French Journal*, in Moses Tyson and Henry Guppy (eds) *The French Journals of Mrs Thrale and Dr Johnson* (Manchester, 1932), pp. 67–166, at p. 129.

248. Ibid., p. 77.

249. See also Hester Piozzi, *Observations and Reflections Made in the Course of a Journey Through France, Italy, and Germany* (2 vols, London, 1789), ii. 238. Hereafter cited in the text.

250. Brownell, *Johnson's Attitude to the Arts*, p. 167.

251. Piozzi, *Welsh Journal*, p. 197.

252. Brownell, *Johnson's Attitude to the Arts*, pp. 159–62.

253. Mrs Piozzi did not read Boswell's journal until the spring of 1775, after the Welsh tour, and Johnson did not mention it in his letter about Inch Keith (see Clifford *Piozzi*, p. 125; *Letters*, ii. 54–55), but the anecdote may well have been known to her by word of mouth.

254. Johnson's words on Cowley are an apt summary: he 'gives inferences instead of images, and shews not what may be supposed to have been seen, but what thoughts the sight might have suggested'. (*Lives*, i. 51).

255. Piozzi, *Welsh Journal*, p. 179.

256. This Miltonic reference may have come to Johnson because of the local folklore about Red Castle Hill at Hawkestone: 'There have been several accounts of this very extraordinary place; the generally received notion, prevalent among all the country people in that neighbourhood, that it was formerly the habitation of two huge Giants named Tarquin and Tarquinus, however absurd and ridiculous in itself, is as perfect correspondent with the style of the place.' T. Rodenhurst, *A Description of Hawkestone* (2nd edn, Shrewsbury, 1784), pp. 39–40.

257. Hester Piozzi, *The Piozzi Letters: correspondence of Hester Lynch Piozzi, 1784–1821 (formerly Mrs Thrale)*, ed. Edward A. and Lillian D. Bloom (6 vols, Newark, NJ, 1989–2003), i. 115–16.

258. Similarly, but less artfully, Piozzi commented in response to rural unrest in Britain in the 1790s on 'a sudden change from dreadful Cold to Warmth – not even seasonable.... I think even the natural World exhibits Signs and Wonders now; while the Commotions in Civil and political Life are such as no History can give Precedent for certainly... 'Tis a wonderful World truly, and displays such Changes of Colour in its last State of Existence, as dazzle and confound an Observer'. Piozzi, *Letters*, ii. 233.

259. Ian Ousby, *The Englishman's England: taste, travel and the rise of tourism* (Cambridge, 1990), p. 9.

260. Anderson, *Over the Alps*, p. 23.

261. J.C.D. Clark, *The Language of Liberty, 1660–1832: political discourse and social dynamics in the Anglo-American world* (Cambridge, 1994).

262. Brownell, *Johnson's Attitude to the Arts*, p. 178.

263. Stafford, *Voyage into Substance*, p. 1. See also p. 18.

264. See Gascoigne, *Banks and the English Enlightenment*; idem, *Cambridge in the Age of the Enlightenment: science, religion and politics from the restoration to the French Revolution* (Cambridge, 1988); and Larry Stewart, *The Rise of Public Science: rhetoric, technology, and natural philosophy in Newtonian Britain, 1660–1750* (Cambridge, 1992).

265. Barbara Stafford, *Artful Science: Enlightenment Education and the Eclipse of Visual Education* (Cambridge, Mass., 1994), p. 76 herself has subsequently realized the need to analyze the interlinkage of aesthetics, science and religion: 'we need to restore the study of religion to the Age of Reason'.

Chapter 8

1. E.L. McAdam, Johnson's Lives of Sarpi, Blake, and Drake, in *Proceedings of the Modern Language Association of America*, 58, 1943, pp. 466–76, at p. 471.
2. For sources, see McAdam, Johnson's Lives, corrected by O.M. Brack, Johnson's Life of Admiral Blake and the development of biographical technique, in *Modern Philology*, 85, 1988, pp. 523–31, and *idem* The Gentleman's Magazine, concealed printing, and the texts of Samuel Johnson's Lives of Admiral Robert Blake and Sir Francis Drake, in *Studies in Bibliography*, 40, 1987, pp. 140–46.
3. Robert DeMaria, *The Life of Samuel Johnson: a critical biography* (Oxford, 1993), p. 77.
4. Paul Langford, *A Polite and Commercial People: England, 1727–73* (Oxford, 1989), p. 51.
5. J.D. Fleeman (ed.) *Early Biographical Writings of Dr Johnson* (Farnborough, 1973), p. 36. All further references are to this edition, in the text as *EBW*.
6. *Gentleman's Magazine (GM)*, vol. x, 1740, pp. 103–105.
7. Ibid., xi, 1741, p. 165.
8. The 'Life of Blake' has less relevance to the present inquiry.
9. Richard Helgerson, *Forms of Nationhood: the making of Elizabethan England* (Chicago, 1992), Ch. 4.
10. Langford, *Polite and Commercial People*, p. 53.
11. Thomas Kaminski, *The Early Career of Samuel Johnson* (New York, 1987), p. 111.
12. Nicholas Bourne, [printer] *Sir Francis Drake Revived* (London, 1653), p. 51. Bourne's work collects four narratives about Drake, of which Johnson's life focuses on two: *Sir Francis Drake Revived* by Philip Nichols (London, 1626), hereafter referred to in the text as *FDR*; and *The World Encompassed* by Sir Francis Drake, Bt. (London, 1628) [New pagination], hereafter referred to in the text as *WE*.
13. Nathaniel Crouch, [printer] *The English Hero: or, Sir Francis Drake Reviv'd* (London, 1716), p. 32. This volume was itself based on Bourne's collection and is hereafter referred to in the text as *EH*. On Crouch, see Robert Mayer, Nathaniel Crouch, bookseller and historian: popular historiography and cultural power in late seventeenth-century England, in *Eighteenth-century Studies*, 27, 1994, pp. 391–420. Johnson also added reflections to the effect that the Symerons were rational to reward useful skills such as ironmaking (cf. *FDR*, p. 49), that skin painting made the Brazilian Indians as comfortable as clothing (*EBW*, p. 54; cf. *WE*, p. 23), and that the hardships of Drake's sailors in a fifty-day storm were no greater than those of the island natives buffeted by the same conditions (*EBW*, pp. 57–58; cf. *WE*, p. 43 and *EH*, p. 80).
14. A point recognized for this period (in the case of Pope) by Edward Said, *Orientalism: Western conceptions of the Orient* (London, 1978), p. 45.
15. Anthony Pagden, *European Encounters with the New World: from renaissance to romanticism* (New Haven, 1993), p. 21. See also Stephen Greenblatt, *Marvelous Possessions: the wonder of the new world* (Oxford, 1991) for similar ideas but a more pessimistic interpretation.

16. A like consideration of the need to contain the exotic depiction of the natural world within the bounds of credibility may well have dictated the omission of a passage on the trees of the island of Maio in the *World Encompassed*: Johnson called the island 'extremely fruitful' (*EBW*, p. 51), but did not add the reflection 'that the Sun doth never withdraw himself farther off from them [the trees], but that with his lively heat he quickeneth and strengthneth the power of the soile and plant; neither ever have they any such frost and cold, or thereby to loose their green hew and appearance.' (*WE*, pp. 8–9). This also removed the personification of the natural world, in line with Johnson's editing of providential passages.

17. To 'God's Will' (*FDR*, p. 72), to what 'pleased God' (*FDR*, p. 43) and to moments when 'God Almighty so provided for us, by giving us [a] good storm' (*FDR*, p. 87).

18. This omission was repeated near the close of the 'Life' for an island which Bourne saw as an 'experience of God's wonderfull wisedome in many rare and admirable creatures' (*WE*, p. 96; cf. *EBW*, pp. 4–5).

19. DeMaria, *Life of Samuel Johnson*, p. 77.

20. O.M. Brack and Thomas Kaminski, Johnson, James and the Medicinal Dictionary, in *Modern Philology*, 81, 1983–84, pp. 378–400, 385.

21. See Chapter 7.

22. James Boswell, *Life of Johnson*, ed. G.B. Hill, rev. L.F. Powell (6 vols, Oxford, 1934–50), i. 68. Hereafter referred to in the text as *Life*.

23. Other scientific biographies in Robert James's *Medicinal Dictionary* were traditionally ascribed to Johnson. See Brack and Kaminski, Johnson, James and the Medicinal Dictionary, for scepticism on Johnson's authorship. They are followed in this scepticism by J.D. Fleeman, *A Bibliography of the Works of Samuel Johnson* (2 vols, continuously paginated, Oxford, 2000), p. 36. In the light of this, I have only referenced these biographies in support of points also made in works more firmly ascribed to Johnson.

24. Ibid., p. 391.

25. Johnson's authorship of this piece has been disputed in ibid., pp. 392–93.

26. The same phrase – bent of his genius – was used in the *Medical Dictionary* to explain Tournefort's vocation (*EBW*, p. 151).

27. Kaminski, *Early Career*, p. 54, referring to Boerhaave.

28. Albert Schultens, *Oratio Academica in Memoriam Hermanni Boerhavii* (London, 1739), p. 19.

29. Cf. Bernard Le Bovier de Fontenelle, 'Eloge de Morin' in his *Histoire de l'Academe Royale des Sciences* (Paris, 1724). John L. Abbott, Dr Johnson, Fontenelle, Le Clerk and six 'French' lives, in *Modern Philology*, 63, 1965–66, pp. 121–27 argues that 'Johnson quite evidently expands and intensifies Fontenelle's rather prosaic commentary' (p. 125). Fontenelle had spoken of reading 'avec un plaisir mêlé d'horreur le récit de leur descente dans la Grotte d'Antiparos' and of 'ces Rocher affreux & presque inaccessibles' (pp. 176, 186). Johnson (?) accurately conveyed the sense of the sublime and the scientist in this translation.

30. Schultens, *Oratio Academica*, p. 31.

31. Ibid., p. 17 and Fontenelle, *Histoire*, p. 310, 311.

32. Schultens, *Oratio Academica*, p. 78. I am grateful to George Boys-Stones for providing an accurate translation. For a full comparison of Schultens and Johnson's redaction, see Richard B. Reynolds, Johnson's 'Life of Boerhaave' in perspective, in *Yearbook of English Studies*, 5, 1975 pp. 115–29.

33. Reynolds, Johnson's Life, p. 117.

34. On the other hand, Kaminski, *Early Career* (p. 51), notes the reverse in Johnson's poem 'To Eliza plucking laurel in Mr Pope's gardens' (1738). In this case the English translation is more elaborate and rhetorical in its garden imagery than in Johnson's own Latin original.

35. *Posthumous Works of the Learned Sir Thomas Browne* (London, 1712), p. ii; and *Religio Medici* (London, 1736), p. xxix.

36. Troas is mentioned in *Acts*, 16.8–11.

37. Kaminski, *Early Career*, p. 171.

38. *GM*, vii. 111–12.

39. Ibid., 622–23.

40. Ibid., vi. 526–27.

41. Ibid., 601–603.

42. R.Y. was probably Richard Yate (R. Yate writes in the *GM* of April 1743). He was clearly a defender of rationalistic Anglicanism, bordering on Deism, writing *A Letter in Defence of Dr Middleton: Being an Argument Proving that Miraculous Powers were Never Truly Wrought but by Persons Divinely Inspired* (London, 1749) and a *Letter* (London, 1739) against enthusiastic Baptist preaching.

43. *GM*, viii. 13; vi. 241–42; and viii. 241–45. On Enlightenment debates concerning the geography of paradise, see C.W.J. Withers, Geography, Enlightenment and the Paradise Question, in David Livingstone and Charles Withers (eds) *Geography and Enlightenment* (Chicago, 1999), pp. 67–92.

44. Kaminski, *Early Career*, pp. 44, 154.

45. Ibid., pp. 60–61.

46. *GM*, viii. 577–78, 635–36; ix. 14–15.

47. Other factual accounts were to appear during the period of Johnson's close connection with the magazine, such as Captain Elton's account of his travels in Russia in January 1742.

48. Similarly, pieces such as the 'Description of the Island of Cuba' (October 1741) contained religious reflections on the lack of improvement of the island, there being 'more Priests than industrious Planters', but also gave information on the island's size, situation and physical geography.

49. Ibid., xii. 216, 263–64, 531–32. On the growing importance of Maps in the *Gentleman's Magazine*, see E.A. Reitan, Expanding Horizons: Maps in the *Gentleman's Magazine*, 1731–54, in *Imago Mundi*, 37, 1985, pp. 54–62. He dates the expansion of interest as starting in 1739. On Packe, see Michael Charlesworth, Mapping, the Body and Desire: Christopher Packe's Chorography of Kent, in Denis Cosgrove (ed.) *Mappings* (London, 1999), pp. 109–24.

50. Ibid., xiii. 542.

51. Which therefore expressed a different view from Johnson's on Pococke, see Chapter 7, pp. 226–28.

52. Arthur Sherbo, The Making of Ramblers 186 and 187, *Proceedings of the Modern Language Association of America*, 67, 1952, pp. 575–80 suggests that Egede's account may have been the source for Johnson's tale of Ajut and Anningait.

53. See Chapter 6, pp. 185–90.

54. See Donald Greene, *The Politics of Samuel Johnson* (2nd edn, Athens, Ga., 1990), p. 141.

55. Brack, Johnson's Life of Blake, p. 523.

56. R.D. Stock, *Samuel Johnson and Neoclassical Dramatic Theory: The Intellectual Context of the Preface to Shakespeare* (Lincoln, Nebraska, 1973), p. 156. See also Arthur Sherbo, *Samuel Johnson, Editor of Shakespeare* (Urbana, Illinois, 1956) Appendix C; and W.B. Carnochan, Johnsonian metaphor and the

'Adamant of Shakespeare', in *Studies in English Literature*,1500–1900, 10, 1970, pp. 541–49.

57. Arthur Sherbo (ed.) *Johnson on Shakespeare* (2 vols, continuously paginated, New Haven, 1968), p. 99. All references, unless otherwise stated, are to this edition, in the text as *JOS.*

58. William Warburton(ed.) *The Works of Shakespear* (8 vols, London, 1747), p. xxv.

59. John Upton, *Critical Observations on Shakespeare* (2nd edn, London, 1748), pp. 14–15.

60. James Barclay, *An Examination of Mr Kenrick's Review of Mr Johnson's Edition of Shakespeare* (London, 1766), p. vi.

61. Stock, *Johnson and Neoclassical Theory*, p. 28.

62. See pp. 275–90.

63. On this and the other category of nature, see Basil Willey, *The Eighteenth-Century Background: Studies in the Idea of Nature in the Thought of the Period* (London, 1940).

64. Stock, *Johnson and Neoclassical Theory*, p. 127.

65. On the cultural history of footnotes, see Anthony Grafton, *The Footnote: A Curious History* (London, 1997).

66. Richard Jones, *Lewis Theobald* (New York, 1919); Robert DeMaria, *Johnson's Dictionary and the Language of Learning* (Oxford, 1986) and idem, *Life of Samuel Johnson*; and J.C.D. Clark, *Samuel Johnson: Literature, Religion and English Cultural Politics from the Restoration to Romanticism* (Cambridge, 1994).

67. Anthony Grafton, *Defenders of the Text: The Traditions of Scholarship in the Age of Science, 1450–1800* (Cambridge, Mass., 1991), p. 5.

68. Alexander Pope, *The Works of Shakespear* (6 vols, London, 1723–5), i. ix.

69. Warburton, *Shakespear*, ii. 333. Referring to *As You Like It*, III. ii. 31–32. References to Shakespeare by line are to *Complete Works*, ed. Stanley Wells *et al.* (Oxford, 1988).

70. Ibid., iv. 319. For the division of souls, see the note to Lewis Theobald, *The Works of Shakespeare* (7 vols, London, 1733), ii. 482.

71. IV. ii. 178.

72. Warburton, *Shakespear*, vii. 316.

73. Ibid., vi. 15. See also Peter Whalley, *An Enquiry into the Learning of Shakespeare* (London, 1748), p. 34.

74. Ibid., viii. 293–94. See also Warburton's 'Dedication to Free-Thinkers' in his *Divine Legation of Moses.*

75. Ibid., vi. 16.

76. Ibid., vi. 20, 22.

77. Peter Seary, *Lewis Theobald and the Editing of Shakespeare* (Oxford, 1990), p. 109.

78. Warburton, *Shakespear*, ii. 246.

79. Theobald, *Shakespeare*, vii. 64.

80. Warburton, *Shakespear*, vii. 426.

81. Ibid., vi. 334.

82. Warburton had similarly cited Varenius to support the suggestion that in *Timon of Athens*, Shakespeare might have been referring to the idea 'that Saltiness of the Sea is caused by several Ranges, or Mounds of Roch-Salt'. The issue of the chronology of geographical knowledge again went undiscussed. Warburton, *Shakespear*, vi. 224.

83. Theobald, *Shakespeare*, iv. 48.This was an anonymous contributor's remark. For a less flattering interpretation of the meaning of 'nook-shotten', more in keeping with its being spoken by the French Duke of Brittany, see *Henry V*, ed. T.W. Craik (London, 1995), p. 226n.

84. Ibid., ii. 282 and vi. 177.
85. Ibid., iv. 96; see also v. 450.
86. Warburton, *Shakespear*, vi. 108.
87. Theobald, *Shakespeare*, vi. 100.
88. William Warburton, *A View of Lord Bolingbroke's philosophy* (2 vols, London, 1754–55), i. 121; see also *idem, A Critical and Philosophical Enquiry into the Causes of Prodigies and Miracles, as Related by Historians* (London, 1727), p. 92.
89. Warburton, *View*, ii. 2.
90. Warburton, *Critical and Philosophical Enquiry*, p. 90 and note to p. 10. See also *idem, Julian. Or, a Discourse Concerning the Earthquake and Firey Eruption, which Defeated the Emperor's Attempt to Rebuild the Temple at Jerusalem* (2nd edn, London, 1751). Richard Hurd's 1794 biography of Warburton also links him with the Cambridge Platonists at a number of points in its argument. The best modern study of Warburton's theology is B.W. Young, *Religion and Enlightenment in Eighteenth-Century England: Theological Debate from Locke to Burke* (Oxford, 1998), Chapter 5.
91. Ibid., p. 2.
92. For early eighteenth-century editorial principles, see Margreta De Grazia, *Shakespeare Verbatim: The Reproduction of Authenticity and the 1790 Apparatus* (Oxford, 1991); Arthur Sherbo *The Birth of Shakespeare Studies: Commentators from Rowe (1709) to Boswell-Malone (1821)* (East Lansing, Michigan, 1986). Pope was similar in his approach: see James R. Sutherland, 'The Dull Duty of an Editor', in *Review of English Studies*, 21, 1945, pp. 202–15; John A. Hart, Pope as scholar-editor, in *Studies in Bibliography*, 23, 1970, pp. 45–59; and Simon Jarvis, *Scholars and Gentlemen: Shakespearian Textual Criticism and Representations of Scholarly Labour, 1725–65* (Oxford, 1995), Chapter 2.
93. Warburton, *Critical and Philosophical Enquiry*, p. 92.
94. Ibid., p. 63.
95. Theobald, *Shakespeare*, iii. 99.
96. The correspondence (mainly Theobald's side of it) was reprinted in John Nichols (ed.) *Illustrations of the Literary History of the Eighteenth Century* (8 vols, London, 1817–58) vol 2. For a discussion, see Seary, *Lewis Theobald*.
97. Thomas Edwards, *The Canons of Criticism, and Glossary, being a Supplement to Mr Warburton's edition of Shakespear* (5th edn, London, 1753), p. 60. See also p. 159, and Benjamin Heath, *A Revisal of Shakespear's Text* (London, 1765), pp. 63, 115.
98. Heath, *Revisal*, p. 375. See also Edwards, *Canons* p. 33, 67.
99. Heath, *Revisal*, p. 123.
100. Ibid., p. 32, 81.
101. Pope, *Shakespear*, i. 29, iv. 18, v. 316, vi. 534.
102. For this, see Seary, *Lewis Theobald*, p. 67.
103. Theobald, *Shakespeare*, i. 14.
104. Ibid., i. 235, 246. Warburton also provided some less philosophical notes on the popularity of travel in Elizabethan England. See Warburton, *Shakespear*, i. 185, and viii. 293–94.
105. Thomas Hanmer, *Works of Shakespear* (6 vols, London, 1745 edn), v. 62 and Theobald, *Shakespeare*, iii. 312–13. Gesner's work came out before Shakespeare's birth (1516–65), and thus the beliefs he referred to must in principle have been available to Shakespeare.
106. See C.S. Lim, Emendation of Shakespeare in the Eighteenth Century: The Case of Johnson, in *Cahiers Elisabethains*, 33, 1988, pp. 23–30.
107. Warburton, *Shakespear*, i. 43.

108. This had, however, been noted before Johnson's edition by Zachary Grey, *Critical, Historical, and Explanatory Notes on Shakespeare* (2 vols, London, 1754), ii. 24.

109. As Rymer had done when he said of *Othello's* 'huge Eclipse / Of Sun and Moon—' (V. ii. 108–109) that 'it wou'd be uncivil to ask Flamstead if the Sun and Moon can both together be so hugely eclipsed'. Thomas Rymer, *A Short View of Tragedy* (London, 1693), p. 141.

110. The Arden edition points out that the case for Shakespeare having made a geographical error here is less clear than Johnson thought: see J.H.P. Pafford, *The Winter's Tale* (London, 1963), p. 66n.

111. W.K. Wimsatt (ed.) *Samuel Johnson on Shakespeare* (London, 1960), p. xxvi.

112. Cf. *JOS* p. 279 and Heath *Revisal*, p. 136. Sherbo, *Johnson, Editor of Shakespeare*, argued Johnson plagiarized Heath, a claim rejected by Arthur M. Eastman, In defence of Johnson, in *Shakespeare Quarterly*, 8, 1957, pp. 493–500. In this case at least, it was so obvious that Warburton had misread, that the independence of the two can be assumed. Indeed, Theobald had in correspondence suggested to Warburton that this was a mistake: see Nichols, *Illustrations*, ii. 627.

113. Samuel Johnson, *Plays of Shakespeare* (8 vols, London, 1765) vol., viii. Appendix to vi. 18.

114. See Peter Martin, *Edmond Malone, Shakespearian Scholar: A Literary Biography* (Cambridge, 1995), pp. 24–25; and David Nichol Smith, *Shakespeare in the Eighteenth Century* (Oxford, 1928), p. 55.

115. John Kenyon, *The History Men: The Historical Profession in England Since the Renaissance* (2nd edn, London, 1993), Ch. 3.

116. Richard Farmer, *An Essay on the Learning of Shakespeare* (Cambridge, 1767), p. 8.

117. Nichols, *Illustrations*, vi. 149.

118. See Samuel Johnson and George Steevens (eds) *Plays of William Shakespeare*, (10 vols, London, 1778), i. 73–74; and Edmond Malone (ed.) *The Plays and Poems of William Shakspeare* (10 vols, London, 1790), i. pt. i. xi.

119. Steevens, in Malone, *Plays*, viii. 354.

120. Ibid., i, pt. ii. 43; i. pt. ii. 185; and iii. 24 respectively.

121. Whalley, *Enquiry*, p. 21; and Steevens in Edmond Malone and James Boswell (the younger) *The Works of Shakespeare* (21 vols, London, 1821), xv. 139.

122. Malone, *Plays*, i. pt. ii. 91.

123. Joseph Ritson, *Remarks, Critical and Illustrative, on the Text and Notes of the Last Editor of Shakespeare* (London, 1783), p. 182.

124. Ibid., p. 188.

125. Gary Taylor, *Reinventing Shakespeare: A Cultural History from the Restoration to the Present* (London, 1990), pp. 129–33.

126. Stock, *Johnson and Neoclassical theory*, p. xxi.

127. See Martin, *Malone*, p. 83, 136–37.

128. Malone, *Plays*, i. pt. i. lxix; and Farmer, *Essay*, p. 2.

129. Malone, *Plays*, i. pt. ii. 19 and iv. 381–84.

130. Ibid., ii. 487.

131. *Boswell-Malone*, x. 211–12.

132. Malone, *Plays*, iv. 213; see also i. 60. In the 1821 edition, Blakeway similarly argued that the *Midsummer Night's Dream's* 'winds piping to us in vain' was a reference 'to the bad weather with which England was visited about this time', citing King's lectures of 1593–4. *Boswell-Malone*, v. 342.

133. Malone, *Plays*, iii. 253; see also i. pt. ii. 195–96. This reference was first used in the second edition of Hanmer's edition (1770–1). The note was by Thomas Warton.

134. *Boswell-Malone*, v. 235.
135. Ibid., xxi. 519–35.
136. Malone, *Plays*, viii. 278–79, for the potato; for this more generally, see Sherbo, *Birth of Shakespeare studies*, p. 58 ff.
137. Malone, *Plays*, i. pt. ii. 85, and *Boswell-Malone*, xv. 156.
138. Keith Thomas, *Religion and the Decline of Magic: studies in popular beliefs in sixteenth-and seventeenth-century England* (Harmondsworth, 1971); and *idem Man and the Natural World: changing attitudes in England, 1500–1800* (Harmondsworth, 1983).
139. This is the general thesis of De Grazia, *Shakespeare Verbatim*.
140. See Charles W.J. Withers, Towards a History of Geography in the Public Sphere, in *History of Science*, 37, 1999, pp. 45–78.
141. See Richard Schwartz, *Samuel Johnson and the New Science* (Madison, 1971).
142. Samuel Johnson, *Lives of the English Poets*, ed. G.B. Hill (3 vols, Oxford, 1905), ii. 94 and i. 434. Hereafter referred to in the text as *Lives*.
143. I.T. [John Toland] *The Life of John Milton* (London, 1699 ed.), p. 26.
144. Toland, *Life* p. 127; Edward Phillips (ed.) *Letters of State, Written by Mr John Milton* (London, 1694), p. xxxvi.
145. Thomas Newton (ed.) *Paradise Lost* (2 vols, 3rd edn, London, 1754), i. xxxvii; Jonathan Richardson, father and son, *Explanatory Notes and Remarks on Milton's Paradise Lost* (London, 1734), p. cxiii.
146. John Milton, *Paradise Lost* (14th edn, London, 1741), p. xxiii; followed by *Biographia Britannica: or, the lives of the most eminent persons who have flourished in Great Britain and Ireland* (6 vols, London, 1747–66), [continuously paginated] p. 3118.
147. Richardson, *Explanatory notes*, p. cxiii.
148. It is worth noting that one of Johnson's sources for the 'Life of Thomson' mentioned that autumn was his favourite season for composition, but Johnson declined to take up the point made in the 'Life of Milton' again. See Patrick Murdock, 'An Account of the Life and Writings of Mr James Thomson', prefaced to *Works* (2 vols, London, 1762), i. xvii.
149. Martin Maner, *The Philosophical Biographer: doubt and dialectic in Johnson's 'Lives of the Poets'* (Athens, Ga., 1988), p. 102.
150. Maner, *Philosophical Biographer*, pp. 131–32.
151. Owen Ruffhead, *The Life of Alexander Pope* (London, 1769), p. 197. See also *Biographia Britannica*, p. 3410.
152. Cited in J.V. Guerinot *Pamphlet attacks on Alexander Pope, 1711–44: a descriptive bibliography* (London, 1969), p. 192. See also p. 155. Br—me is Pope's co-translator of the *Odyssey*, Broome.
153. Ibid., p. 306. See also p. 256.
154. Theophilus Cibber, *The Lives of the Poets of Great Britain and Ireland from the Time of Dean Swift* (5 vols, London 1753), ii. 52. On the relationship between Shiels and Johnson, see William R. Keast, Johnson and 'Cibber's' 'Lives of the Poets', 1753, in Carroll Camden (ed.) *Restoration and Eighteenth-Century Literature: essays in honour of Alan Dugald McKillop* (Chicago, 1963), pp. 89–101; and James Battersby, Johnson and Shiels: biographers of Addison, in *Studies in English Literature, 1500–1900*, 9, 1969, pp. 521–37.
155. Sprat's 'Account of the Life and Writings of Mr Abraham Cowley' is prefixed to Richard Hurd (ed.) *Select Works in Verse and Prose of Mr A Cowley* (2 vols, Dublin, 1772). Quote at i. 36.
156. Ibid., i. 36, 37.

157. Robert Potter, *The Art of Criticism; as exemplified in Dr Johnson's lives* (London, 1789), p. 155.

158. Hurd (ed.) *Select Works*, i. 40.

159. Robert Potter, *An Inquiry into Some Passages in Dr Johnson's Lives of the Poets* (London, 1783), p. 8.

160. Some have argued that Johnson was inaccurate on this point of competition. For evidence of Shenstone's dislike of Lyttelton's gardening at Hagley, see David Coffin, *The English Garden: meditation and memorial* (Princeton, 1994), p. 40, 210.

161. G.B. Hill, *Johnsonian Miscellanies* (2 vols, Oxford, 1897), ii. 210. The comment was made by Thomas Percy.

162. For Johnson's sources, see James H. Leicester, Johnson's Life of Shenstone: some observations on the sources, in Magdi Wahba (ed.) *Johnsonian Studies* (Cairo, 1962), pp. 189–222.

163. Richard Graves, *Recollection of Some Particulars in the Life of the Late William Shenstone, Esq* (London, 1788) p. 55, 23, 53.

164. Ibid., pp. 108–109.

165. Potter, *Art of Criticism*, pp. 9–10, 110–11, 172–74. Also James Callendar, *A Critical Review of the works of Dr Samuel Johnson* (Edinburgh, 1783), p. 60.

166. Robert Folkenflik, *Samuel Johnson, biographer* (Ithaca, 1978), p. 35.

167. Potter, *Art of Criticism*, pp. 155, 173.

168. Leopold Damrosch, *The Uses of Johnson's Criticism* (Charlottesville, 1976), p. 136.

169. Oliver Sigworth, Johnson's *Lycidas*: The End of Renaissance Criticism, in *Eighteenth-Century Studies*, 1, 1967, pp. 159–68.

170. See Jean H. Hagstrum, *Samuel Johnson's Literary Criticism* (Chicago, 1952 [1967 edn]), p. 68.

171. See George Sewell (ed.) *Poems by Mr John Philips* (2nd edn, London, 1715), p. 27.

172. Jean Pierre de Crousaz, *A Commentary on Mr Pope's Principles of Morality, or Essay on Man* [trans., Samuel Johnson] (London, 1742), notes to pp. 40, 37–38.

173. William Warburton (ed.) *Works of Alexander Pope* (9 vols, London, 1752 edn), iii. 25–26.

174. Nalini Jain, *The Mind's Extensive View: Samuel Johnson as a critic of poetic language* (Strathtay, 1991), p. 65.

175. William Tindal, *Remarks on Dr Johnson's Life, and Critical Observations on the Works of Mr Gray* (London, 1782), pp. 38–39.

176. Richardson, *Explanatory Notes*, p. ciii.

177. Patrick Hume in Newton *Paradise Lost*, i. 256.

178. This was a critical commonplace about *Cyder* after Sewell, *Poems*, pp. 7–8, 18.

179. Francis Hutcheson, *An Inquiry into the Origin of our Ideas of Beauty and Virtue* (3rd edn, London, 1729), pp. 9–10.

180. For Dennis, see John Barnard (edn) *Pope: the critical heritage* (London, 1973), p. 89; and [Joseph Warton] *An Essay on the Genius and Writings of Pope* (2 vols, London, 1756–82), i. 20.

181. Cited in Barnard, *Pope*, p. 409.

182. Yet Johnson had commented in these terms about Savage's *The Wanderer* in his *Life* in 1744.

183. Murdock, *Works*, i. ix.

184. Percival Stockdale (ed.) *The Seasons* (London, 1793) Notes, not paginated, note to Spring v. 346. For Stockdale's relationship with Johnson, see Howard Weinbrot, Samuel Johnson, Percival Stockdale, and bric-bats from Grubstreet: some later response to the *Lives of the Poets*, in *Huntington Library Quarterly*, 56, 1993, pp. 105–34.

185. There was a parallel between Johnson's opinion and that of Shiels in Cibber's *Lives*. Although the 'Life of Thomson' may be one case where Shiels was not dependent on Johnson, for this specific point 'since Shiels was working from no known written source, it may well have been Johnson by whom the poem 'has been judged defective in point of plan'.' Hilbert H. Campbell, Shiels and Johnson: biographers of Thomson, in *Studies in English literature, 1500–1900*, 12, 1972, pp. 535–44, p. 541. Perhaps Johnson generalized the point he made of Savage in 1744.

186. William Gilpin, *Observations on the river Wye* (London, 1782), p. 59, 61.

187. John Scott, *Critical Essays on some of the Poems, of Several English Poets* (London, 1785), p. 112.

188. For more recent comments, see John Barrell, *The Idea of Landscape and the Sense of Place, 1730–1840: an approach to the poetry of John Clare* (Cambridge, 1972).

189. See Hagstrum, *Johnson's Literary Criticism*, p. 113.

190. Folkenflik, *Samuel Johnson, Biographer*, p. 10.

191. See Stephen Daniels, Marxism, Culture and Duplicity of landscape, in R. Peet and N. Thrift (eds) *New Models in Geography* (2 vols, London, 1989), ii. 196–219; John Berger, *Ways of Seeing* (London, 1972); idem, *The White Bird and Other Essays* (London, 1985); and Simon Schama, *Landscape and Memory* (London, 1995).

192. Samuel Johnson, *Poems*, ed. E.L. McAdam with George Milne (New Haven, 1964), p. 13, ll. 21–24. Translation of Horace, *Odes*, II. xiv.

193. Ibid., p. 4 ll. 18–20.

194. Ibid., p. 4, ll. 23–27.

195. See Walter Bate, *The Achievement of Samuel Johnson* (Oxford, 1955), p. 9. Also powerful on this theme is John Wain, *Samuel Johnson* (London, 1974).

196. See H.F. Hallett, Dr Johnson's refutation of Bishop Berkeley, in *Mind*, 56, 1947, pp. 132–47.

197. Kaminski, *Early Career*, p. 89.

198. See James Boswell, *London Journal, 1762–3*, ed. Frederick A. Pottle (Edinburgh, 1991 edn), p. 330; and *James Boswell's Life of Johnson: an edition of the original manuscript in four volumes*, ed. Marshall Waingrow (4 vols, Edinburgh, 1994–), i. 323, n. 3, 4.

199. Hester Piozzi, *Anecdotes of the Late Samuel Johnson, LLD., During the Last Twenty Years of his Life*, ed. Arthur Sherbo (London, 1974), p. 151.

200. Samuel Johnson, *Letters*, ed. Bruce Redford (5 vols, Oxford, 1992–94), iv. 191. Hereafter referred to in the text as *Letters*.

201. Hill (ed.) *Johnsonian Miscellanies*, ii. 353.

202. *Letters*, i. 249, n. 1 records that by 1765 Taylor was building an octagonal drawing room in Adam style.

203. John Hawkins, *The Life of Samuel Johnson* (2nd edn, London, 1787), p. 530, also mentioned this garden behind Johnson's house at Bolt Court as one 'which he took delight in watering'.

204. See Stephen Daniels, The Political Iconography of Woodland in later-Georgian England, in Denis Cosgrove and Stephen Daniels (eds) *The Iconongraphy of Landscape: Essays on the Symbolic Representation, Design and Use of Past Environments* (Cambridge, 1988), pp. 43–82.

205. See Chester Chapin, Samuel Johnson's 'wonderful' experience, in Wahba (ed.) *Johnsonian Studies*, pp. 51–60.

206. Hill (ed.) *Johnsonian Miscellanies*, ii. 395–96.

207. Johnson, *Poems*, p. 343, ll. 1–8.

208. Ibid., pp. 342–43. For a full discussion, see Barry Baldwin, *The Latin and Greek Poems of Samuel Johnson: text, translation and commentary* (London, 1995), pp. 157–62. He points out that 'there is a good deal of Horatian language in the poem' (p. 159).

209. Ibid., pp. 348–49. This translation is the Reverend James Henry's (1787). Baldwin *Latin and Greek Poems*, p. 147, translates it more literally as 'with a bountiful hand sow in me the seeds of virtue, that an ample harvest of goodness shall result'. It is interesting that Henry chose to translate 'messis' as 'prospect', and perhaps shows the stronger religious resonance of the term at the time (as in Johnson's *Dictionary*, see Chapter 5), which made him sacrifice literal fidelity of translation to the retention of mood.

Chapter 9

1. See Chapter 3.
2. For this point in the literature of the period, see R.D. Stock, *The Holy and the Daemonic from Sir Thomas Browne to William Blake* (Princeton, 1982); and for a later era, see Robert Ryan, *The Romantic Reformation: religious politics in English literature, 1789–1824* (Cambridge, 1997).
3. See Chapter 5.
4. William Wordsworth, Tintern Abbey (1798), ll. 80–82.
5. See John Barrell, *The Idea of Landscape and the Sense of Place, 1730–1840: An Approach to the Poetry of John Clare* (Cambridge, 1972); idem, *The Dark Side of the Landscape: the rural poor in English painting, 1730–1840* (Cambridge, 1980); and Ann Bermingham *Landscape and Ideology: the English rustic tradition, 1740–1860*. Although all three claim to extend back into the eighteenth century, their model of the politics of landscape is in each case more convincing for the later years. James Turner, *The Politics of Landscape: rural scenery and society in English poetry, 1630–1660* (Oxford, 1979), deals with a much earlier period, but it is scarcely more convincing to extrapolate forwards to the eighteenth century from the Civil War patterns of the politics of landscape than to extrapolate back the discursive connections of the post-revolutionary period.
6. See Charles W.J. Withers, Geography, Natural History and the Eighteenth-Century Enlightenment, putting the world in place, in *History Workshop Journal*, 39,1995, pp. 137–63.
7. See Larry Stewart, *The Rise of Public Science: rhetoric, technology, and natural philosophy in Newtonian Britain, 1660–1750* (Cambridge, 1992).
8. See Chapters 6 and 8.
9. On this process, see Geoffrey Galt Harpham, *The Ascetic Imperative in Culture and Criticism* (Chicago, 1987), esp. p. 71.
10. See Chapter 7.
11. See Chapter 8.
12. See Chapter 3, pp. 65–67.
13. John Hutchinson, *Philosophical and Theological Works*, ed. Robert Spearman and Julius Bate (3rd edn, 12 vols, London, 1748–9), v. 84–85.
14. Hutchinson, *Works*, iii. p. ix.
15. See for example his discussion of the olive tree, in *Works*, viii. 119–27.
16. Hutchinson, *Works*, x. 1.
17. Ibid., v. 103.
18. Ibid., xii. 259–359.

19. Ibid., xi. 159.
20. William Jones, *Theological, Philosophical and Miscellaneous Works* (12 vols, London, 1801), iv. 6.
21. William Jones, *The Book of Nature; or, The True Sense of Things Explained and Made Easy to the Capacities of Children* (5th edn, London 1803), pp. xiii–xiv.
22. Jones, *Works*, vi. 281.
23. William Jones, *Physiological Disquisitions; or, Discourses on the Natural Philosophy of the Elements* (London, 1781), p. 480.
24. Jones, *Physiological Disquisitions*, p. 527.
25. James Boswell, *Life of Johnson*, ed. G.B. Hill, rev. L.F. Powell (6 vols, Oxford, 1934–50), ii. 468.
26. William Jones, *Observations in a Journey to Paris by Way of Flanders, in the Month of August 1776* (2 vols, London, 1777), i. p. vii.
27. Jones, *Observations*, i. 38–39.
28. *Life*, iii. 53 has a reference to Hutchinson by Johnson, but I agree with L.F. Powell's note that Johnson is referring to Francis Hutcheson. The passage G.B. Hill referred to in Hutchinson's works (xii. 38 ff.) does not contain the argument Johnson is expounding at that point.
29. See Chapter 3.
30. See Chapter 3 and 4.
31. For Knight see Chapter 6; for Hanway, Boswell and Piozzi see Chapter 7; and for Warburton, see Chapter 8.
32. See Paul Fussell, *The Rhetorical World of Augustan Humanism: ethics and imagery from Swift to Burke* (Oxford, 1965).

Bibliography

Part I: Works by Samuel Johnson

The Yale Edition of the Works of Samuel Johnson. General editor, John Middendorf (New Haven, 1958 to the present).
Diaries, Prayers, Annals, ed. E.L. McAdam, with Donald and Mary Hyde (New Haven, 1958).
Idler and Adventurer, eds W.J. Bate, J.M. Bullitt and L.F. Powell (New Haven, 1963).
Rambler, eds W.J. Bate and Albrecht B. Strauss (3 vols, New Haven, 1964).
Poems, ed. E.L. McAdam, with George Milne (New Haven, 1964).
Johnson on Shakespeare, ed. Arthur Sherbo, with introduction by Bertrand H. Bronson (2 vols, continuously paginated, New Haven, 1968).
Political Writings, ed. Donald J. Greene (New Haven, 1977).
Sermons, eds Jean H. Hagstrum and James Gray (New Haven, 1978).
Voyage to Abyssinia, ed. Joel J. Gold (New Haven, 1735).
Rasselas and other tales, ed. Gwin J. Kolb (New Haven, 1990).

Other Editions

A Commentary on Mr Pope's Principles of Morality, or Essay on Man, by Jean Pierre de Crousaz, trans. with notes by Samuel Johnson (London, 1742).
A Course of Lectures on the English Law, 1767–73, by Robert Chambers with Samuel Johnson, ed. Thomas Curley (2 vols, Oxford, 1986).
A Dictionary of the English Language (London, 1755 [Longman Facsimile, 1990]),
A Dictionary of the English Language (4th edn, London, 1773).
Early Biographical Writings, ed. J.D. Fleeman (Farnborough, 1973).
Journey to the Western Islands of Scotland, ed. J.D. Fleeman (Oxford, 1775).
Letters, ed. Bruce Redford (5 vols, Oxford, 1992–94).
Lives of the Poets, ed. G.B. Hill (3 vols, Oxford, 1905).
The Latin and Greek Poems of Samuel Johnson: Text, Translation and Commentary by Barry Baldwin, (London, 1995).
Plays of Shakespeare (8 vols, London, 1765).
Plays of William Shakespeare (10 vols, London, 1778), Samuel Johnson and George Steevens (eds).
Samuel Johnson on Shakespeare, ed. W.K. Wimsatt (London, 1960).
Prefaces and Dedications, ed. Allen T. Hazen (New Haven, 1937).
Review of 'The civil and natural history of Jamaica', by Patrick Browne, in *Literary Magazine, or Universal Review*, 1756, IV pp. 176–85.
Review of 'Essay on the description of China, in two volumes folio'. From the French of Pere Du Halde, in *Gentleman's Magazine*, 1742, XII, pp. 320–23, 353–57 and 484–86.
Review of 'A Free Enquiry into the nature and origin of evil', by Soame Jenyns, in *Literary Magazine, or Universal Review*, 1756–7, XIII, pp. 171–75, XIV, pp. 251–53 and XV, pp. 301–306.
Review of 'A Journal of eight days journey ... To which is added an Essay on Tea' by Jonas Hanway, in *Literary Magazine, or Universal Review*, 1756–7, VII, pp. 335–42 and XIII, pp. 161–67.

Review of 'Observations on the ancient and present state of the Isles of Scilly' by William Borlase, in *Literary Magazine, or Universal Review*, 1756, II, pp. 91–97.
Review of 'Philosophical Transactions, Volume xlix', in *Literary Magazine, or Universal Review*, 1756, IV, pp. 193–97.

Part II: Primary sources

Adams, George *A Treatise Describing and Explaining the Construction and Use of New Celestial and Terrestrial Globes* (London, 1766).
Addison, Joseph *A Letter from Italy* (London, 1701).
——*Remarks on Several Parts of Italy, &c. in the Years 1701, 1702, 1703* (5th edn, London, 1736).
——*The Works of the right honourable Joseph Addison* (6 vols, London, 1854–6).
——*The Drummer; or the Haunted House* (1716) in *The Miscellaneous Works of Joseph Addison: Volume 1: Poems and Plays*, ed. A.C. Guthkelch (London, 1914), pp. 431–90.
——*The Freeholder*, ed. James Leheny (Oxford, 1979).
Addison, Joseph and Richard Steele *The Spectator*, ed. Donald F. Bond (5 vols, Oxford, 1965).
Alison, Archibald *Essays on the Nature and Principles of Taste* (Dublin edn, 1792).
Anonymous *Chatsworth. A Poem* (n.p. 1788?) [Bod. S.M. Gough Derby 2(8)].
——*The Rise and Progress of the Present Taste in Planting Parks, Pleasure Grounds, Gardens, etc.* [1767], in John Dixon Hunt and Peter Willis (eds). *The Genius of the Place: the English Landscape Garden, 1620–1820* (Paperback edn, Cambridge, Mass., 1988), pp. 299–300.
——*A Trip to North Wales* [by a barrister of the Temple], in John Torbuck (ed.) *A Collection of Welsh Travels and Memoirs of Wales* (London, 1738).
——Review of *The Mysteries of Udolpho*, *The Gentleman's Magazine*, 64, 1794, p. 834.
Aristophanes *Lysistrata/The Acharnians/The Clouds*, trans. Alan R. Sommerstein (Harmondsworth, 1973).
Bailey, Nathan *Dictionarium Britannicum: Or, a more Complete Universal Etymological Dictionary than Any Extant* (London, 1730).
Bailey, Nathan, rev. Joseph Nichol Scott *A New Universal Etymological Dictionary* (new edn, London, 1764).
Baker, Thomas *Reflections upon Learning, wherein is shewn the insufficiency thereof, in its several particulars in order to evince the usefulness and necessity of revelation* (3rd edn, London, 1700).
Baretti, Giuseppe *An Account of the Manner and Customs of Italy: with observations on the Mistakes of some Travellers with Respect to that Country* (2 vols, London, 1768).
——*A Journey from London to Genoa through England, Portugal, Spain and France* (4 vols, London, 1770).
Barclay, James *An Examination of Mr Kenrick's review of Mr Johnson's Edition of Shakespeare* (London, 1766).
Barnard, John (ed.) *Pope: The Critical Heritage* (London, 1973).
Barrow, Isaac *Works* (3 vols, 5th edn, London, 1741).
Baxter, Richard *The Divine Life* (London, 1664).
——*The Reasons of the Christian Religion* (London, 1667).
Bell, John *Travels from St. Petersburgh in Russia, to various parts of Asia* (2 vols, Edinburgh, 1788 [1st edn, 1763]).
Bentley, Richard *Eight Sermons preach'd at the honourable Robert Boyle's Lecture, in the first year, MDCXCII* (6th edn, Cambridge, 1735).

Berkeley, George *Philosophical Works*, ed. M.R. Ayers (rev. edn, London, 1989).

Bevan, Edward *Box-Hill; a descriptive poem* (London, 1777).

Biographia Britannica: or, the lives of the most eminent men who have flourished in Great Britain and Ireland (6 vols, continuously paginated, London, 1747–66).

Blackmore, Richard *The Creation; A Philosophical Poem* (London, 1712).

Blainville, P. de *Travels through Holland, Germany, Switzerland, but especially Italy* (trans., 3 vols, London, 1757).

Borlase, William *Observations on the Ancient and Present State of the Islands of Scilly* (Oxford, 1756).

Boswell, James *An Account of Corsica* (Glasgow, 1768).

——*Correspondence and other papers relating to the making of the 'Life of Johnson'*, ed. Marshall Waingrow (London, 1969).

——*Correspondence of James Boswell with David Garrick, Edmund Burke and Edmond Malone*, ed. Peter Baker *et al.* (London, 1968).

——*Journal of a Tour to the Hebrides*, eds Frederick A. Pottle and Charles H. Bennett (London, 1936).

——*Life of Johnson*, ed. G.B. Hill, rev. L.F. Powell (6 vols, Oxford, 1934–50).

—— *Life of Johnson: an edition of the original manuscript in four volumes*, ed. Marshall Waingrow (4 vols, Edinburgh 1994–).

——*On the Grand Tour: Germany and Switzerland, 1764*, ed. Frederick Pottle (London, 1953).

——*On the Grand Tour: Italy, Corsica, and France, 1765–6*, eds Frank Brady and Frederick Pottle (London, 1955).

Boulton, James (ed.) *Johnson: the Critical Heritage* (London, 1971).

Bourne, Nicholas [Printer] *Sir Francis Drake Revived* (London, 1653).

Bowden, Samuel *Poetical Essays on Several Occasions* (2 vols, London, 1733–35).

Boyle, Robert *A Free Enquiry into the Vulgarly Receiv'd Notion of Nature: Made in an Essay address'd to a Friend* (London, 1685).

——*General heads for the Natural History of a Country, Great or Small; drawn out for the use of Travellers and Navigators* (London, 1692 edn).

——*Works*, ed. Thomas Birch (5 vols, London, 1744).

Brown, John *A Description of the Lake at Keswick* (London, 1770).

Browne, Henry *An Illustration of Stonehenge and Abury, in the County of Wilts., pointing out their Origin and Character, through Considerations hitherto unnoticed* (Salisbury, 1823).

Browne, Patrick *The Civil and Natural History of Jamaica* (2nd edn, London, 1789).

Browne, Thomas *Pseudodoxia Epidemica: or, enquiries into the very many received tenets, and common, presumed truths* (5th edn, London, 1669).

——*Posthumous Works of the learned Thomas Browne* (London, 1712).

——*Religio Medici* (London, 1736 edn).

Brydone, Patrick *A Tour through Sicily and Malta, in a series of letters to William Beckford, Esq.* (2 vols, continuously paginated, Dublin, 1773).

Bunyan, John *The Pilgrim's Progress*, ed. Roger Sharrock (Harmondsworth, 1965).

Burke, Edmund *A Philosophical Enquiry into the Origin of our Ideas of the Sublime and Beautiful* [2nd edn, 1759], ed. James T. Boulton (2nd edn, Oxford, 1987).

——*Pre-Revolutionary Writings*, ed. Ian Harris (Cambridge, 1993).

Burnet, Thomas *The Sacred Theory of the Earth* (2 vols, 4th edn, London, 1719).

Burney, Charles *The Present State of Music in France and Italy; or, the Journal of a Tour taken through these countries* (2nd edn, London, 1773).

Butler, Joseph *The Analogy of Religion, Natural and Revealed* (1736 [London 1906]).

Byng, John *The Torrington Diaries. Containing the Tours through England and Wales of the Hon. John Byng*, ed. C. Bruyn Andrews (4 vols, London 1934–38).
Callendar, James *A Critical Review of the Works of Dr Samuel Johnson* (Edinburgh, 1783).
Campbell, John *The Travels of Edward Brown, Esq.* (2 vols, London, 1753).
Castell, Robert *The Villas of the Ancients Illustrated* (London, 1728).
Cato, Marcus *De Agri Cultura*, trans., William D. Hooper (London, 1935).
The Censor (3 vols, 2nd edn, London, 1717).
Chalmers, Alexander *Works of the English Poets* (19 vols, London, 1810).
Chambers, William *A Dissertation on Oriental Gardens* (2nd edn, London, 1773).
The Champion (2 vols, London, 1741).
Champion, Anthony *Miscellanies in Verse and Prose*, ed. William Henry (London, 1801).
Chandler, Richard *Travels in Asia Minor: or, an account of a tour made at the expense of the Society of Dilettanti* (Dublin, 1775).
Cheyne, George *Philosophical Principles of Religion: Natural and Reveal'd* (2nd edn, London, 1716).
Cibber, Theophilus *The Lives of the Poets of Great Britain and Ireland from the time of Dean Swift* (5 vols, London, 1753).
Clarke, Samuel *Sermons*, ed. John Clarke (10 vols, 3rd edn, London, 1732).
Cockin, William *Ode to the Genius of the Lakes in the North of England* (London, 1780).
Colman, George and David Garrick *The Clandestine Marriage* (London, 1766).
Columella *De Rei Rustica*, trans., H.B. Ash (3 vols, London, 1941).
The Connoisseur (4 vols, 5th edn, Oxford, 1767).
Cooper, Anthony Ashley, Third Earl of Shaftesbury *Characteristics of Men, Manners, Opinions, Times* (3 vols, 6th edn, London, 1737).
Cooper, John Gilbert *The Power of Harmony* (London, 1749).
Coventry, Henry *Philemon to Hydaspes: relating a conversation with Hortensius upon the Subject of False Religions* (London, 1736).
Cowper, William *Poetical Works*, ed. H.S. Milford (Oxford, 1967).
Craddock, Joseph *An Account of some of the most Romantic Parts of North Wales* (London, 1777).
——*Letters from Snowdon: descriptive of a Whole Tour through the Northern Counties of Wales* (London, 1770).
——*Village Memoirs: in a series of letters between a Clergyman and his Family in the Country and his Son in Town* (3rd edn, London, 1775).
Crawford, Charles *Richmond Hill. A Poem* (London, 1777).
Crouch, Nathaniel [Printer] *The English Hero: or, Sir Francis Drake Revived* (London, 1716).
Defoe, Daniel *Caledonia. A Poem in Honour of Scotland, and the Scots Nation* (London, 1707).
——*A Tour Through the Whole Island of Great Britain*, ed. Pat Rogers (Harmondsworth, 1971).
Denham, John *Expans'd Hieroglyphicks: A Critical Edition of Sir John Denham's 'Cooper's Hill'*, ed. Brendan O'Hehir (Berkeley, 1969).
Derham, William *Physico-Theology: or, a Demonstration of the Being and Attributes of God, from His Works of Creation* (4th edn, London, 1716).
Dodsley, Robert *Public Virtue: A Poem* (Dublin, 1754).
——(ed.) *The Museum: or, the Literary and Historical Register* (3 vols, London, 1746).
Drake, Francis *The World Encompassed* (London, 1628).

Duck, Stephen *Poems on Several Occasions* (London, 1736).
——*Caesar's Camp: or, St. George's Hill. A Poem* (London, 1755).
Dyer, John Poems, in *The Works of the English Poets. With Prefaces, Biographical and Critical, by Samuel Johnson* (London, 1790), vol. 58.
Edwards, John *The Tour of the Dove; or, a visit to Dovedale, &c. A Poem* (2nd edn London [1825]).
Edwards, Thomas *The Canons of Criticism, and Glossary, being a Supplement to Mr Warburtons's Edition of Shakespear* (5th edn, London, 1753).
Elwood, Ann *Memoirs of the Literary Ladies of England* (2 vols, London, 1843).
'English, John' [Pseud?] *Travels through Scotland* (London, 1762 [?]).
Erasmus *Praise of Folly*, trans. Betty Radice (rev. edn, Harmondsworth, 1993).
Farmer, Richard *An Essay on the Learning of Shakespeare* (Cambridge, 1767).
Fenton, Elijah (ed.) *Milton's Paradise Lost* (14th edn, London, 1741).
Fielding, Henry *The History of Tom Jones* (London, 1749).
——*The History of Tom Jones, A foundling*, eds John Bender and Simon Stern (Oxford, 1996).
——*The Journal of a Voyage to Lisbon* (London, 1755).
Fiennes, Celia *Journeys* [1685–1712], ed. Christopher Morris (rev edn, London, 1949).
Finch, Anne, Countess of Winchelsea *Poems*, ed. Myra Reynolds (Chicago, 1903).
Fontenelle, Bernard Le Bovier de *Histoire de l'Academe Royale des Sciences* (Paris 1724).
Forster, J. Georg *A Voyage Round the World in his Britannic Majesty's sloop Resolution, commanded by Capt. James Cook, during the Years 1772, 3, 4 and 5* (2 vols, London, 1777).
——*The Freethinker: or, Essays of Wit and Humour* (3 vols, 3rd edn, London, 1739 [for nos. 1–159; then individual nos. uncollected]).
Garnett, Thomas *Observations on a Tour through the Highlands and part of the Western Isles of Scotland* (2 vols, London, 1800).
Gentleman's Magazine (London, 1731 et seq.).
Gerard, Alexander *An Essay on Taste* (London, 1759).
Gibbons, Thomas *Juvenalia: Poems on Various Subjects of Devotion and Virtue* (London, 1750).
Gilpin, William *Observations, relative chiefly to Picturesque Beauty made in the Year 1776, on several parts of Great Britain; Particularly the High-Lands of Scotland* (2 vols, 2nd edn, London, 1792).
——*Observations on the River Wye* (London, 1782).
——*Observations . . . of Cumberland and Westmoreland* (2 vols, London, 1786).
——*Remarks on forest scenery* (2 vols, London, 1791).
——*Three Essays: on Picturesque Beauty; on Picturesque Travel; and on Sketching Landscape* (London, 1792).
——*Sermons preached to a country congregation* (3 vols, Lymington, 1799–1804).
——*Three Dialogues on the Amusements of Clergymen* (London, 1796).
——*Observations on the Western Parts of England* (London, 1798).
——*Moral Contrasts: or, the Power of Religion Exemplified under Different Characters* (Lymington, 1798).
——*Observations on the coasts of Hampshire, Sussex and Kent* (London, 1804).
——*Dialogues on Various Subjects* (London, 1807).
——*Observations on several parts of the counties of Cambridge, Norfolk, Suffolk, and Essex. Also on several parts of North Wales* (London, 1809).
——'An account of the Revd. Mr Gilpin, vicar of Boldre in New Forest, written by himself' in *Gilpin Memoirs*, ed. William Jackson for the *Cumberland and Westmoreland Antiquarian and Archaeological Society* (London, 1879).

Glanville, Joseph *Scepsis Scientifica: or, confest ignorance, the way to science* (London, 1665).

Goldsmith, Oliver *Poems and Plays*, ed. Tom Davis (2nd edn, London, 1990).

——*The Citizen of the World* (2 vols, London, 1762).

Graves, Richard *Columella: or, the Distressed Anchoret* (2 vols, London, 1779).

——*Euphrosyne: or, Amusements on the Road of Life* (London, 1776).

——*Recollections of some particulars of the life of the late William Shenstone, esq.* (London, 1788).

——*The Spiritual Quixote: or, the Summer's Ramble of Mr Geoffry Wildgoose. A Comic Romance* (2 vols, Dublin, 1774).

Grey, Zachary *A Chronological and Historical Account of the most Memorable Earthquakes that have happened in the World, from the Beginning of the Christian Period to the Present Year, 1750* (Cambridge, 1750).

——*Critical, Historical, and Explanatory notes on Shakespeare* (2 vols, London, 1754).

Grew, Nehemiah *Cosmologia Sacra: or, a Demonstration of the Universe as it is the Creature and Kingdom of God* (London, 1701).

The Grumbler (London, 1715).

Gwynn, John *An Essay on Design* (London, 1749).

Hale, Matthew *The Primitive Origination of Mankind, Considered and Examined According to the Light of Nature* (London, 1677).

Hall, Samuel An Attempt to Shew, that a Taste for the Beauties of Nature and the Fine Arts, has no Influence Favourable to Morals, in *Memoirs of the Literary and Philosophical Society of Manchester*, 1, 1785, pp. 223–40.

Hammond, Henry *A Practical Catechism* (London, 1677).

Hanmer, Thomas *Works of Shakespear* (6 vols, London, 1745 edn).

Hanway, Jonas *An Historical Account of the British Trade over the Caspian Sea: with a Journal of travels* (4 vols, London, 1753).

——*A Journal of Eight Days' Journey from Portsmouth to Kingston Upon Thames. With Miscellaneous Thoughts, Moral and Religious* (2 vols, 2nd edn, London, 1757).

Hanway, Mary *A Journey to the Highlands of Scotland, with occasional remarks on Dr Johnson's Tour* (London, 1775).

Harris, James *Three Treatises* (London, 1744).

Hawkins, John *The Life of Samuel Johnson* (2nd edn, London, 1787).

Haywood, Eliza *The Female Spectator* (3 vols, London, 1745).

Heath, Benjamin *A Revisal of Shakespear's Text* (London, 1765).

Heely, Joseph *Letters on the Beauties of Hagley, Envil, and the Leasowes* (2 vols, London, 1777).

Herodotus *Histories*, trans. David Greene (Chicago, 1987).

Herring, Thomas *Letters from the Late Most Reverend Thomas Herring* (London, 1777).

Hervey, James *Works* (6 vols, London, 1807).

——*Meditations and Contemplations* (2 vols, 3rd edn, London, 1748).

Hill, G.B. (ed.) *Johnsonian Miscellanies* (2 vols, Oxford, 1897).

Home, Henry, Lord Kames *Elements of Criticism* (3 vols, 2nd edn, Edinburgh, 1763).

Hooker, Richard *Of the Laws of Ecclesiastical Polity* (3 vols, London, 1830 [1st edn, 1593–1661]).

Horace *Odes and Epodes*, trans. C.E. Bennett (London, 1938).

——*Satires, Epistles, Ars Poetica*, trans. H.R. Fairclough (London, 1929).

Horne, George *A Fair, Candid and Impartial State of the Case between Sir Isaac Newton and Mr Hutchinson* (Oxford, 1753).

Howell, James *Instructions and Directions for Forreine Travell* (London, 1650 edn).

Hume, David *Dialogues concerning natural religion* (London, 1779).

——*An Enquiry Concerning the Principles of Morals* [1751], eds L. Selby-Bigge and P.H. Nidditch (3rd edn, Oxford, 1975).

——*Essays: Moral, Political and Literary*, ed. E.F. Miller (2nd edn, Indianapolis, 1987).

——*A Treatise of Human Nature* [1739/40], ed. L. Selby-Bigge and P.H. Nidditch (2nd edn, Oxford, 1978).

Humphreys, Samuel *Cannons. A Poem* (London, 1728).

Hurn, William *Heath Hill: A Descriptive Poem in Four Cantos* (Colchester, 1777).

Hutcheson, Francis *An Inquiry into the Origin of our Ideas of Beauty and Virtue* (3rd edn, London, 1729).

Hutchinson, John *Philosophical and Theological Works*, eds Robert Spearman and Julius Bate (12 vols, 3rd edn, London 1748–49).

Hutchinson, John *Tour through the High Peak of Derbyshire* (Macclesfield, 1809).

Hutton, James Theory of the Earth, in *Trans. of the Royal Society of Edinburgh*, 1, 1788, pp. 209–304.

Jago, Richard *Edge-Hill, or, the Rural Prospect Delineated and Moralized* (London, 1767).

Jones, William *The Book of Nature: or, the true sense of things explained and made easy to the capacities of children* (5th edn, London 1803).

——*Observations in a Journey to Paris by way of Flanders, in the month of August 1776* (2 vols, London, 1777).

——*Physiological Disquisitions: or, Discourses on the Natural Philosophy of the Elements* (London, 1781).

——*Theological, Philosophical and Miscellaneous Works* (12 vols, London, 1801).

Kavanagh, Julia *English Women of Letters* (2 vols, London, 1863).

Keate, George *Poetical Works* (2 vols, London, 1780).

Kennedy, John *A Complete System of Astronomical Chronology, unfolding the Scriptures* (London, 1762).

Keysler, John George *Travels through Germany, Bohemia, Hungary, Switzerland, Italy and Lorrain* (trans., 4 vols, London, 1756).

Kirkpatrick, James *The Sea-Piece. A Narrative. Philosophical and Descriptive* (London, 1750).

Knight, Ellis Cornelia *Dinarbas; a tale: being a continuation of Rasselas, Prince of Abyssinia*, ed. Lynne Meloccaro (London, 1994).

Knight, Richard Payne *The Landscape. A Didactic Poem* (2nd edn, London, 1795).

Knox, John *A Tour through the Highlands of Scotland, and the Hebrides Islands in MDCCLXXXVI* (London, 1787).

Knox, Vicesimus *Essays, Moral and Literary* (2 vols, London, 1782).

Lawrence, John *The Clergyman's Recreation: Shewing the Pleasure and Profit of the Art of Gardening* (4th edn, London, 1716).

——*The Gentleman's Recreation: or, the Second Part of the Art of Gardening Improved* (2nd edn, London, 1717).

Leland, John *Leland's Itinerary: travels in Tudor England*, ed. John Chandler (Stroud, 1993).

Leigh, Charles *The Natural History of Lancashire, Cheshire, and the Peak in Derbyshire* (Oxford, 1700).

Lloyd, Robert *Poems* (London, 1762).

——*Loyd A Month's Tour in North Wales, Dublin and its Environs* (London, 1781) [Bod. S.M. Gough Wales 20(3)].

Lyttelton, George An Account of a Journey into Wales [1756], in Henry Windham *A Gentleman's Tour Through Monmouthshire and Wales* (new edn, London, 1781).

M'Nichol, Donald *Remarks on Dr Samuel Johnson's Journey to the Hebrides* (London, 1779).

Macbean, Alexander *A Dictionary of Ancient Geography* (London, 1773).

Macky, John *A Journey Through Scotland* (London, 1723).

Mallett, David *The Excursion; a Poem* (London, 1728).

Malone, Edmond (ed.) *The Plays and Poems of Shakspeare* (10 vols, London, 1790).

Malone, Edmond and James Boswell (the younger) (eds) *The Works of Shakespeare* (21 vols, London, 1821).

Marshall, Joseph *Travels* (4 vols, 2nd edn, London, 1773).

Marshall, William *Planting and Rural Ornament* (2nd edn, London, 1796).

Martin, Benjamin *Lingua Britannica Reformata: or, a new universal English Dictionary* (2nd edn, London, 1754).

Martin, Martin *A Description of the Western Islands of Scotland* (London, 1703).

Mason, George *An Essay on Design in Gardening* (2nd edn, London, 1795).

Mason, William *The English Garden* (London, 1782).

——(ed.) *Memoirs of the Life and Writings of William Whitehead* (York, 1788).

Maude, Thomas *Verbeia; or, Wharfdale, A Poem Descriptive and Didactic* (York, 1782).

——*Wensley-Dale; or, Rural Contemplations; A Poem* (3rd edn, London, 1780).

Milton, John *Complete Poems*, ed. John T. Shawcross (rev. edn, New York, 1971).

Moore, Henry *Picturesque Excursions from Derby to Matlock Bath, and its vicinity* (Derby, 1818).

Munro, Donald Description of the Western Isles of Scotland, called Hybrides [1594], in *Miscellania Scotica: A Collection of Tracts Relating to the History, Antiquities, Topography and Literature of Scotland*, vol II (Glasgow, 1818).

Murdock, Patrick (ed.) *Works of James Thomson* (London, 1762).

Newton, Thomas (ed.) *Milton's Paradise Lost* (2 vols, 3rd edn, London, 1754).

Newman, John Henry *Idea of a University, Defined and Illustrated* [1852], ed. Martin Svaglic (Notre Dame, 1982).

Nichols, John (ed.) *Illustrations of the Literary History of the Eighteenth Century* (8 vols, London, 1817–58).

Nichols, Philip *Sir Francis Drake Revived* (London, 1626).

Nourse, Timothy *Campania Foelix: or, a Discourse of the Benefits and Improvements of Husbandry* (London, 1700).

Olla Podrida (2nd edn, London, 1788).

Paine, Thomas *Rights of Man, Common Sense and other political writings*, ed. Mark Philp (Oxford, 1995).

Paley, William *Works* (London, 1827).

Patrick, Simon *A Brief Account of the New Set of 'Latitude Men'* (London, 1662).

Patterson, Samuel *Another Traveller! or, Cursory Remarks and Critical Observations made on a Journey through part of the Netherlands, by Coriat Junior* (London, 1767).

Pearch, George (ed.) *A Collection of Poems in Four Volumes by Several Hands* (4 vols, London, 1770).

Pennant, Thomas *A Tour in Scotland MDCCLXIX* (2nd edn, London, 1772).

——*A Tour in Scotland and Voyage to the Hebrides, MDCCLXXII* (2 vols, Chester, 1774).

——*A Tour in Wales* (2 vols, London, 1784).

Percival, Thomas *Moral and Literary Dissertations* (Warrington, 1784).

Phillips, Edward (ed.) *Letters of State, written by Mr John Milton* (London, 1694).

Piozzi, Hester *Anecdotes of the Late Samuel Johnson, LLD., during the last twenty years of his Life*, ed. Arthur Sherbo (Oxford, 1974).

——French Journal, ed. Moses Tyson and Henry Guppy, in *The French Journals of Mrs Thrale and Dr Johnson* (Manchester, 1932).

——Journal of a Tour in Wales with Dr Johnson, in A.M. Broadley *Dr Johnson and Mrs Thrale* (London, 1910).

——*Observations and Reflections made in the Course of a Journey through France, Italy and Germany* (2 vols, London, 1789).

——*The Piozzi Letters: Correspondence of Hester Lynch Piozzi, 1784–1821 (formerly Mrs Thrale)*, ed. Edward A. and Lillian D. Bloom (6 vols, Newark, N.J., 1989–2003).

——*Thraliana: the diary of Mrs Hester Lynch Thrale (later Mrs Piozzi), 1776–1809*, ed. Katharine Balderston (2 vols, continuously paginated, Oxford, 1959).

The Plain Dealer (2 vols, London, 1730).

Pococke, Richard *A Description of the East, and some other Countries* (2 vols, London, 1743).

Pope, Alexander *Windsor Forest*, ed. E. Audra and A. Williams (London, 1961).

——(ed.) *Works of Shakespear* (6 vols, London, 1723–25).

Potter, Robert *The Art of Criticism; as Exemplified in Dr Johnson's Lives* (London, 1789).

——*An Inquiry into some Passages in Dr Johnson's Lives of the Poets* (London, 1783).

Price, Richard *Political Writings*, ed. D.O. Thomas (Cambridge, 1991).

Price, Uvedale *An Essay on the Picturesque* (London, 1796 edn).

Quintilian *Institutio Oratoria*, trans. H.E. Butler (4 vols, London, 1921).

Radcliffe, Ann *A Sicilian Romance*, ed. Alison Milbank (Oxford, 1993).

——*The Mysteries of Udolpho*, ed. Bonamy Dobrée (Oxford, 1966).

——*The Italian, Or The Confessional of the Black Penitents. A Romance*, ed. Frederick Garber (Oxford, 1981).

——*The Romance of the Forest*, ed. Chloe Chard (Oxford, 1986).

——*A Journey made in the Summer of 1794, Through Holland and the Western Frontier of Germany, with a return down the Rhine: To which are added Observations during a tour to the Lakes of Lancashire, Westmoreland, and Cumberland* (London, 1795).

Ramsay, Allan *The Investigator* (London, 1762).

Ray, John *The Wisdom of God as Manifested in the Works of the Creation* (London, 1691).

Repton, Humphrey Sketches and Hints on Landscape Gardening [1795], in John Nolen (ed.) *The Art of Landscape Gardening* (London, 1907).

——Theory and Practice of Landscape Gardening [1804], in John Nolen (ed.) *The Art of Landscape Gardening* (London, 1907).

Reynolds, Frances *An Enquiry concerning the Principles of Taste, and the Origins of our Ideas of Beauty, &c.* (London, 1789).

Reynolds, Joshua *Discourses on Art*, ed. Robert R. Wark (New Haven, 1975).

Richardson, Jonathan *An Essay on the Theory of Painting* (London, 1725 edn).

Richardson, Jonathan, father and son *Explanatory Notes and Remarks on Milton's Paradise Lost* (London, 1734).

Richardson, Samuel *Sir Charles Grandison* (London, 1754).

——*Clarissa, or the history of a young lady*, ed. Angus Ross (Harmondsworth, 1985).

Ritso, George *Kew Gardens. A Poem* (London, 1753).

Ritson, Joseph *Remarks, Critical and Illustrative, on the Text and Notes of the Last Editor of Shakespeare* (London, 1783).

Rodenhurst, T *A Description of Hawkestone* (2nd edn, Shrewsbury, 1784).

Rogers, John *Nineteen Sermons on several occasions* (London, 1735).

Rolt, Richard *Cambria. A Poem, Illustrated with Historical, Critical and Other Explanatory Notes* (London, 1749).

Ruffhead, Owen *The Life of Alexander Pope* (London, 1769).

Rymer, Thomas *A Short View of Tragedy* (London, 1693).

Sacheverell, William *An Account of the Isle of Man* (London, 1702).

——*An Account of the Isle of Man*, ed. J.G. Cumming (Douglas, 1859).

Savage, Richard *The Wanderer; a Vision in Five Cantos* (London, 1729).

Schultens, Albert *Oratio Academica in Memoriam Hermanni Boerhavii* (London, 1739).

Scott, John *Critical Essays on Some of the Poems of Several English Poets* (London, 1785).
Scott, Walter *Prose Works* (Edinburgh, 1834–36).
Sewell, George (ed.) *Poems by Mr John Philips* (2nd edn, London, 1715).
Shakespeare, William *Complete Works*, eds Stanley Wells *et al*. (Oxford, 1988).
——*Henry V*, ed. T.W. Craik (London, 1995).
——*The Winter's Tale*, ed. J.H.P. Pafford (London, 1963).
Shaw, Thomas *Travels, or Observations relating to the Barbary and the Levant* (2nd edn, London, 1757).
Simond, Louis *Journal of a Tour and Residence in Great Britain, During the Years 1810 and 1811* (2 vols, Edinburgh, 1815).
Skrine, Henry *Two Successive Tours through the whole of Wales* (London, 1795).
Smith, Adam *The Theory of Moral Sentiments* [1759], eds D.D. Raphael and A.L. Macfie (Oxford, 1976).
Smollett, Tobias *The Expedition of Humphrey Clinker* (London, 1771).
——*Travels through France and Italy* (2 vols, London, 1766).
South, Robert *Twelve Sermons preached on several occasions* (6 vols, 4th edn, London, 1718).
Spence, Joseph *Crito: or, a Dialogue on Beauty by 'Sir Harry Beaumont'* (London, 1752).
Sprat, Thomas Account of the Life and Writings of Mr Abraham Cowley [1668], in Richard Hurd (ed.) *Select Works in verse and prose of Mr A. Cowley* (2 vols, Dublin, 1772).
Steele, Richard *et al The Guardian* (Philadelphia, 1831).
——*The Guardian* ed. John Stephens (Lexington, Ky., 1982).
——*The Tatler*, ed. Donald F. Bond (3 vols, Oxford, 1987).
Sterne, Laurence *The Life and Opinions of Tristram Shandy, Gentleman*, eds Melvyn and Joan New (3 vols, Gainsville, Florida, 1978).
——*Sermons*, ed. Melvyn New (Gainesville, Florida, 1996).
——*A Sentimental Journey through France and Italy by Mr Yorick* (2 vols, London, 1768).
——*A Sentimental Journey*, ed. Ian Jack (Oxford, 1984).
Stockdale, Percival (ed.) *The Seasons by James Thomson* (London, 1793).
Story, Thomas *A Journal of the Life of Thomas Story* (London, 1747).
Strabo *The Geography of* Strabo, trans. H.L. Jones (8 vols, London, 1917–32).
Swift, Jonathan (?) The Briton Described: or, a Journey thro' Wales, in John Torbuck (ed.) *A Collection of Welsh Travels and Memoirs of Wales* (London, 1738).
Switzer, Stephen *Ichnographia Rustica: or, the Nobleman, Gentleman, and Gardener's Recreation* (London, 1718).
Talfourd, T.N. 'Memoir of the Life and Writings of Mrs Radcliffe' in *Gaston de Blondeville, or, The Court of Henry III. Keeping Festival in Ardenne, A Romance* (4 vols, London, 1826).
Taylor, Jeremy *A Course of Sermons for all the Sundays of the Year* (London, 1826 [1st edn, 1653]).
Theobald, Lewis (ed.) *The Works of Shakespeare* (7 vols, London, 1733).
Thomson, James *The Seasons*, ed. James Sambrook (Oxford, 1981).
——*The Seasons and the Castle of Indolence*, ed. James Sambrook (Oxford, 1972).
Tillotson, John *Works* (3 vols, 5th edn, London, 1735).
Tindal, William *Remarks on Dr Johnson's Life, and Critical Observations on the works of Mr Gray* (London, 1782).
Toland, John *The Life of John Milton* (London, 1699 edn).
Topham, Edward *Letters from Edinburgh; written in the Years 1774 and 1775* (London, 1776).
Tucker, Josiah *Instructions for Travellers* (Dublin, 1758).
Turnbull, George *A Treatise on Ancient Painting* (London, 1740).

Turner, Thomas *The Diary of Thomas Turner, 1754–65*, ed. David Vaisey (Oxford, 1984 [Reprint, East Hoathley, 1994]).

Twiss, Richard *Travels through Portugal and Spain, in 1772 and 1773* (London., 1775).

Upton, John *Critical Observations on Shakespeare* (2nd edn, London, 1748).

Ussher, James *Clio: or, a Discourse on Taste* (3rd edn, London, 1772).

Varro, Marcus Terrentius *Rerum Rusticorum*, trans., William D. Hooper (London, 1935).

Vitruvius *De Architectura*, trans. F. Granger (2 vols, London, 1934).

Walker, John The Rev. *Dr Walker's Reports on the Hebrides of 1764 and 1771*, ed. Margaret McKay (Edinburgh, 1980).

Walpole, Horace The History of Modern Taste in Gardening, in *Anecdotes of Painting in England* (4 vols, Strawberry Hill, 1771), vol. 4.

——Journal of Visits to Country Seats, &c., ed. Paget Toynbee, in *Walpole Society*, 66, 1927–8, pp. 9–80.

Wansleben, John *The Present State of Egypt; or, a New Relation of a Late Voyage into that Kingdom* (London, 1678).

Warburton, William *A Critical and Philosophical Enquiry into the Causes of Prodigies and Miracles, as related by Historians* (London, 1727).

——*Julian. or, a discourse concerning the arthquake and Firey Eruption, which defeated the Emperor's attempt to rebuild the Temple at Jerusalem* (2nd edn, London, 1751).

——*A View of Lord Bolingbroke's Philosophy* (2 vols, London, 1754–5).

——(ed.) *Works of Alexander Pope* (9 vols, London, 1752 edn).

——(ed.) *Works of Shakespeare* (8 vols, London, 1747).

Warner, Richard *A Tour through the Northern Counties of England, and the Borders of Scotland* (2 vols, Bath, 1802).

Warton, Joseph *The Enthusiast; or, The Lover of Nature* (London, 1744).

——*An Essay on the Genius and Writings of Pope* (2 vols, London, 1756–82).

Warton, Thomas *Correspondence*, ed. David Fairer (Athens, Ga., 1995).

Webb, Daniel *An Inquiry into the Beauties of Painting* (2nd edn, London, 1761).

Webster, William *The Weekly Miscellany, by Richard Hooker* (2 vols, 2nd edn, London, 1738).

Wesley, John *Journal* ed. Nehemiah Curnock (8 vols, London, 1909–16).

——*Serious Thoughts Occasioned by the late Earthquakes in Lisbon* (London, 1755).

——*A Survey of the Wisdom of God in the Creation* (3 vols, Bristol, 1763–70).

Whalley, Peter *An Enquiry into the Learning of Shakespeare* (London, 1748).

White, Gilbert *The Natural History and Antiquities of Selbourne*, ed. Richard Mabey (Harmondsworth, 1977).

Wilkins, John *Of the Principles and Duties of Natural Religion* (7th edn, London, 1715).

Windham, Henry *A Gentleman's Tour through Monmouthshire and Wales* (new edn, London, 1781).

Woodhouse, James *Poems on Several Occasions* (2nd edn, London, 1766).

Wordsworth, William *Poetical Works*, ed. Thomas Hutchinson, rev. E. de Selincourt (Oxford, 1936).

——*"Tintern Abbey" (1798)* in *Complete Poetical works*, ed. Thomas Hutchinson and Ernest de Selincourt (Oxford, 1969).

——*The World* (6 vols, London, 1755).

Worthington, William *The Scripture Theory of the Earth* (London, 1773).

Wraxall, Nathaniel *A Tour Through Some of the Northern parts of Europe* (Dublin, 1776).

Wright, Edward *Some Observations made Travelling through France, Italy, &c. In the Years 1722, 3, 4 and 5* (London, 1730).

Yate, Richard *A Letter in defence of Dr Middleton: being an Argument Proving that Miraculous Powers were Never Truly Wrought but by Persons Divinely Inspired* (London, 1749).
——*Letter to a Gentleman in Stourbridge* (London, 1739).
Young, Arthur *A Six Months Tour through the North of England* (4 vols, 2nd edn, London, 1771).
——*A Six Weeks Tour, through the Southern Counties of England and Wales* (2nd edn, London, 1769).
Young, Edward *Night Thoughts*, ed. S. Cornford (Cambridge, 1989).
——*The correspondence of Edward Young*, ed. Henry Pettit, (Oxford, 1971).
——*Conjectures on Original Composition*, in *Works* (6 vols, London, 1774–78).

Part III: Secondary writings relating to Samuel Johnson

Abbott, John L. Dr Johnson, Fontenelle, LeClerk and six 'French' Lives, in *Modern Philology*, 63, 1965–6, pp. 121–27.
The Age of Johnson Volumes 7 and 8 (1996–7).
Atkinson, A.D. Dr Johnson and some physico-theological themes, in *Notes and Queries*, 197, 1952, pp. 16–18, 162–65 and 249–53.
Alkon, Paul Robert South, William Law and Samuel Johnson, in *Studies in English Literature, 1500–1900*, 6, 1966, pp. 499–528.
——*Samuel Johnson and Moral Discipline* (Evanston, 1967).
Balderston, Katharine Dr Johnson and William Law, in *Proceedings of the Modern Language Association of America*, 75, 1960, pp. 382–94.
Bate, Walter J. *The Achievement of Samuel Johnson* (New York, 1955).
——*Samuel Johnson* (London, 1975).
Battersby, James Johnson and Shiels: biographers of Addison, in *Studies in English Literature, 1500–1900*, 9, 1969, pp. 521–37.
Bloom, Edward A. *Samuel Johnson on Grub Street* (Providence, 1957).
——Symbolic Names in Johnson's Periodical Essays, in *Modern Language Quarterly*, 13, 1952, pp. 333–52.
Boyd, D.V. Vanity and Vacuity: a reading of Johnson's Verse Satires, in *English Literary History*, 39, 1972, pp. 387–403.
Brack, O.M. The *Gentleman's Magazine*, concealed printing, and the texts of Samuel Johnson's Lives of Admiral Robert Blake and Sir Francis Drake, in *Studies in Bibliography*, 40, 1987, pp. 140–46.
——Johnson's Life of Admiral Blake and the development of biographical technique, in *Modern Philology*, 85, 1988, pp. 523–31.
Brack, O.M. and Thomas Kaminski, Johnson, James and the *Medicinal Dictionary*, in *Modern Philology*, 81, 1983–84, pp. 140–46.
Bronson, Bertrand H. *Johnsonian Agonistes and Other Essays* (Berkeley, 1965).
Brownell, Morris *Samuel Johnson's Attitude to the Arts* (Oxford, 1989).
Butt, John Johnson's Practice in Poetical Imitation, in F.W. Hilles (ed.) *New Light on Dr Johnson. Essays on the Occasion of his 250th Birthday* (New Haven, 1959), pp. 19–34.
Campbell, Charles C. Image and Symbol in *Rasselas*: narrative form and the Flux of Life, in *English Studies in Canada*, 16, 1990, pp. 263–78.
Campbell, Hilbert H. Shiels and Johnson: biographers of Thomson, in *Studies in English Literature, 1500–1900*, 12, 1972, pp. 535–44.
Cannon, John *Samuel Johnson and the Politics of Hanoverian England* (Oxford, 1994).
Carnochan, W.B. Johnsonian Metaphor and the 'Adamant of Shakespeare', in *Studies in English Literature, 1500–1900*, 10, 1970, pp. 541–49.

Chapin, Chester Religion and the Nature of Johnson's Toryism, in *Cithara Essays in the Judaeo-Christian Tradition*, 29, 1990, pp. 38–54.

——*The Religious Thought of Samuel Johnson* (Ann Arbor, 1968).

——Samuel Johnson's 'Wonderful Experience', in Magdi Wahba (ed.) *Johnsonian Studies* (Cairo, 1962), pp. 51–60.

Clark, J.C.D. *Samuel Johnson: Literature, Religion and English Cultural Politics from the Restoration to Romanticism* (Cambridge, 1994).

Clark, Jonathan and Howard Erskine-Hill (eds) *Samuel Johnson in Historical Context* (London, 2002).

Clifford, James L. *Dictionary Johnson: Samuel Johnson's Middle Years* (London, 1979).

——*Young Samuel Johnson* (London, 1955).

Curley, Thomas *Samuel Johnson and the Age of Travel* (Athens, Ga., 1976).

Damrosch, Leopold *The uses of Johnson's Criticism* (Charlottesville, 1976).

DeMaria, Robert *Johnson's Dictionary and the Language of Learning* (Oxford, 1986).

——*The Life of Samuel Johnson: A Criticial Biography* (Oxford, 1993).

——*Samuel Johnson and the Life of Reading* (Baltimore, 1997).

——*The Politics of Johnson's Dictionary*, in *Proceedings of the Modern Language Association of America*, 104, 1989, pp. 64–74.

——*The Theory of Language in Johnson's Dictionary*, in Paul Korshin (ed.) *Johnson after Two Hundred Years* (Philadelphia, 1986), pp. 159–74.

Ditchfield, G.M. Samuel Johnson and the Dissenters, in *Bulletin of the John Rylands Library*, 68, 1986, pp. 373–409.

Eastman, Arthur M. In defence of Johnson, in *Shakespeare Quarterly*, 8, 1957, pp. 493–500.

Einbond, Bernard *Samuel Johnson's Allegory* (The Hague, 1971).

Elder, A.T. Irony and Humour in the *Rambler*, in *University of Toronto Quarterly*, 30, 1960, pp. 57–71.

Erskine-Hill, Howard The Political Character of Samuel Johnson, in Isobel Grundy (ed.) *Samuel Johnson: New Critical Essays* (London, 1984), pp. 107–36.

Fleeman, J.D. *A Bibliography of the Works of Samuel Johnson* (2 vols, continuously paginated, Oxford, 2000).

Folkenflik, Robert Samuel Johnson and Art, in Paul Alkon and Robert Folkenflik *Samuel Johnson: Pictures and Words* (Los Angeles, 1984).

——*Samuel Johnson, Biographer* (Ithaca, 1978).

Fussell, Paul *Samuel Johnson and the Life of Writing* (London, 1972).

Gilmore, Thomas B. Implicit Criticism of Thomson's *Seasons* in Johnson's *Dictionary*, in *Modern Philology*, 86, 1988–9, pp. 265–73.

Gold, Joel J. Johnson's Translation of Lobo, in *Proceedings of the Modern Language Association of America*, 80, 1965, pp. 51–61.

Gray, James *Johnson's Sermons: a study* (Oxford, 1972).

Greene, Donald J. Pictures to the Mind: Johnson and Imagery, in Mary Lascelles, *et al.* (eds) *Johnson, Boswell and their Circle: Essays Presented to Lawrence Fitzroy Powell* (Oxford, 1965), pp. 135–58.

——*The Politics of Samuel Johnson* (2nd edn, Athens, Ga., 1990).

——Response, in *Essays in Criticism*, 10, 1960, pp. 476–80.

——Samuel Johnson, Journalist, in Donavan H. Bond and W. Reynolds McLeod (eds) *Newsletters to Newspapers: Eighteenth-Century Journalism* (West Virginia, 1977), pp. 87–101.

——Samuel Johnson and Natural Law, in *Journal of British Studies*, 2, 1962, pp. 59–75.

Grundy, Isobel *Samuel Johnson and the Scale of Greatness* (Leicester, 1986).

Hagstrum, Jean H. *Samuel Johnson's Literary Criticism* (Chicago, 1952).

Hansen, Marlene The Happy Valley: a version of Hell and a Version of Pastoral, in *New Rambler*, 14, 1973, pp. 24–30.

Hardy, John P. Johnson's *London*: the Country vs. the City, in R.F. Brissenden (ed.) *Studies in the Eighteenth-Century. Papers Presented to the David Nichol Smith Memorial Seminar* (Canberra, 1968), pp. 251–68.

——*Samuel Johnson: A Critical Study* (London, 1979).

Hart, Francis Johnson as Philosophic Traveller: the Perfection of an Idea, in *English Literary History*, 36, 1969, pp. 679–95.

Hart, Jeffrey Johnson's *A Journey to the Western Islands*: History as art, in *Essays in Criticism*, 10, 1960, pp. 44–59.

Hart, Kevin Economic Acts: Johnson in Scotland, in *Eighteenth-Century Life*, 16, 1992, pp. 94–110.

Hedrick, Elizabeth Fixing the Language: Johnson, Chesterfield, and the Plan of the *Dictionary*, in *English Literary History*, 55, 1988, pp. 421–42.

——Locke's theory of Language and Johnson's *Dictionary*, in *Eighteenth-Century Studies*, 20, 1986–7, pp. 422–44.

Hickey, Alison 'Extensive Views' in Johnson's *Journey to the Western Islands of Scotland*, in *Studies in English Literature, 1500–1900*, 32, 1992, pp. 537–53.

Hill, G.B. *Footsteps of Dr Johnson* (London, 1890).

Hilles, Frederick Johnson's Poetic Fire, in F.W. Hilles and Harold Bloom (eds) *From Sensibility to Romanticism: Essays Presented to Frederick Pottle* (Oxford, 1965), pp. 67–77.

Horgan, A.D. *Johnson on Language: An Introduction* (London, 1994).

Hudson, Nicholas *Samuel Johnson and Eighteenth-Century Thought* (Oxford, 1988).

——Three Steps to Perfection: *Rasselas* and the Philosophy of Richard Hooker, in *Eighteenth-Century Life*, 14, 1990, pp. 29–39.

Hutchings, William and William Ruddick, Samuel Johnson and Landscape, in Nalini Jain (ed.) *Re-Viewing Samuel Johnson* (London, 1991), pp. 57–81.

Jackson, H.J. Johnson and Burton: The *Anatomy of Melancholy* and the *Dictionary of the English Language*, in *English Studies in Canada*, 5, 1979, pp. 36–48.

Jain, Nalini *The Mind's Extensive View: Samuel Johnson as a Critic of Poetic Language* (Strathtay, 1991).

Jemielity, Thomas Dr Johnson and the Uses of Travel, in *Philological Quarterly*, 51, 1972, pp. 448–59.

——'More in Notions than in Facts': Samuel Johnson's *Journey to the Western Islands*, in *Dalhousie Review*, 49, 1969, pp. 319–30.

——Thomas Pennant's Scottish Tours and *A Journey to the Western Islands of Scotland*, in Prem Nath (ed.) *Fresh Reflections on Samuel Johnson: Essays in Criticism* (New York, 1987), pp. 312–27.

Jenkins, Ralph E. 'And I Travelled after him': Johnson and Pennant in Scotland, in *Texas Studies in Literature and Language*, 14, 1972, pp. 445–62.

Jones, Emrys The Artistic Form of *Rasselas*, in *Review of English Literature*, N.S. 18, 1967, pp. 387–401.

Kaminski, Thomas *The Early Career of Samuel Johnson* (New York, 1987).

Kaul, R.P. The Philosopher of Nature in *Rasselas* XXII, in *Indian Journal of English Studies*, 3, 1962, pp. 116–20.

Keast, William R. Johnson and 'Cibber's' *Lives of the Poets*, 1753, in Carroll Camden (ed.) *Restoration and Eighteenth-Century Literature: Essays in Honour of Alan Dugald McKillop* (Chicago, 1963), pp. 89–101.

Ketton-Cremer, R.W. Johnson and Countryside, in Mary Lascelles *et al.* (eds) *Johnson, Boswell and their Circle: Essays Presented to Lawrence Fitzroy Powell* (Oxford, 1965), pp. 65–75.

Kolb, Gwin J. Johnson's Dissertation on Flying, in F.W. Hilles (ed.) *New Light on Dr Johnson. Essays on the Occasion of his 250th Birthday* (New Haven, 1959), pp. 91–106.

——The 'Paradise' in Abyssinia and the 'Happy Valley' in *Rasselas*, in *Modern Philology*, 56, 1958, pp. 10–16.

——Rousseau and the Background to the 'Life led according to Nature' in Chapter 22 of *Rasselas*, in *Modern Philology*, 73, 1976 (Supplement), pp. S.66-S.73.

——The Structure of *Rasselas*, in *Proceedings of the Modern Language Association of America*, 65, 1951, pp. 698–717.

Kolb, Gwin J. and Ruth A. The Selection and use of illustrative quotations in Dr Johnson's *Dictionary*, in Howard Weinbrot (ed.) *New Aspects of Lexicography: Literary Criticism, Intellectual History and Social Change* (Carbondale, 1972), pp. 61–72.

Koper, Peter T. Samuel Johnson's Rhetorical Stance in the *Rambler*, in *Style*, 12, 1978, pp. 23–34.

Korshin, Paul J. Johnson and the Renaissance Dictionary, in *Journal of the History of Ideas*, 35, 1975, pp. 300–12.

Landa, Louis A. Johnson's Feathered Man: 'A Dissertation on the Art of Flying' Considered, in W.H. Bond (ed.) *Eighteenth-Century Studies in Honour of Donald Hyde* (London, 1970), pp. 161–78.

Lascelles, Mary Johnson and Juvenal, in F.W. Hilles (ed.) *New Light on Dr Johnson. Essays on the Occasion of his 250th Birthday* (New Haven, 1959), pp. 35–55.

Leicester, James H. Johnson's Life of Shenstone: some observations on the sources, in Magdi Wahba (ed.) *Johnsonian Studies* (Cairo, 1962), pp. 189–222.

Lim, C.S. Emendation of Shakespeare in the Eighteenth Century: the case of Johnson, in *Cahiers Elisabethains*, 33, 1988, pp. 23–30.

Lipking, Lawrence Learning to Read Johnson: *The Vision of Theodore* and *The Vanity of Human Wishes*, in *English Literary History*, 43, 1749, pp. 517–37.

Lockhart, Donald 'The Fourth Son of the Mighty Emperor': the Ethiopian Background of Johnson's *Rasselas*, in *Proceedings of the Modern Language Association of America*, 78, 1963, pp. 516–28.

Lombardo, Agostino The Importance of Imlac, in Magdi Wahba (ed.) *Bicentenary Essays on Rasselas* (Cairo, 1959), pp. 31–49.

Lynn, Stephen Johnson's *Rambler* and Eighteenth-Century Rhetoric, in *Eighteenth-Century Studies*, 19, 1986, pp. 461–79.

McAdam, E.L. *Dr Johnson and the English Law* (Syracuse, 1951).

——Johnson's Lives of Sarpi, Blake and Drake, in *Proceedings of the Modern Language Association of America*, 58, 1943, pp. 466–76.

McGuffie, Helen *Samuel Johnson in the British Press, 1749–84. A Chronological Checklist* (New York, 1976).

McIntosh, Carey *The Choice of Life: Samuel Johnson and the World of Fiction* (New Haven, 1973).

McLaverty, James From Definition to Explanation: Locke's influence on Johnson's *Dictionary*, in *Journal of the History of Ideas*, 47, 1986, pp. 377–94.

Maner, Martin *The Philosophical Biographer: Doubt and Dialectic in Johnson's 'Lives of the Poets'* (Athens. Ga., 1988).

Manzalaoui, Mahmoud *Rasselas* and some Mediaeval Ancillaries, in Magdi Wahba (ed.) *Bicentenary Essays on Rasselas* (Cairo, 1959), pp. 59–73.

Mayhew, Robert *Geography and Literature in Historical Context: Samuel Johnson and Eighteenth-Century English Conceptions of Geography* (Oxford, 1997).

Meier, T.K. Pattern in Johnson's *A Journey to the Western Islands*, in *Studies in Scottish Literature*, 5, 1967, pp. 185–93.

O'Brien, Karen Johnson's View of the Scottish Enlightenment in *A Journey to the Western Islands of Scotland*, in *The Age of Johnson: A scholarly Annual*, Volume 4 (New York, 1991), pp. 59–82.

O'Flaherty, Patrick Johnson's *Idler*: the Equipment of a Satirist, in *English Literary History*, 37, 1970, pp. 211–25.

——Johnson in the Hebrides: Philosopher Becalmed, in *Studies in Burke and his Time*, 13, 1971, pp. 1986–2001.

——Towards an Understanding of Johnson's *Rambler*, in *Studies in English Literature, 1500–1900*, 18, 1978, pp. 523–36.

Olson, Robert C. *Motto, Context, Essay. The Classical Background of Samuel Johnson's Rambler and Adventurer Essays* (New York, 1984).

Pagliaro, Harold Structural Patterns of Control in *Rasselas*, in John Middendorf (ed.) *English Writers of the Eighteenth Century* (New York, 1971), pp. 208–29.

Pierce, Charles *The Religious Life of Samuel Johnson* (London, 1983).

Pittock, Murray Johnson and Scotland, in Jonathan Clark and Howard Erskine-Hill (eds) *Samuel Johnson in Historical Context* (London, 2002), pp. 184–96.

Preston, Thomas The Biblical Context of Johnson's *Rasselas*, in *Proceedings of the Modern Language Association of America*, 84, 1969, pp. 274–81.

——Homeric Allusion in *A Journey to the Western Islands of Scotland*, in *Eighteenth-Century Studies*, 5, 1971–72, pp. 545–58.

Quinlan, Maurice *Samuel Johnson: A Layman's Religion* (Madison, 1964).

Radner, John B. The Significance of Johnson's Changing Views of the Hebrides, in John Burke and Donald Kay (eds) *The Unknown Samuel Johnson* (Madison, 1983), pp. 131–49.

Reddick, Allen *The Making of Johnson's Dictionary, 1746–73* (Cambridge, 1990).

Reynolds, Richard B. Johnson's 'Life of Boerhaave' in perspective, in *Yearbook of English Studies*, 5, 1975, pp. 115–29.

Riely, John C. The Pattern of Imagery in Johnson's Periodical Essays, in *Eighteenth-Century Studies*, 3, 1970, pp. 384–97.

Roberts, S.C. *Dr Johnson and Others* (Cambridge, 1958).

Rogers, Pat *Johnson and Boswell: the Transit of Caledonia* (Oxford, 1995).

Sachs, Ariel *Passionate Intelligence: Imagination and Reason in the Work of Samuel Johnson* (Baltimore, 1967).

San Juan, E. The Actual and the Ideal in the Making of Johnson's *Dictionary*, in *University of Toronto Quarterly*, 34, 1964–65, pp. 146–58.

Schalit, Ann Literature as Product and Process: two different accounts of the same trip, in *Serif: the Kent State University Quarterly*, 4, 1967, pp. 10–17.

Scherwatzky, Steven Johnson, *Rasselas* and the Politics of Empire, in *Eighteenth-Century Life*, 16, 1992, pp. 103–13.

Schwartz, Richard B. Johnson's *Journey*, in *Journal of English and Germanic Philology*, 69, 1970, pp. 292–303.

——Johnson's Philosopher of Nature: *Rasselas*, Chapter 22, in *Modern Philology*, 74, 1976, pp. 196–200.

——*Samuel Johnson and the New Science* (Madison, 1971).

——*Samuel Johnson and the Problem of Evil* (Madison, 1975).

Sherbo, Arthur Johnson's Intent in the *Journey to the Western Islands of Scotland*, in *Essays in Criticism*, 16, 1966, pp. 382–97.

——Dr Johnson's Revision of his *Dictionary*, in *Philological Quarterly*, 31, 1952, pp. 372–82.

——The Making of Ramblers 186 and 187, in *Proceedings of the Modern Language Association of America*, 67, 1952, pp. 575–80.

——*Samuel Johnson, editor of Shakespeare* (Urbana, Illinois, 1956).

——Some Animadversions to Patrick O'Flaherty's Journey to the Western Islands of Scotland, in *Studies in Burke and his Time*, 13, 1971, pp. 2119–27.

Siebert, Donald Johnson as Satirical Traveller: *A Journey to the Western Islands of Scotland*, in *Tennessee Studies in Literature*, 19, 1974, pp. 137–48.

Sigworth, Oliver Johnson's *Lycidas*: the end of Renaissance Criticism, in *Eighteenth-Century Studies*, 1, 1967, pp. 159–68.

Sledd, James H. and Gwin J. Kolb *Dr Johnson's Dictionary: Essays in the Biography of a Book* (London, 1955).

Stewart, Keith Samuel Johnson and the Ocean of Life: variations on a Commonplace, in *Papers in Language and Literature*, 23, 1987, pp. 303–17.

Stock, R.D. *Samuel Johnson and Neoclassical Dramatic Theory: the Intellectual context of the Preface to Shakespeare* (Lincoln, Nebraska, 1973).

Sudan, Rajani Foreign Bodies: contracting identity in Johnson's *London* and the *Life of Savage*, in *Criticism*, 34, 1992, pp. 173–92.

Tillotson, Geoffrey Time in *Rasselas*, in Magdi Wahba (ed.) *Bicentenary Essays on Rasselas* (Cairo, 1959), pp. 97–103.

Tomarken, Edward *Johnson, Rasselas and the Choice of Criticism* (Lexington, 1989).

——Travels into the Unknown: *Rasselas* and *A Journey to the Western Islands of Scotland*, in John Burke and Donald Kay (eds) *The Unknown Samuel Johnson* (Madison, 1983), pp. 150–67.

Tucker, Susie and Henry Gifford, Johnson's Poetic Imagination, in *Review of English Studies*, N.S. 8, 1957, pp. 241–8.

Vance, John A. *Samuel Johnson and the Sense of History* (Athens, Ga., 1986).

Wain, John Dr Johnson's Poetry, in his *A House for Truth: Critical Essays* (London, 1972), pp. 105–29.

——*Samuel Johnson* (London, 1974).

Waingrow, Marshall The Mighty Moral of *Irene*, in F.W. Hilles and Harold Bloom (eds) *From Sensibility to Romanticism: Essays Presented to Frederick Pottle* (Oxford, 1965), pp. 79–92.

Walker, Robert *Eighteenth-Century Arguments for Immortality and Johnson's Rasselas* (Victoria, 1977).

——Johnson and the 'Age of Evidences', in *Huntington Library Quarterly*, 44, 1980–81, pp. 27–42.

——Johnson and the Trees of Scotland, in *Philological Quarterly*, 61, 1982, pp. 98–101.

Wasserman, Earl Johnson's *Rasselas*: Implicit Contexts, in *Journal of English and Germanic Philology*, 74, 1975, pp. 1–25.

Weinbrot, Howard Johnson's *London* and Juvenal's Third Satire: the Country as Ironic Norm, in *Modern Philology*, 73, 1976 (Supplement) S.56–65.

——Samuel Johnson, Percival Stockdale, and bric-bats from Grub Street: some later responses to the *Lives of the Poets*, in *Huntington Library Quarterly*, 56, 1993, pp. 105–34.

Weitzman, Arthur More Light on *Rasselas*: the Background to the Egyptian episodes, in *Philological Quarterly*, 48, 1969, pp. 42–58.

Wiltshire, John *Samuel Johnson and the Medical World: the Doctor and the Patient* (Cambridge, 1991).

Wimsatt, W.K. In Praise of *Rasselas*: four notes (converging), in Maynard Mack and Ian Gregor (eds) *Imagined Worlds: Essays on English Novels and Novelists in Honour of John Butt* (London, 1968).

——Johnson's *Dictionary*, in F.W. Hilles (ed.) *New Light on Dr Johnson. Essays on the occasion of his 250th Birthday* (New Haven, 1959), pp. 65–90.

——*Philosophic Words: a Study of Style and Meaning in the Rambler and Dictionary of Samuel Johnson* (New Haven, 1948).

——*The Prose Style of Samuel Johnson* (New Haven, 1941).

——Samuel Johnson and Dryden's *DuFresnoy*, in *Studies in Philology*, 48, 1951, pp. 26–39.

Woodruff, James Johnson's *Rambler* in contemporary context, in *Bulletin of Research in the Humanities*, 85, 1982, pp. 27–64.

——*Rasselas* and the Tradition of Menippean Satire, in Isobel Grundy (ed.) *Samuel Johnson: New Critical Essays* (London, 1984), pp. 158–85.

Part IV: Other secondary sources

Adams, Percy *Travellers and Travel Liars, 1660–1800* (Berkeley, 1962).

Allen, Brian Jonathan Tyers's Other Garden, in *Journal of Garden History*, 1, 1981, pp. 215–38.

Almond, Philip *Heaven and Hell in Enlightenment England* (Cambridge, 1994).

Anderson, Patrick *Over the Alps: Reflections on Travel and Travel Writing, with Special Reference to the Grand Tours of Boswell, Beckford, and Byron* (London, 1969).

Andrews, Malcolm *The Search for the Picturesque: Landscape Aesthetics and Tourism in Britain, 1760–1800* (Aldershot, 1989).

Arthos, J. *The Language of Natural Description in Eighteenth-Century Poetry* (London, 1949).

Ashcraft, Richard Latitudinarianism and toleration: historical myth versus political history in Richard Kroll *et al.* (eds) *Philosophy, Science and Religion in England, 1640–1700* (Cambridge, 1992), pp. 151–77.

Ashton, Nigel Horne and Heterodoxy: The defence of Anglican beliefs in the late Enlightenment, in *English Historical Review*, 108, 1993, pp. 899–907.

Atkinson, R.E. *Knowledge and Explanation in History: an Introduction to the Philosophy of History* (London, 1978).

Aubin, R.A. *Topographical Poetry in XVIIIth century England* (London, 1936).

Auspitz, J.L. Michael Joseph Oakeshott (1901–1990), in A. Jesse Norman (ed.) *The Achievement of Michael Oakeshott* (London, 1993), pp. 1–25.

Ballantyne, Andrew *Architecture, Landscape and Liberty: Richard Payne Knight and the Picturesque* (Cambridge, 1997).

Barbier, Carl Paul *William Gilpin: his Drawings, Teaching, and Theory of the Picturesque* (Oxford, 1963).

Bareham, Tony 'Paths to Guide Imagination's Flight': Some Eighteenth-Century Poets Abroad, in John McVeagh (ed.) *All Before Them: English Literature and the Wider World, 1660–1780* (London, 1990), pp. 247–62.

Barnard, Toby Art, Architecture, Artifacts and Ascendancy, in *Bullán*, 1.2, 1994, pp. 17–34.

Barrell, John *The Birth of Pandora and the Division of Knowledge* (London, 1992).

——*The Dark Side of the Landscape: the rural poor in English Painting, 1730–1840* (Cambridge, 1980).

——*English Literature in History, 1730–80. An Equal, Wide Survey* (London, 1983).

——Geographies of Hardy's Wessex, in *Journal of Historical Geography*, 8, 1982, pp. 347–61.

——*The Idea of Landscape and the Sense of Place, 1730–1840: An Approach to the Poetry of John Clare* (Cambridge, 1972).

——*The Political Theory of Painting from Reynolds to Hazlitt: 'the Body of the Public'* (New Haven, 1986).

——The Public Prospect and the Private View, in Simon Pugh (ed.) *Reading Landscape: Country – City – Capital* (Manchester, 1990), pp. 19–40.

Batten, Charles *Pleasurable Instruction: Form and Convention in Eighteenth-Century Travel Literature* (Berkeley, 1978).

Battestin, Martin *The Providence of Wit: Aspects of Form in Augustan Literature and the Arts* (Oxford, 1974).

——Tom Jones: the argument of design' in Henry Knight Miller *et al.*, eds., *The Augustan Milieu: essays presented to Louis A. Landa* (Oxford, 1970), pp. 289–319.

Battestin, Martin with Ruthe Battestin *Henry Fielding: A Life* (London, 1989).

Baxandall, Michael *Patterns of Intention: on the Historical Explanation of Pictures* (New Haven, 1985)

Beattie, John *Crime and the Courts in England, 1660–1800* (Oxford, 1986).

Berdoulay, V. The Contextual Approach, in D.R. Stoddart (ed.) *Geography, Ideology and Social Concern* (Oxford, 1981), pp. 8–16.

Berger, John *Ways of Seeing* (London, 1972).

——*The White Bird and other Essays* (London, 1985).

Berlin, Isaiah The Concept of Scientific History, in his *Concepts and Categories: Philosophical Essays* (Oxford, 1978), pp. 103–42.

Bermingham, Ann *Landscape and Ideology: the English Rustic Tradition, 1740–1860* (London, 1986).

Bevir, Mark The Errors of Linguistic Contextualism, in *History and Theory*, 31, 1992, pp. 276–98.

Black, Jeremy *The British Abroad: The Grand Tour in the Eighteenth Century* (Stroud, 1992).

——*The English Press in the Eighteenth Century* (Philadelphia, 1987).

Bloom, Edward A. and Lillian D. *Joseph Addison's Sociable Animal: in the Market Place, on the Hustings, in the Pulpit* (Providence, 1971).

Boucher, David The Creation of the Past: British Idealism and Michael Oakeshott's Philosophy of History, in *History and Theory*, 23, 1984, pp. 193–214.

Bradley, F.H. *The Presuppositions of Critical History* (Oxford, 1874).

Bradley, James *Religion, Revolution and English Radicalism: Non-Conformity in Eighteenth-Century Politics and Society* (Cambridge, 1990).

Brauer, George C. *The Education of a Gentleman: Theories of Gentlemanly Education in England, 1660–1775* (New York, 1959).

Brent, Richard God's Providence: Liberal political economy as natural theology at Oxford, 1825–62, in Michael Bentley (ed.) *Public and Private Doctrine: essays in British History presented to Maurice Cowling* (Cambridge, 1993), pp. 85–107.

Bronson, Bertrand H. When was Neoclassicism? in his *Facets of the Enlightenment: studies in English Literature and its Contexts* (Berkeley, 1968), pp. 1–25.

——The pre-Romantic or post-Augustan mode in his *Facets of the Enlightenment* (Berkeley, 1968), pp. 159–72.

Brownell, Morris *Alexander Pope and the Arts of Georgian England* (Oxford, 1978).

Butler, Janet The garden: early symbol of Clarissa's complicity in *Studies in English Literature*, 24, 1984, pp. 527–44.

Butler, Marilyn *Jane Austen and the War of Ideas* (2nd edn, Oxford, 1987).
Butlin, Robin Ideological Contexts and the Reconstruction of Biblical Landscapes in the seventeenth and early-eighteenth-centuries: Dr Edward Wells and the Historical Geography of the Holy Land, in Alan Baker and Gideon Biger (eds) *Ideology and Landscape in Historical Perspective: essays on the Meanings of some places in the Past* (Cambridge, 1992), pp. 31–62.
Butterfield, Herbert *The Whig Interpretation of History* (Cambridge, 1931).
Carnochan, W.B. *Confinement and Flight: an Essay on English Literature of the Eighteenth Century* (Berkeley, 1977).
Charlesworth, Michael Mapping, the Body and Desire: Christopher Packe's Chorography of Kent, in Denis Cosgrove (ed.) *Mappings* (London, 1999), pp. 109–24.
Clark, Henry *The History of English Nonconformity, Volume 2* (London, 1913).
Clark, J.C.D. *The Dynamics of Change: the Crisis of the 1750s and the English Party Systems* (Cambridge, 1982).
——*English Society, 1688–1832: Ideology, Social Structure and Political Practice during the Ancien Régime* (Cambridge, 1985).
——*The Language of Liberty, 1660–1832: Political Discourse and Social Dynamics in the Anglo-American World* (Cambridge, 1994).
Clery, E.J. *The rise of supernatural fiction, 1762–1800* (Cambridge, 1999).
Clifford, James L. *Hester Lynch Piozzi (Mrs Thrale)* (2nd edn, corrected, Oxford, 1968).
Clingham, Greg *Boswell's The Life of Johnson* (Cambridge, 1992).
Coffin, David *The English Garden: Meditation and Memorial* (Princeton, 1994).
Cohen, Murray *Sensible Words: Linguistic Practice in England, 1640–1785* (Baltimore, 1977).
Cohen, Ralph (ed.) *Studies in Eighteenth-Century British Art and Aesthetics* (Berkeley, 1985).
Colley, Linda *In Defiance of Oligarchy: the Tory Party, 1714–60* (Cambridge, 1982).
Collingwood, R.G. *The Idea of History* (rev edn, Jan van der Dussen, Oxford, 1993).
Conant, Martha *The Oriental Tale in Eighteenth-Century England* (New York, 1908).
Condren, Conal *The Language of Politics in Seventeenth-Century England* (London, 1994).
Copley, Stephen and Peter Garside (eds) *The Politics of the Picturesque: Literature, Landscape and Aesthetics since 1770* (Cambridge, 1994).
Corbin, Alain *The Lure of the Sea: the Discovery of the Seaside in the Western World, 1750–1850* (trans., Harmondsworth edn, 1995).
Corfield, P. Class by Name and Number in Eighteenth-Century Britain, in *History*, 72, 1987, pp. 38–61.
Cosgrove, Denis Landscape Studies in Geography and Cognate Fields of the Humanities and Social Sciences, in *Landscape Research*, 15.3, 1900, pp. 1–6.
——The Myth and the Stones of Venice: An Historical Geography of a Symbolic Landscape, in *Journal of Historical Geography*, 8, 1982, pp. 145–69.
——*The Palladian Landscape: Geographical Change and its Cultural Representations in Sixteenth-Century Venice* (Leicester, 1993).
——Place, Landscapes and the Dialectics of Cultural Geography, in *Canadian Geographer*, 22, 1978, pp. 66–72.
——Prospect, Perspective and the Evolution of the Landscape Idea, in *Transactions of the Institute of British Geographers*, N.S. 10, 1985, pp. 45–62.
——*Social Formation and Symbolic Landscape* (Beckenham, 1984).
Cosgrove, Denis and Peter Jackson, New Directions in Cultural Geography, in *Area*, 19, 1987, pp. 95–101.

Cruickshanks, Eveline *Political Untouchables: the Tories and the '45*(London, 1979).

Curley, Thomas Boswell's Liberty-Loving Account of Corsica, in Greg Clingham (ed.) *New Light on Boswell: Critical and Historical Essays on the Occasion of the Bicentenary of 'The Life of Johnson'* (Cambridge, 1991), pp. 89–103.

Curtius, Ernst *European Literature in the Latin Middle Ages* (trans., London, 1953).

Damisch, Hubert *The Origin of Perspective* (trans., Cambridge, Mass., 1994).

Daniels, Stephen *Humphry Repton: Landscape Gardening and the Geography of Georgian England* (New Haven, 1999).

——DeLoutherbourg's Chemical Theatre: *Coalbrookdale By Night*, in John Barrell (ed.) *Painting and the Politics of Culture: New Essays on British Art, 1700–1850* (Oxford, 1992), pp. 195–230.

——*Fields of Vision: Landscape Imagery and National Identity in England and the United States* (Cambridge, 1993).

——The Implications of Industry: Turner and Leeds, in Simon Pugh (ed.) *Reading Landscape: Country – City – Capital* (Manchester, 1990), pp. 66–80.

——Marxism, Culture and Duplicity of Landscape, in R. Peet and N. Thrift (eds) *New Models in Geography* (2 vols, London, 1989), ii. 196–219.

——The Political Iconography of Woodland in later-Georgian England, in Denis Cosgrove and Stephen Daniels (eds) *The Iconography of Landscape: Essays on the Symbolic Representation, Design and Use of Past Environments* (Cambridge, 1988), pp. 43–82.

Daniels, Stephen and Denis Cosgrove, Spectacle and Text: Landscape Metaphors in Cultural Geography, in James Duncan and David Ley (eds) *Place/Culture/Representation* (London, 1993), pp. 57–77.

Danto, Arthur C. *Analytical Philosophy of History* (Cambridge, 1965).

Danziger, Marlies K. Self-Restaint and Self-Display in the Authorial Comments in the *Life of Johnson*, in Greg Clingham (ed.) *New Light on Boswell: Critical and Historical Essays on the occasion of the Bicentenary of 'The Life of Johnson'* (Cambridge, 1991), pp. 162–73.

De Bolla, Peter *The Education of the Eye: Painting, Landscape, and Architecture in Eighteenth-Century Britain* (Stanford, 2003).

——*The Discourse of the Sublime: Readings in History, Aesthetics and the Subject* (Oxford, 1989).

De Grazia, Margreta *Shakespeare Verbatim: the Reproduction of Authenticity and the 1790 Apparatus* (Oxford, 1991).

Dilke, O.A.W. *Greek & Roman Maps* (Baltimore, 1985).

Dobson, Michael and Nicola Watson *England's Elizabeth: An Afterlife in Fame and Fantasy* (Oxford, 2002).

Downey, James *The Eighteenth-Century Pulpit: A Study of the Sermons of Butler, Berkeley, Secker, Sterne, Whitefield and Wesley* (Oxford, 1969).

Driver, Felix The Historicity of Human Geography, in *Progress in Human Geography*, 12, 1988, pp. 497–506.

Duckworth, Alistair Fiction and some uses of the country house setting from Richardson to Scott, in D.C. Streatfield and Alistair Duckworth, *Landscape in the Gardens and Literature of Eighteenth-Century England* (Los Angeles, 1981), pp. 89–128.

Dunn, John *Political Obligation in Historical Context: Essays in Political Thought* (Cambridge, 1980).

——*The Political Thought of John Locke: an historical approach to the argument of 'The Two Treatises of Government'* (Cambridge, 1969).

Eco, Umberto *et al. Interpretation and Overinterpretation* (Cambridge, 1992).

Edwards, Philip *The Story of the Voyage: Sea-Narratives in Eighteenth-Century England* (Cambridge, 1994).

Everett, Nigel *The Tory View of Landscape* (New Haven, 1994).

Fabricant, Carole The Aesthetics and Politics of Landscape in the Eighteenth Century, in Ralph Cohen (ed.) *Studies in Eighteenth-Century British Art and Aesthetics* (Berkeley, 1985), pp. 49–81.

——Binding and Dressing Nature's Loose Tresses: The Ideology of Augustan Landscape Design, in Roseann Runte (ed.) *Studies in Eighteenth-Century Culture, Vol. 8* (Wisconsin, 1979), pp. 109–35.

——The Literature of Domestic tourism and the Public Consumption of Private Property, in Felicity Nussbaum and Laura Brown (eds) *The New Eighteenth Century: Theory – Politics – English Literature* (London, 1987), pp. 254–75.

——*Swift's Landscapes* (Baltimore, 1982).

Feingold, Richard *Nature and Society: Later-Eighteenth Century Uses of the Pastoral and Georgic* (Hassocks, Sussex, 1978).

Femia, J.V. An Historicist Critique of 'Revisionist' Methods for Studying the History of Ideas, in James Tully (ed.) *Meaning and Context: Quentin Skinner and his Critics* (Cambridge, 1988), pp. 156–75.

Fitter, Chris *Poetry, Space, Landscape: toward a new theory* (Cambridge, 1995).

Fitzpatrick, Martin Latitudinarianism at the Parting of the Ways: a suggestion in John Walsh *et al., The Church of England, c. 1689–1833: From Toleration to Tractarianism* (Cambridge, 1993), pp. 209–27.

Force, Pierre *Self-Interest before Adam Smith: A Genealogy of Economic Science* (Cambridge, 2003).

Forster, Howard *Edward Young: the Poet of the Night Thoughts, 1683–1765* (Alburgh, Norfolk, 1986).

Foucault, Michel *The Archaeology of Knowledge* (trans., London, 1972).

——*The Order of Things* (trans., London, 1970).

Fruchtman, Jack *Thomas Paine and the Religion of Nature* (Baltimore, 1993).

Frye, Northrop Towards defining an age of sensibility in *English Literary History*, 23, 1956, pp. 144–52.

Fumagalli, Vito *Landscapes of Fear: Perceptions of Nature and the City in the Middle Ages* (trans., Cambridge, 1994).

Fussell, Paul Patrick Brydone: the eighteenth-century traveller as representative man, in *Bulletin of the New York Public Library*, 66, 1962, pp. 349–63.

——*The Rhetorical World of Augustan Humanism: Ethics and Imagery from Swift to Burke* (Oxford, 1965).

Gascoigne, John *Cambridge in the Age of the Enlightenment: Science, Religion and Politics from the Restoration to the French Revolution* (Cambridge, 1989).

——*Joseph Banks and the English Enlightenment: Useful Knowledge and Polite Culture* (Cambridge, 1994).

Gerrard, Christine *The Patriot Opposition to Walpole: Politics, Poetry, and National Myth, 1725–42* (Oxford, 1994).

Gilley, Sheridan Christianity and Enlightenment: an Historical Survey, in *History of European Ideas*, 1, 1981, pp. 103–21.

Glacken, Clarence *Traces on the Rhodian Shore: Nature and Culture in Western Thought from Ancient times to the end of the Eighteenth Century* (Berkeley, 1967).

Gombrich, Ernst *Symbolic Images: Studies in the Art of the Renaissance II* (London, 1972).

Goodman, Nelson *Ways of Worldmaking* (Indianapolis, 1978).

Grafton, Anthony *Defenders of the Text: the Traditions of Scholarship in the Age of Science,1450–1800* (Cambridge, Mass., 1991).
——*The Footnote: A Curious History* (London, 1997).
Graham, Walter *English Literary Periodicals* (New York, 1930).
Grant, Robert *Oakeshott* (London, 1990).
Greaves, R.M. Religion in the University of Oxford, 1715–1800, in L.S. Sutherland and L.G. Mitchell (eds) *The History of the University of Oxford*, Vol V, *The Eighteenth Century* (Oxford, 1986), pp. 401–24.
Greene, Jack P. *Peripheries and Center: Constitutional Development in the Extended Polities of the British Empire and the United States, 1607–1788* (Athens, Ga., 1986).
Green, Nicholas Looking at the Landscape: Class Formation and the Visual, in Eric Hirsch and Michael O'Hanlon (eds) *The Anthropology of Landscape: Perspectives on Place and Space* (Oxford, 1995), pp. 31–42.
Greenblatt, Stephen *Marvelous Possessions: the Wonder of the New World* (Oxford, 1991).
Guerinot, J.V. *Pamphlet Attacks on Alexander Pope, 1711–44: A Descriptive Bibliography* (London, 1969).
Hagstrum, Jean H. *The Sister Arts: the Tradition of Literary Pictorialism and English Poetry from Dryden to Gray* (Chicago, 1958).
Hammond, Lansing *Laurence Sterne's Sermons of Mr Yorick* (New Haven, 1948).
Harlan, David Intellectual History and the Return of Literature, in *American Historical Review*, 94, 1989, pp. 581–609.
——Reply to David Hollinger, in *American Historical Review*, 94, 1989, pp. 622–27.
Harpham, Geoffrey Galt *The Ascetic Imperative in Culture and Criticism* (Chicago, 1987).
Harris, Ian,Paine and Burke: God, Nature and Politics, in Michael Bentley (ed.) *Public and Private Doctrine: Essays in British History Presented to Maurice Cowling* (Cambridge, 1993), pp. 34–62.
Hart, John A. Pope as Scholar-Editor, in *Studies in Bibliography*, 23, 1970, pp. 45–59.
Haskell, Francis *History and its Images: Art and the Interpretation of the Past* (New Haven, 1993).
Havens, R.D. *The Influence of Milton on English Poetry* (Cambridge, Mass., 1922).
Hay, D. Property, Authority and the Criminal Law, in D. Hay *et al.* (eds) *Albion's Fatal Tree: Crime and Society in Eighteenth-Century England* (London, 1975), pp. 17–63.
——War, Dearth and Theft in the Eighteenth Century: the record of the English courts, in *Past and Present*, 95, 1982, pp. 117–60.
Haydon, Colin *AntiCatholicism in Eighteenth-Century England, c. 1714–80: A Political and Social study* (Manchester, 1993).
Helgerson, Richard *Forms of Nationhood: the Elizabethan Writing of England* (Chicago, 1992).
Hilton, Boyd *The Age of Atonement: the Influence of Evangelicalism on Social and Economic Thought, 1795–1865* (Oxford, 1988).
Hemmingway, Andrew *Landscape Imagery and Urban Culture in Early-Nineteenth-Century-Britain* (Cambridge, 1992).
Hill, Christopher *The English Bible and the Seventeenth-Century Revolution* (Harmondsworth, 1993).
Hirsch, E.D. *Validity in Interpretation* (New Haven, 1967).
Hirsch, Eric, Introduction: Landscape: Between Place and Space, in Eric Hirsch and Michael O'Hanlon (eds) *The Anthropology of Landscape: Perspectives on Place and Space* (Oxford, 1995), pp. 1–30.
Hirschman, A.O. *The Passions and the Interests: Political Arguments for Capitalism Before its Triumph* (Princeton, 1977).

——*The Rhetoric of Reaction: Perversity, Futility, Jeopardy* (Cambridge, Mass., 1991).

Horne, T.A. *Property Rights and Poverty: Political Argument in Britain, 1605–1834* (Chapel Hill, 1990).

Howkins, Alun J.M.W. Turner at Petworth: Agricultural Improvement and the Politics of Landscape, in John Barrell (ed.) *Painting and the Politics of Culture: New Essays on British Art, 1700–1850* (Oxford, 1992), pp. 231–51.

Hunt, John Dixon *The Figure in the Landscape: Poetry, Painting and Gardening during the Eighteenth Century* (Baltimore, 1976).

——*Gardens and the Picturesque: Studies in the History of Landscape Architecture* (Cambridge, Mass., 1992).

Hunt, John Dixon and Peter Willis (eds) *The Genius of the Place: the English Landscape Garden, 1620–1820* (Paperback edn, Cambridge, Mass., 1988).

Hunter, J. Paul *Before Novels: the Cultural Contexts of Eighteenth-Century English Fiction* (New York, 1990).

Hunter, Michael Latitudinarianism and the "ideology" of the early Royal Society in Richard Kroll *et al.*, *Philosophy, science and religion in England, 1640–1700* (Cambridge, 1992), pp. 199–229.

Hurlbutt, Robert *Hume, Newton and the Design Argument* (Lincoln, Ne., 1965).

Hussey, Christopher *The Picturesque: Studies in a Point of View* (London, 1927).

Hylson-Smith, Kenneth *High Churchmanship in the Church of England: from the Late-Sixteenth Century to the Late-Twentieth Century* (Edinburgh, 1993).

Jackson, J.R. de J. *Historical Criticism and the Meaning of Texts* (London, 1989).

Jacob, Margaret *The Newtonians and the English Revolution 1689–1720* (Hassocks, Sussex, 1976).

Jacques, David *Georgian Gardens: in the Realm of Nature* (London, 1983).

Janowitz, Anne The Chartist Picturesque, in Stephen Copley and Peter Garside (eds) *The Politics of the Picturesque: Literature, Landscape and Aesthetics since 1770* (Cambridge, 1994), pp. 261–80.

——*England's Ruins: Poetic Purpose and the National Landscape* (Oxford, 1990).

Janssen, P.L. Political Thought as Traditionary Action: the Critical Response to Skinner and Pocock, *in History and Theory*, 24, 1985, pp. 115–46.

Jarvis, Simon *Scholars and Gentlemen: Shakespearian Textual Criticism and Representations of Scholarly Labour, 1725–65* (Oxford, 1995).

Jones, Richard *Lewis Theobald* (New York, 1919).

Joost, Nicholas The Authorship of the Freethinker, in Richmond P. Bond (ed.) *Studies in the Early English Periodical* (Chapel Hill, 1957), pp. 103–34.

Keener, Frederick *The Chain of Becoming: the Philosophic Tale, the Novel, and the Neglected Realism of the Enlightenment: Swift, Montesquieu, Voltaire, Johnson and Austen* (New York, 1983).

Kelsall, Malcolm The Iconography of Stourhead, in *Journal of the Warburg and Courtauld Institutes*, 46, 1983, pp. 135–43.

Kenyon, John *The History Men: the Historical Profession in England since the Renaissance* (2nd edn, London, 1993).

Kermode, Frank *Forms of Attention* (Chicago, 1985).

King, P.J.M. Decision-Makers and Decision-Making in the English Criminal Law, 1750–1800, in *Historical Journal*, 27, 1984, pp. 25–58.

Klonk, Charlotte *Science and the Perception of Nature: British Landscape Art in the Late Eighteenth and Early Nineteenth Centuries* (New Haven, 1996).

Kramnick, Isaac *Bolingbroke and his Circle: the Politics of Nostalgia in the Age of Walpole* (2nd edn, Ithaca, 1992).

Kripke, Saul *Wittgenstein on Rules and Private Language* (Oxford, 1982).
Kroll, Richard Introduction in Richard Kroll *et al.* (eds) *Philosophy, Science and Religion in England, 1640–1700* (Cambridge, 1992), pp. 1–28.
LaCapra, Dominick *Rethinking Intellectual History: Texts, Contexts, Language* (Ithaca, 1983).
Langbein, J.H. Albion's Fatal Flaws, in *Past and Present*, 98, 1983, pp. 96–120.
Langford, Paul *A Polite and Commercial People: England, 1727–73* (Oxford, 1989).
——*Public Life and the Propertied Englishman, 1698–1798* (Oxford, 1991).
Le Mahieu, D.L. *The mind of William Paley: A Philosopher and His Age* (Lincoln, Ne., 1976).
Lessenich, Rolf P. *Elements of Pulpit Oratory in Eighteenth-Century England (1660–1800)* (Cologne, 1972).
Levine, Joseph *The Battle of the Books: History and Literature in the Augustan Age* (Ithaca, 1991).
Lipking, Lawrence *The Ordering of the Arts in Eighteenth-Century England* (Princeton, 1970).
Livingstone, David N. *The Geographical Tradition: Episodes in the History of a Contested Enterprise* (Oxford, 1992).
Lonsdale, Roger *Dr Charles Burney: A Literary Biography* (Oxford, 1965).
Lovejoy, A.O. *The Great Chain of Being: A Study in the History of an Idea* (Cambridge, Ma., 1936).
Loveridge, Mark *Laurence Sterne and the Argument about Design* (London, 1982).
McCoy, Gerard 'Patriots, Protestants and Papists': religion and Ascendancy, 1714–60, in *Bullán*, 1.1, 1994, pp. 105–18.
McGann, J.J. *Historical Studies and Literary Criticism* (Wisconsin, 1985).
McGuire, J.E. Boyle's Conception of Nature, in *Journal of the History of Ideas*, 33, 1972, pp. 523–42.
MacIntyre, Alasdair *After Virtue: A Study in Moral Theory* (2nd edn, London, 1985).
McKillop, Alan Local Attachment and Cosmopolitanism – the Eighteenth-Century Pattern, in F.W. Hilles and Harold Bloom (eds) *From Sensibility to Romanticism: Essays Presented to Frederick Pottle* (Oxford, 1965), pp. 191–218.
Macpherson, C.B. *The Political Theory of Possessive Individualism: Hobbes to Locke* (Oxford, 1962).
Mack, Maynard *The Garden and the City: Retirement and Politics in the Later Poetry of Pope, 1731–43* (Toronto, 1969).
Mączak, Antoni *Travel in Early Modern Europe* (trans., Cambridge, 1995).
Manwaring, Elizabeth *Italian Landscape in Eighteenth-Century England* (New York, 1925).
Martin, Peter *Edmond Malone, Shakespearian Scholar: A Literary Biography* (Cambridge, 1995).
——'Pursuing Innocent Pleasures': the Gardening World of Alexander Pope (Hamden, Connecticut, 1984).
Mather, F.C. *High-Church Prophet: Bishop Samuel Horsley (1733–1806) and the Caroline Tradition in the later Georgian Church* (Oxford, 1992).
Maxwell, Constantia *The English Traveller in France, 1698–1815* (London, 1932).
Mayer, Robert Nathaniel Crouch, bookseller and historian: popular historiography and cultural power in late seventeenth-century England, in *Eighteenth-century Studies*, 27, 1994, pp. 391–420.
Mayhew, Robert *Enlightenment Geography: The Political Languages of British Geography, 1650–1850* (London, 2000).
——The Denominational Politics of Travel Writings: The Case of Tory Anglicans in the 1770s, in *Studies in Travel Writing*, 3, 1999, pp. 47–81.
Mendyk, Stan A.E. *'Speculum Britanniae': Regional Study, Antiquarianism, and Science in Britain to 1700* (Toronto, 1989).

Michasiw, Kim Ian Nine revisionist theses on the Picturesque in *Representations*, 38, 1992, pp. 76–100.

Miles, Robert *Ann Radcliffe: The Great Enchantress* (Manchester, 1995).

Miles, Rogers B. *Science, religion and belief: the clerical virtuosi of the Royal Society of London, 1663–1687* (New York, 1992).

Miller, Peter N. *Defining the Common Good: Empire, Religion and Philosophy in Eighteenth-Century Britain* (Cambridge, 1994).

Minogue, Kenneth Method in Intellectual History: Quentin Skinner's *Foundations*, in James Tully (ed.) *Meaning and Context: Quentin Skinner and his Critics* (Cambridge, 1988), pp. 176–93.

Mitchell, W. Fraser *English Pulpit Oratory: from Andrewes to Tillotson* (London, 1932).

Mitchell, W.J.T. Gombrich and the Rise of Landscape, in Ann Bermingham and John Brewer (eds) *The Consumption of Culture, 1600–1800: Image, Object, Text* (London, 1995), pp. 103–18.

——Imperial Landscape, in W.J.T. Mitchell (ed.) *Landscape and Power* (Chicago, 1994), pp. 5–34.

Moir, Esther *The Discovery of Britain: the English Tourist, 1540 to 1840* (London, 1964).

Morgan, Majorie *Manners, Morals and Class in England, 1774–1858* (London, 1994).

Mukerji, Chandra Reading and Writing with Nature: a Materialist Approach to French Formal Gardens, in John Brewer and Roy Porter (eds) *Consumption and the World of Goods* (London, 1993), pp. 439–61.

Mullan, John *Sentiment and Sociability: the language of feeling in the eighteenth century* (Oxford, 1988).

Neale, R.S. *Class in English History, 1650–1800* (Oxford, 1981).

Neeson, J.M. *Commoners: Common Right, Enclosure and Social Change in England, 1700–1820* (Cambridge, 1993).

New, Melyvn *Laurence Sterne as Satirist: A reading of Tristram Shandy* (Gainesville, Florida, 1969).

Nicholson, Margaret *Mountain Gloom and Mountain Glory: the Development of the Aesthetics of the Infinite* (Ithaca, 1959).

Nockles, Peter B. Church Parties in the Pretractarian Church of England, 1750–1833: the 'orthodox' – some problems, in John Walsh *et al.* (eds) *The Church of England, c. 1689 to c. 1833: from Toleration to Tractarianism* (Cambridge, 1993), pp. 334–59.

——*The Oxford Movement in Context: Anglican High Churchmanship, 1760–1857* (Cambridge, 1994).

O'Gorman, Frank *Voters Patrons and Parties: the Unreformed Electorate in Hanoverian England, 1734–1832* (Oxford, 1989).

Oakeshott, Michael *Experience and its Modes* (Cambridge, 1933).

——History and the Social Sciences, in Institute of Sociology, *The Social Sciences: Their Relation in Theory and in Teaching* (London, 1936), pp. 71–81.

——The Importance of the Historical Element in Christianity, in his *Religion, Politics, and the Moral Life* (ed. Timothy Fuller, New Haven,1993), pp. 63–73.

——*On History and other Essays* (Oxford, 1983).

——*Rationalism in Politics and other Essays* (rev. edn, Indianapolis, 1991).

——*The Voice of Liberal Learning: Michael Oakeshott on Education*, ed. Timothy Fuller (New Haven, 1991).

Ousby, Ian *The Englishman's England: Taste, Travel and the Rise of Tourism* (Cambridge, 1990).

Pagden, Anthony *European Encounters With the New World: from Renaissance to Romanticism* (New Haven, 1993).

Paulson, Ronald *Breaking and Remaking: Aesthetic Practice in England, 1700–1820* (New Brunswick, 1989).

Pears, Iain *The Discovery of Painting: the Growth of Interest in the Arts in England, 1680–1768* (New Haven, 1988).

Pocock, J.G.A. A Discourse of Sovereignty: Observations on the Work in Progress, in Nicholas Phillipson and Quentin Skinner (eds) *Political Discourse in Early Modern Britain* (Cambridge, 1993), pp. 377–428.

——*The Machiavellian Moment: Florentine Political Thought and the Atlantic Republican Tradition* (Princeton, 1975).

——Political thought in the English-speaking Atlantic, 1760–1790: (ii) Empire, Revolution and the End of Early Modernity, in J.G.A. Pocock *et al.* (eds) *The Varieties of British Political Thought, 1500–1800* (Cambridge, 1993), pp. 283–317.

——*Politics, Language and Time: Essays in Political Thought and History* (New York, 1971).

——*Virtue, Commerce and History: Essays on Political Thought and History, Chiefly in the Eighteenth Century* (Cambridge, 1985).

——Within the Margins: the definition of orthodoxy, in R.D. Lund (ed.) *The Margins of Orthodoxy: Heterodox Writing and Cultural Response, 1660–1750* (Cambridge, 1995), pp. 33–53.

Popper, Karl *The Poverty of Historicism* (London, 1957).

Porter, Roy Review of Barrell, *Painting and the Politics of Culture*, in *Journal of Historical Geography*, 19, 1993, pp. 244–45.

Pottle, Frederick *James Boswell: the Earlier Years, 1740–69* (London, 1966).

Poovey, Mary Ideology and *The Mysteries of Udolpho*, *Criticism*, 21, 1979, pp. 307–30.

Powell, L.F. Boswell's original journal of his tour to the Hebrides and the printed Version, in *Essays and Studies*, 23, 1938, pp. 58–69.

Prince, Hugh Art and Agrarian Change, 1710–1815, in Denis Cosgrove and Stephen Daniels (eds) *The Iconography of Landscape: essays on the Symbolic Representation, Design and Use of Past Environments* (Cambridge, 1988), pp. 98–118.

Rack, Henry D *Reasonable Enthusiast: John Wesley and the Rise of Methodism* (2nd edn. London, 1992).

Raven, James *Judging New Wealth: Popular Publishing and Responses to Commerce in England, 1750–1800* (Oxford, 1992).

Redford, Bruce *The Converse of the Pen: Acts of Intimacy in the Eighteenth-century Familiar Letter* (Chicago, 1986).

Reitan, E.A. Expanding Horizons: Maps in the *Gentleman's Magazine*, 1731–54, in *Imago Mundi*, 37, 1985, pp. 54–62.

Relph, Edward *Place and Placelessness* (London, 1976).

——*Rational Landscapes and Humanistic Geography* (London, 1981).

Rivers, Isabel *Reason, grace and sentiment: A Study of the Language of Religion and Ethics in England 1660–1780, Volume 1: From Whichcote to Wesley* (Cambridge, 1991).

Rogers, Pat *The Text of Great Britain: Theme and Design in Defoe's Tour* (Newark, NJ., 1998).

Rosenthal, Michael *British Landscape Painting* (London, 1982).

Roston, Murray *Changing Perspectives in Literature and the Visual Arts, 1650–1820* (Princeton, 1990).

Røstvig, Maren-Sofie *The Happy Man: Studies in the Metamorphisis of a Classical Ideal, vol 2, 1700–60* (Oslo, 1958).

Rupp, E. Gordon *Religion in England, 1688–1791* (Oxford, 1986).

Ryan, Robert *The Romantic Reformation: Religious Politics in English Literature, 1789–1824* (Cambridge, 1997).

St John Bliss, Isabel Young's *Night Thoughts* in relation to contemporary Christian apologetics in *Proceedings of the Modern Language Association of America*, 49, 1949, pp. 37–70.

Sack, James J. *From Jacobite to Conservative: Reaction and Orthodoxy in Britain, c. 1760–c. 1832* (Cambridge, 1993).

Saglia, Diego Looking at the other: cultural difference and the traveller's gaze in *The Italian*, *Studies in the Novel*, 28, 1996 pp. 12–37.

Said, Edward *Orientalism: Western Conceptions of the Orient* (London, 1978).

Sambrook, James *James Thomson, 1700–1748: A life* (Oxford, 1991).

Schaffer, Simon The Consuming Flame: Electrical Showmen and Tory Mystics in the World of Goods, in John Brewer and Roy Porter (eds) *Consumption and the World of Goods* (London, 1993), pp. 489–526.

Schama, Simon *Landscape and Memory* (London, 1995).

Schopenhauer, Arthur On History, in his *The World and Will and Representation*, trans., E.F.T. Payne (2 vols, New York, 1966), ii. 439–46.

Scruton, Roger *The Philosopher on Dover Beach: Essays* (Manchester, 1990).

Seary, Peter *Lewis Theobald and the Editing of Shakespeare* (Oxford, 1990).

Shapin, Steven *A Social History of Truth: Civility and Science in Seventeenth-Century England* (Chicago, 1994).

——Social Uses of Science, in George Rousseau and Roy Porter (eds) *The Ferment of Knowledge: Studies in the Historiography of Eighteenth-Century Science* (Cambridge, 1980), pp. 93–139.

Sher, Richard B. *Church and University in the Scottish Enlightenment: The Moderate Literati of Edinburgh* (Princeton, 1985).

——Scottish Divines and Legal Lairds: Boswell's Scots Presbyterian Identity, in Greg Clingham (ed.) *New Light on Boswell: Critical and Historical Essays on the Occasion of the Bicentenary of 'The Life of Johnson'* (Cambridge, 1991), pp. 162–73.

Sherbo, Arthur *The Birth of Shakespearian Studies: Commentators from Rowe (1709) to Boswell-Malone (1821)* (East Lansing, Michigan, 1986).

Simpson, A.W.B. *An Introduction to the History of the Land Law* (Oxford, 1961).

Sitter, John *Literary loneliness in mid-eighteenth-century England* (Ithaca, 1982).

Skinner, Quentin *The Foundations of Modern Political Thought* (2 vols, Cambridge, 1978).

——*Reason and Rhetoric in the Philosophy of Hobbes* (Cambridge, 1996).

Smith, David Nichol *Shakespeare in the Eighteenth Century* (Oxford, 1928).

Smithers, Peter *The Life of Joseph Addison* (Oxford, 1954).

Solkin, David *Painting for Money: The Visual Arts and the Public Sphere in Eighteenth-Century England* (New Haven, 1993).

——*Richard Wilson: The Landscape of Reaction* (London, 1982).

Spellman, W. M. *The Latitudinarians and the church of England, 1660–1700* (Athens, Ga., 1993).

Spurr, John *The Restoration Church of England, 1646–89* (New Haven, 1991).

Stafford, Barbara *Artful Science: Enlightenment entertainment and the Eclipse of Visual Education* (Cambridge, Mass., 1994).

——*Body Criticism: Imaging the Unseen in Enlightenment Art and Medicine* (Cambridge, Mass., 1991).

——*Voyage into Substance: Art, Science and the Illustrated Travel Account, 1760–1840* (Cambridge, Mass., 1984).

Starnes, De Witt T. and Gertrude Noyes *The English Dictionary from Cawdrey to Johnson, 1604–1755* (Chapel Hill, 1946).

Stewart, Larry *The Rise of Public Science: Rhetoric, Technology and Natural Philosophy in Newtonian Britain, 1660–1750* (Cambridge, 1992).

Stock, R.D. *The Holy and The Daemonic from Sir Thomas Browne to William Blake* (Princeton, 1982).

Streatfield, D.C. Art and Nature in the English Landscape Landscape Garden: Design theory and practice, 1700–1818, in D.C. Streatfield and Alistair Duckworth, *Landscape in the Gardens and Literature of Eighteenth-Century England* (Los Angeles, 1981), pp. 1–87.

Sutherland, James R. 'The Dull Duty of an Editor', in *Review of English Studies*, 21, 1945, pp. 202–15.

Sykes, Norman *From Sheldon to Secker: Aspects of English Church History 1660–1768* (Cambridge, 1959).

Taylor, Gary *Reinventing Shakespeare: A Cultural History from the Restoration to the Present* (London, 1990).

Taylor, James *Jonas Hanway: Founder of the Marine Society: Charity and Policy in Eighteenth-Century Britain* (London, 1985).

Templeman, William D. *The Life and Work of William Gilpin (1724–1804). Master of the Picturesque and Vicar of Boldre*, in *University of Illinois Studies in Language and Literature*, 24, 1939.

Thomas, Keith *Man and the Natural World: Changing Attitudes in England, 1500–1800* (Harmondsworth, 1983).

——*Religion and the Decline of Magic: Studies in Popular Beliefs in Sixteenth- and Seventeenth-Century England* (Harmondsworth, 1971).

Thompson, E.P. Rough Music, in his *Customs in Common* (London, 1991), pp. 467–538.

Todorov, T. *Genres in Discourse* (Cambridge, 1990).

Toews, J.E. Intellectual History after the Linguistic Turn: the Autonomy of Meaning and the Irreducibility of Experience, in *American Historical Review*, 92, 1987, pp. 879–907.

Tribe, Keith *Land, Labour and Economic Discourse* (London, 1978).

Tuck, Richard *Natural Rights Theories: Their Origin and Development* (Cambridge, 1979).

Tucker, Susie I. *Protean Shape. A Study in Eighteenth-Century Vocabulary and Usage* (London, 1967).

Tully, James *A Discourse on Property: John Locke and his Adversaries* (Cambridge, 1980).

——(ed.) *Meaning and Context: Quentin Skinner and his Critics* (Cambridge, 1988).

Turner, James *The Politics of Landscape: Rural Scenery and Society in English Poetry, 1630–1660* (Oxford, 1979).

——The Structure of Henry Hoare's Stourhead, in *Art Bulletin*, 61, 1979, pp. 68–77.

Tuveson, Ernst Space, Deity and the 'Natural Sublime', in *Modern Language Quarterly*, 12, 1951, pp. 20–38.

Veeser, H. Aram (ed.) *The New Historicism* (London, 1989).

Wall, Cynthia *The Literary and Cultural Spaces of Restoration London* (Cambridge, 1998).

Walsh, John and Stephen Taylor, Introduction: the Church and Anglicanism in the 'Long' Eighteenth Century, in John Walsh *et al.* (eds) *The Church of England, c. 1689 to c. 1833: from Toleration to Tractarianism* (Cambridge, 1993), pp. 1–64.

Warnke, Martin *Political Landscape: An Art History of Nature* (trans., London, 1994).

Wasserman, Earl Nature Moralized: the Divine Analogy in the Eighteenth Century, in *English Literary History*, 20, 1953, pp. 39–76.

Waterman, A.M.C. *Revolution, Economics and Religion: Christian Political Economy, 1798–1833* (Cambridge, 1991).

Weinbrot, Howard *The Formal Strain: Studies in Augustan Imitation and Satire* (Chicago, 1969).

Wellech, S 'Class versus Rank': the transformation of Eighteenth-Century English Social terms and theories of production, in *Journal of the History of Ideas*, 47, 1986, pp. 409–31.

Wells, Harwell The Philosophical Michael Oakeshott, in *Journal of the History of Ideas*, 55, 1994, pp. 124–45.

Whale, John Romantics, Explorers and Picturesque Travellers, in Stephen Copley and Peter Garside (eds) *The Politics of the Picturesque: Literature, Landscape and Aesthetics since 1770* (Cambridge, 1994), pp. 175–95.

White, Hayden *The Content of the Form: Narrative Discourse and Historical Representation* (Baltimore, 1987).

Williams, Raymond *The Country and the City* (London, 1973).

Williamson, Tom *Polite Landscapes: Gardens and Society in Eighteenth-Century England* (Baltimore, 1995).

Willey, Basil *The Eighteenth-Century Background: Studies in the Idea of Nature in the Thought of the Period* (London, 1940).

Winch, Donald *Adam Smith's Politics: An Essay in Historiographical Revision* (Cambridge, 1978).

——*Riches and Poverty: An Intellectual History of Political Economy in Britain, 1750–1834* (Cambridge,1996).

Withers, Charles W.J. Geography, natural history and the eighteenth-century Enlightenment, putting the world in place, in *History Workshop Journal*, 39,1995, pp. 137–63.

——Towards a History of Geography in the Public Sphere, in *History of Science*, 37, 1999, pp. 45–78.

——Geography, Enlightenment and the Paradise Question, in David Livingstone and
·   Charles Withers (eds) *Geography and Enlightenment* (Chicago, 1999), pp. 67–92.

Wittgenstein, Ludwig *Philosophical Investigations*, trans G.E.M. Anscombe (3rd edn, Oxford, 1967).

Wojcik, Jan *Robert Boyle and the Limits of Reason* (Cambridge, 1997).

Wollheim, Richard *Art and Its Objects* (2nd edn, Cambridge, 1980).

——*Painting as an Art* (London, 1987).

Woodbridge, Kenneth Henry Hoare's Paradise, in *Art Bulletin*, 47, 1965, pp. 83–116.

Woodward, John *An Essay Toward a Natural History of the Earth* (London, 1695).

Wyatt, John *Wordsworth and the Geologists* (Cambridge, 1995).

Yeo, Richard *Defining Science: William Whewell, Natural Knowledge and Public Debate in Early Victorian Britain* (Cambridge, 1993).

——*Encyclopaedic Visions: Scientific Dictionaries and Enlightenment Culture* (Cambridge, 2001).

Young, B.W. *Religion and Enlightenment in Eighteenth-Century England: Theological Debate from Locke to Burke* (Oxford, 1998).

Index

Printed in the United States
32262LVS00001B/13-18

9 780333 993088